5000044197

D1348398

On Growth, Form and Computers

To my parents
SK

On Growth, Form and Computers

Edited by Sanjeev Kumar and Peter J. Bentley
Department of Computer Sciences, University College London, UK

Amsterdam • Boston • Heidelberg • London • New York • Oxford • Paris
San Diego • San Francisco • Singapore • Sydney • Tokyo

This book is printed on acid-free paper

Elsevier Academic Press
84 Theobald's Road, London WC1X 8RR, UK
http://www.academicpress.com

Elsevier Academic Press
525 B Street, Suite 1900, San Diego, California 92101-4495, USA
http://www.academicpress.com

ISBN 0-12-428765-4

Library of Congress Catalog Number: 2003104592

A catalogue record for this book is available from the British Library

Printed and bound in Great Britain

03 04 05 06 07 08 B 9 8 7 6 5 4 3 2 1

Contents

About the editors *vii*
Foreword *ix*
List of contributors *xiii*
Preface *xix*
Acknowledgements *xxi*

1 An introduction to computational development 1
Sanjeev Kumar and **Peter J Bentley**

Section 1 DEVELOPMENTAL BIOLOGY **45**

2 Relationships between development and evolution 47
Lewis Wolpert

3 The principles of cell signalling 64
John T Hancock

4 From genotype to phenotype: looking into the black box 82
Jessica A Bolker

5 Plasticity and reprogramming of differentiated cells in amphibian regeneration 92
Jeremy P Brockes and **Anoop Kumar**

Section 2 ANALYTICAL MODELS OF DEVELOPMENTAL BIOLOGY **107**

6 Qualitative modelling and simulation of developmental regulatory networks 109
Hidde de Jong, Johannes Geiselmann and **Denis Thieffry**

7 Models for pattern formation and the position-specific activation of genes 135
Hans Meinhardt

8 Signalling in multicellular models of plant development 156
Henrik Jönsson, Bruce E Shapiro, Elliot M Meyerowitz and **Eric Mjolsness**

9 Computing an organism: on the interface between informatic and dynamic processes — 162
Paulien Hogeweg

Section 3 THE ROLE OF PHYSICS IN DEVELOPMENT — **179**

10 Broken symmetries and biological patterns — 181
Ian Stewart

11 Using mechanics to map genotype to phenotype — 203
Mark A Miodownik

12 How synthetic biology provides insights into contact-mediated lateral inhibition and other mechanisms — 220
Kurt W Fleischer

Section 4 DEVELOPMENTAL BIOLOGY INSPIRED COMPUTATION — **237**

13 The evolution of evolvability — 239
Richard Dawkins

14 Artificial genomes as models of gene regulation — 256
Torsten Reil

15 Evolving the program for a cell: from French flags to Boolean circuits — 278
Julian F Miller and **Wolfgang Banzhaf**

16 Combining developmental processes and their physics in an artificial evolutionary system to evolve shapes — 302
Peter Eggenberger Hotz

17 Evolution of differentiated multi-threaded digital organisms — 319
Tom Ray and **Joseph Hart**

Section 5 APPLICATIONS OF BIOLOGICALLY INSPIRED DEVELOPMENT — **337**

18 Artificial life models of neural development — 339
Angelo Cangelosi, Stefano Nolfi and **Domenico Parisi**

19 Evolving computational neural systems using synthetic developmental mechanisms — 353
Alistair G Rust, Rod Adams, Maria Schilstra and **Hamid Bolouri**

20 A developmental model for the evolution of complete autonomous agents — 377
Frank Dellaert and **Randall D Beer**

21 Harnessing morphogenesis — 392
Nick Jakobi

22 Evolvable hardware: pumping life into dead silicon — 405
Pauline Haddow, Gunnar Tufte and **Piet van Remortel**

Glossary — *424*
Index — *431*
Colour Plates are between pages 138 and 139

About the editors

Sanjeev Kumar

Sanjeev is a doctorate researcher at the Department of Computer Science, University College London, focusing on artificial life, computational development and evolutionary computation. His research has included the use of evolutionary algorithms for satellite design, and modelling biological gradient formation and maintenance as part of positional information theory with Michel Kerszberg and Lewis Wolpert. For the past four years he has worked with Peter Bentley on biologically plausible computational models of development. He has been a member of the programme committee for numerous conferences in the area, including the International Conference on Evolvable Systems and the International Conference on Evolvable Hardware.

Peter J. Bentley

Peter is an Honorary Research Fellow at the Department of Computer Science, University College London, and Visiting Fellow of University of Kent at Canterbury. He is known for his prolific research covering all aspects of evolutionary computation, including computational development, artificial immune systems, creative problem solving, and more, and applied to diverse applications including biological modelling, robot control, floor-planning, fraud detection and music composition. He is a regular keynote speaker at international conferences and is a consultant, convenor, chair and reviewer for workshops, conferences, journals and books on evolutionary design and evolutionary computation. He has been guest editor of special issues in journals on subjects such as evolutionary design, creative evolutionary systems and artificial immune systems. He served as chair of the Council of Editors for the International Society of Genetic and Evolutionary Computation. He is editor of the books *Evolutionary Design by Computers* and *Creative Evolutionary Systems* and author of the popular science book *Digital Biology*.

Foreword

Though there were precursors, it is reasonable to date the conception of modern, computer-based studies of artificial life from studies of self-reproduction by von Neumann and Ulam (von Neumann, 1966). More than a quarter of a century later, Chris Langton was both the midwife and the person who named the offspring (Langton, 1989). From the outset, reproduction and, implicitly, adaptation – the basis of Darwin's great theory – were features of artificial life. This book, published half a century after von Neumann's unfinished manuscript became available, underlines the centrality of these ideas. Though the field is still far from its goals, the excitement remains. Whatever the long-term outcome, we can only agree with Eddington's statement, 'The contemplation in natural science of a wider domain than the actual leads to a far better understanding of the actual' (Levy, 1992).

In both life and artificial life, perpetual novelty is a feature that continually comes to the fore. Of course, perpetual novelty in this context means much more than simple random variation. Still, it is not as mysterious or as difficult to attain as it might seem. Fewer than a dozen rules suffice to define the game of chess, but no two games of chess are alike (barring deliberate repeats) and, after centuries of study, we still discover new principles for playing the game well. Much the same can be said of Euclidean geometry, with new, non-trivial theorems being discovered after more than two millennia of close study. Is this kind of perpetual novelty related to that of life and artificial life?

My answer is yes and, moreover, I think the relation is close. Rules, axioms and computer instructions serve as generators that define a set of possible configurations, be they legal arrangements of pieces on a game board, strings of symbols that define theorems, or programs that direct a general-purpose computer. Similarly, molecular genetics makes it clear that interacting genes, defined by the 4-nucleotide DNA code, act much as a program with many conditionals, organizing the substrate into a network of reactions that maintains and defines a biological cell. Of course, there are many levels of organization in a cell and several different dynamics. That is what makes genomics and proteomics so complex, but most scientists believe that the perpetual novelty we observe in biology can be explained in terms of these generators. And, however complex the models reported here, they are all generated by programs based upon a small 'alphabet' of computer instructions.

What should we expect, then, of these computer-based models? Most computer-based models originate in an attempt to answer some question or cluster of questions. The art of model construction is to choose the mechanisms or rules defining the model – its generators – so that the configurations they define map clearly onto the area defined by the questions. In the case of living systems, the questions are so difficult that it is often a major triumph to discover *any* set of mechanisms sufficient to generate the phenomenon of interest. Von Neumann's paper is a case in point. Until that paper appeared, most philosophers held that self-reproduction was a defining characteristic of life, using a *recursus ad infinitum* argument that a machine could not reproduce itself: the machine would have to have a plan of its organization, but that plan would have to include the plan itself, and so on. Von Neumann's model offered an existence proof of a self-reproducing machine, thereby neatly collapsing sophisticated philosophical arguments to the contrary.

Much of the most productive work in artificial life has this proof-of-principle character. Without attempting to match living systems in detail, such models show that some set of mechanisms or rules is sufficient to generate the phenomena of interest. Because it is difficult to find any set of sufficient mechanisms for most lifelike phenomena, this is already a useful step. It is not 'just' a theoretical step either. Theories tell us where to look and so it is with these existence proof models. We can use experimental methods to see whether such mechanisms have counterparts or variants in real systems. And even if the living system does not have counterparts, the differences may well suggest a new set of mechanisms that do have counterparts.

Computer-based models are important in this endeavour because of the generated nature of most living phenomena. A fertilized egg provides a wonderful example. The rules and mechanisms embedded in that single cell can direct the generation of a complex metazoan, such as a primate consisting of billions of cells. A great variety of organisms are generated in this fashion, ranging from sponges to elephants. It is not easy to build analytic, equation-based models to describe this procedure. Even when it can be done, say with finite difference equations, the usual tools for analysing equation-based models, such as determination of fixed points or statistical analyses offer few insights.

Again, the game of chess offers a simple illustration of the difficulty of such analyses: a statistical analysis of the moves in a set of games will give few, if any, clues about the network of conditional decisions that defines a good strategy. Moreover, the fixed point of the game, the minimax value, is unknown, and of itself would tell us little about good strategies. Another metaphor strengthens the point: consider a porous meteorite plunged into a bath of liquid under high pressure. The fixed point in this case is a uniform distribution of the fluid throughout the cavities and pores of the meteorite, but the process of getting to this uniform distribution – with the near-surface, porous cavities being penetrated first, and so on – is a different matter entirely. The fixed point tells little about the trajectory of changes leading to it. In the case of living systems, the endpoint is death, which tells us little about the process of living. We need an entirely different kind of analysis to determine the characteristics of living systems.

What do computer-based models offer that is different from analytic equation-based models? They are executable. That is, we can *observe* the generated trajectory as it unfolds. Of equal importance, we can restart the model from selected, well-defined starting points, observing the changes imposed by modifications of the generators and rules of interaction. Though we lose the generality of good analytic models, such as the Lotka-Volterra equations, we do not lose rigour, and we gain in return great amounts of information about the trajectory. We also gain new insights into macro-phenomena such as robustness, speciation and the origins of autocatalytic reactions – phenomena that are critical to an understanding of living systems.

Most importantly, computer-based models, when successful, clearly demonstrate the emergence of macro-phenomena from micro-phenomena. Here we come to a question that is currently a subject of much debate in science and the philosophy of science: can we explain living phenomena with a reductionist approach? This question actually requires answers to two closely related questions:

(i) Can observable macro-phenomena be produced by a relatively simple set of generators and interactions?
(ii) Can we provide well-defined ways for macro-phenomena to influence the interactions of the generators?

There are agent-based models that give a proof-of-principle that both questions can be answered in the affirmative (see, for example, Arthur *et al.*, 1997). Indeed, we do not yet have any clear examples of questions for which reduction must fail. Under the circumstances, it seems reasonable to me to press as hard as we can with this kind of reduction until we are stopped in our tracks by formidable barriers. Even then, given the great successes of reduction, it would seem reasonable to retain some optimism that new variants of the tools of reduction will find ways through such barriers. In any case, in the study of life and artificial life, we are far from 'end of science', even when it is interpreted as coextensive with reduction.

In model-based studies, it is important to avoid premature criticism centred on 'what has been left out'. The art of model building is similar to the art of cartooning: the object is to select, and even exaggerate, salient points in order to concentrate on the questions of interest. The well-known Crick-Watson-Franklin story of modelling DNA's structure makes the point (Maddox, 2002). And, as Ian Stewart says in his chapter in this volume, 'It is not useful to criticize these models on the grounds of what they leave out: the test is what they predict, and how well it fits the *appropriate* aspects of reality, with what they leave in'. Only in this way can we begin to build the grand interaction between theory and experiment that will let us tame the floods of data being produced by the new tools of biology.

There are many fundamental questions in the study of life and artificial life that are still open. Several of them are the subject of papers in this volume. I will look at just three examples:

(i) Ian Stewart makes a strong case for the relevance of a theory of pattern. From the days of D'Arcy Thompson (Thompson, 1961) and Weyl (1952) we have known the critical role of pattern in understanding complex systems. But, for living systems, there is as yet little in the way of theory that lets us pick out significant pattern-generating mechanisms – the counterparts of patterns like gambits, forks, discovered checks, and the like, that enable us to understand chess.
(ii) John Hancock emphasizes the *sine qua non* role of signalling in biological cells. As he says, '... cell signalling events are in control of the activity of the cell ... (which is) crucial for survival of the cell ... and is vital for the development of cells within multicellular organisms ...'. A general framework for studying signal networks, with accompanying analytic techniques if possible, is much to be desired.
(iii) Angelo Cangelosi and his co-authors tackle the difficult problem of generating a complex structure using a limited set of instructions. As they say: 'The nervous system is part of the phenotype which is derived from the genotype through a process called development. ... The phenotype is progressively built by executing the inherited growing instructions. ... (The objective is to) encode repeated structures (such as network composed of several sub-networks with similar local connectivity) in a compact way'. It is of particular interest to be able to subject these instructions for growing to evolutionary processes, the artificial life counterpart of the current studies of 'evo-devo' (evolutionary development) in biology.

In closing, I would like to emphasize one final question that seems to me pivotal to the whole artificial life enterprise: can we build open-ended computer-based models of evolution? As in the case of perpetual novelty, open-ended does *not* mean simple random variation that goes on *ad infinitum*. It is easy enough to build a non-repeating random number generation process where the numbers continually increase in length. Open-ended means generating agents that are increasingly adapted, in some reasonable sense, to an environment that offers increasingly complex challenges. For example, it is conceivable that an agent-based model can be designed in which the agents adapt to each other in 'arms races' with increasing numbers of dimensions. The dimensions increase because the agents, in adapting to each other, form 'niches' which offer opportunities for still other kinds of agents. I have great confidence that such open-ended processes can be designed, though we have no tested models yet. If we can do so, then we will have ways to seek the resolution of another 'great debate': does evolution exhibit progress? With an open-ended model we can go on to try for a proof-of-principle that there are progressive evolutionary systems.

John H. Holland
Fridhem, 2002

References

Arthur, W.B., Holland, J.H., LeBaron, R., Palmer, R. and Taylor, P. (1997) Asset pricing under endogenous expectations in an artificial stock market. *Economic Notes*, **26**(2), 297–330.

Langton, C.G. (ed.) (1989) *Artificial Life*. Addison-Wesley.

Levy, S. (1992) *Artificial Life*. Pantheon Books.

Maddox, B. (2002) *Rosalind Franklin: The Dark Lady of DNA*. Harper-Collins.

Thompson, D'Arcy W. (1961) *On Growth and Form* (ed. Bonner, T.J.) Cambridge University Press.

von Neumann, J. (1966) *Theory of Self-Reproducing Automata*. (Completed by A.W. Burks). University of Illinois Press.

Weyl, H. (1952) *Symmetry*. Princeton University Press.

Contributors

Rod Adams
Department of Computer Science, Faculty of Engineering and Information Sciences, University of Hertfordshire, College Lane, Hatfield, Hertfordshire AL10 9AB, UK
Email: r.g.adams@herts.ac.uk

Wolfgang Banzhaf
Department of Computer Science, LS 11, University of Dortmund, D-44221 Dortmund, Germany
Email: banzhaf@cs.uni-dortmund.de

Randall D Beer
Department of Electrical Engineering and Computer Science, Case Western Reserve University, Cleveland, OH 44106, USA
Email: beer@eecs.cwru.edu,rxb9@po.cwru.edu

Peter J Bentley
Department of Computer Science, University College London, Gower Street, London WC1E 6BT, UK
Email: p.bentley@cs.ucl.ac.uk

Jessica A Bolker
Department of Zoology & Institute for the Development and Evolution of Wet Animals, University of New Hampshire, Durham, NH 03824, USA
Email: jbolker@cisunix.unh.edu

Hamid Bolouri
Institute for Systems Biology, 1441 North 34th Street, Seattle, USA
Email: HBolouri@systemsbiology.org

Jeremy P Brockes
Department of Biochemistry & Molecular Biology, University College London, Gower Street, London WC1E 6BT, UK
Email: j.brockes@ucl.ac.uk

Angelo Cangelosi
School of Computing, University of Plymouth, Plymouth, PL4 8AA, UK
Email: acangelosi@plymouth.ac.uk

Richard Dawkins
University Museum of Natural History, University of Oxford, Oxford OX1 3PW, UK

Frank Dellaert
College of Computing, Georgia Institute of Technology, Atlanta, Georgia 30332-0280, USA
Email: frank@cc.gatech.edu

Kurt W Fleischer
Pixar Animation Studios, 1200 Park Avenue, Emeryville, California 94608, USA
Email: kurt@pixar.com

Johannes Geiselmann
Laboratoire adaptation et Pathogénie des Microorganismes, CNRS FRE2620, Université Joseph Fourier, Bât. CERMO, 460, rue de la Piscine, BP53, 38041 Grenoble Cedex 9, France
Email: hans.geiselmann@ujf-grenoble.fr

Pauline Haddow
Norwegian University of Science and Technology, Sem Sælands vei 7-9, NO-7491 Trondheim, Norway
Email: pauline.haddow@idi.ntnu.no

John T Hancock
Faculty of Applied Sciences, School of BioSciences, University of the West of England, Bristol BS16 1QY, UK
Email: john.hancock@uwe.ac.uk

Joseph Hart
ATR Human Information Science Laboratories, 2-2-2 Hikaridai, Seika-cho, Soraku-gun, Kyoto 619-0288, Japan
jhart@atr.co.jp

Paulien Hogeweg
Theoretical Biology and Bioinformatics Group, Utrecht University, Padualaan 8, 3584, CH Utrecht, The Netherlands
Email: P.Hogeweg@bio.uu.nl

John H Holland
Psychology Department, University of Michigan, Department of Psychology, Ann Arbor, MI 48109, USA
Email: jholland@umich.edu

Peter Eggenberger Hotz
Artificial Intelligence Laboratory, Department of Information Technology, University of Zurich, Winterthurerstrasse 190, CH-8057, Zurich, Switzerland
Email: eggen@ifi.unizh.ch

Nick Jakobi
Managing Director MASA UK, Sussex Innovation Centre, Science Park Square, Brighton BN19SB, UK
Email: nick.jakobi@animaths.com

Hidde de Jong
Institut National de Recherche en Informatique et en Automatique (INRIA), Unite de recherche Rhône-Alpes, 655 avenue de l'Europe, Montbonnot, 38334 Saint Ismier CEDEX, France
Email: Hidde.de-Jong@inrialpes.fr

Henrik Jönsson
Complex Systems Division, Department of Theoretical Physics, Lund University, Sölvegatan 14A, 223 62 Lund, Sweden
Email: henrik@thep.lu.se

Anoop Kumar
Department of Biochemistry & Molecular Biology, University College London, Gower Street, London WC1E 6BT, UK
Email: anoop.kumar@ucl.ac.uk

Sanjeev Kumar
Department of Computer Science, University College London, Gower Street, London, WC1E 6BT, UK
Email: s.kumar@cs.ucl.ac.uk

Hans Meinhardt
Max-Planck-Institut für Entwicklungsbiologie, Spemannstr. 37–39 1 IV, D-72076 Tübingen, Germany
Email: hans.meinhardt@tuebingen.mpg.de

Elliot M Meyerowitz
Division of Biology, California Institute of Technology, 1200 E. California Blvd, Pasadena, CA 91125, USA
Email: meyerow@cco.caltech.edu

Julian F Miller
Department of Computer Science, University of Birmingham, Edgbaston, Birmingham B15 2TT, UK
Email: j.miller@cs.bham.ac.uk

Mark A Miodownik
Department of Mechanical Engineering, King's College London, Strand, London WC2R 2LS, UK
Email: mark.miodownik@kcl.ac.uk

Eric Mjolsness
Department of Information and Computer Science and Institute for Genomics and Bioinformatics, University of California, Irvine, CA 92697-3425, USA
Email: emj@uci.edu

Stefano Nolfi
Neural Systems & Artificial Life Div., National Research Council of Italy, Viale Marx 15, I-00137 Rome, Italy
Email: nolfi@ip.rm.cnr.it

Domenico Parisi
Neural Systems & Artificial Life Div., National Research Council of Italy, Viale Marx 15, I-00137 Rome, Italy
Email: parisi@ip.rm.cnr.it

Tom Ray
Department of Zoology, 730 Van Vleet Oval, Room 314, University of Oklahoma, Norman, Oklahoma 73019, USA
Email: tray@ou.edu, ray@his.atr.co.jp, ray@santafe.edu

Torsten Reil
Department of Zoology, University of Oxford, Oxford OX1 3PS, UK
Email: reil@naturalmotion.com

Piet van Remortel
COMO lab, Dept. of Computer Science, VUB – Vrije Universiteit Brussel, Pleinlaan 2, B-1050 Brussel, Belgium
Email: pvremort@vub.ac.be

Alistair G Rust
Institute for Systems Biology, 1441 North 34th Street, Seattle, USA
Email: arust@systemsbiology.org

Maria Schilstra
Neural Systems Group, Science & Technology Research Centre, University of Hertfordshire, College Lane, Hatfield, Hertfordshire AL10 9AB, UK
Email: m.j.1.schilstra@herts.ac.uk

Bruce E Shapiro
Jet Propulsion Laboratory, California Institute of Technology, M/S 126–347, 4800 Oak Grove Drive, Pasadena, CA 91109, USA
Email: bshapiro@jpl.nasa.gov

Ian Stewart
Mathematics Institute, University of Warwick, Coventry CV4 7AL, UK
Email: ins@maths.warwick.ac.uk

Denis Thieffry
Laboratoire de Génétique et Physiologie du Développement (LGPD), Parc Scientifique de Luminy, 13288 Marseille Cedex 9, France
Email: thieffry@lgpd.univ-mrs.fr

Gunnar Tufte
Norwegian University of Science and Technology, Sem Sælands vei 7-9, NO-7491 Trondheim, Norway
Email: Gunnar.Tufte@idi.ntnu.no

Lewis Wolpert
Department of Anatomy and Developmental Biology, University College, Gower Street, London WC1E 6BT, UK
Email: l.wolpert@cs.ucl.ac.uk

Preface

On Growth and Form, the classic by the great D'Arcy Wentworth Thompson provides the general inspiration for this book. D'Arcy was not one to run with the herd; he was an original thinker and brilliant classicist, mathematician and physicist, who provided a source of fresh air to developmental biology by examining growth and form in the light of physics and mathematics, courageously ignoring chemistry and genetics. Despite this omission of what are now regarded as the main sciences in understanding growth and form, D'Arcy's message is not in the least bit impaired. Instead, in today's biochemistry dominant world D'Arcy's work highlights, as it did in his own time, the role physics plays in growth and form. This book takes its name from D'Arcy's *magnum opus* and hopes to do it justice.

Developmental biology is a fascinating subject. It has a long history – stemming as far back as the times of Hippocrates and Aristotle – rich with its fair share of failures, successes and controversies. From preformationism and epigenesis to Mendel's laws, to Haeckel's idea of ontogeny recapitulating phylogeny, to the discovery of the structure of DNA and the discovery of the cell. All have been either great achievements or failures. Nevertheless, developmental biology has influenced much thought and lines of inquiry in many disciplines.

What, however, does developmental biology have to offer computer science? The answer is *construction*. Construction provides the unifying theme between the two subjects. Developmental biology seeks to understand how organisms are constructed. Computer science (and especially artificial life and evolutionary computation) seeks to learn how to construct complex technology capable of adaptive, robust self-organization.

Encouraging the exchange of ideas between two subjects is always an ambitious undertaking and this book is no exception. In starting this project, our main goal was to convey something of the beauty of developmental biology to artificial life researchers. It soon became clear, however, that something of the beauty of artificial life could, and should, be conveyed to biologists too. The sophisticated developmental processes and mechanisms of construction evolved by natural selection provide the general source of inspiration for the field of computational development, and consequently, the subject matter of this book.

This book has two main aims. First, to provide a textbook and source of inspiration for new students to the subject of computational development; and secondly, to encourage greater dialogue between biologists and computer science researchers and to show how computer scientists can learn from developmental biology to improve our technology.

This volume provides an authoritative introduction to developmental biology and computational development; it makes clear key biological and computational concepts and ideas. The volume stands as a compilation of essays on important topics in developmental biology and computational development from some of the most respected researchers in the fields. It acts as an introductory guide to two very complex disciplines.

The book is organized into five sections, and for ease of reference ordered according to their content, such that chapters focusing mostly on biology are at the beginning, while chapters focusing mostly on computer science are at the end. The first section (Chapters 2, 3, 4 and 5) deals with developmental biology, exploring topics such as relationships between development and evolution, pattern formation, cell signalling and plasticity. The second section (Chapters 6, 7, 8 and 9) begins to formalize ideas through mathematical and computational models of developmental regulatory networks, pattern formation, plant development and *Dictyostelium discoideum* development. The third section (Chapters 10, 11 and 12) explores the relationships between physics and development and illustrates the importance of effects such as surface tension and contact-mediated inhibition through examples and computer models. In the fourth section (Chapters 13, 14, 15, 16 and 17) we begin to move away from models designed to explain biological processes and towards algorithms designed to be useful for problem solving in computer science. Here we investigate development-inspired computer algorithms to examine the evolution of evolvability, gene regulatory networks, pattern generation and differentiation, and A-Life based developmental systems. The final section (Chapters 18, 19, 20, 21 and 22) focuses on practical technological applications of development-inspired computation. Explorations of neural development, agent and robot control and evolvable hardware (electronic circuit design) are provided. A glossary is included at the back to help out with all the terminology.

And finally, from both of us, we hope you find this book as interesting and inspiring as we have. Enjoy.

Sanjeev Kumar
Peter Bentley

Acknowledgements

We would both like to thank:

Lisa Tickner and all the guys and girls at Academic Press for making this book what it is.

Denise Penrose, Emilia Thiuri and everyone at Morgan Kaufmann for helping to market the book to all those difficult computer scientists.

All the contributors to the book for their excellent work. Their efforts have made this book better than we dared hoped for.

Our friends and colleagues Max Christian, Jungwon Kim, Bill Langdon, Anthony Ruto, Tom Quick, Supiya Ujjin, Dave Yoo-Foo and everyone in rooms 301 and 303 and at University College, London.

Sanjeev's acknowledgements

I would like to thank my co-editor, friend and PhD supervisor Peter Bentley for his support, words of encouragement and advice, excellent supervision and for believing in me; without him, this book would not have been possible.

Additionally, I am deeply indebted to Lewis Wolpert for his supervision, unwavering support, constant encouragement and for teaching me how to think about biology. Lewis' enthusiasm for biology is positively infectious; he is an inspiration. I thank him for his much-valued friendship.

A further debt of gratitude must be expressed to Michel Kerszberg for teaching me how to model genetics and whose unequivocal support, understanding and words of encouragement have always been an inspiration. His knowledge of biology, computer science and physics has provided much reassurance.

Additionally, I would like to thank Clinton W. Kell III for his much-valued and continued support of my research.

It is a pleasure to thank many dear friends:

Tom Quick for all the great discussions, helpful suggestions, criticisms, and last minute proof reading, in particular for his advice on the embodied development section. Jasbir, Satnam and Hardeep Flora for keeping me sane and providing welcome distractions; also Jagjeet and Sarabjeet Hayer, Ian Oszvald, Samantha Payne, Nicky Quick and Piet van Remortel.

My heartfelt appreciation to Edalin Paltoo for her support and encouragement throughout; and to my sisters Savita and Anju Kumar for their suggestions, tireless proof-reading and endless encouragement; and to my dear parents, Abnash and Santosh Kumar, for their steadfast support in all my endeavours.

Peter's acknowledgements

I would also like to thank my co-editor and friend Sanj for his enthusiasm and his great efforts as commissioning editor – without him we would not have had the impressive array of contributors that we have. He is also due some well-deserved thanks for putting up with my direct manner and the pressure of deadlines I piled upon his shoulders.

Thanks also to Paul O'Higgins – the first development biologist I ever went to see, many years ago. Luckily it was a very rewarding experience and through him I found Lewis (and where would any of us be without him).

Thanks to all of my PhD students, past and present, for their excitement, drive and fascinating research that helps me with my own explorations of the world. They are: David Basanta, Ramona Behravan, Katie Bentley, Tim Gordon, Jungwon Kim, Siavash Mahdavi, Rob Shipman, Supi Ujjin, (and of course, Sanj). Also many thanks to all those pals at UCL and around the world who I enjoy stimulating conversations with; too many to name, but you know who you are.

Thanks to my friends and family for providing support and welcome distractions, as ever.

And finally (as usual) I would like to thank the cruel and indifferent, yet astonishing creative process of natural evolution for providing the inspiration for my work. Long may it continue to do so.

Cover acknowledgements

Top left: Embryo with differentiated cells and diffusing proteins developed by Sanjeev Kumar using the Evolutionary Developmental System.

Top right: To create this image of a thorn-covered bust, we used thousands of interacting geometric elements constrained to lie on a surface defined by a polygonal dataset. Each element tries to match the orientation of its neighbours, creating a flow field. The elements are then rendered as thorns (on the head) or patches (on the neck and chest). Note that the size of each thorn is relative to the local feature size on the dataset. Smaller thorns appear around the mouth and nose. Cindy Ball helped with modeling this object, and Erik Winfree wrote some important sections of the code. The image was rendered with John Snyder's ray-tracer, with parallel extensions written by Mark Montague. Many thanks to Barbara Meier and members of the Caltech Graphics Group for other software and support. This image first appeared on the cover of *Proceedings of ACM SIGGRAPH 1995* and in the article 'Cellular Texture Generation' by Fleischer, Laidlaw, Currin, Barr, pp. 239–248, copyright © 1995, reprinted with permission from ACM, Inc.

Bottom left: A man-made microchain and gears, fabricated by Sandia National Laboratories. Reproduced with permission of Sandia National Laboratories.

Bottom right: An adult fly, a complex multicellular organism. Reproduced with permission of FlyBase (http://flybase.bio.indiana.edu).

An introduction to computational development

1

SANJEEV KUMAR and **PETER J BENTLEY**

1.1 Introduction

Man has walked on the moon, explored the deepest oceans and built atomic-scale processing devices. With our technological prowess, we like to think that we know everything. But there are still some fundamental questions that remain unanswered. They are so fundamental that our children often ask them of us, to our bewilderment and confusion. A common example is, 'where did I come from?'

The answer to this simple question is so complex, that hundreds of years of research and the work of tens of thousands of scientists have not fully answered it. We know the answer has something to do with evolution, a molecule called DNA, little machines called cells, special chemicals called proteins and a set of processes that follow the instructions in the DNA to develop a hugely complicated organism. We know quite a bit about evolution. But the developmental processes that created you and me from a single fertilized egg are still proving a challenge to understand. One of the aims of this book is to improve our ability to answer questions like this.

Embryology has come a long way since the times of Hippocrates, Aristotle and even Roux (Needham, 1959; Wolpert, 1969, 1998). With the amalgamation of disciplines such as cellular biology, molecular biology and genetics, embryology is, as Bolker observes, no longer a 'black box' and cannot be ignored (Chapter 4). Now better known as developmental biology, the field has seen dramatic progress. Genetics, for example, has promised profound insights into development with the Human Genome Project completing its initial quest to sequence the entire human genome and, more recently, the complete sequencing of the mouse genome has strengthened such promises.

But even with knowledge of genes, proteins and cells, how can we determine their function during development? Remove a gene and you will see a change in the fully developed organism; remove a person from a crowd of friends and their conversation might take a different course. Does that mean that the gene *caused* that phenotypic change, or that the missing person *caused* a change of conversational topic? Not necessarily. Calculating the impact of each element within a complex interacting system is very difficult. To do it, we need new tools. We need computer models.

 ISBN 0-12-428765-4

This book is about development and computers. It is about using computers to understand development further. It is also about using our understandings of development to create new computational techniques that improve our technology. In this chapter we introduce this new combination of computers and development, termed *computational development*.

Computational development promises further insights and a deeper understanding into the processes of development, genetics and cellular and molecular biology through computational modelling. Even rather complex topics in developmental biology are benefiting from this field. Developmental biologists would, for example, be surprised to discover that topics such as *heterochrony* (the study of the effect of changes in timing and rate of development in an evolutionary context) can be modelled in some detail (as described by Angelo Cangelosi in Chapter 18 and Cangelosi, 1999). Also, topics as hotly debated as Haeckel's idea that ontogeny recapitulates phylogeny (Gould, 1974; Raff, 1996) have been diligently covered in the computational development literature (Nolfi and Parisi, 1995).

Computational development also promises new capabilities for our computer software: self-designing, adapting, self-repairing and *developing* solutions to problems. We anticipate that it will enhance the capabilities of evolutionary computation, neural computation and artificial life.

1.1.1 What's to come

This chapter is divided into three major sections. In the first (1.2), we provide an overview of the biology behind development. If you are new to biological ideas of morphogenesis, differentiation, pattern formation, cells, proteins and genes, we strongly recommend that you start with this section. The second major section (1.3) covers the new field of computational development, or how we use computers to model developmental processes. It explains why we are interested in doing this, what computational development is, what it entails and an overview of some of the literature on this topic. Finally, the third major section (1.4) presents two case studies: an example of modelling complex interactions between proteins and an example of a testbed able to evolve and develop simple multicellular forms by computer.

1.2 An overview of development

Natural evolution has evolved countless organisms of varied morphologies. But before this astonishing diversity of life could evolve, evolution had to create the process of development.

So, what is development? In the words of our colleague Lewis Wolpert at University College London, 'Development is essentially the emergence of organized structures from an initially very simple group of cells' (Wolpert, 1998). It is a process of construction that emerges from the interplay between proteins, genes, cells and the environment, resulting in the formation of an organism. It is the reason why you are able to read these words instead of floating around an ocean looking like an amoeba. Evolution designs life, but development builds it.

1.2.1 Developmental processes

Central to development is construction and self-organization. The production of a massively complex form, comprising trillions of cells all working in harmony with each other, all from a single initial cell, is truly one of the wonders of evolution. How does a single cell give rise to a multicellular organism?

All cellular behaviour is controlled by proteins, which are produced by genes. But we will first examine development at the level of the cell. There are five main processes involved in biological development:

- cleavage divisions
- pattern formation
- morphogenesis
- cellular differentiation and
- growth.

Cleavage division

Cleavage division involves the zygote (fertilized cell) undergoing a series of rapid divisions to create more cells. Unlike *cellular proliferation* where cells grow after dividing, during cleavage there is no increase in cellular mass between each division. The result of cleavage is a hollow ball of cells, known as the blastula.

Pattern formation

Pattern formation is the process by which 'a spatial and temporal pattern of cell activities is organized within the embryo so that a well-ordered structure develops' (Wolpert, 1998). Pattern formation comprises two main stages: the process by which the initial body plan is laid down and the allocation of cells to different *germ layers* (primary cell layers in an embryo). The first of these stages – the laying down of the body plan – results in the setting up of a coordinate system. This is achieved through two axes, which define the anterior and posterior ends and the dorsal and ventral sides of the body. Both axes are at right angles to each other (Figure 1.1). The second stage of pattern formation is also responsible for creating the different germ layers, namely: the *ectoderm* (external layer of cells), *mesoderm* (middle layer) and *endoderm* (inner layer of cells).

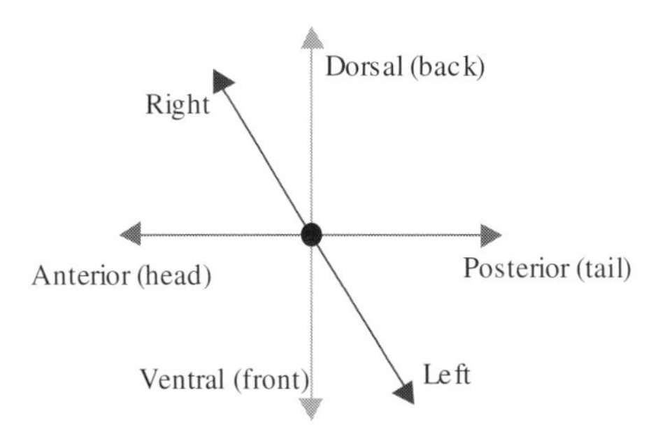

Figure 1.1 Initial body plan. Two axes defining anterior and posterior ends, dorsal and ventral sides (a third axis showing left and right is also shown).

One important theory related to pattern formation was developed by Lewis Wolpert in 1969. *Positional information* is a theory of how cells acquire positional identity and value related to their position along a line with respect to boundaries at both ends of the line. The line is specified by a diffusing morphogen (Wolpert, 1998). Cells can have their position specified by a number of mechanisms. A popular mechanism is that of a diffusing morphogen gradient. For a more detailed exposition of positional information the reader is referred to Chapter 2. In addition, Chapter 15 by Miller provides a lovely illustration of the ideas behind positional information and size regulation using a computer simulation.

Lateral inhibition is another important developmental mechanism. During development, differentiating cells can emit a signal molecule that acts locally by inhibiting the nearest neighbour cells from developing similarly (Wolpert, 1998). This can give rise to regularly spaced patterns, such as that exhibited by the pattern of feathers on the skin of birds. Again, Chapter 15 shows pattern formation in a computer that appears to emerge in a process akin to lateral inhibition.

Morphogenesis

Morphogenesis involves incredible change in the three-dimensional form of the developing embryo as a result of cell movement and conformational changes that generate forces. Extensive cell migration (movement of cells) can also occur. The most dramatic change during morphogenesis, as Wolpert points out, is *gastrulation*. Typically, all animal embryos undergo the dramatic changes of gastrulation, or gut formation. In general, gastrulation involves cells on the outside of the embryo folding and moving inwards to form the gut. In the case of the sea urchin, gastrulation can carve out a hole through the middle of the blastula forming a *gastrula* (multilayered embryo with a cup-shape indentation in it). Chapter 16 shows an excellent computer simulation of invagination.

Cell differentiation

Cell differentiation, the fourth process, is a gradual process by which cells acquire different structure and function from one another, resulting in the emergence of distinct cell types, for example, neurons or skin cells. Differentiation is fundamentally about the different proteins cells contain. If a cell has become terminally differentiated, it continues to produce these proteins due to a change in gene expression that causes a stable pattern of gene activity, else the cell may continue differentiating over successive cell divisions. Differentiation is influenced therefore, by at least the following two processes:

- cell signalling – intercellular communication and
- asymmetric division – division that results in the asymmetric apportioning of factors (proteins) in the parent cell, causing parent and daughter cells to acquire different developmental fates. It also acts as a symmetry-breaking mechanism (Turing, 1952; Wolpert, 1998; Stewart, Chapter 10 this volume).

Growth

The final process, growth, involves an increase in size due to one of a number of methods: cell proliferation, in which cells multiply; a general increase in cell size and the accretion of extracellular materials, such as bone (see Chapter 2 for a description of *Drosophila* development). Note that these processes do not necessarily occur sequentially, but overlap.

1.2.2 Cells

So the development of an organism relies upon its cells undergoing cleavage divisions, pattern formation, morphogenesis, cellular differentiation and growth. But how can cells perform such diverse functions? What is a cell?

Cells are complicated entities; they are the atomic constructional unit of organisms. Cells come in two specific classes: *prokaryotic* (bacteria and blue-green algae) and *eukaryotic* (plants and animals). The DNA in a prokaryotic cell is not encased within a nuclear envelope. Eukaryotic cells enclose their DNA within a membrane, conferring an additional opportunity to control gene regulation (they also contain *organelles* such as the chloroplast and mitochondria).

Generally, (in the terminology of computer science) cells can be likened to autonomous agents in that they have:

- sensors (in the form of protein-based receptors that bind signals, or ion-channels that permit signals through the cell membrane that receive information from the environment)
- internal logic that integrates this information (the genome) and
- effectors (synthesized proteins) able to perturb the environment.

As should be evident, cells are fundamentally protein-processing machines, sensing protein signals, being controlled by proteins and outputting new proteins for other cells to sense.

Cell membrane

The cell takes great pains to separate itself from its immediate environment: that is to say the cell has a very well defined boundary – the cell membrane. The immediate purpose of the membrane is to prevent proteins from seeping away. At the same time, however, proteins need to be able to enter and leave the cell in order to affect it. This passage of proteins is not random. Cells exercise specificity: they can select which proteins enter and leave them. To this end, the cell membrane is semipermeable (or selectively permeable, as it is also known), allowing only certain protein molecules through.

Cell signalling

Groups of cells can influence the development of another group of cells by emitting signals. This process is termed *induction*. Inductive signals provide instructions to cells on how to behave. In essence, the inductive signal 'selects' a single cellular response from an already limited number of responses (Wolpert, 1998).

Cell signalling enables cells to detect and respond to conditions within the extracellular environment. In the case of multicellular organisms, cells need to be able to communicate over short and long distances in order to shape the developing organism. Without signalling, only asymmetric distribution of factors during cleavage could cause cells to become different from one another. The importance of cell signalling in influencing an otherwise unchanging mass of cells to vary in identity and function (differentiation) was highlighted by Spemann's famous organizer experiment, in which a signalling centre was able to organize new axes in an amphibian embryo (Wolpert, 1998).

Types of cell signalling in biology fall into four main categories:

- endocrine (long range, e.g. hormones)
- paracrine (short range)
- autocrine (internal self-signalling) and
- synaptic (long-range effects through changes in the electrical potential across the membrane of the cell; used in the nervous system).

(There is also a fifth: juxtacrine signalling.) Chapter 3 provides further definitions and explanations. First, cells may signal by emitting long or short range protein molecules. A target cell then recognizes the protein and binds it. Secondly, membrane-bound surface proteins on the signalling cell are recognized and bound by membrane-bound surface receptors on a target cell. The term cell signalling is normally used to describe interactions between cells, not intracell signalling (autocrine), where a cell emits proteins to affect itself.

Cells can communicate by transmitting a signal in three main ways, by the:

- release of molecules by one cell being detected by receptors on another cell
- detection of membrane proteins on one cell by receptors on another cell, and
- transfer of small molecules through gap junctions.

The binding of a protein to a cell's receptor triggers many internal cell reactions that relay information to the genome. Such reactions constitute *signal transduction* pathways.

Cell division

Cells multiply by duplicating their contents and then dividing into two cells, parent and daughter. This forms a cell division cycle, which is a complicated process that consists of a number of stages (Alberts *et al.*, 1994):

- interphase (where DNA replicates and proteins are synthesized before and after mitosis)
- mitosis (nuclear division, which itself consists of a number of stages) and
- cytokinesis (the division of the cytoplasm of a cell following the division of the nucleus).

Discrete cell behaviours such as division and death occur against a backdrop of continuous growth.

Cell division is a crucial aspect of development. Two types of cell division occur: symmetric and asymmetric. The symmetry or asymmetry is in relation to cytoplasmic factors sequestered within the cell. Symmetric division occurs when the plane of cleavage divides the cell into equal sizes with equal proportions of cytoplasmic proteins. Asymmetric division occurs when the plane of cleavage divides the cell into unequal sizes with daughter and parent cell containing different cytoplasmic factors. The orientation or direction of cell division is not random, it is controlled and directed. However, exactly what determines the direction in which a cell is to divide is still under investigation in cell biology.

1.2.3 Proteins, enzymes, polypeptides and catalysts

As we have seen, development relies on proteins to control all of the complex behaviour of cells. Proteins are macromolecules derived from genes in DNA. Each cell contains several thousand different types of proteins, performing numerous functions. Some of the roles proteins adopt are:

- structural components of cell and tissues
- general housekeeping, i.e. involved in the transport and storage of small molecules, e.g. haemoglobin and oxygen
- transmission of information between cells, e.g. protein hormones and
- defending against infection (e.g. antibodies).

The most fundamental property of a protein is its ability to act as an enzyme. Enzymes are essentially catalysts, responsible for catalysing the vast majority of chemical reactions in biological systems. Catalysts increase the rate of chemical reactions by lowering the activation energy of reactions without altering the chemical equilibrium between reactants and products and without themselves being consumed or permanently altered.

Proteins are polymers comprised of twenty different amino acids. These amino acids are joined together by peptide bonds forming *polypeptides*. Polypeptides are thus linear chains of amino acids, typically hundreds or thousands of amino acids in length. Each protein comprises a unique sequence of amino acids, determined by the order of nucleotides in a gene. Proteins take on distinct 3D conformations that are critical to their function. The shapes of proteins are determined by their amino acid sequences.

Genes specify the order in which amino acids are to be incorporated into a protein. The order of nucleotides in a gene specifies the amino acid sequence of a protein via translation (see below) in which messenger RNA (mRNA) acts as a template for protein synthesis.

1.2.4 Genes

Proteins may specify the behaviour of cells, but it is the genome that controls proteins. The genome specifies when and where proteins are synthesized. By controlling the temporal and spatial synthesis and decay of proteins, the genome is able to control the overall form of an organism with high precision.

There is a common and often dangerous misconception that there exists a single gene for any single behavioural trait, for example, a gene for blue eyes. The implication that there exists a one-to-one correspondence between gene and phenotypic trait is simply incorrect and perhaps solely a consequence of our need to reduce phenomena to simple cause and effect. There are no genes for eye colour, for leg length or for any other phenotypic trait. There are only genes for proteins. Thus, genes only have one purpose – they specify proteins to be synthesized. These proteins, by virtue of their chemical properties and the complicated conformations they adopt, construct highly intricate and complex networks of interactions. The presence or absence of different proteins within these networks in turn regulate the synthesis and decay of further proteins. The process of regulation is of paramount importance during development. Proteins regulate proteins, which regulate other proteins, which regulate genes, whose products regulate other genes, including themselves, and so on.

DNA

DNA (*deoxyribonucleic acid*) forms the genetic material for almost all organisms on the planet (with the exception of some viruses that replace DNA with RNA or *ribonucleic acid* as the genetic material) (Lewin, 1999). The discovery of the structure of DNA by Watson and Crick in 1953, represented a milestone in modern genetics. They proposed that DNA had a double helical structure that contained two *polynucleotide* chains, forming the backbone of the molecule.

Nucleic acids consist of polynucleotide chains. Nucleic acids contain four types of *bases*, two *purines*: *adenine* (A), *guanine* (G) and two *pyrimidines*: *cytosine* (C) and *thymine* (T). In RNA thymine is replaced by *uracil* (U). The DNA molecule is built up as follows: the bases extend inwards from each nucleotide chain, with purines pairing with pyrimidines, resulting in the following *base pairings*: guanine pairs with cytosine (G-C) and adenine pairs with thymine (A-T) or (A-U in RNA). These base pairings are termed *complementary*.

The sequence of nucleotides in DNA is important as it codes for *amino acids* that constitute the building blocks of the corresponding *polypeptide* or *protein*. In order to construct the corresponding protein the sequence of nucleotides is read three at a time – the nucleotide triplet is termed a *codon*. Each codon corresponds to a single amino acid. The genetic code is read in separate triplets or non-overlapping triplets with successive codons represented by successive tri-nucleotides. Sixty-one define amino acids (although only 20 amino acids exist) while the remaining three represent stop signals for protein synthesis. The code is *unambiguous*, i.e. one codon does not specify more than one amino acid. However, the code is *degenerate* in that more than one codon can specify each amino acid.

Information in DNA and RNA is conveyed by the order of the bases in the polynucleotide chains. A consequence of complementary base pairing is that one strand of DNA or RNA, can act as a template for the synthesis of a complementary strand. Nucleic acids are therefore uniquely capable of directing their own self-replication.

DNA does not directly control protein synthesis. Protein synthesis occurs in the cell cytoplasm under the control of RNA synthesized from the DNA template. So RNA is responsible for conveying information from DNA to the sites of protein synthesis in the cell, known as *ribosomes*.

Gene expression and regulation

In eukaryotic genetics, genes are flanked by a region of DNA that is rich in G-C nucleotide bonds, commonly referred to as the *TATA box* (Lewin, 1999). Its presence informs the cell transcription machinery that a gene is present. Once informed, the machinery (in the form of transcription factor TFIID) binds to a region extending upstream of the TATA box. A protein, known as the TATA-binding protein (TBP), is responsible for recognizing the TATA box. The region that TFIID binds to is known as the *promoter* (Lewin, 1999) (the term is now fast being replaced by a new, more accurate term, the *cis-regulatory region*). This region contains transcription factor target sites that proteins bind to, thus influencing transcription inhibition or activation (Davidson, 2001). These transcription factor target sites can be viewed as preconditions that need to be satisfied to various degrees in order for the gene to activate. Once activated, the gene transcribes the region adjacent to the cis-regulatory region, known as the coding region, responsible for specifiying which protein is to be synthesized. The process by which a gene is transcribed and gives rise to a protein is known as *gene expression*.

The expression of genes is controlled by regulatory proteins. These proteins are emitted by genes and in turn control and affect the expression of other genes. Regulation can lead to long chains of regulatory control: with genes regulating genes that in turn regulate other genes, and so on. In so doing, development can control the synthesis of proteins over space and time.

The process of controlling protein synthesis is termed *regulation* and is crucial to development (Lewin, 1999). (Note that for developmental biologists the term regulation also means the ability of the embryo to develop normally even if parts of it are rearranged or removed.) There are many methods of achieving regulation, these vary depending on organismal cell type, for example, prokaryotic cells have no nuclear envelope and so RNA synthesis and protein synthesis do not occur individually, but simultaneously. This is in contrast to eukaryotic cells, which do have a nuclear envelope enabling mRNA and protein synthesis to occur separately. This separation in both space and time can cause the RNA to degrade before protein synthesis (translation) can occur; thus providing eukaryotic cells with an additional method of regulation. Apart from this, there are two main types of regulation: positive and negative. Positive regulation can be likened to amplification. With initially small amounts of protein present, regulatory proteins can bind to genes causing additional synthesis of the same protein, thus amplifying what was there already. Negative regulation is the opposite: a regulatory protein binds to the cis-site of the gene and represses activation, resulting in no expression of the gene.

1.2.5 Summary of development

This first section of the chapter has given a very brief overview of natural development, or how multicellular organisms are built. We have seen that cells undergo five important processes: cleavage divisions, pattern formation, morphogenesis, cellular differentiation and growth. We examined some of the complexity of cells: how they sense and react to their environments, how they divide and how they are controlled by proteins and use proteins to signal their companions. We briefly looked at proteins: how they act as catalysts, how their molecular structure forces them to fold into specific shapes which then affects their function and how they are used for everything from structural components of cells and tissues, to signals. Finally, we examined the genome: how genes within DNA are transcribed with the help of RNA to make proteins and how the transcription is regulated by the presence or absence of other proteins, thus forming complex gene regulatory networks.

Chapters 2 to 5 of this volume continue this exploration of development further. Chapters 6 to 9 begin to explain how some of our understandings of the biology can be formalized and explained mathematically and through computer modelling.

1.3 Computational development

The second major section of the chapter examines *computational development*: how computers have been used to model development and what those models do. However, first we need to understand why we are interested in computational development and what it is.

1.3.1 Why development?

The advantages that a good computational model of development can bring to biology are almost self-evident. Computers enable us to experiment with different hypotheses of how organisms develop and they enable us to analyse that development in far greater depth than is possible in 'wet experiments'. They also allow us to run more experiments than are normally possible and at almost no expense. But this book is not just aimed at biologists who wish to see how computers can be used to help them. It is also aimed at computer scientists who wish to exploit some of the power of development in our technology.

Why is development of any importance to fields such as evolutionary computation and artificial life? The answer is construction.

The main goal of developmental and evolutionary biologists is to understand construction. Indeed as Fontana *et al.* remark, 'The principal problem in evolution is one of construction' (Fontana *et al.*, 1994). In their quest, developmental biologists are aided by the plethora of organisms (or systems) that make up our ecosystems. Through studying developmental systems in organisms such as the chick, *Drosophila* (fruit fly), the mouse, the nematode worm and *Xenopus*, biologists have built up a good picture of the mechanisms at work.

The field of computer science (specifically artificial life and evolutionary computation) are equally obsessed with construction. Whether it is the construction of life-like entities or complex designs or solutions, construction is the unifying theme between the two subjects.

Why do we need development at all? In technological fields, the dream of complex technology that can design itself requires a new way of thinking. We cannot go on building knowledge-rich systems where human designers dictate what should and should not be possible. Instead, we need systems capable of building up complexity from a set of low-level components. Such systems need to be able to learn and adapt in order to discover the most effective ways of assembling components into novel solutions. And this is exactly what developmental processes in biology do, to great effect.

So why bother with development when evolutionary algorithms (computer models of evolution used for problem solving) permit the evolution of solutions to our problems? The trouble with applying traditional evolutionary algorithms (EAs) to problems is that, generally, as our problems get more complex so do the types of solutions they demand. This is particularly troublesome for traditional EAs as typically there is a one-to-one relationship between the genotype and the corresponding solution description (the equivalent of one gene for each strand of hair on a person's body). As solutions to problems become more complex, the length of the genome encoding the solution typically increases. This leads to the silly situation in which a problem demanding a solution with n constituents requires at least n genes, where n could easily run into the thousands or millions. The standard answer to such problems has been the addition of problem-specific knowledge into genetic representations. We can do this by telling the computer that subroutines, loops, symmetry or subdesigns are necessary. However, for very complex solutions we may not have any knowledge about the best way to solve the problem. And even if we do, by adding our own ideas about how solutions should be constructed, we constrain evolution and prevent it from finding alternatives.

Clearly, biology does not handle organismal construction in such an inefficient manner. Instead, evolution by natural selection evolved a highly intricate, non-linear method of mapping genotype to phenotype: development. Almost every organism on earth exists as a result of the correct unfolding of events due to its developmental programme. Biology is able to construct the smallest organisms to the largest, from ants to blue whales, all using similar fundamental processes of development. Evolution has not had an external God to provide knowledge about the best way to develop life. It had learnt that certain developmental processes result in the formation of modularity, providing complex solutions that contain the equivalent of subroutines, loops, symmetry and subdesigns. Development has enabled evolution to learn how to create complexity.

As is listed in the 'advantages' in the next section, these processes have enormous potential. Technology that can design itself, build itself, repair itself and adapt to its environment would be a revolution. The combination of evolution with development has the potential to bring about this revolution. And while it may still sound ambitious or akin to science fiction, researchers are actively working towards all of these goals using ideas from development right now.

Advantages and disadvantages

As with any type of computational technique, algorithms based on development can create their own problems. For example, they can be:

- difficult to evolve by computer
- difficult to analyse
- difficult to create by hand, and
- computationally expensive.

It may be argued, however, that the potential advantages they confer far outweigh their disadvantages. Advantages include:

- reduction of the genotype
- automatic emergence of complexity
- compact genotypes defining complex phenotypes
- repeated structure (subroutining, symmetry, segmentation)
- adaptability
- robustness to noise (fault tolerance)
- regenerative capabilities
- regulatory capabilities
- able to help in understanding real biological processes and mechanisms.

The other sections of this chapter (and all of the other chapters in this book) explore these ideas further.

Now we have examined why we are interested in the topic, we must try to define it.

1.3.2 What is computational development?

Evolutionary algorithms

Much recent research on computational development originates from computer scientists in the field of evolutionary computation (a field that overlaps with artificial life). This field of computer science studies a class of population-based, stochastic search algorithms known as evolutionary algorithms (EAs). There are four main types of EA: the genetic algorithm (GA) (Holland, 1975), genetic programming (GP) (Koza, 1992), evolutionary programming (EP) (Fogel *et al.*, 1966), and evolutionary strategies (ES) (Rechenberg, 1973). All are highly successful 'problem-solvers' used to evolve solutions for hundreds of different applications: everything from design optimization and robot control to music composition (Bentley, 1999; Bentley and Corne, 2001). Figure 1.2 illustrates the algorithm for the simple GA.

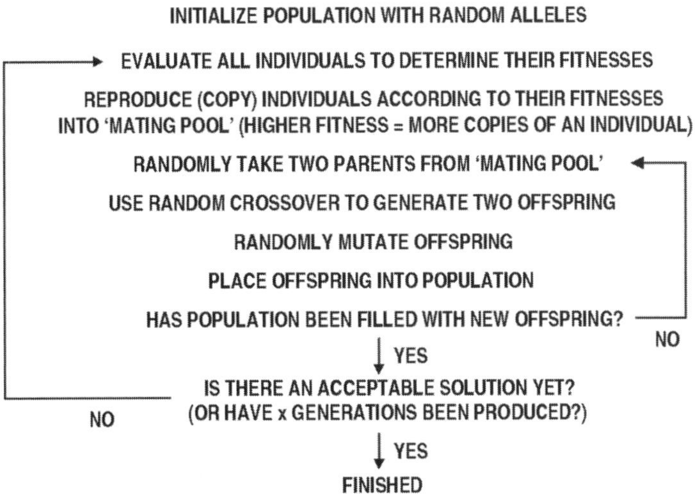

INITIALIZE POPULATION WITH RANDOM ALLELES
EVALUATE ALL INDIVIDUALS TO DETERMINE THEIR FITNESSES
REPRODUCE (COPY) INDIVIDUALS ACCORDING TO THEIR FITNESSES INTO 'MATING POOL' (HIGHER FITNESS = MORE COPIES OF AN INDIVIDUAL)
RANDOMLY TAKE TWO PARENTS FROM 'MATING POOL'
USE RANDOM CROSSOVER TO GENERATE TWO OFFSPRING
RANDOMLY MUTATE OFFSPRING
PLACE OFFSPRING INTO POPULATION
HAS POPULATION BEEN FILLED WITH NEW OFFSPRING? NO
YES
IS THERE AN ACCEPTABLE SOLUTION YET? (OR HAVE x GENERATIONS BEEN PRODUCED?) NO
YES
FINISHED

Figure 1.2 The simple genetic algorithm (GA).

Traditionally, the GA was the only EA to make the explicit separation between genotype and phenotype. For historical reasons, rather than including an explicit development stage between genotype and phenotype, evolutionary algorithms of all classes have focused on modelling evolution and omitted complex genotype to phenotype mappings. Typically, in EAs, the relationship between a gene and its phenotypic effect is one-to-one, as is illustrated in Figure 1.3a. There is a linear relationship between genotype and phenotype – no development!

The recent trends towards the addition of development in EAs has often stemmed from purely practical concerns. Evolutionary algorithms are very efficient search techniques in computer science, but with a linear mapping from genotype to phenotype, the only way to evolve complex solutions is to have complex genotypes – which often cannot be evolved by computer without the addition of domain-specific knowledge. It is anticipated that the addition of development may overcome such limitations to current EAs.

Alternative forms of development

Development is often seen as being unique to biology, but this not always a useful way of viewing it. Here, we argue that development is about fundamental, self-organizing processes and mechanisms of construction. It is not about the specific type of material performing the construction. A system with the appropriate properties will undergo development, whether made from protein molecules, cells and cartilage, or from binary digits, subroutines and data types. Just as we can evolve solutions in computers using evolutionary algorithms (EAs) and both solve our problems and gain improved understandings of natural evolution, so we can *develop* solutions in computers. If materials were all that mattered then all computational (and mathematical) models are ultimately flawed, including any conclusions drawn from them. Fortunately, it is the processes that matter.

Developmental algorithms cannot follow every aspect of development to the letter; simplifying assumptions must be made. This applies also to the scale at which modelling is performed. Modelling at too fine a level of granularity (e.g. at atomic scales) can result in large computational overheads and can render the problem computationally intractable, even with the most powerful of computers. Modelling at too coarse a level (e.g. ignoring protein–cell and cell–cell interactions) may prevent many of the developmental processes that are desirable.

Determining the level of abstraction at which to model development is often difficult. Fortunately for the computer scientists, computational development is not just about modelling developmental processes to learn more about biology. From a 'complex systems' point of view, the construction of complex solutions to problems is of paramount importance. The ability to self-organize would be advantageous for numerous technological disciplines, such as evolutionary design, evolutionary robotics and evolvable hardware. Consider, for example, the young field of *nanotechnology*, which promises tiny, nanoscale robots able to rescue ageing and dying cells, or even give the immune system a helping hand by detecting and destroying foreign pathogens. This is easier said than done. An immediate problem facing nanotechnology is one of construction. Robots at the nanoscale are too small to construct using conventional methods, so new methods are required. One of these is to employ self-organizing processes as in developmental biology.

So computational development is not merely a tool biology can employ to understand the processes and mechanisms underlying organismal construction. It is also a tool for researchers in technological fields to help construct complex adaptive systems. Just as natural selection evolved intricate processes of construction, computational development is in a position to investigate what can only be described as alternative forms of development for our own technological, constructional problems. There is much to be learned from developmental biology.

Definitions

So computational development has great potential. But what is it? What does it mean to *develop*? Why are some processes developmental and others not?

As we have seen, most evolutionary algorithms do not have complex mapping stages from genotype to phenotype. Most EAs are designed to evolve solutions in computers, not to develop them. Because of this, they often have direct and linear mappings from gene to parameter (phenotypic effect) (Figure 1.3a). In contrast, development relies upon highly indirect and non-linear mappings.

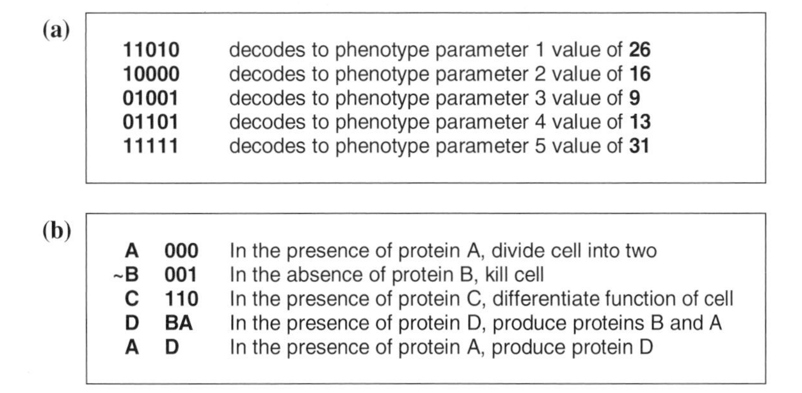

Figure 1.3 (a) An example of a traditional genotype in computer models of evolution: genes map directly to parameter values. (b) An example of a developmental encoding required in models of development: genes act as instructions.

In biology, rather than the genome containing a full description of the phenotype, as in one-to-one mappings, it contains a set of instructions to construct the phenotype. In computer science, when the genotype is a set of such instructions it is termed a *generative program* or *developmental encoding*. So for development to occur in our computer models, we require a developmental encoding, i.e. our genes must act like instructions (Figure 1.3b). Development then becomes the process(es) of executing those instructions and dealing with the highly parallel interactions between them and the structures they create.

But we can go further than this in our definition of computational development. In biology there seems to be a clear idea of separate, interacting entities. For example, entities known as genes interact with other genes, through the transcription of proteins. Proteins interact with each other and with genes and cells. Cells interact with each other, with substances they secrete, with collections of cells known as tissues and with proteins and genes. Tissues interact with each other, as does cartilage and bone, eventually forming organisms. And, of course, organisms interact with each other and with other organisms as they continue to develop and live their lives.

Development requires the idea of separate entities or modules, hierarchies of such modules and multiple interactions between each other and between the different levels of the hierarchy (Figure 1.4). This is an important computational tool used in development – the separation of function into separate modules helps isolate, preserve and protect those functions. Their contents are less likely to be damaged by evolution or by neighbouring modules (we use subroutine functions in computer programming languages for the same kinds of reasons). The multiple interactions provide an extraordinary capability of 'emergence' of self-organization, in much the

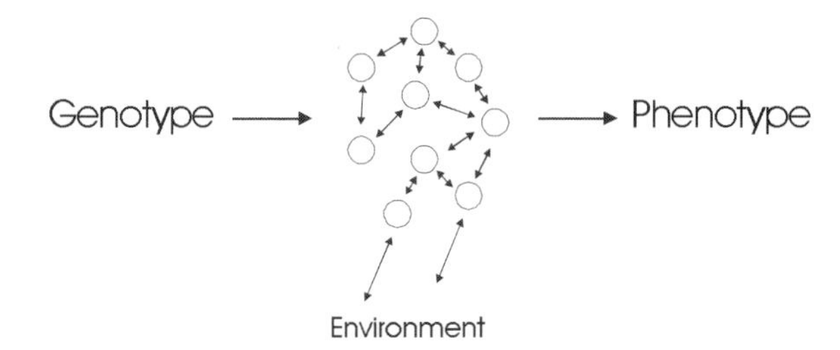

Figure 1.4 Developmental mapping from genotype to phenotype (only a single set of modules and their interactions is shown). Note that the separation of genotype, developmental mapping and phenotype is shown purely for convenience of explanation. A more appropriate viewpoint would merge all three together, for the genetic interactions are an integral part of development and the phenotype is nothing but the 'result so far' of the non-stop developmental processes that continue throughout the life of the organism.

same way that multiple interactions of an ant colony can result in complex, organized behaviour (Holland, 1998; Bentley, 2002).

To this end, we define a module to mean *any discrete constructional unit able to sense its environment, process its sensed inputs and affect its environment, as controlled by the developmental encoding* (Figure 1.5).

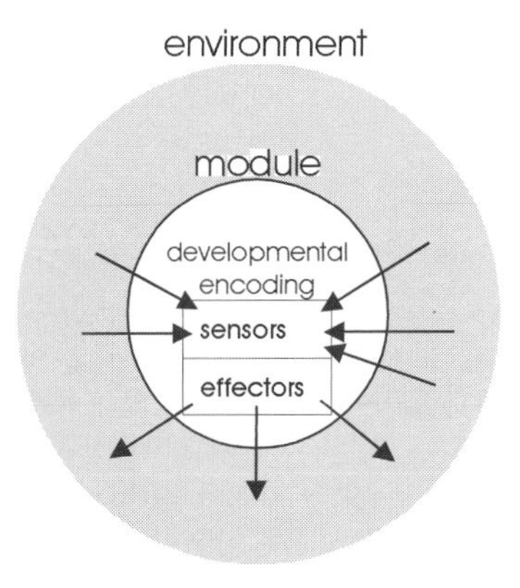

Figure 1.5 A module is able to sense its environment, process its sensed inputs and affect its environment, as controlled by the developmental encoding.

This allows us to capture the idea of genes and proteins and their interactions with each other. The environment of a protein is a molecular-scale universe where proteins from a trillion different sources could affect it through molecular interactions. It also allows us to capture the idea of higher-level modules: cells that interact with each other. The environment of a cell is a horribly complex world of other cells, molecules, surface tensions, compression and expansion forces. From interacting genes to interacting organisms, we can develop them if we have a developmental encoding that controls the construction and behaviour of modules (and modules of modules, and so on).

Finally, while modules may comprise most of the environments of their companions, development does require a more fundamental environment – a set of physical laws such as forces, molecular bonds and electromagnetic radiation, to enable the parsing of the development encoding and the creation and interactions of modules. Of course in our computers we may pick and choose which of the laws of physics we wish to include. But computational development requires its instructions to be read and its modules to form and interact. Without some physical laws, nothing would develop.

With this idea of developmental encodings, hierarchies of interacting modules and environments, we can assess our computational models of development. A system that has a direct one-to-one mapping from gene to parameter, has no interacting modules and thus has zero development occurring. A system with only one type of module and interactions between those modules, may perform weak development (see Figure 1.4). A system with multiple interacting modules (e.g. genes, cells, tissues, organs, etc) may perform strong development (Figure 1.6). Of course, the strongest form of development is biological development in nature.

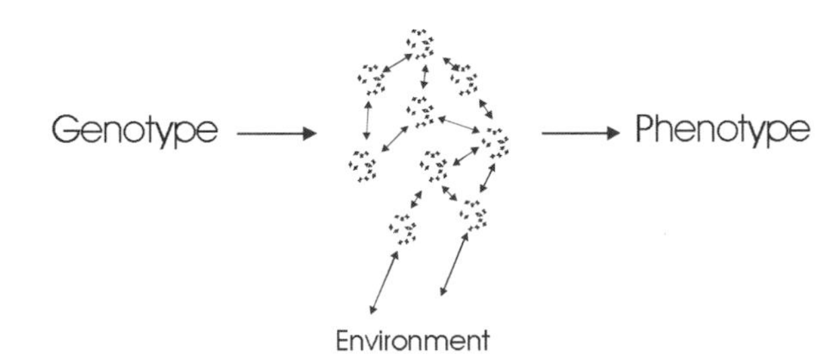

Figure 1.6 Stronger developmental mapping from genotype to phenotype (two sets of modules and their interactions are shown).

1.3.3 Embodied development: development and the environment

Organisms do not exist in a vacuum. They are *embodied* within a complex, dynamic environment with which they interact. Maturana and Varela term the interaction between the embodied system and environment, *structural coupling* (Maturana and Varela, 1992). When an organism is embodied in its environment, its response to environmental events or perturbations is determined by its structure (its components and the relationships between them) and vice versa. At the same time, the structure of organism and environment are each affected and altered by perturbations from the other. All biological systems are intimately coupled to their environment. For example, humans have evolved a range of sensors to detect those aspects of their environment that affect them most and have also evolved extraordinarily diverse methods of interacting with their environments.

Structural coupling requires *mutual perturbatory channels* (Quick *et al.*, 1999a, b) between system and environment (Beer, 1995; Pfeifer, 1998). To be embodied within an environment, an entity must be perturbed by and must perturb its environment. Clearly, genes, proteins, cells, tissues and so on, are all highly structurally coupled to their respective environments for development to work. In contrast, most man-made devices have very few (if any) sensors and few ways in which they can perturb their environments. Such comparisons enable us to think of embodiment as a measure on a sliding scale: the more perturbatory channels between an entity and an environment, the more structually coupled, or embodied, it is.

Biological systems are not the only type of system that this measure applies to. Its appeal is that it is not limited (unique) to biology or any other medium. Quick *et al.* have termed this matter-independent quality, *ontology independence* (Quick *et al.*, 1999a). All that counts is that a system be embodied within an environment and there be some degree of structural coupling between the two. So, whether biology, silicon, or software-based, the raw material is immaterial. This is important to the field of artificial life and its quest to create systems comparable to biology, as it represents an important step forward in being able to compare in a meaningful manner the complexity of a myriad of disparate systems.

Fleischer's work on the multiple mechanisms of morphogenesis (see Chapter 12) serves to illustrate the concepts of ontology-independent structural coupling and perturbatory bandwidth (Fleischer and Barr, 1993). His model is an example of a software-based medium that has an environment and embodied developing system within it. The environment has physics and the developing system is connected to it through sensors and effectors. Thus, the system and environment are able to interact and perturb each other. Other excellent examples of such work include that of Sims (1999) and Hornby *et al.* (2001) which use developmental encodings to evolve both creature morphology and neural network controllers while directly interacting with an environment. These are examples of richness of system-environment interaction (increased perturbatory bandwidth) that are often overlooked because they are embodied in software, i.e. a simulation. A more obvious candidate for such concepts is the pioneering work of Adrian Thompson on evolvable hardware (Thompson, 1996a, b), in which highly unconventional circuits were evolved in real FPGA chips.

Thompson's work on evolvable hardware serves to illustrate the concept of degrees of freedom (Thompson, 1996b). It highlights the complexity achieved through rich system–environment interaction, i.e. increased perturbatory bandwidth available to the evolving system: silicon. Thompson initially had trouble understanding how the evolved solution achieved the desired behaviour. Analysis soon revealed that evolution had taken advantage of the increased perturbatory bandwidth open to the silicon, such as temperature, and the intrinsic physical properties including resistance, capacitance and inductance to achieve the desired aim of evolving a two-tone discriminator.

Unfortunately, embodiment and increased perturbatory bandwidths are meaningless without a rich environment with which to interact. The work of Thompson illustrates how evolutionary computation can harness the richness and complexity of real physics, just as natural evolution has done within its developmental processes. The work was impressive, for it showed how perturbatory channels could be increased: blurring the boundaries between computer model and physical world. But computational models are often entirely based within computers. Having entities embodied within environments is very simple in a computer model. The difficulty lies in creating a sufficiently rich and complex environment. This is the reason for the success of Fleischer's, Sims's and Hornby's research – their environments were rich with dynamics such as surface tension, gravity, friction, inertia, moments and energy dissipation. It is also the reason for the success of many of the models of development presented in this book. If you can get the laws of physics right in your model, then your modules (be they genes or cells) will be given intrinsic properties such as adhesion, diffusion, friction and mass. Evolution (or you) can then design developmental processes that exploit these intrinsic properties of modules and their environment. A simple environment reduces the capabilities of development. A rich and complex environment enhances the potential for development. (See Chapters 4, 10, 11, 12 and 16 for a detailed discussion of the role of the environment on biological development.)

1.3.4 Evolvability

Evolution is lazy. It exploits every conceivable (and inconceivable) way to make the minimum necessary adaptation. Evolution's greatest and unrivalled skill is its ability to find and use every avenue to achieve almost nothing. Like a river flowing down a rugged landscape, it always finds the easiest path. In nature it is able to do this trick often enough to generate ingenuity and complexity beyond our comprehension.

Using evolutionary algorithms in our computers, we evolve good solutions to many problems all the time. The trouble is, to make an evolutionary algorithm evolve, you often need to be trained in the black arts of genetic representation wrangling and fitness function fiddling. In our computers, evolution often gets stuck – the population converges on a bad solution and then never improves.

Reasons for such a lack of evolvability permeate the literature on evolutionary computation. For a long time it was believed that the fitness function held the key. If the function was multimodal with many local maxima, the population might converge on a local maxima unrelated to the global optimum and (with nothing but a virtually impossible mutation able to find a better solution) become stuck there. Some researchers discovered that, contrary to this view, a converged population could evolve over many hundreds or thousands of generations, by following 'neutral pathways'. Genetic drift would randomly push the population around until it would reach a point where small mutations to better solutions became possible. The key to this seemed to be in the right kind of genetic representation.

Ever since the beginnings of evolutionary computation, we have known about good and bad genetic representations. Traditionally, a representation should have strong 'causality' – a change to a gene should make a similar change to the phenotype. Genetic representations with strong causality are much easier to evolve. So representations that have a one-to-one mapping from gene to parameter value, and no development, are easier to evolve. Adding a non-linear mapping such as development from genotype to phenotype reduces the causality, or makes it highly variable and inconsistent. Now effects such as *pleiotropy* (a single gene affecting more than one character of the phenotype: one-to-many), and *polygeny* (one or more genes combining to affect a single character of the phenotype: many-to-one) become common. Even worse, complex *epistasis* (linkage between genes) occurs, with the effects of genes expressed later in development dependent on the effects of genes expressed earlier. So compared to one-to-one mappings, where any gene can be changed and always produce a predictable and similar change to the phenotype, in developmental systems changing any gene can often completely disrupt the whole developmental process, resulting in dramatic changes to the phenotype. Evolution likes small changes – it does not jump from one species to another. In our computers, development often turns a small genetic change into a massive (and disruptive) phenotypic alteration, thus halting evolution in its tracks.

Some argue that the ability of development to enable major phenotypic changes may be responsible for some of the significant landmarks in evolutionary history, e.g. the creation of segmentation (see Chapter 13 by Dawkins). As such, development would behave like a fine-tuneable bias for evolution, making the future evolution of some phenotypes more likely, others less likely (Gordon and Bentley, 2002). This is likely to be true in nature, but in our computers we still struggle to give evolution control of the biases. All too often, the bias seems to force evolution away from the phenotypes we might prefer and towards simpler ones that use development less.

Developmental processes turn out to be one of the biggest headaches that we know of for evolutionary computation. While we can hand-design them with some success, creating an evolvable developmental process has proven an enormous challenge. We know that natural evolution also suffers from similar problems by the existence of the Hox genes (see Chapter 2). These are ancient genes used by most of life early in development. As such, the effects of most genes expressed later in development rely on the Hox genes, so evolution is unable to make significant changes to these genes without destroying the organism. Our problem in computational development is that far too many of our digital genes behave like this – leaving almost nothing for evolution to modify.

The problems may be partially due to the discrete methods of encoding commonly used in many computational development systems. Treating genes as binary 'on/off' switches, or proteins as 'on/off' flags, prevents evolution from fine-tuning systems. A good way to remove a gene is slowly to reduce its effect over many generations, until a mutation to remove it no longer damages the phenotype. Discrete 'on/off' encodings do not allow such gradual changes. In the same way, binary coding (often used to represent values of genes) and binary matching of proteins to genes is undesirable.

Although we can never hope for a simple, linear causality in developmental systems, we can achieve *massively gradual* non-linear systems. If every part of the developmental system can be fine tuned by extremely small amounts, then the effects of all the interacting modules can be fine tuned by evolution. Methods to achieve this include the use of real coding instead of binary, different concentrations of proteins instead of 'on/off' and probabilistic expression of genes instead of deterministic. Experiments have shown that these approaches do enable evolution to fine tune developmental systems successfully (Bentley, 2003a, b, c). Massively gradual development seems to be the key to evolvability.

1.3.5 Literature related to computational development

Reaction-diffusion systems

Computational development was, arguably, founded by mathematician Alan Turing in the 1950s with his seminal work on the chemical basis of morphogenesis (Turing, 1952). Turing's work involved investigating pattern due to the breakdown of symmetry and homogeneity in initially homogeneous continuous media. Through the use of chemicals, which he termed morphogens (meaning generators of form), he showed how small discrepancies in chemical concentrations could be amplified and during the course of development act as symmetry-breaking mechanisms that give rise to patterns. The patterning mechanisms Turing proposed (termed *reaction-diffusion* systems) were captured by a set of partial differential equations. These equations capture the dynamics of autocatalytic and antagonistic chemical reactions with diffusion, in which a compound B undergoes an autocatalytic reaction, thus synthesizing more of itself. At the same time compound A begins synthesizing compound B, which inhibits the formation of A. Compounds A and B have different rates of diffusion through the medium (Meinhardt, 1998). The patterns generated by such dynamical systems are termed *Turing patterns*.

Activator-inhibitor models

Since Turing's work there has been much research into such dynamical systems as generators of pattern in development, most notably the work of Hans Meinhardt (Meinhardt, 1982, 1998). Meinhardt and Gierer have proposed and extensively studied a class of pattern forming reaction-diffusion system, known as an *activator-inhibitor* model (Meinhardt, 1998), which uses the concept of local autocatalysis and long range inhibition. Pattern is the result of strong positive feedback from small discrepancies in an otherwise homogeneous system.

Meinhardt illustrates the model with an example of star formation in which 'a local increase in matter attracts more cosmic material – the self enhancing process'. The opposing, antagonistic effect is provided by the depletion of the dust. Meinhardt points out that in this example there is an additional active antagonistic effect produced by the star itself: 'the emitted light exerts a so-called light pressure that repels dust particles'.

The behaviour of such dynamical systems is often difficult to predict. This can be attributed to (as Meinhardt observes) fluctuations in initial conditions rendering accurate predictions impossible. Meinhardt presents some of his recent work in Chapter 7 of this book.

Random Boolean networks

In 1969, Stuart Kauffman developed a model of genetic regulatory networks, which he termed *random Boolean networks* (RBNs) (Kauffman, 1969). RBNs are Boolean in essence, i.e. a gene in an RBN can either be 'on' or 'off', there is no middle ground. Networks consist of N genes each able to regulate K other genes. Kauffman noticed that these RBNs spontaneously displayed some rather stable, ordered behaviour in randomly generated networks. The inherent stability and order in random networks led him to conclude that evolution by natural selection need not have evolved highly complex, ordered entities from scratch, but rather the inherent 'order for free' in such networks provided an array of choices for natural selection to pick amongst.

Many biologists fear that RBNs are oversimplified and do not capture the complex subtleties of real genetics. Indeed, much genetics research is highlighting the intricate, complex nature of genetic regulatory networks (GRNs) in biology. One need only look at some of the highly complex and intricate schematics of gene regulatory structure to realize that the biologists' fears may be warranted. Current emerging genetics data show how the activity of just a single gene is regulated by many factors. This translates to a high K value in the RBN model. Additionally, K varies for different genes; this is not the case in RBNs, where K is fixed for every gene and is typically low – around 4 (Reil, 2000). Nevertheless, RBNs are intended as an idealized Boolean abstraction model of genetic regulatory networks that captures some important aspects of real genetics. As Ian Stewart observes, 'It is not useful to criticize these models on the grounds of what they leave out: the test is what they predict, and how well it fits the *appropriate* aspects of reality, with what they leave in' (see Chapter 10).

Lindenmayer systems

In 1968, Aristid Lindenmayer proposed a formalism based on string rewriting grammars as a model of development for multicellular organisms, now better known as *L-systems* (Lindenmayer, 1968). Since then, L-systems have seen much popularity, with Smith importing computer graphics techniques to visualize the processes and structures being modelled, it was not long before L-systems were successfully applied to the modelling and visualization of plant development. Since then, L-systems have become synonymous with models of plant development. Visualizing the results of L-systems never fails to impress; they are strikingly similar to the structure of trees and plants. Since Lindenmayer, Prusinkiewicz and his team have conducted much research into the area of modelling and visualization of plant development (Prusinkiewicz and Lindenmayer, 1990; Prusinkiewicz *et al.*, 1993, 1995, 1996).

L-systems are a powerful formalism that come in different classes, such as parametric, differential and map L-systems. L-systems are essentially a set of parallel re-writing rules or productions that replace predecessor (parent) modules by successor (daughter) modules starting from an initial condition or axiom. As Prusinkiewicz observes, the term *module* in the context of L-systems '... denotes any discrete constructional units that are repeated as the plant develops'. An example of a different application of L-systems is given in Chapter 22 of this book.

Cellular encoding

In recent years, other researchers have examined grammars as means to model development. Although biologically inspired, the motivation behind Gruau's work was not to model biology to the letter, but to investigate the use of a developmental encoding to construct neural networks in an efficient modular manner (Gruau, 1994). To this end, Gruau developed a technique he termed cellular encoding.

Cellular encoding takes the form of a *grammar-tree*, used to encode a developmental process. The language upon which the grammar is defined uses instructions that correspond to local graph transformations controlling cell division (the genotype). Each cell retains a copy of the entire grammar-tree, but has a pointer pointing to different sub-trees of the main tree. Development begins with a single cell, which is subjected to cell division by duplication, according to the instructions in the genome. The result of this process is the phenotype, a neural network.

Cellular encoding is an example of a *context free system*, in which neighbour state is not taken into consideration (it does not use cell signalling). Additionally, the model does not make use of diffusing chemicals, such as proteins. Symbolic knowledge is used in the grammar to permit hierarchies, symmetry and problem decomposition into sub-problems.

Evolutionary neurogenesis

In contrast to these approaches to computational development, much recent research originates from computer scientists in the field of evolutionary computation. Kitano and his group were among the first to conduct research into evolving computational development (Kitano, 1990; Kitano *et al.*, 1998). In particular, Kitano has had great success in evolving large neural networks using ideas from development. Kitano extended his initial work and developed neurogenetic learning (NGL) which exhibited better scaling and convergence properties than had previously been shown. This work also exhibited phenomena such as differentiation. In addition, Kitano *et al.* conducted a fascinating project to simulate *C. elegans*, and more recently Kyoda *et al.* (2000) have developed a generalized simulator for multicellular development. The system is used to investigate the spatiotemporal patterning in single and multicellular organsims.

Evolutionary 2D morphogenesis

Fleischer's excellent work on the multiple mechanisms of morphogenesis represents one of the best works on artificial life to date. Fleischer developed a very sophisticated development testbed (Fleischer and Barr, 1993), with which he examined, in particular, the effect of mechanical forces, such as surface tension between cells during morphogenesis. Fleischer's initial motivation was to evolve neural networks, but his work soon took a departure from the norm in that his motivation changed to investigate the multiple mechanisms (i.e. chemical, mechanical and electrical) development uses during morphogenesis.

One aspect of development Fleischer investigated with his testbed was long-range contact inhibition, i.e. the spread of an inhibitory signal through cell-to-cell contact alone (see Chapter 12 for a more detailed exposition of this work). Fleischer showed that it was possible for an inhibitory signal to be propagated without diffusion, but by contact inhibition alone, i.e. he managed to evolve systems in which an inhibitory signal was passed from cell to cell.

Fleischer developed a real-coded representation to be used with an evolutionary algorithm. The representation encoded parameters for a set of conditional differential equations that lay at the heart of the model. These equations controlled the activity of discrete cells (in particular, continuous processes such as growth and movement) and discrete events such as divide and die. The testbed incorporated important developmental concepts and phenomena such as cellular movement, differentiation, chemical diffusion, receptor-ligand binding and cell-signalling, substances were also capable of reacting.

For an example of evolutionary 3D morphogenesis, see Chapter 16 by Eggenberger. Chapters 18 to 21 provide other examples of neurogenesis.

Other applications and models

Computational development as a subject has matured steadily since Turing. Over the last decade, the resurgence of interest has fuelled much research centred on the study of computational models of development and we do not have the room to review all the work in this area (especially since the other chapters of this book cover much of it). The applications of these models range from testing hypotheses and making new predictions in biology to the construction of complex adaptive systems in computer science (specifically, artificial life) and the sciences of complexity. A wide diversity of topics have been covered in such research. Examples include:

- modelling the activity of single genes (Gibson and Mjolsness, 2001).
- modelling gene regulatory networks (Kauffman, 1993; Reil, 2000; Bongard, 2002a; Bentley, 2003a, b and Chapter 14 of this book).
- models of neurogenesis for the construction of neural networks for robot controllers (see Chapters 12, 18, 19, 20, 21) (Boers and Kuiper, 1992; Fleischer and Barr, 1993; Dellaert and Beer, 1994; Gruau, 1994; Kodjabachian and Meyer, 1994, 1995; Jakobi, 1995; Vaario *et al.*, 1995; Rust *et al.*, 1996; Astor and Adami, 1998; Cangelosi, 1999).
- models of aspects of biological development, such as:
  - lateral inhibition mediated through the Notch-Delta pathway in *Drosophila* (Mjolsness *et al.*, 1995; Gibson and Mjolsness, 2001)
  - slug morphogenesis (see Chapter 9 for a review of the underlying principles of this work) (Saville and Hogeweg, 1997)
  - metabolism and differentiation (Kitano, 1995)
  - plant development (Lindenmeyer, 1968 and Chapter 8 of this book)
  - visual models of morphogenesis (Prusinkiewicz *et al.*, 1996).
- the use of generative encodings for the construction of artificial organisms in simulated physical environments and real robots (Sims, 1999; Bongard and Paul, 2000; Hornby *et al.*, 2001).
- the use of development for evolvable hardware and electronic circuit design (Koza *et al.*, 1999; Miller and Thomson, 2000; Canham and Tyrrell, 2002; Gordon and Bentley, 2002 and Chapter 22).
- the use of lateral inhibition for the placing of cellular phone masts (Tateson, 1998).
- the connectionist approach to modelling, proposed by Eric Mjolsness (Mjolsness *et al.*, 1995).
- the large scale modelling of development for morphogenesis in evolutionary design (Fleischer and Barr, 1993; Eggenberger, 1997; Kumar and Bentley, 2002).

1.3.6 Summary of computational development

In the second major section of the chapter we have examined computational development. Researchers are investigating this area for two reasons: to improve our understanding of development in biology and to improve the capabilities of our technology by incorporating developmental processes. We went on to define computational development, showing how it is necessary to use a developmental encoding for genotypes, which controls the behaviour and interaction of modules. The importance of environment was then stressed – development relies on strong embodiment of its modules and a rich and complex environment. Issues of evolvability were discussed, including the idea that massively gradual developmental systems may improve our abilities to create evolvable computational development. Finally, we gave an overview of some of the literature in this area.

1.4 Case studies

In the third and final major section of this chapter, we present two models to illustrate how developmental systems can be modelled by computer.

1.4.1 Case study 1: modelling complex chemistries

It is often very difficult to model the exact behaviour of specific genes and proteins in biology – the data are usually unavailable or the dynamics too complex. Luckily such accurate modelling is often unnecessary, especially when the behaviour of interest is, for example, the ability of evolution to create new gene regulatory networks or other higher level behaviours.

In this first case study, we look at one approach to modelling the interaction of proteins and genes and of exploring issues of evolvability of such systems. The research described here focuses on methods to enrich the genetic space in which evolution searches. In this work, a biologically plausible model of gene regulatory networks is constructed through the use of genes that are expressed into fractal proteins – subsets of the Mandelbrot set that can interact and react according to their own fractal chemistry (Bentley, 2003a, b, c).

The motivations behind this work are twofold. First, the aim was to create a highly complex chemistry (i.e. a rich environment) that would enable more possibilities for developmental processes. The second aim was to improve the evolvability of developmental processes by removing the 'discreteness' of traditional representations. Using ideas from evolutionary biology and neutral network theory, most evolutionary progress is made through countless small changes to the genotype, many of which make no change to the phenotype. Discrete binary representations often prevent such gradual change, so in this work all parameters are real-coded and can be modified by large or small amounts; the discrete on/off switching boundaries are smoothed by a non-linear probabilistic function or an amplification/reduction function. Both the structure of the fractal search space and the movement through the space using creep mutations appear to increase evolvability with great success (Bentley, 2003c).

Before we look at how fractals can be used to represent proteins, let us review the Mandelbrot set.

Mandelbrot set

Given the equation $x_{t+1} = x_t^2 + c$ where x_t and c are imaginary numbers, Benoit Mandelbrot wanted to know which values of c would make the length of the imaginary number stored in x_t stop growing when the equation was applied for an infinite number of times. He discovered that if the length ever went above 2, then it was unbounded – it would grow forever but, for the right imaginary values of c, sometimes the result would simply oscillate between different lengths less than 2.

Mandelbrot used his computer to apply the equation many times for different values of c. For each value of c, the computer would stop early if the length of the imaginary number in x_t was 2 or more. If the computer had not stopped early for that value of c, a black dot was drawn. The dot was placed at coordinate (m, n) using the numbers from the value of c: $(m + ni)$ where m was varied from −2.4 to 1.34 and n was varied from 1.4 to −1.4, to fill the computer screen. The result was the infinite complexity of the 'squashed bug' shape we know so well today (Mandelbrot, 1982).

Representation

The key to this model is the idea of modelling the complex shapes produced by protein folding in nature as fractal shapes. In this case study we will focus on the fractal proteins. Other work describes how gene regulatory networks can be evolved using these ideas (Bentley, 2003a, b, c).

The representation comprises:

- fractal proteins, defined as subsets of the Mandelbrot set
- environment, which can contain one or more fractal proteins (expressed from the environment gene(s)), and one or more cells
- cell, which contains a genome and cytoplasm and which has some behaviours
- cytoplasm, which can contain one or more fractal proteins
- genome, which comprises structural genes and regulatory genes. In this work, the structural genes are divided into different types: cell receptor genes, environment genes and behavioural genes
- regulatory gene, comprising operator (or promoter) region and coding (or output) region
- cell receptor gene, a structural gene with a coding region which acts like a mask, permitting variable portions of the environmental proteins to enter the corresponding cell cytoplasm
- environment gene, a structural gene which determines which proteins (maternal factors) will be present in the environment of the cell(s)
- behavioural gene, a structural gene comprising operator region and a cellular behaviour region.

Defining a fractal protein

In more detail, a fractal protein is a finite square subset of the Mandelbrot set, defined by three 'codons' (x,y,z) that form the coding region of a gene in the genome of a cell. Each (x,y,z) triplet is expressed as a protein by calculating the square fractal subset with centre coordinates (x,y) and sides of length z (Figure 1.7). In this way, it is possible to achieve as much complexity (or more) compared to natural protein folding in nature.

Figure 1.7 Example of a fractal protein defined by ($x = 0.132541887$, $y = 0.698126164$, $z = 0.468306528$).

In addition to shape, each fractal protein represents a certain concentration of protein (from 0 meaning 'does not exist' to 200 meaning 'saturated'), determined by protein production and diffusion rates.

Fractal chemistry

Cell cytoplasms and the environment usually contain more than one fractal protein. In an attempt to harness the complexity available from these fractals, multiple proteins are merged. The result is a product of their own 'fractal chemistry' which naturally emerges through the fractal interactions.

Fractal proteins are merged (for each point sampled) by iterating through the fractal equation of all proteins in 'parallel', and stopping as soon as the length of any is unbounded (i.e. greater than 2). Intuitively, this results in black regions being treated as though they are transparent and paler regions 'winning' over darker regions (Figure 1.8).

Figure 1.8 Two fractal proteins (left, middle) and the resulting merged fractal protein combination (right).

Calculating concentration levels

The total concentration of two or more merged fractal proteins is the mean of the different concentrations seen in their merged product. For example, Figure 1.8 shows the shape of two merged fractal proteins. Figure 1.9 illustrates the resultant areas of different concentration in the product. When being compared to the (*xp*, *yp*, *zp*) promoter region of a gene, the concentration seen on that promoter is described by all those regions that 'fall under' the promoter (Figure 1.10). In other words, the merged product is masked by the promoter fractal and the total concentration on the promoter is the mean of the resulting concentrations (Figure 1.10).

Updating protein concentration levels

At every time step the new concentration of each protein is calculated. This is formed by summing two separate terms: the previous concentration level after diffusion (*diffusedconc*) and the new concentration output by a gene (*geneoutputconc*). These two terms model the reduction in concentration of proteins over time and the production of new proteins over time, respectively (Bentley, 2003b), where:

Figure 1.9 The different concentrations of the two fractal proteins and the concentration levels in their merged product.

$$diffusedconc = prevconcentration - prevconcentration \,/\, \text{PROTEINDEC} + 0.2)$$

(PROTEINDEC is a constant normally set to 5,
the final addition of 0.2 ensures a minimum level of diffusion)

and:

$$geneoutputconc = totalconc \times tanh((totalconc - ct) \,/\, \text{CWIDTH}) \,/\, \text{CINC}$$

where: *totalconc* is the mean concentration seen at the promoter,
ct is the concentration threshold from the gene promoter
CWIDTH is a constant (normally set to 30)
CINC is a constant (normally set to 2).

Figure 1.10 The shape of the desired protein as defined by a promoter and the concentration levels seen on that promoter (total concentration is taken as mean). Note that although the second protein may decrease affinity (similarity) to the promoter, should the second protein have a higher concentration level to the first, it will boost overall concentration seen by the promoter, i.e. act like a catalyst to speed up (or slow down, if lower) the 'reaction'.

Genes

All genes in the genome contain 7 real-coded values:

xp	*yp*	*zp*	*Affinity threshold*	*Concentration threshold*	*x*	*y*	*z*	*type*

where *xp, yp, zp, Affinity threshold, Concentration threshold* define the promoter (cis-region, operator or precondition) for the gene and *x,y,z* define the coding region of the gene. The type value defines which type of gene is being represented and can be one or all of the following: environment, receptor, behavioural or regulatory. This enables the type of genes to be set independently of their position in the genome, enabling variable-length genomes. It also enables genes to be multifunctional, i.e. a gene might be expressed both as an environmental protein (maternal factor) and a cell behaviour.

When *Affinity threshold* is a positive value, one or more proteins must match the promoter shape defined by (*xp,yp,zp*) with a difference equal to or lower than *Affinity threshold* for the gene to be activated. When *Affinity threshold* is a negative value, one or more proteins must match the promoter shape defined by *xp,yp,zp* with a difference equal to or lower than |*Affinity threshold*| for the gene to be repressed (not activated).

To calculate whether a gene should be activated, all fractal proteins in the cell cytoplasm are merged (including the masked environmental proteins, see later) and the combined fractal mixture is compared to the promoter region of the gene.

The similarity between two fractal proteins (or a fractal protein and a merged fractal protein combination) is calculated by sampling a series of points in each and summing the difference between all the resulting values. (Black regions of fractals are ignored.) Given the similarity matching score between cell cytoplasm fractals and gene promoter, the activation probability of a gene is given by:

$$activationprob = (1 + \tanh((matchnum - Affinity\ threshold - \mathrm{Ct}) / \mathrm{Cs})) / 2$$

where: *matchnum* is the matching score,
Affinity threshold is the matching threshold from the gene promoter
Ct is a threshold constant (normally set to 50)
Cs is a sharpness constant (normally set to 50).

'Regulatory' gene. Should a regulatory gene be activated by other protein(s) in the cytoplasm (which have concentrations above 0) matching its promoter region, its corresponding coding region (*x,y,z*) is expressed (by calculating the subset of the Mandelbrot set) and new concentration level calculated. To do this, the concentration of the resulting protein is modified by incrementing with *geneoutputconc*, the result of a function of the concentration threshold (*Ct*) and the mean total concentration seen at the gene promoter (*totalconc*), as given earlier. In this way, higher concentrations of protein on the promoter will cause an increased rate of output protein concentration growth, while lower concentrations (below the *Ct* threshold) will increase the diffusion rate of the output protein. The cell cytoplasm, which holds all current proteins, is updated at the end of the developmental cycle.

'Cell receptor' gene. At present, the promoter region of the cell receptor gene is ignored and this gene is always activated. As usual, the corresponding coding region (*x,y,z*) is expressed by calculating the subset of the Mandelbrot set. However, the resultant fractal protein is treated as a mask for the environmental proteins, where all black regions of the mask are treated as opaque and all other regions treated as transparent (for an example, see Figure 1.11). If there is more than one receptor gene, only the first in the genome is used.

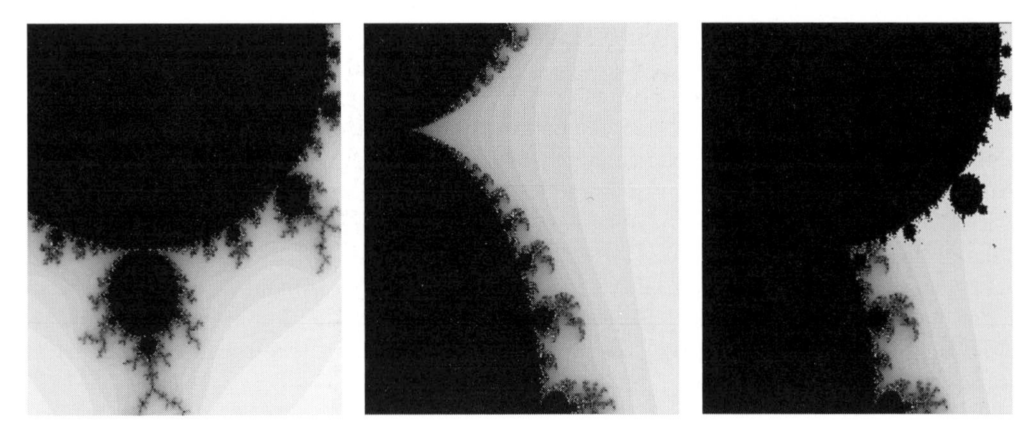

Figure 1.11 Cell receptor protein (left), environment protein (middle), resulting masked protein to be combined with cytoplasm (right).

'Environment' gene. Like the cell receptor gene, this gene is always activated. It produces environmental factors for all cells: fractal proteins of concentration 200. If there is more than one environmental gene, the resulting environmental proteins are merged before being masked by the receptor protein.

'Behavioural' gene. A behavioural gene is activated when other protein(s) in the cytoplasm match its promoter region and the overall concentration is above its *Concentration threshold* value. Instead of the coding region (x,y,z) being expressed as a protein, these three real values are decoded to specify a range of different cellular functions, depending on the application. If there are more behavioural genes than are required, only the first encountered in the genome are used.

Fractal sampling

Fractal proteins are normally stored in memory as lists of constructor (x,y,z) values rather than as bitmaps. All fractal calculations (masking, merging, comparisons) are performed at the same time, by sampling the fractals at a resolution of 15×15 points. Note that the comparison is normally performed between the single fractal defined by (xp,yp,zp) of a gene and the merged combination of all other proteins currently in the cytoplasm. The fractal being compared is treated a little like the cell receptor mask – only those regions that are not black are actually compared with the contents of the cytoplasm.

Development

An individual begins life as a single cell in a given environment. To develop the individual from this zygote into the final phenotype, fractal proteins are iteratively calculated and matched against all genes of the genome. Should any genes be activated, the result of their activation (be it a new protein, receptor or cellular behaviour) is generated at the end of the current cycle. Development continues for d cycles, where d is dependent on the problem. Note that if one of the cellular behaviours includes the creation of new cells, then development will iterate through all genes of the genome in all cells.

Evolution

The genetic algorithm used in this work has been used extensively elsewhere for other applications (including GADES (Bentley, 1999)). A dual population structure is employed, where child solutions are maintained and evaluated and then inserted into a larger adult population, replacing the least fit. The fittest *n* are randomly picked as parents from the adult population. The degree of negative selection pressure can be controlled by modifying the relative sizes of the two populations. Likewise the degree of positive selection pressure is set by varying *n*. When child and adult population sizes are equal, the algorithm resembles a canonical or generational GA. When the child population size is reduced, the algorithm resembles a steady-state GA. Typically the child population size is set to 80% of the adult size and $n = 40\%$. (For further details of this GA, refer to Bentley (1999).)

Unless specified, alleles are initialized randomly, with (*xp,yp,zp*) and (*x,y,z*) values between −1.0 and 1.0 and *thresh* between −10 000 and 10 000. The ranges and precision of the alleles are limited only by the storage capacity of *double* and *long* 'C' data types – no range constraints were set in the code.

Genetic operators. Genes are real-coded, but genomes may comprise variable numbers of genes. Given two parent genomes, the crossover operator examines each gene of parent 1 in turn, finding the most similar gene of the same type in parent 2. Similarity is measured by calculating the differences between values of operator and coding regions of genes. One of the two genes is then randomly allocated to the child. If the genome of parent 2 is shorter, the child inherits the remaining genes from parent 1. If the genomes are the same length, this crossover acts as uniform crossover.

Mutation is also interesting, particularly since these genes actually code for proteins in this system. There are four main types of mutation used here:

1. Creep mutation, where (*xp,yp,zp*) and (*x,y,z*) values are incremented or decremented by a random number between 0 and 0.5, *Affinity threshold* is incremented or decremented by a random number between 0 and 16 384 and *Concentration threshold* is incremented or decremented by a random number between 0 and 200.
2. Duplication mutation, where a (*xp,yp,zp*) or (*x,y,z*) region of one gene randomly replaces a (*xp,yp,zp*) or (*x,y,z*) of another gene. (This permits evolution to create matching promoter regions and coding regions quickly.)
3. Gene mutation, where a random gene in the genome is either removed or a duplicate added.
4. Sign flip mutation, where the sign of *Affinity threshold* is reversed.

Crossover is always applied; all mutations occur with probability 0.01 per gene.

Examples of results

Extensive experiments have shown that fractal proteins enable evolution to create specific, desirable gene regulatory networks (Bentley, 2003a, b) and, if permitted to evolve further, will even improve these solutions, making them efficient and robust against damage (Bentley, 2003c). Evolution achieves this by designing specific genes that code for proteins that interact (according to their fractal chemistries) in useful and coherent ways. Figure 1.12 shows some of the shapes of the evolved proteins. Figure 1.13 shows an example of a perfect GRN pattern for a specific evaluation function. Figures 1.14 and 1.15 (see page 30) illustrate how the shapes of the proteins also enable complex changes in protein concentrations over time as they react with each other.

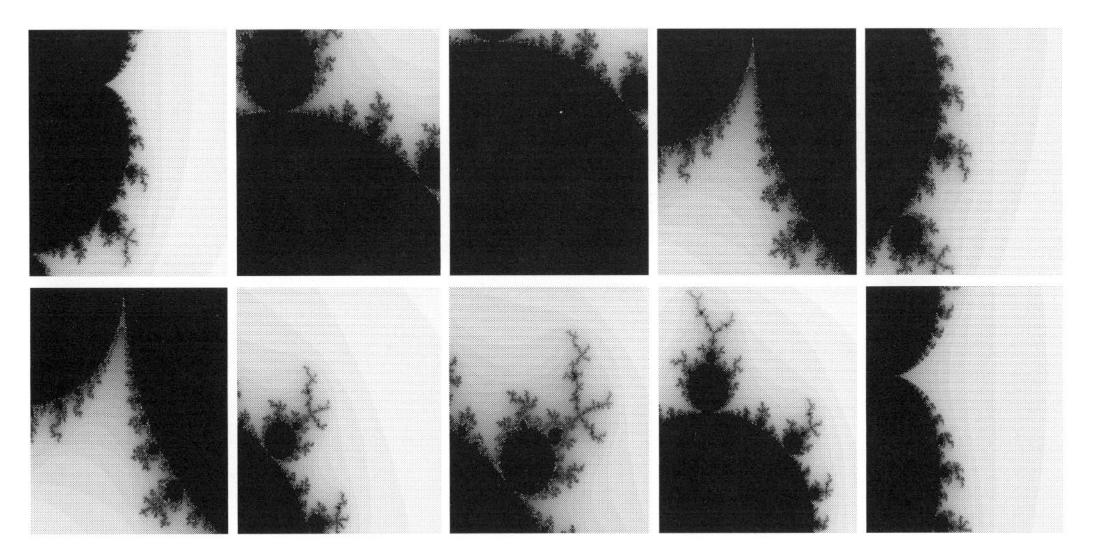

Figure 1.12 Examples of fractal protein shapes evolved by computer, designed to interact with each other in a GRN (Bentley, 2003b).

1.4.2 Case study 2: an evolutionary developmental model

In the second case study we illustrate a developmental testbed, known as an *evolutionary developmental system* (EDS), designed by Kumar and Bentley for the investigation of multicellular developmental processes and mechanisms for evolutionary design. The system uses analogues of proteins, genes, receptors and cells coupled to a genetic algorithm to evolve robust adaptive developmental programmes that specify the construction of varying three-dimensional morphologies, such as spheres, cubes, rectangles. The purpose of the work is to study the processes, mechanisms and pathways evolved to generate such forms.

This system is intended to model biological development very closely in order to discover the key components of development and their potential for computer science. This section is divided

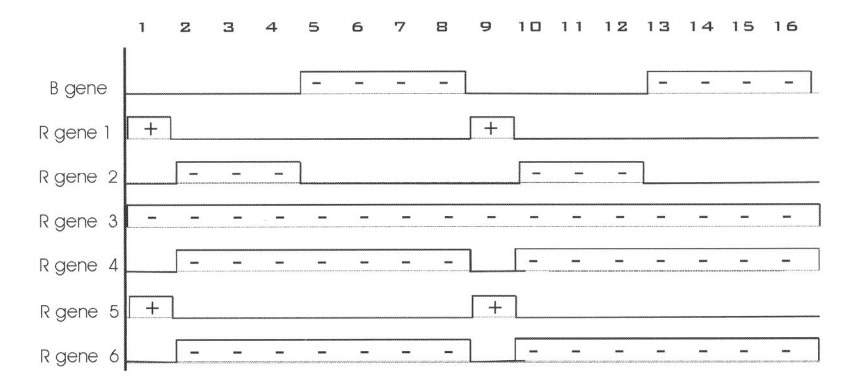

Figure 1.13 GRN pattern of a perfect solution. A high '–' indicates a gene active because it has not been repressed. A high '+' indicates a gene activated by a match to its promoter. A low indicates the gene is inactive (Bentley, 2003b).

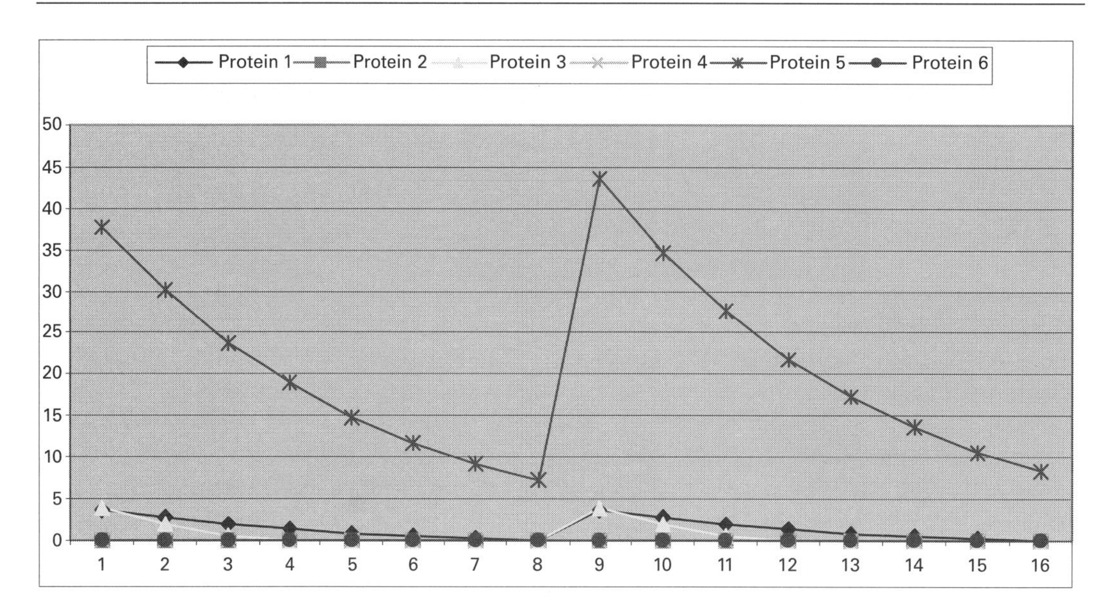

Figure 1.14 The concentrations (y-axis) of each protein in the cytoplasm every time step (x-axis) for the GRN in Figure 1.13. Note how proteins 1 and 3 begin at almost identical concentration levels, but the concentration of protein 3 decreases to zero in exactly 4 time steps. This is achieved by the action of gene 3 which suppresses concentration levels (i.e. speeds up diffusion) of protein 3, compared to the normal diffusion rate shown by protein 1 (Bentley, 2003b).

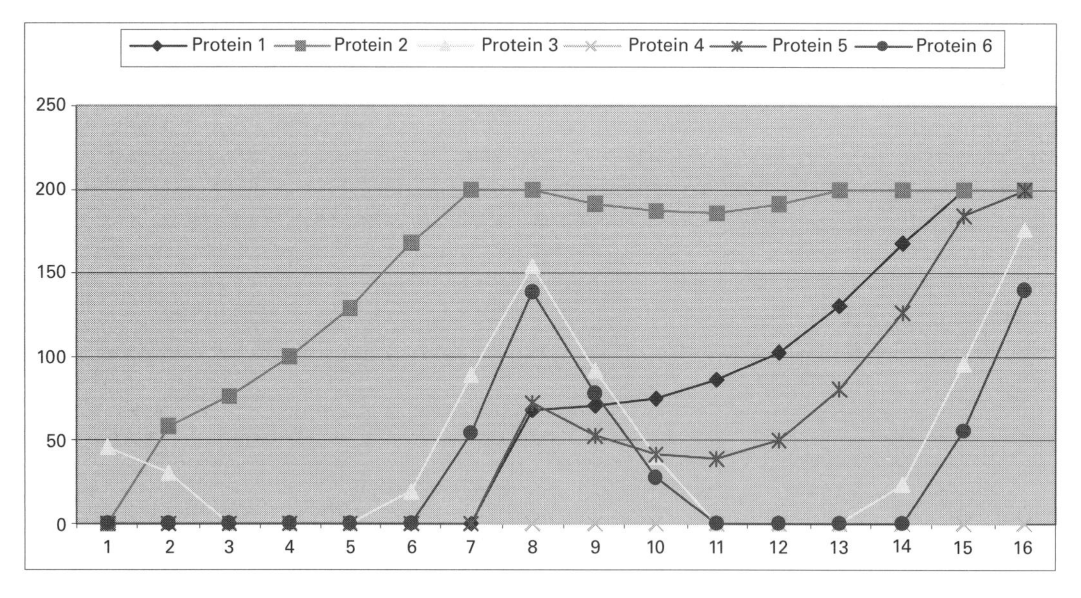

Figure 1.15 Protein concentration levels (y-axis) during development of a different evolved solution each time step (x-axis). In this solution, proteins interact and behave as catalysts to ensure that specific genes are active or inactive at specific times (Bentley, 2003b).

into subsections covering different aspects of the evolutionary developmental system (EDS). It begins with an overview of the entire system, followed by sections detailing individual components in isolation. These individual components are then drawn together and how they work as part of the overall developmental system is detailed as well as the role of evolution and how the genetic algorithm is wrapped around the developmental core. Finally, we present some examples of results generated during on-going experiments.

The evolutionary developmental system (EDS)

The evolutionary developmental system is a software testbed designed to encapsulate key developmental processes within a computer model. At the heart of the EDS lies the developmental core. This implements concepts such as embryos, cells, cell cytoplasm, cell wall, proteins, receptors, transcription factors (TFs), genes and cis-regulatory regions. Genes and proteins form the atomic elements of the system. A cell stores proteins within its cytoplasm and its genome (comprising rules that collectively define the developmental programme) in the nucleus. The overall embryo is the entire collection of cells (and proteins emitted by them) in some final conformation attained after a period of development. Finally, a genetic algorithm is wrapped around the developmental core. This provides the system with the ability to evolve genomes for the developmental machinery to execute.

Components of the EDS

The following sections describe the main components of the developmental model: proteins, genes and cells.

Proteins

The EDS captures the concept of a protein as an object. Each protein has an ID tag, which is simply an integer number. The EDS uses eight proteins (although number of proteins used is a user-defined variable in the system). Protein objects contain both a current and a new state object (at the end of each developmental cycle all protein new states are swapped with current states to provide 'parallel' protein behaviour). These protein state objects house important protein-specific information, for example, the protein diffusion coefficient.

Protein creation, initialization and destruction In the EDS, proteins do not exist in isolation; they are created and owned by cells. Thus, during protein construction each protein is allocated spatial coordinates inherited from the cell creating the protein. Handling protein coordinate initialization using this method overcomes the problem of knowing which cell created which proteins.

A protein lookup table (extracted from the genome, see next section) holds details about all proteins and is used to initialize each protein upon creation. It has the following details for each protein:

Rate of synthesis	amount by which the protein is synthesized
Rate of decay	amount by which the protein decays
Diffusion coefficient	amount by which the protein diffuses
Interaction strength	strength of protein interaction, i.e. activation or inhibition
Protein type	ID tag, e.g. long-range hormone, or short-range receptors

Additionally, each protein keeps the following variables:

Bound?	whether or not a receptor protein is currently bound (only operational in receptor proteins)
Protein source concentration	the current concentration of the protein
Spatial coordinates	the position of the source of the protein

Protein destruction in the EDS is implemented by simply setting the protein's source concentration to zero: if the concentration is zero there can be no diffusion, unless more of the protein is synthesized.

Protein diffusion Diffusion is the process by which molecules spread or wander due to thermal motions (Alberts *et al.*, 1994). When molecules in liquids collide, the result is random movement. Protein molecules are no different: they diffuse.

The average distance that a molecule travels from its starting point is proportional to the square root of the time taken to do so. For example, if a molecule takes on average 1 second to move 1 μm, it will take 4 seconds to move 2 μm, 9 seconds to move 3 μm, and 100 seconds to move 10 μm. Diffusion represents an efficient method for molecules to move short distances, but an inefficient method to move over large distances. Generally, small molecules move faster than large molecules (Alberts *et al.*, 1994).

Protein diffusion in the EDS models this behaviour. Diffusion is implemented by using a Gaussian function centred on the protein source. The use of the Gaussian assumes proteins diffuse equally in all directions from the cell.

In more detail: the source concentration records the amount of the current protein. Every iteration, its value is decremented by the corresponding 'rate of decay' parameter. If expressed by a gene, its value is also incremented by the corresponding 'rate of synthesis' parameter. To calculate the concentration of a protein at a distance x from the protein source:

$$concentration = s \times \mathbf{e}^{\frac{-x^2}{2d^2}}$$

where: d is the diffusion coefficient of the current protein
x is the distance from protein source to current point
s is the current protein source concentration.

Figure 1.16 illustrates the way protein concentration changes according to the three variables: distance, diffusion coefficient and source concentration.

Genes

The EDS employs two genomes. The first contains protein specific values (e.g. synthesis, decay, diffusion rates, see above). These are encoded as real floating-point numbers. The second describes the architecture of the genome to be used for development; it describes which proteins are to play a part in the regulation of different genes. It is this second genome that is copied into every cell during development; the information evolved on the first genome is only needed to initialize proteins with their respective properties.

In nature, genes can be viewed as comprising two main regions: the cis-regulatory (Davidson, 2001) region and the coding region. Cis-regulatory regions are located just before (upstream of) their associated coding regions and effectively serve as switches that integrate signals received (in the form of proteins) from both the extracellular environment and the cytoplasm. Coding regions specify a protein to be transcribed upon successful occupation of the cis-regulatory

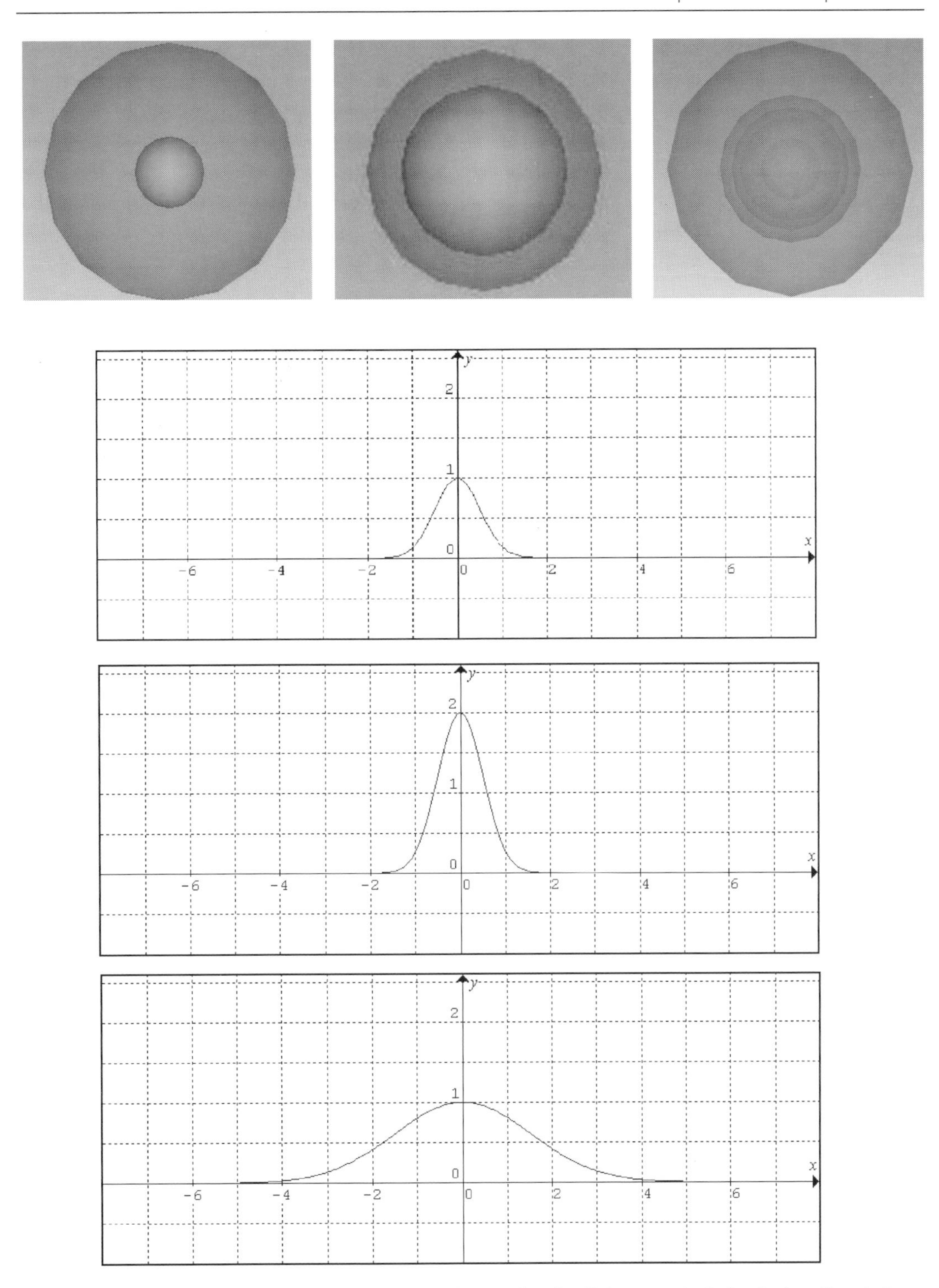

Figure 1.16 (top) Examples of proteins with their associated cell (at centre). Left: single cell emitting a long-range hormone-type protein. Middle: single cell emitting a short range (local) protein. Right: single cell emitting four proteins of various spread, reflected by the radius of each protein sphere.
(bottom) Plot of protein concentration against distance from source, where: $d = 0.5$ and $s = 1.0$ (top), $d = 0.5$ and $s = 2.0$ (middle), and $d = 1.5$ and $s = 1.0$ (bottom).

region by assembling transcription machinery. Currently, the EDS's underlying genetics model assumes a 'one gene, one protein' (Lewin, 1999) simplification rule (despite biology's ability to construct multiple proteins); this aids in the analysis of resulting genetic regulatory networks. To this end, the activation of a single gene, in the EDS, results in the transcription of a single protein. This is, currently, ensured by imposing the following structure over genes: each gene comprises both a cis-regulatory region and a consequent protein-coding region.

A novel genome representation (based on eukaryotic genetics) was devised as part of ongoing research into the use of development for evolutionary design using the EDS testbed (Kumar and Bentley, 2003). The genome is represented as an array of gene objects (Figure 1.17). Genes are objects containing two members: a regulatory region and a protein-coding region. The cis-regulatory region contains an array of TF target sites; these sites bind TFs in order to regulate the activity of the gene.

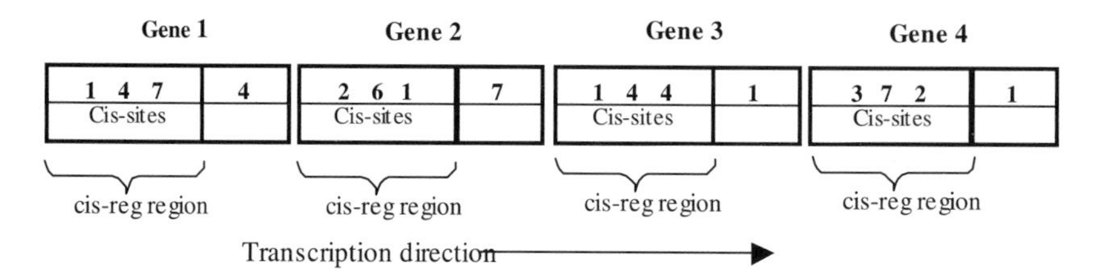

Figure 1.17 An arbitrary genome created by hand. Genes consist of two objects: a cis-regulatory region and a coding region. Cis-regulatory regions consist of transcription factor target sites that bind TFs, triggering transcription of the adjacent coding region. Each number denotes a protein.

The gene then integrates these TFs and either switches the gene 'on' or 'off'. Integration is performed by summing the products of the concentration and interaction strength (weight) of each TF, to find the total input of all TFs occupying a single gene's cis-regulatory region:

$$input_j = \sum_{i=1}^{d} conc_i * weight_{ij}$$

where: $input_j$ represents the total input, jth gene, i, is the current TF,
d is the total number of TF proteins visible to the current gene,
$conc_i$ is the concentration of ith TF at the centre of the current cell,
$weight_{ij}$ is the interaction strength between TF i and gene j.

This sum provides the input to an equation containing a sigmoid threshold function (a hyperbolic tangent function), which provides a probability (between 0 and 1) of gene activation.

$$activity_j = \frac{input_j - THRESHOLD\_CONSTANT}{SHARPNESS\_CONSTANT}$$

where: $activity_j$ represents the total activity of the jth gene,
$input_j$ is the total input to the jth gene,
$THRESHOLD\_CONSTANT$ is a constant normally set to 0.1,
$SHARPNESS\_CONSTANT$ is a constant normally set to 0.1.

The *activation probability* is the probability of the *jth* gene activating given by:

$$activation\_probability_j = \frac{1 + \tanh(activity_j)}{2}$$

where: $activation\_probability_j$ is the probability of activation for the *jth* gene, $activity_j$ represents the total activity of the *jth* gene.

Figure 1.18 illustrates this sigmoid-based calculation used to determine whether a gene is activated and produces its corresponding transcription factor or not.

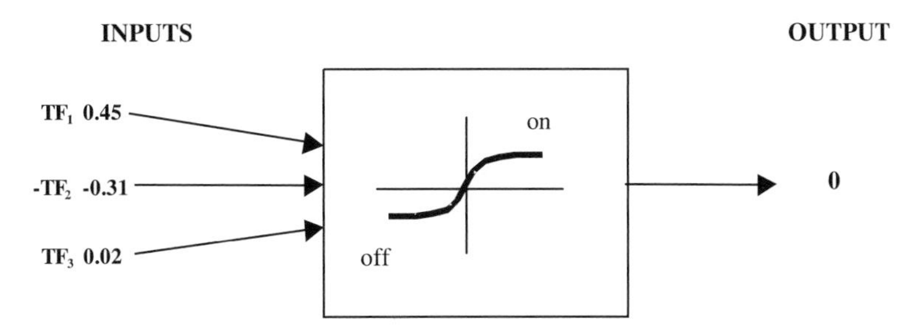

Figure 1.18 A gene showing the various positive and negative inputs received in the form of transcription factors, with their respective interaction strengths (weights) and concentrations of 0.24, 0.87 and 0.11 respectively. Internally, the gene integrates these TFs and produces a gene activation probability (in this case 0, and the gene is not transcribed). TF_1 and TF_3 are both activators, whereas TF_2 is a repressor, denoted by a '−' symbol.

Cells

Cells can be viewed as autonomous agents. These agents have sensors in the form of surface receptors able to detect the presence of certain molecules within the environment. Additionally, the cell has effectors in the form of hundreds and thousands of protein molecules transcribed from a single chromosome able to affect other genes in other cells. Cells resemble multitasking agents, able to carry out a range of behaviours. For example, cells are able to multiply, differentiate and die.

Like protein objects, cell objects in the EDS have two states: *current* and *new*. During development, the system examines the current state of each cell, depositing the results of the protein interactions on the cell's genome in that time step into the new state of the cell. After each developmental cycle, the current and new state of each cell is swapped ready for the next cycle.

The EDS supports a range of different cell behaviours, triggered by the expression of certain genes. These are currently: division (when an existing cell 'divides', a new cell object is created and placed in a neighbouring position), differentiation (where the function of a cell is fixed, e.g. colour = 'red' or colour = 'blue') and apoptosis (programmed cell death).

The EDS uses an *n-ary* tree data structure to store the cells of the embryo, the root of which is the zygote. As development proceeds cell multiplication occurs. The resulting cells are stored as child nodes of the root in the tree. Proteins are stored within each cell. When a cell needs to examine its local environment to determine which signals it is receiving, it traverses the tree, checks the state of the proteins in each cell against its own and integrates the information.

Evolution

A genetic algorithm (GA) is 'wrapped around' the developmental model. The GA represents the driving force of the system. Its main roles are to:

1. provide genotypes for development
2. provide a task or function and hence a measure of success and failure
3. search the space of genotypes that give rise to developmental programmes capable of specifying embryos, correctly and accurately according to the task or function.

Individuals within the population of the genetic algorithm comprise a genotype, a phenotype (in the form of an embryo object) and a fitness score. After the population is created, each individual has its fitness assessed through the process of development. Each individual is permitted to execute its developmental programme according to the instructions in the genome. After development has ended a fitness score is assigned to the individual based upon the desired objective function.

The EDS uses a generational GA with tournament selection (typically using $\frac{1}{4}$ of population size) and real coding. Crossover is applied with 100% probability. Creep mutation is applied with a Gaussian distribution (small changes more likely than large changes), with probability between 0.01 and 0.001 per gene.

Coordinates and visualization The underlying coordinate system used by the EDS is isospatial. All coordinate systems have inherent biases towards different morphologies; the isospatial system is no different. However, the isospatial system bias results in what can only be described as more natural (biologic) morphologies than its Cartesian counterpart (Frazer, 1995). Isospatial coordinates permit a single cell to have up to twelve equidistant neighbours defined by 6 axes (Figure 1.19), Cartesian coordinates only permit 6 neighbours.

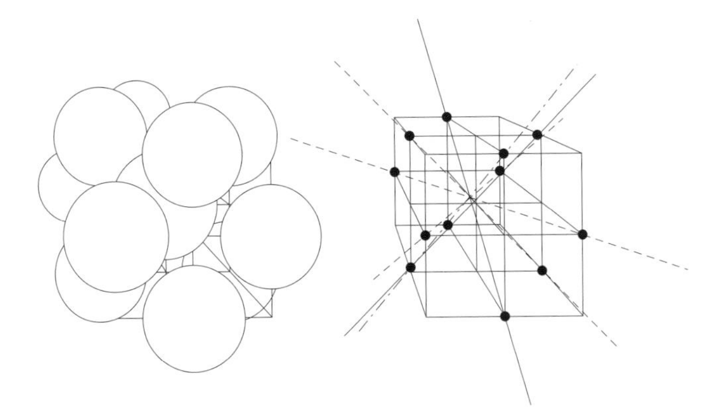

Figure 1.19 Isospatial coordinates permit twelve equidistant neighbours for each cell (left) and are plotted using six axes (right).

The EDS automatically writes VRML files of developed embryos, enabling three-dimensional rendered cells and proteins to be visualized. Cells are represented by spheres of fixed radius; proteins are shown as translucent spheres of radius equal to the extent of their diffusion from their source cells. In order to place a cell in VRML its Cartesian coordinates need to be defined: to this end, isospatial coordinates are converted to Cartesian. Figure 1.16 (top) (see page 33) illustrates how cells and proteins appear when rendered.

Experiments

Because of the complexity of the system, numerous experiments can be performed to assess behaviour and capabilities. Here we briefly outline two:

1. The ability of genes and proteins to interact and form genomic regulatory networks within a single cell.
2. The evolution of a 3D multicellular embryo with form as close to a prespecified shape as possible.

Genetic regulatory networks In order to assess the natural capability of the EDS to form GRNs independently of evolution, genomes of five random genes were created and allowed to develop in the system for ten developmental steps. The cell was seeded with a random set of eight proteins (maternal factors).

Figure 1.20 (top) shows an example of the results of this experiment. The pattern shows gene four exhibiting autocatalytic behaviour having initially bound to protein zero. (Gene four is activated when in the presence of protein zero and produces protein zero when activated.)

Figure 1.20 (bottom) shows an example of the pattern that results when the initial random proteins (initial conditions) are varied very slightly, but the genome is kept constant. Again, gene four shows the same autocatalytic behaviour, but now the GRN has found an alternative pattern of activation. These two runs illustrate the difference the initial proteins can make on the resulting GRN.

Morphogenesis: evolving a spherical embryo In addition to GRNs, the other important capability of the EDS is cellular behaviour. The second experiment focuses on morphogenesis, i.e. the generation of an embryo with specific form, constructed through appropriate cellular division and placement, from an initial single zygote. For this experiment, the genetic algorithm was set up as described previously, with the fitness function providing selection pressure towards spherical embryos of radius 2 (cells have a radius of 0.5).

Figure 1.21 shows examples of the initially random embryos with their corresponding proteins produced by the GRNs. Figure 1.22 shows two examples of final 'spherical' embryos. As well as having appropriate forms, it is clear that the use of proteins has been reduced by evolution. Interestingly, analysis indicates that evolution did not require complex GRNs to produce such shapes. It seems likely that it is the natural tendency of the EDS to produce near-spherical balls of cells, hence evolution simply did not need to evolve intricate GRNs for this task. Colour plate 1.1 shows four embryos with differentiated cells.

1.4.3 Summary of case studies

The third and final major section of this chapter has provided two detailed case studies of computational development. The first focused on modelling complex chemistries by the use of fractal proteins, designed to be evolvable through the high exploitability of their infinitely variable shapes and through their combination into fractal 'compounds'. The second described an evolutionary developmental system, based on an object-oriented model of proteins, genes and cells, capable of intricate genomic regulatory networks and spherical embryos constructed from balls of cells.

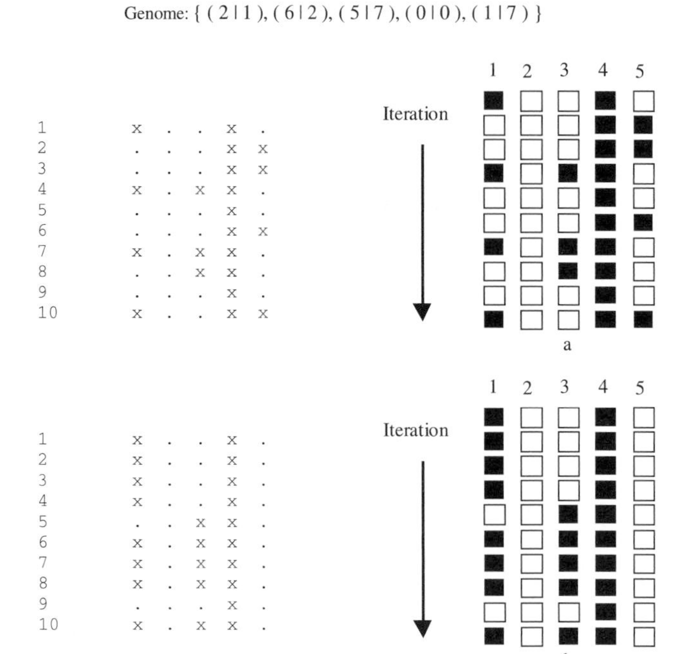

Figure 1.20 Gene expression patterns for a run of a randomly created genome seeded with a random subset of proteins. The left side shows the raw output from the system where an 'x' means the gene in that column is 'on' and '.' means the gene is 'off'. The right side depicts this text pattern as a graphical output viewed as a 1D CA iterated over ten time-steps. Note, gene 4, i.e. (0 | 0) is autocatalytic.

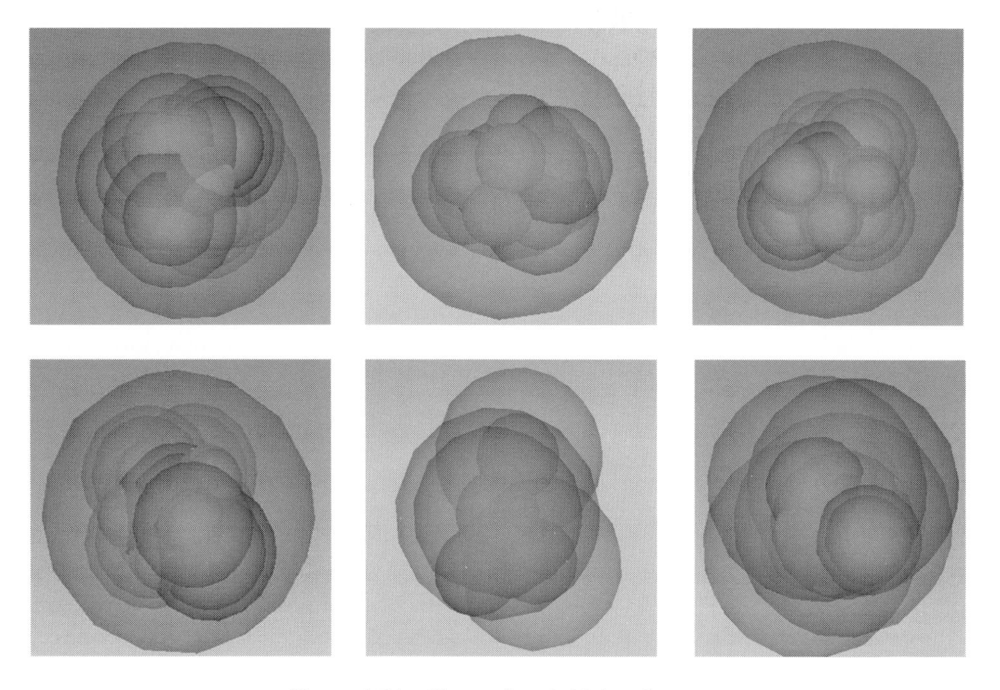

Figure 1.21 Six random initial embryos.

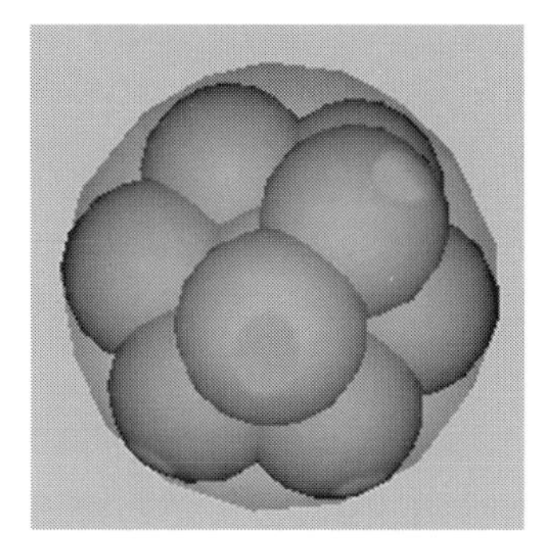
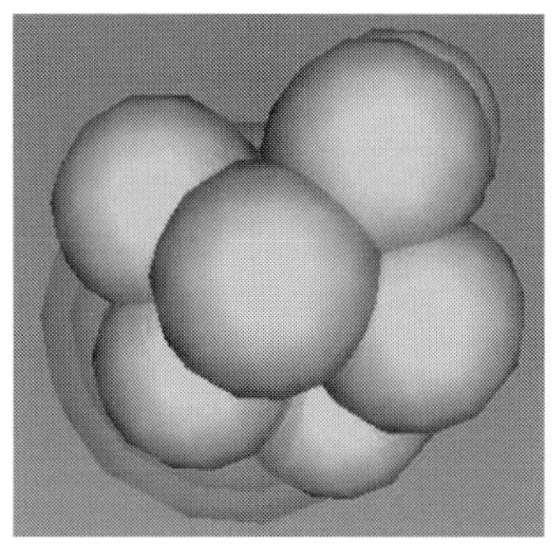

Figure 1.22 Two 'spherical' embryos. Using the equation of a sphere as a fitness function with sphere of radius 2.0.

These are just two ways in which development can be modelled by computer. There are many other ideas and approaches, as you will discover in the other chapters of this book.

1.5 Chapter summary

Development is one of the miracles of nature – it is extremely impressive and we do not really understand it. Computational development provides us with a new way of investigating developmental processes, through computer modelling. It also provides the technologists new and exciting methods for using computers to solve our problems.

This chapter has provided an introduction to computational development, but in a sense, the whole of this book is designed as an introduction to this exciting new field. From detailed explorations of the biology, mathematical formalizations of gene–gene or cell–cell interactions and the role of physics in development, to detailed computer models of development and applications of developmental algorithms to specific problems, this book attempts to give you, our reader, the best introduction possible. We hope you enjoy it!

Recommended reading

Davidson, E.H. (2001) ***Genomic Regulatory Systems: Development and Evolution.*** Academic Press.
Gould, S.J. (1974) ***Ontogeny and Phylogeny.*** The Belknap Press of Harvard University Press.
Lawrence, P. (1992) ***The Making of a Fly: the genetics of animal design.*** Blackwell Science.
Lewin, B. (1999) ***Genes VI.*** Oxford University Press.
Meinhardt, H. (1982) ***Models of biological pattern formation.*** Academic Press.
Meinhardt, H. (1998) ***The Algorithmic Beauty of Seashells.*** Springer-Verlag.
Murray, J.D. (1993) ***Mathematical Biology.*** **2nd Edn.** Springer-Verlag.
Needham, J. (1959) ***A History of Embryology.*** Cambridge University Press.
Prusinkiewicz, P. and Lindenmayer, A. (1990) ***The Algorithmic Beauty of Plants.*** Springer-Verlag.
Raff. R. (1996) ***The Shape of Life: Genes, Development, and the evolution of animal form.*** The University of Chicago Press.
Smith, J.M. and Szathmary, E. (1995) ***The Major Transitions in Evolution.*** Oxford University Press.

Thompson, D'arcy (1961) ***On Growth and Form***. Cambridge University Press.
Wolpert, L. (1991) ***The Triumph of the Embryo.*** Oxford University Press.
Wolpert, L. (1998) ***The Principles of Development***. Oxford University Press.

References

Alberts, B., Johnson, A., Lewis, J., Raff, M., Roberts, K. and Walter, P. (1994) *Molecular Biology of the Cell*. Garland Publishing.

Astor, C. and Adami, C. (1998) Development and evolution of neural networks in an artificial chemistry. In Wilke, C., Altmeyer, S. and Martinetz, T. eds. *Proceedings of the Third German Workshop on Artificial Life*, pp. 15–30. Verlag Harri Deutsch.

Beer, R.D. (1995) A dynamical system perspective on agent-environment interaction. *Artificial Intelligence*, **72**(1–2), 173–215.

Bentley, P.J. (ed.) (1999) *Evolutionary Design by Computers*. Morgan Kaufmann Pub.

Bentley, P.J. (2002) *Digital Biology*. Simon and Schuster.

Bentley, P. J. (2003a) Evolving Fractal Proteins. In *Proc. of ICES '03, the 5th International Conference on Evolvable Systems: From Biology to Hardware. LNCS*, **2606**, 81–92. Springer-Verlag.

Bentley, P.J. (2003b) Fractal Proteins. To appear in *Genetic Programming and Evolvable Machines Journal*.

Bentley, P.J. (2003c) *Evolving Beyond Perfection: An Investigation of the Effects of Long-Term Evolution on Fractal Gene Regulatory Networks* (in press).

Bentley, P.J. and Corne, D.W. (eds) (2001) *Creative Evolutionary Systems*. Morgan Kaufmann Pub.

Bentley, P.J. and Kumar, S. (1999) Three ways to grow designs: A comparison of three embryogenies for an evolutionary design problem. In *Proceedings of Genetic and Evolutionary Computation* 1999, pp. 35–43. Morgan Kaufmann.

Boers, E.J.W. and Kuiper, H. (1992) *Biological Metaphors and the Design of Modular Artificial Neural Networks*. Masters Thesis Leiden University.

Bongard, J.C. (2002) Evolving modular genetic regulatory networks. In *Proceedings of the IEEE 2002 Congress on Evolutionary Computation (CEC2002)*, pp. 1872–1877. IEEE Press.

Bongard, J.C. and Paul, C. (2000) Investigating Morphological Symmetry and Locomotive Efficiency Using Virtual Embodied Evolution. In Meyer, J.-A. *et al.* eds. *From Animals to Animats: Proceedings of the Sixth International Conference on the Simulation of Adaptive Behavior*: MIT Press.

Cangelosi A. (1999) Heterochrony and adaptation in developing neural networks. In Banzhaf, W. *et al.* eds. *Proceedings of GECCO99 Genetic and Evolutionary Computation Conference*, pp. 1241–1248. Morgan Kaufmann.

Canham, R.O. and Tyrrell, A.M. (2002) A multilayered immune system for hardware fault tolerance within an embryonic array. In Timmis, J. and Bentley, P.J. eds. *Proc. of the First Int. Conf. on Artificial Immune Systems*, pp. 3–11. University of Kent Printing Unit.

Chaplain, M.A.J., Singh, G.D. and McLachlan, J.C. (eds) (1999) *On Growth and Form: Spatio-temporal Pattern Formation in Biology*. John Wiley Publishers.

Davidson, E.H. (2001) *Genomic Regulatory Systems: Development and Evolution*. Academic Press.

Dellaert, F. and Beer, R.D. (1994) Toward an evolvable model of development for autonomous agent synthesis. In Brooks, R. and Maes, P. eds. *Proceedings of the Fourth Conference on Artificial Life*, MIT Press.

Dellaert, F. and Beer, R.D. (1996) A developmental model for the evolution of complete autonomous agents. In Maes, P. *et al.* eds. *From Animals to Animats: Proceedings of the Fourth International Conference on Simulation of Adaptive Behavior*, pp. 393–401. MIT Press.

Eggenberger, P. (1996) Cell interactions as a control tool of developmental processes for evolutionary robotics. In Maes, P. *et al.* eds. *Proceedings of the Fourth International Conference on Simulation of Adaptive Behavior*, pp. 440–448. MIT Press.

Eggenberger. P. (1997) Evolving morphologies of simulated 3D organisms based on differential gene expression. *Proceedings of the Fourth European Conference on Artificial Life*, pp. 205–213. Springer-Verlag.

Fleischer, K. and Barr, A. (1993) A simulation testbed for the study of multicellular development: The multiple mechanisms of morphogenesis. In Langton, C. ed., *Artificial Life III*, pp. 389–416. Addison-Wesley.

Fogel, L.J., Owens, A.J. and Walsh, M.J. (1966) *Artificial Intelligence through Simulated Evolution*. Wiley.

Fontana, W., Wagner, G. and Buss, L.W. (1994) Beyond digital naturalism, *Artificial Life*, **1**, 211–227.

Frazer, J. (1995) *An Evolutionary Architecture*. Architectural Assoc.

Gibson, M.A. and Mjolsness, E. (2001) Modelling the activity of single genes. In Bower, J.M. and Bolouri, H. eds. *Computational Modelling of Genetic and Biochemical Networks*, pp. 1–48. MIT Press.

Gordon, T.W. and Bentley, P.J. (2002) Towards development in evolvable hardware. In *Proc. of the 2002 NASA/DoD Conference on Evolvable Hardware* (EH-2002, Washington DC, July 15–18, 2002), pp. 241–250. IEEE Press.

Gould, S.J. (1974) *Ontogeny and Phylogeny*. The Belknap Press of Harvard University Press.

Gruau, F. (1994) *Neural Network Synthesis using Cellular Encoding and the Genetic Algorithm*. PhD thesis, Laboratoire de l'Informatique du Parallelisme, Ecole Normale Superieure de Lyon, France.

Holland, J.H. (1975) *Adaptation in Natural and Artificial Systems*. University of Michigan Press.

Holland, J.H. (1998) *Emergence: from chaos to order*. Oxford University Press.

Hornby, G.S., Lipson, H. and Pollack, J.B. (2001) Evolution of generative design systems for modular physical robots. In *Intl. Conf. on Robotics and Automation*, pp. 4146-4151. IEEE-RAS Pub.

Jakobi, N. (1995) Harnessing morphogenesis. In *Proceedings of International Conference on Information Processing in Cells and Tissues*, pp. 29–41.

Kauffman, S.A. (1969) Metabolic stability and epigenesis in randomly constructed genetic nets. *Journal of Theoretical Biology*, **22**, 437–467.

Kauffman, S.A. (1993) *The Origins of Order: self-organisation and selection in evolution*. Oxford University Press.

Kerszberg, M. and Changeux, J-P. (1998) A simple molecular model of neurulation. *BioEssays* **20**: 758–770.

Kitano, H. (1990) Designing neural networks using genetic algorithms with graph generation system. *Complex Systems*, **4**, 461–476.

Kitano, H. (1995) A simple model of neurogenesis and cell differentiation based on evolutionary large-scale chaos. *Artificial Life*, **2** (1), 79–99.

Kitano, H., Hamahashi, S. and Luke, S. (1998) The perfect *C. elegans* project: an initial report. *Artificial Life*, **4**(2),141–156.

Kodjabachian, J. and Meyer, J.-A. (1994) Development, learning and evolution in animats. In *Perception to Action Conference Proceedings*, Gaussier, P. and Nicoud, J.-D. eds, pp. 96–109. IEEE Computer Society Press.

Kodjabachian, J. and Meyer, J. (1995) Evolution and development of control architectures in animats. *Robotics and Autonomous Systems*, **16**(2–4), 161–182.

Koza, J. (1992) *Genetic Programming: On the Programming of Computers by Means of Natural Selection*. MIT Press.

Koza, J., Bennett III, F.H., Andre, D. and Keane, M.A. (1999) *Genetic Programming III. Darwinian Invention and Problem Solving*. Morgan Kaufmann.

Kumar, S. and Bentley, P.J. (1999) The ABCs of Evolutionary Design: Investigating the Evolvability of Embryogenies for morphogenesis. A Late-breaking paper submitted to *Genetic and Evolutionary Computation Conference* (GECCO '99), July 14–17, 1999, Orlando.

Kumar, S. and Bentley, P.J. (2002) Computational embryology: past, present and future. In Ghosh, A. and Tsutsui, S. eds. *Theory and Application of Evolutionary Computation: Recent Trends*. Springer-Verlag.

Kumar, S. and Bentley, P.J. (2003) Biologically inspired evolutionary development. In *Proceedings of the International Conference on Evolvable Systems: from biology to hardware (ICES 2003)*. Trondheim, Norway.

Kyoda, K., Muraki, M., and Kitano, H. (2000) Construction of a generalized simulator for multicellular organisms and its application to Smad signal transduction, *Proc. Pacific Symposium on Biocomputing 2000*, pp. 317–328. World Scientific Pub.

Lawrence, P. (1992) *The Making of a Fly: the genetics of animal design*. Blackwell Science.

Lewin, B. (1999). *Genes VI*. Oxford University Press.

Lindenmeyer, A. (1968) Mathematical models for cellular interaction in development, parts I and II. *Journal of Theoretical Biology*, **18**, 280–315.

Mandelbrot, B. (1982) *The Fractal Geometry of Nature*. W.H. Freeman & Company.

Maturana, H.R. and Varela, F.J. (1992) *The Tree of Knowledge: the Biological Roots of Human Understanding*. Revised edition. Shambala Publications, Inc.

Meinhardt, H. (1982) *Models of Biological Pattern Formation*. Academic Press.

Meinhardt, H. (1998) *The Algorithmic Beauty of Seashells*. Springer-Verlag.

Miller, J.F. and Thomson, P. (2000) Cartesian genetic programming. In *Proceedings of the Third European Conference on Genetic Programming. LNCS*, **1802**, 121–132.

Mjolsness, E., Sharp, D.H. and Reinitz, J. (1995) A connectionist model of development. *Journal of Theoretical Biology*, **176**, 291–300.

Murray, J.D. (1993) *Mathematical Biology*. 2nd ed. Springer-Verlag.

Needham, J. (1959) *A History of Embryology*. Cambridge University Press.

Nolfi, S. and Parisi, D. (1993) *Phylogenetic Recapitulation in the Ontogeny of Artificial Neural Networks*. Technical Report PCIA-18–93, Institute of Psychology, C.N.R.- Rome.

Nolfi, S. and Parisi, D. (1995) Evolving artificial neural networks that develop in time. In Mor'an, F., Moreno, A., Merelo, J.J. and Chac'on, P. eds. *Advances in Artificial Life, Proceedings of the Third European Conference on Artificial Life*, Granada, pp. 353–367. Springer.

Odell, G., Oster, G., Alberch, P. and Burnside, B. (1981) The mechanical basis of morphogenesis. *Developmental Biology*, 85.

Pfeifer, R. (1998) Embodied system life. In *Proc. of the International Symposium on System Life*, Tokyo, July. IEEE Press.

Prusinkiewicz, P. and Lindenmayer, A. (1990) *The Algorithmic Beauty of Plants*. Springer-Verlag.

Prusinkiewicz, P., Hammel, M. and Mjolsness, E. (1993) Animation of plant development. *Proceedings of SIGGRAPH 93*. Anaheim, California, August 1–6. ACM Pub.

Prusinkiewicz, P., James, M., Mech, R. and Hanan, J. (1995) The Artificial Life of Plants. *Course Notes of SIGGRAPH '95*, **I**, 1–38. ACM Pub.

Prusinkiewicz, P., Hammel, M., Hanan, J. and Mech. R. (1996) Visual models of plant development. In Rozenberg, G. and Salomaa, A. eds. *Handbook of Formal Languages*. Springer-Verlag.

Quick, T., Dautenhahn, D., Nehaniv, C. and Roberts, G. (1999a) On bots and bacteria: ontology independent embodiment. In *Advances in Artificial Life: 5th European Conference on Artificial Life (ECAL'99)*, Lausanne, Switzerland, pp. 339–343. Springer.

Quick, T., Dautenhahn, K., Nehaniv, C. and Roberts, G. (1999b) The essence of embodiment: A framework for understanding and exploiting structural coupling between system and environment. In *Proc. of Third Int. Conf. on Computing Anticipatory Systems* (CASYS'99), Symposium 4 on Anticipatory, Control and Robotic Systems, Liege, Belgium, pp. 16–17. AIP Conference Proceedings 517.

Raff. R. (1996) *The Shape of Life: Genes, Development, and the Evolution of Animal Form*. The University of Chicago Press.

Rechenberg, I. (1973) *Evolutionstrategie: Optimierung Technischer Systeme nach Prinzipien der Biolischen Evolution*. Frommann-Holzboog Verlag.

Reil, T. (2000) Models of Gene Regulation – A Review. In Maley, C.C. and Boudreau, E. eds. *Artificial Life 7 Workshop Proceedings*, pp. 107–113, MIT Press.

Rust, A.G., Adams, R. George, S. and Bolouri, H. (1996) *Artificial Evolution: Modelling the Development of the Retina*. Technical Memorandum ERDC/1996/0015, ERDC, University of Hertfordshire.

Saville, N. and Hogeweg, P. (1997) The modelling of slug morphogenesis. *Journal of Theoretical Biology*, **22**, 437–467.

Sims, K. (1999). Evolving three dimensional morphology and behaviour. In Bentley, P.J. ed. *Evolutionary Design by Computers*. Morgan Kaufmann Pub.

Smith, J.M. and Szathmary, E. (1995) *The Major Transitions in Evolution*. Oxford University Press.

Tateson, R. (1998) Self-organising pattern formation: fruit flies and cell phones. In *Proceedings of Parallel Problem Solving From Nature (PPSN 1998)*. pp. 732–744. Springer LNCS 1498.

Thompson, A. (1996a) Evolutionary Techniques for Fault Tolerance. *Proc. UKACC Int. Conf. on Control* (CONTROL'96), IEE Conference Publication No. 427, pp. 693–698.

Thompson, A. (1996b) An evolved circuit, intrinsic in silicon, entwined with physics. *International Conference on Evolvable Systems*, pp. 390–405.

Thompson, D'arcy (1961) *On Growth and Form*. Cambridge University Press.

Turing, A. (1952). The chemical basis of morphogenesis. *Phil. Trans. R. Soc. London B*, **237**, 37–72.

Vaario, J., Hori, K. and Ohsuga, S. (1995) Toward evolutionary design of autonomous systems. *Int. Journal in Computer Simulation*, **5**, 187–206. Springer LNCS 1259.

Wolpert, L. (1969) Positional information and the spatial pattern of cellular dfferentiation. *Journal of Theoretical Biology*, **25**, 1–47.

Wolpert, L. (1991) *The Triumph of the Embryo*. Oxford University Press.

Wolpert, L. (1998) *The Principles of Development*. Oxford University Press.

Yuh, C.-H., Bolouri, H. and Davidson, E. H. (1998) Genomic cis-regulatory logic: functional analysis and computational model of a sea urchin gene control system. *Science*, **279**, 1896–1902.

Section 1

Developmental Biology

Relationships between development and evolution

2

LEWIS WOLPERT

2.1 Evolutionary developmental biology

In one sense, DNA is rather boring and passive. It is the proteins produced from the DNA code that control the behaviour of the cells of the embryo. They carry out the key processes in the cell from production of energy to specific cell functions as seen in the variety of cell types. All changes in animal form and function during evolution are due to changes in their DNA. Such changes determine which proteins are made, where they are made and when, during embryonic development. One of the most important concepts in evolutionary developmental biology is that any developmental model for a structure must be able to account for the development of earlier forms in the ancestors. This means that multicellular organism development is almost inextricably linked to evolution (Carroll *et al.*, 2001).

It has been suggested that nothing in biology makes sense unless viewed in the light of evolution. Certainly it would be very difficult to make sense of many aspects of development without an evolutionary perspective. Every structure has two histories that relate to how it developed: ontogeny (its complete development to maturity) and phylogeny (its evolutionary history). Ontogeny does not recapitulate phylogeny but embryos often pass through stages that their evolutionary ancestors passed through. For example, in vertebrate development despite different modes of very early development, all vertebrate embryos develop to a rather similar phylotypic stage after which their development diverges (see Figure 2.1). This shared phylotypic stage, which is the embryonic stage after neurulation (formation of the neural tube, which later develops into the central nervous system) and the formation of the somites (body segments), is probably a stage through which some distant ancestor of the vertebrates passed. It has persisted ever since, to become a fundamental characteristic of the development of all vertebrates, whereas the stages before and after the phylotypic stage have evolved differently in different organisms.

In evolution, changes in organs usually involve modification of the development of existing structures – tinkering with what is already there. Good examples are the evolution of the jaws from the pharyngeal arches of jawless ancestors and the incus and stapes of the middle ear from bones originally at the joint between upper and lower jaws. However, it is possible that new structures could develop, as has been suggested for the digits of the vertebrate limb, but the developmental mechanisms would still be similar. It is striking how conserved developmental

 ISBN 0-12-428765-4

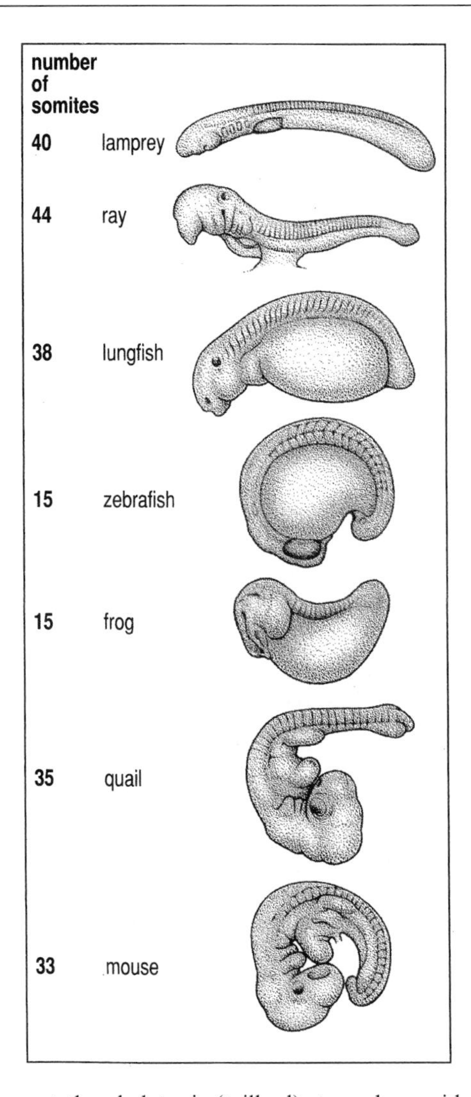

Figure 2.1 Vertebrate embryos at the phylotypic (tailbud) stage show wide variation in somite number.

mechanisms are in pattern formation both with respect to the genes involved and the intercellular signals. For example, many systems use the same positional information but interpret it differently. One of the ways the developmental programmes have been changed is by gene duplication, which allows one of the two genes to diverge and take on new functions – Hox genes are an example. Another mechanism for change involves the relative growth rates of parts of a structure.

The changes in genes are thought to be mainly in the so called cis-regulatory region of the genes rather than in the nature of the protein for which the gene codes. The cis-regulatory region is where protein factors – transcription factors – bind and determine whether the gene will be transcribed or not (Figure 2.2). Such regions can be very complex and, for example, the pair rule stripes in *Drosophila* (fruit fly) development are due to the genes having cis-regulatory regions such that the gene is expressed in seven distinct stripes across developing cells. Again, in the sea-urchin, the cis-regulatory region for a gene expressed in the gut is complex, and controls the

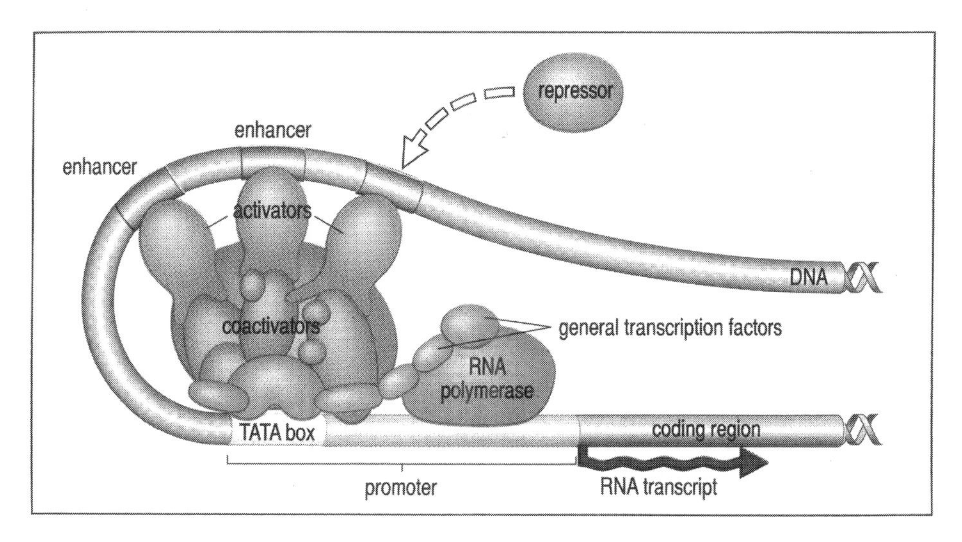

Figure 2.2 Gene expression is regulated by the coordinated action of gene regulatory proteins that bind to cis-regulatory control regions in DNA. The transcription machinery, composed of RNA polymerase and the general transcription factors that bind to the promoter region, is common to many cells. Other regulatory regions, which have to be bound by gene regulatory proteins (activators and repressors) before transcription can occur, are located adjacent to the promoter and usually also further downstream and sometimes upstream, of the gene. The upstream regulatory regions, such as enhancers, may be many kilobases away from the start point of transcription.

gene's spatial expression during development. The evolution of cis-regulatory regions can involve gene duplication, modification of existing regions and the co-options of other regulatory elements (Davidson, 2001).

2.1.1 Development of pattern

As an example of the developmental processes that have evolved I will consider pattern formation, which is a key process in embryonic development (Wolpert *et al.*, 2002). Many organisms use similar processes. When the fertilized egg divides, the cells it gives rise to become different and give rise to a variety of structures such as the main body and its appendages. These structures have a well-defined spatial organization – the vertebrate limb is a good example – and pattern formation is the process by which this organization is brought about. There are three other related processes: *cell differentiation* is the process by which different specialized adult cell types such as muscle, cartilage and neurons are generated; *morphogenesis*, or change in form, such as neural tube formation; and *growth* – a general feature of pattern formation that it is initially specified for any organ on a small scale rarely greater than one hundred cell diameters. These processes are related to one another but, in general, pattern formation is primary in that it specifies the other processes. Thus patterning in vertebrate limb development specifies where cartilage will form and how it will grow; it also specifies ectodermal cells such as the apical ridge which controls the overall form of the limb.

Positional information

In vertebrate development the main mechanism for pattern formation involves cell-to-cell interactions by cell signalling. One class of patterning mechanism is based on the generation by some cells (which may be zygotic or maternal) of positional information. Cells acquire their positional value and then interpret this according to their genetic constitution and developmental history. Studies on regeneration of newt limbs and insect tibia show very clearly that even adult cells can retain their positional values and generate new ones (Wolpert *et al.*, 2002).

How do cells acquire their positional values? Positional fields, that is the set of cells which have positional values with respect to the same boundaries, are small, less than about thirty cells along any axis. Thus patterning at any one time occurs with rather few cells and growth gives rise to the larger structures seen in the adult. A widely proposed model for specifying positional values is based on a graded distribution of a diffusible molecule, a morphogen. This morphogen would be released by cells at a boundary and diffuse over the positional field. Positional values would then be specified by the local concentration of the morphogen, whose concentration would decrease with distance from the source (Kersberg, 1999).

Unfortunately, this simple model of diffusion from a source as the way position is specified neglects the presence of receptors for the morphogens. These affect the diffusion and the effect of the morphogen on the cells. Computer simulations of the process show that receptors close to the source may be expected to bind the morphogen and render it unable to propagate further before they become saturated. This gives rise not to a gradient but a flat distribution of activated receptor near the source and then a sharp decrease at a distance. However, if the binding of the morphogen to the receptor has a low kinetic binding constant, only a small fraction of the morphogen will at first be retained by the receptor and a smoother gradient of activated receptor can form. There is also now substantial evidence that a variety of other factors can influence the shape of the gradient by the morphogen regulating receptor concentration. For example the gradient in decapentaplegic in the patterning of the *Drosophila* wing is modified by the receptors for the morphogen; in one case the receptors are induced to increase in number by high concentrations of the morphogen and this slows down its lateral diffusion.

Another model is based on a mechanism in which different receptor subtypes cooperate in passing the morphogen molecules along the cell membrane and then from cell to cell. There is also the possibility of the morphogen being passed from cell to cell in vesicles. Yet another very different mechanism for setting up a gradient is based on measuring time on a region like a progress zone. For example if a group of cells are proliferating in a well-defined region such as the progress zone at the tip of the limb or the regressing node region during chick gastrulation, cells are continually leaving the region. Then if the cells can measure the time T they have spent in the defined region, and this becomes frozen as they leave the progress zone, there will be a trail of cells with a gradient of increasing T.

While a variety of candidate morphogens have been proposed for specifying positional information, in most cases the evidence is still not conclusive. The best evidence comes from the development of the early embryo of *Drosophila* but here the gradients are established in effectively a single cell with many nuclei and there are no individual cell membranes. Very good examples are the gradients in decapentaplegic, a member of the TGF beta family and wingless in patterning the *Drosophila* wing. In vertebrates, good candidates are activin-like molecules in the early *Xenopus* embryo and sonic hedgehog in the developing limb and neural tube.

Interpretation of positional information requires the cells to respond in a specific manner to different positional values (Gurdon and Bourillot, 2001). If positional value is specified by the concentration of a morphogen then different levels of the morphogen must result in different cell behaviour including different patterns of gene expression. This implies that cells can respond accurately to threshold levels of morphogens. Just how this is done is not yet known but there is good evidence that cells can respond to different levels of a morphogen.

This has several important implications for evolution. It means that a major change in development of the embryo comes from changes in interpretation of positional information, that is the cells' response to signals. In fact there are a rather limited number of families of signalling molecules in most embryos – these include the TGF beta family, FGFs, sonic hedgehog, Wnts, Notch-delta, the ephrins and EGFs. Evolution is both conservative and lazy, using the same signals again and again both within the same embryo and in other distantly related species; most of the key genes in vertebrate development are similar to those in *Drosophila* (fruit fly). Patterning using positional information allows for highly localized changes in the interpretation of position at particular sites. It is also a feature of development that the embryo at an early stage is broken up into largely independent 'modules' of a small size which are under separate genetic control. There is also very good evidence that many structures make use of the same positional information but interpret it differently because of their developmental history. A classic case is that of the antenna and leg of *Drosophila*. A single mutation can convert an antenna into a leg and by making genetic mosaics it was shown that they use the same positional information but interpret it differently because of their developmental history – the antenna is in the anterior region of the body. Similar considerations apply to the fore and hind limbs of vertebrates. These differences in interpretation involve the Hox genes.

Hox genes

Hox genes are members of the homeobox gene family, which is characterized by a short 180 base pair motif, the homeobox, which encodes a helix-turn-helix domain that is involved in transcriptional regulation (Wolpert *et al.*, 2002). Two features characterize all known Hox genes: the individual genes are organized into one or more gene clusters or complexes and the order of expression of individual genes along the anteroposterior axis is usually the same as their sequential order in the gene complex (Figure 2.3).

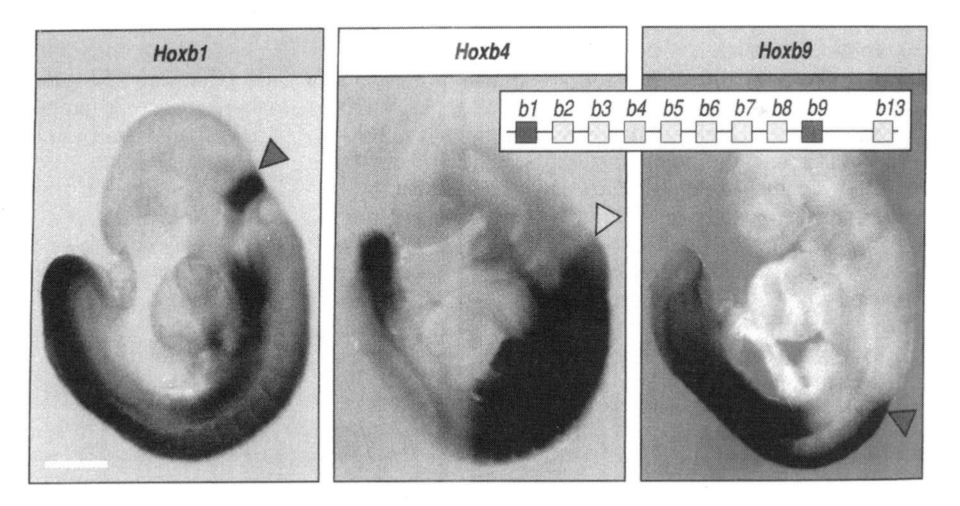

Figure 2.3 Hoxb gene expression in the mouse embryo. The arrowheads indicate the anterior boundary of expression of each gene within the neural tube. The position of the three genes within the Hoxb gene complex is indicated (inset). Scale bar = 0.5 mm.

Hox genes are key genes in the control of development and are expressed regionally along the anteroposterior axis of the embryo. The apparent universality of Hox genes, and certain other genes, in animal development has led to the concept of the zootype. This defines the pattern of expression of these key genes along the anteroposterior axis of the embryo, which is present in all animals.

The role of the Hox genes is to specify positional identity in the embryo rather than the development of any specific structure. These positional values are interpreted differently in different embryos to influence how the cells in a region develop into, for example, segments and appendages. The Hox genes exert this influence by their action on the genes controlling the development of these structures. Changes in the downstream targets of the Hox genes can thus be a major source of change in evolution. In addition, changes in the pattern of Hox gene expression along the body can have important consequences. An example is a relatively minor modification of the body plan that has taken place within vertebrates. One easily distinguishable feature of pattern along the anteroposterior axis in vertebrates is the number and type of vertebrae in the main anatomical regions – cervical (neck), thoracic, lumbar (between lower rib and pelvis), sacral, and caudal (tail). The number of vertebrae in a particular region varies considerably among the different vertebrate classes – mammals have seven cervical vertebrae, whereas birds can have between 13 and 15. How does this difference arise? A comparison between the mouse and the chick shows that the domains of Hox gene expression have shifted in parallel with the change in number of vertebrae. For example, the anterior boundary of Hoxc6 expression in the mesoderm in mice and chicks is always at the boundary of the cervical and thoracic regions. Moreover, the Hoxc6 expression boundary is also at the cervical–thoracic boundary in geese, which have three more cervical vertebrae than chicks, and in frogs, which only have three or four cervical vertebrae in all. The changes in the spatial expression of Hoxc6 correlate with the number of cervical vertebrae. Other Hox genes are also involved in the patterning of the anteroposterior axis, and their boundaries also shift with a change in anatomy.

Thus a major feature of evolution relates to the downstream targets of the Hox genes. Unfortunately, these are largely unknown but are a major research area. The number of downstream targets may be very large.

There is thus the conservation of some developmental mechanisms at the cellular and molecular level among distantly related organisms. The widespread use of the Hox gene complex and of the same few families of protein signalling molecules provide excellent examples of this. It seems that when a useful developmental mechanism evolved, it was used again and again. Bird wings and insect wings have some rather superficial similarities and have similar functions, yet are very different in their structure. The insect wing is a double-layered epithelial structure (outer tissue and membranes), whereas the vertebrate limb develops mainly from a mesenchymal core (undifferentiated cells derived from the mesoderm) surrounded by ectoderm. However, despite these great anatomical differences, there are striking similarities in the genes and signalling molecules involved in patterning insect legs, insect wings and vertebrate limbs.

All these relationships suggest that, during evolution, a mechanism for patterning and setting up the axes of appendages appeared in some common ancestor of insects and vertebrates. Subsequently, the genes and signals involved acquired different downstream targets so that they could interact with different sets of genes, yet the same set of signals retain their organizing function in these very different appendages. The individual genes involved in specifying the limb axes are probably more ancient than either insect or vertebrate limbs.

Drosophila *development*

The development of the embryo of the fruitfly *Drosophila* provides a key model for understanding pattern formation (Figure 2.4). In the unfertilized egg, bicoid mRNA is localized in the anterior end having being laid down there by maternal genes during oogenesis (the production and growth of the egg). After fertilization it is translated and the bicoid protein diffuses from the anterior end and forms a concentration gradient along the anteroposterior axis. This provides the positional information required for further patterning along this axis. Historically,

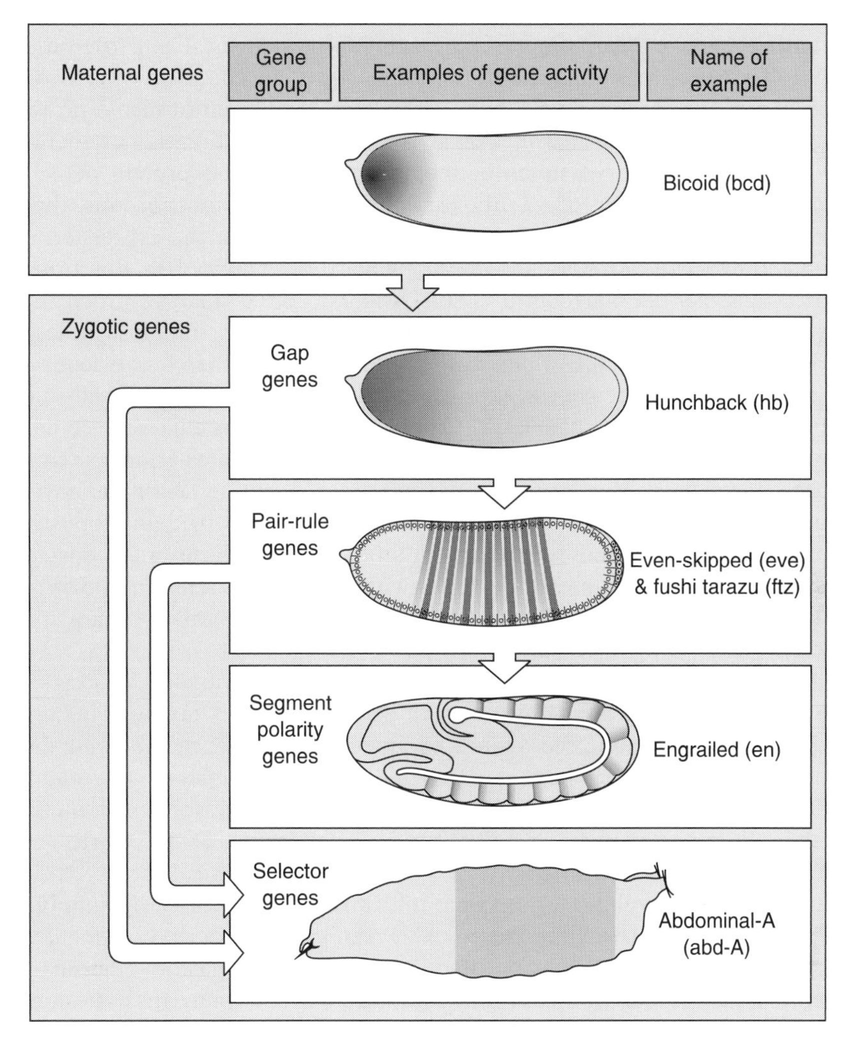

Figure 2.4 The sequential expression of different sets of genes establishes the body plan along the anteroposterior axis. After fertilization, maternal gene products laid down in the egg, such as *bicoid* mRNA, are translated. They provide positional information which activates the zygotic genes. The four main classes of zygotic genes acting along the anteroposterior axis are the gap genes, the pair-rule genes, the segment-polarity genes and the selector or homeotic genes. The gap genes define regional differences that result in the expression of a periodic pattern of gene activity by the pair-rule genes, which define the parasegments and foreshadow segmentation. The segment-polarity genes elaborate the pattern in the segments and segment identity is determined by Hox selector genes.

the bicoid protein gradient provided the first reliable evidence for the existence of the morphogen gradients that had been postulated to control pattern formation. However, it is unusual as it is a transcription factor and can only act as a morphogen as there are no cell membranes separating the nuclei of the early embryo – it is essentially a single cell with many nuclei.

As the bicoid protein diffuses along the *Drosophila* embryo it also breaks down (it has a half-life of about 30 minutes) and this breakdown is important in establishing the anteroposterior concentration gradient. It switches on certain zygotic genes at different threshold concentrations, so initiating a new pattern of gene expression along the axis. The gap genes are the first zygotic genes to be expressed along the anteroposterior axis, and they too code for transcription factors. Their expression is initiated by the anteroposterior gradient of bicoid protein while the embryo is still a syncytial blastoderm, that is the nuclei have divided but there are no cell membranes dividing the embryo up into individual cells. Bicoid protein activates anterior expression of the gap gene *hunchback* above a threshold concentration and the hunchback protein, which is then translated, is instrumental in switching on the expression of the other gap genes, including giant, *Kruppel* and *knirps* which are expressed in this order along the anteroposterior axis. As the blastoderm is still acellular at the stage at which the gap genes are expressed the gap gene proteins can diffuse away from their sites of synthesis. They are short-lived proteins with a half-life of minutes. Their distribution therefore extends only slightly beyond the region in which the gene is expressed and this typically gives a bell-shaped protein concentration profile. The hunchback protein is exceptional in this respect as its gene is expressed over a broad anterior region and it has a steep anteroposterior protein gradient. It is only switched on when the bicoid protein, which is a transcription factor, is above a threshold concentration. The dorsoventral axis, which is at right angles to the anteroposterior axis, is specified by a set of maternal genes separate from those that specify the anteroposterior axis and another gradient is set up, this time in the nuclei with the concentration of dorsal being highest ventrally. Another key system in which gradients in positional information play a key role is in the development of the wing imaginal disc.

Patterning of tissues by gradients has been found in a variety of tissues. What is striking is the similarity in both the principles involved as well as the molecules involved in signalling. The same sets of protein signals are used again and again and include members of the TGF beta family, the hedgehog and wingless families. Particularly clear examples in vertebrates come from studies on limb development where there is very good evidence for a morphogen gradient patterning the anteroposterior axis and from the patterning of the early mesoderm in *Xenopus* embryos.

From the viewpoint of evolution, positional information provides a very suitable system for the generation of novel patterns. Since there is little relation between the positional values of the cells and the expressed pattern of cellular differentiation, the same positional mechanisms can be used to generate an enormous variety of different patterns. The change in pattern would reflect the changes in interpretation of positional values rather than signalling between the cells. Thus the complexity of patterning comes from the cellular response to common signals.

2.2 The evolution of development

2.2.1 Evolution of new structures

Comparison of embryos of related species has suggested an important generalization: those characteristics shared by all members of a group of animals appear earlier in evolution. In the vertebrates, a good example of such a general characteristic would be the notochord (a skeletal rod of tissue enclosed by a firm sheath), which is common to all vertebrates and is also found in other chordate embryos. Paired appendages, such as limbs, which develop later, are special characters that are not found in other chordates and which differ in form among different vertebrates. All vertebrate embryos pass through a common phylotypic stage, which then gives rise to the diverse forms of the different vertebrate classes. However, the development of the different vertebrate classes before the phylotypic stage is also highly divergent, because of their very different modes of reproduction; some developmental features that precede the phylotypic stage are evolutionarily highly advanced, such as the formation of a trophoblast (the differentiated outer layer of cells) and inner cell mass by mammals.

An embryo's development reflects the evolutionary history of its ancestors. Structures found at a particular embryonic stage have become modified during evolution into different forms in the different groups. In vertebrates, one good example of this is the evolution of the branchial arches and clefts that are present in all vertebrate embryos, including humans. These are not the relics of the gill arches and gill slits of an adult fish-like ancestor, but of structures that would have been present in the embryo of the fish-like ancestor. During evolution the branchial arches have given rise both to the gills of primitive jawless fishes and, in a later modification, to jaws (Figure 2.5). When the ancestor of land vertebrates left the sea, gills were no longer required but the embryonic structures that gave rise to them persisted. With time they became modified and in mammals, including humans, they now give rise to different structures in the face and neck. The cleft between the first and second branchial arches provides the opening for the Eustachian tube and endodermal cells in the clefts give rise to a variety of glands, such as the thyroid and thymus (Figure 2.6).

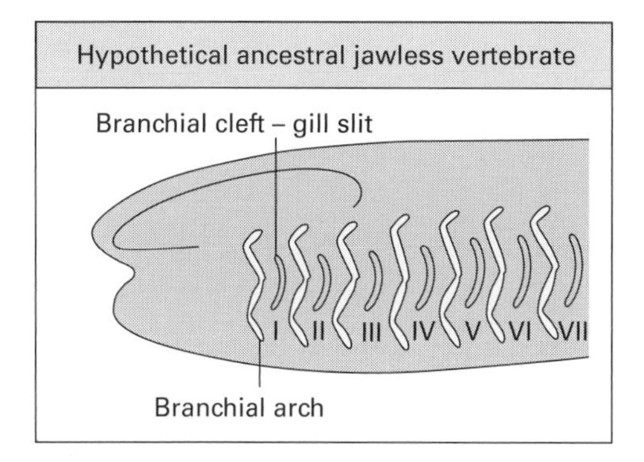

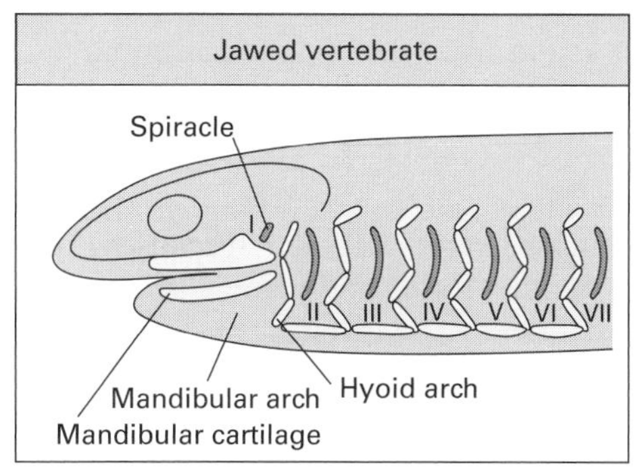

Figure 2.5 Modification of the branchial arches during the evolution of jaws in vertebrates. The ancestral jawless fish had a series of at least seven gill slits – branchial clefts – supported by cartilaginous or bony arches. Jaws developed from a modification of the first arch to give the mandibular arch, with the mandibular cartilage of the lower jaw and the hyoid arch behind it.

Evolution rarely generates a completely novel structure out of the blue. New anatomical features usually arise from modification of an existing structure. A nice example is provided by the evolution of the mammalian middle ear. This is made up of three bones that transmit sound from the eardrum (the tympanic membrane) to the inner ear. In the reptilian ancestors of mammals, the joint between the skull and the lower jaw was between the quadrate bone of

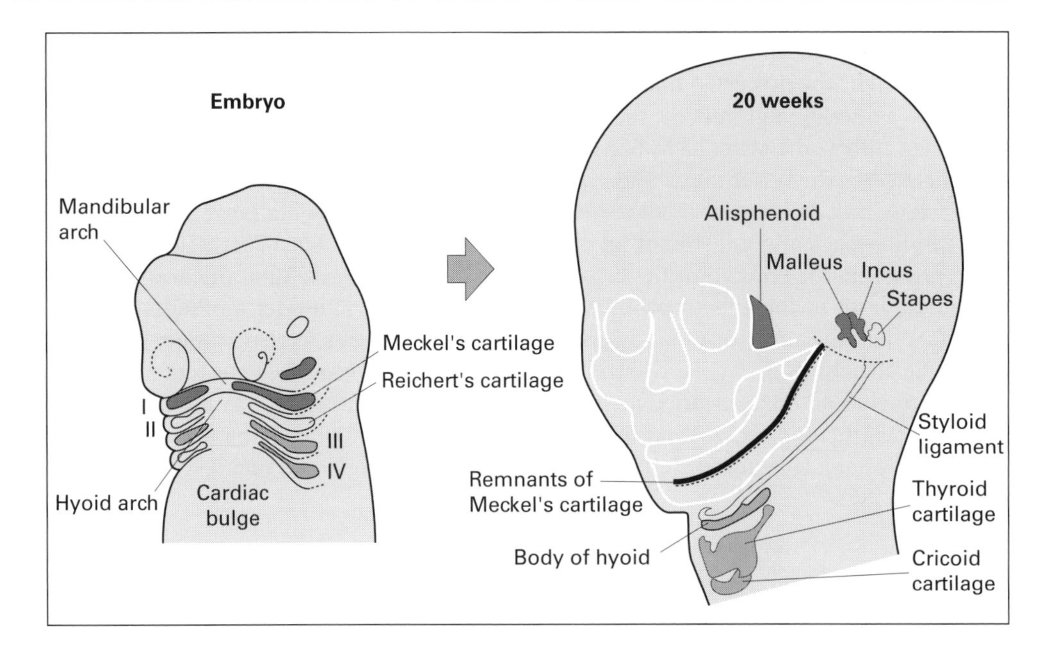

Figure 2.6 Fate of branchial arch cartilage in humans. In the embryo, cartilage develops which gives rise to elements of the three auditory ossicles, the hyoid, and the pharyngeal skeleton.

the skull and the articular bone of the lower jaw, which were also involved in transmitting sound. During mammalian evolution, the lower jaw became just one bone, the dentary, with the articular no longer attached to the lower jaw. By changes in their development, the articular and the quadrate bones in mammals were modified into two bones, the malleus and the incus, whose function was now to transmit sound from the tympanic membrane to the inner ear.

The skull bones of fish remain unfused and retain the segmental series of the gill arches.

Limbs

The limbs of tetrapod (four-limbed) vertebrates are special characters that develop after the phylotypic stage. Amphibians, reptiles, birds and mammals have limbs, whereas fish have fins. The limbs of the first land vertebrates evolved from the pelvic and pectoral fins of their fish-like ancestors. The basic limb pattern is highly conserved in both the forelimbs and hind limbs of all tetrapods, although there are some differences both between forelimbs and hind limbs and between different vertebrates.

The fossil record suggests that the transition from fins to limbs occurred in the Devonian period, between 400 and 360 million years ago. The transition probably occurred when the fish ancestors of the tetrapod vertebrates living in shallow waters moved onto the land. Proximal skeletal elements corresponding to the humerus, radius and ulna of the tetrapod limb are present within the fins of Devonian lobe-finned fishes, but there are no structures corresponding to digits. How did digits evolve? Some insights have been obtained by examining the development of fins in a modern fish, the zebra fish.

The fin buds of the zebra fish embryo are initially similar to tetrapod limb buds, but important differences soon arise during development. The proximal part of the fin bud gives rise to skeletal elements, which are homologous to the proximal skeletal elements of the tetrapod limb. There are four main proximal skeletal elements in a zebra fish fin which arise from the subdivision of a cartilaginous sheet. The essential difference between fin and limb development is in the distal skeletal elements. In the zebra fish fin bud, an ectodermal fin fold develops at the distal end of the bud and fine bony fin rays are formed within it. These rays have no relation to anything in the vertebrate limb.

If zebra fish fin development reflects that of the primitive ancestor, then tetrapod digits are novel structures, whose appearance is correlated with a new domain of Hox gene expression. However, they may have evolved from the distal recruitment of the same developmental mechanisms and processes that generate the radius and ulna. There are mechanisms in the limb for generating periodic cartilaginous structures such as digits. This may be based on reaction-diffusion mechanisms but the evidence is not yet persuasive. Nevertheless, it is likely that such a mechanism was involved in the evolution of digits by an extension of the region in which the embryonic cartilaginous elements form, together with the establishment of a new pattern of Hox gene expression in the more distal region.

Gene duplication

A major general mechanism of evolutionary change has been gene duplication. Tandem duplication of a gene, which can occur by a variety of mechanisms during DNA replication, provides the embryo with an additional copy of the gene. This copy can diverge in its nucleotide sequence and acquire a new function and regulatory region, so changing its pattern of expression and downstream targets without depriving the organism of the function of the original gene. The process of gene duplication has been fundamental in the evolution of new proteins and new patterns of gene expression; it is clear, for example, that the different haemoglobins (red oxygen-carrying elements in red blood cells) in humans have arisen as a result of gene duplication.

One of the clearest examples of the importance of gene duplication in developmental evolution is provided by the Hox gene complexes (Figure 2.7). Comparing the Hox genes of a variety of species, it is possible to reconstruct the way in which they are likely to have evolved from a simple set of six genes in a common ancestor of all species. *Amphioxus*, which is a vertebrate-like chordate, has many features of a primitive vertebrate: it possesses a dorsal hollow nerve cord, a notochord, and segmental muscles that derive from somites. It has only one Hox gene cluster, and one can think of this cluster as most closely resembling the common ancestor of the four vertebrate Hox gene complexes – Hoxa, Hoxb, Hoxc and Hoxd. It is possible that both the vertebrate and *Drosophila* Hox complexes evolved from a simpler ancestral complex by gene duplication.

Growth and timing

Many of the changes that occur during evolution reflect changes in the relative dimensions of parts of the body. Growth can alter the proportions of the human baby after birth, as the head grows much less than the rest of the body. The variety of face shapes in the different breeds of dog, which are all members of the same species, also provides a good example of the effects of differential growth after birth. All dogs are born with rounded faces; some keep this shape but in others the nasal regions and jaws elongate during growth. The elongated face of the baboon is also the result of growth of this region after birth.

Because structures can grow at different rates, the overall shape of an organism can be changed substantially during evolution by heritable changes in the duration of growth that leads to an increase in the overall size of the organism. In the horse, for example, the central digit of the ancestral horse grew faster than the digits on either side, so that as the animal got larger it ended up longer than the lateral digits.

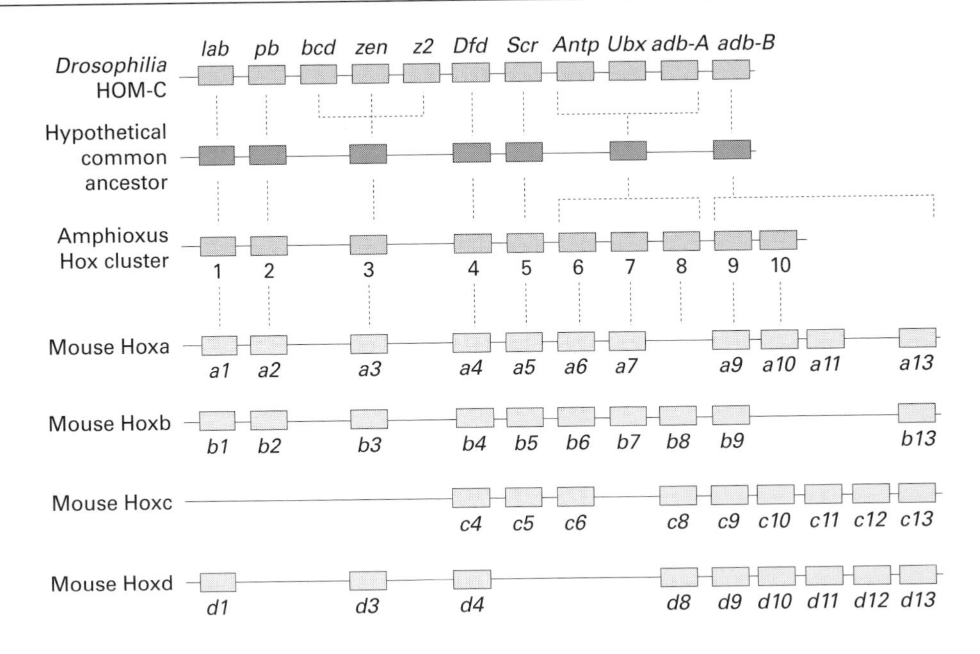

Figure 2.7 Gene duplication and Hox gene evolution. A suggested evolutionary relationship between the Hox genes of a hypothetical common ancestor and *Drosophila* (an arthropod), *Amphioxus* (a cephalochordate) and the mouse (a vertebrate). Duplications of genes of the ancestral set could have given rise to the additional genes in *Drosophila* and *Amphioxus*. Two duplications of the whole cluster in a chordate ancestor of the vertebrates could have given rise to the four separate Hox gene complexes in vertebrates. There has also been a loss of some of the duplicated genes in vertebrates.

Differences among species in the time at which developmental processes occur relative to one another can have dramatic effects on structures. For example differences in the feet of salamanders reflect changes in timing of limb development; in an arboreal (tree-dwelling) species the foot seems to have stopped growing at an earlier stage than in the terrestrial species and, in legless lizards and some snakes, the absence of limbs is due to development being blocked at an early stage.

2.2.2 The origins of multicellularity

The relationship between evolution and development is a very fashionable field at the moment. My own approach is rather different as I focus on how the embryo and multicellularity evolved. I do not know the answers so it is in a sense a just-so story. Rudyard Kipling wrote wonderful just-so stories. There is, for example, a tale of how the camel got its hump and how the elephant got its nose – the crocodile pulled it. My story, which I shall tell for the remainder of this chapter, is rather of that category; it is a just-so story. The supporting evidence is not very strong.

The basic organization and functions shared by all eukaryotic cells (which make up all plants and animals) but not prokaryotes (which make up bacteria), must have been present at least 2 billion years ago, before single-celled eukaryotes diverged. This conservation would include their large size – 1000 times the volume of the prokaryotic cell – their dynamic membranes capable of endocytosis and exocytosis, their membrane-bounded organelles like the nucleus, mitosis and meiosis, sexual reproduction by cell fusion, a cdk/cyclin-based cell cycle, actin- and tubulin-based dynamic cytoskeletons, cilia and flagella, and histone/DNA chromatin complexes. These ancient processes which evolved in the single-celled prokaryotes and early eukaryotes long before metazoa, constitute the core biochemical, genetic and cell biological processes of metazoa (multicellular organisms with distinct tissues and nervous system).

These eukaryotic cells were doing very well. Why did they bother to get together? And what had to be invented to make the embryos? Let me make my position clear. The miracle (and I do not mean it in the religious sense, I mean it in the evolutionary sense), the miracle of evolution, is the cell. While there are theories involving an RNA world and self-organization, it remains a mystery. Once you had the eukaryotic cell, from the point of view of evolution and development, it was downhill all the way: very, very easy.

Development requires turning genes on and off, cell signalling and transduction and cell motility. The ancestral cells had these. Lower eukaryotes, such as flagellates, slime moulds, ciliates and yeast cells, have many control mechanisms known from metazoa. Cell differentiation depends on different genes being active in different cells and the cell cycle can be thought of as a developmental programme. There were kinases (enzymes) turning processes on and off and also genes being turned on and off. There was also signal transduction of stimuli arriving at the cell membrane.

Signalling in unicellular eukaryotes was believed to be confined to mating factors in, for example, ciliates and yeast cells. It is now evident that unicellular eukaryotes depend on extensive signalling systems for their existence.

There are also many similarities of the intracellular transduction systems in uni- and multicellular organisms. Eukaryotic cells had motility and chemotaxis (an ability to move themselves towards or away from a specific chemical). While the slime moulds have nothing to do with the origin of the embryo, they branched from the metazoan line shortly before plants and fungi, and have cell–cell signalling involving several components shared by metazoa, such as cAMP, G-protein linked receptors, a variety of protein kinases and JAK/STAT transcriptional control. From their unicellular past, early metazoa had a lot to draw upon in the evolution of intercellular signalling.

Single cell organisms have molecular motors and these could provide the forces for morphogenesis. Chemotaxis in the slime mould *Dictyostelium* provides an important model for cytoskeletal organization and signal transduction. Chemotaxis is also important in its own right. The chemotactic cell is polarized and polarity is fundamental to many developmental processes. Ligand binding leads to rearrangement of the cytoskeleton – actin polymerization at the anterior end results in filopodial extension, while myosin at the rear contracts to bring it forward. This illustrates how complex the cytoskeleton already was.

Eukaryotic cells had everything (Szathmary and Wolpert, 2003). Among the basic components required for development, I can think of virtually nothing that eukaryotic cells did not have which is required for the developmental processes. And so, the real question is why did they bother to get together and what was the adaptive advantage?

There are arguments that being multicellular allows an organism to have division of labour and specialized functions. But that cannot have been the original selective advantage, for originally they all performed the same functions. So, our argument goes like this. There was mutation in a single cell so that when it divided the cells stuck together. Further division resulted in a loose colony of cells. Now this might have been an advantage if there were other sorts of unicellular predators around – they were more difficult to eat. But what was the real advantage? In hard times when there was no food around and single cells could not survive, some cells in the colony could then eat each other and so survive. If some cells died, then other cells could eat them and survive. That is a major advantage. It is also the origin of cell death (Kersberg and Wolpert, 1998).

There is current evidence to support this idea. If you take planaria or hydra and you starve them, they get smaller and keep their normal form. The way they do this is by the cells eating each other. The way the sponge egg develops is by phagocytosing (engulfing and eating) neighbouring cells. In certain fish and annelids what happens at the time of reproduction is that the adult eats an enormous amount of food and devotes almost all of its body to feeding the egg. In fact, muscle cells are actually broken down and phagocytosed, the eggs are laid and the animal dies.

At a later stage in the evolution of this simple multicellular organism, it was an advantage to identify the cell which was going to be fed by the others and become large. This is the origin of the egg. The cell might have been under different external conditions, for example near the centre of the colony, such that it would enable it not to die, and therefore to feed on the neighbours. The other advantage of having the egg was that it avoided general conflict. While a colony of independently reproducing cells could have been successful, mutations in all the individual lineages would have occurred and accumulated. This would have had two severe disadvantages. The first would have been at the level of how cells interacted in the colony. The cells would acquire different genetic constitutions and this would have led to competition rather than cooperation between the lineages. Secondly, it would have been difficult for the colonies to lose deleterious mutations or mutations in general, including those reverting to unicellular state. The solution to these problems lay in the evolution of the egg; if the various colonies arose from a few germ-like cells with low mutation rate, then the competition and the mutation problems would both disappear. It is not too difficult to imagine a series of mutations which would have given the inner cells an advantage with respect to eating their neighbours so that in hard times the outer cells died. Our origin, I claim, lies in what one might think of as altruistic cannibalism.

2.2.3 Evolvability

Attempts at various forms of multicellularity have been made by evolution at least twenty times independently. There are essentially two ways to make a simple multicellular entity out of single cells: either the single cells divide and the offspring stick together, or a number of solitary cells aggregate to form the colony. Reduction of the propagule (the primitive spoor or egg) to minimal size (i.e. one cell) is advantageous because it reduces conflicts due to increased kinship.

There may be another factor in play. Development from a single cell, an egg, is fundamental and essential for the evolution of multicellular organisms. A feature of evolution of multicellular organisms is the issue of evolvability; that is what cellular properties are necessary for organisms to evolve, as they have, into an enormous variety of different forms. Could an asexual form of reproduction, involving budding (somatic embryogenesis) as in hydra, evolve to give complex new forms? Asked in this way this is not a question of kin selection or competition between cells in the multicellular organism, but an issue that relates to developmental processes that generate the form of the organism. And the answer is no, as essentially all cells must obey the same set of rules. During evolution it is the change in genes that leads to new patterns forming during development. Once the pattern has been set up, as in hydra, it is no longer possible to evolve significant changes for two reasons. First, it is not possible to go through a developmental sequence, and secondly mutations in individual cells mean that they all no longer have the same rules for behaviour. It is only via a coherent developmental programme that organisms can evolve and this requires an egg. There are multicellular organisms like the cellular slime moulds which do not develop from an egg but by aggregation, but their patterning for that reason has remained very simple for hundreds of millions of years – they could not evolve complex patterns of cell behaviour. The strength of the evolvability constraint on development calling for an egg should increase with the complexity of the organism (Wolpert and Szathmary, 2002).

Ontogeny recapitulates phylogeny?

Haeckel really played a very important role in thinking about evolution and development. He had the idea about ontogeny recapitulating phylogeny, which turned out to be incorrect. Ontogeny does not recapitulate phylogeny, the reason why we have something like fish gill slits in our embryonic development is that we partly retain ancestral early embryonic stages. However, on one evolutionary change I think he was right and that is in relation to gastrulation (an important early stage in development where an internal cup shape forms by infolding). He has near the bottom of his evolutionary tree what he calls the gastraea which evolves from the simple blastea. Here there is real evidence that ontogeny really does recapitulate phylogeny. All animals pass through a gastraea-like stage, they gastrulate. Why is gastrulation so similar in all animals? I want to argue that it does actually recapitulate an ancient ancestor.

There is a very simple organism, *Trichoplax*, which is made up of just a single layer of cells and a hollow interior. It is rather like Haeckel's blastea. What is remarkable in *Trichoplax* is that it undergoes a change similar to early gastrulation while feeding. Particles of food or microorganisms that it is going to eat are moved into a digestive chamber.

The basic idea is that a two-layered primitive organism fed on the bottom and it formed an infolding to aid feeding. This basic idea comes from Jaegerstern (1972). In a blastea-like organism, a hollow organism made of a single layer of cells, the feeding was encouraged by currents from cilia. Living on the bottom it formed an invagination to sweep the food into a primitive gut where the cells would engulf the food (Figure 2.8). It takes no stretch of the imagination to see that all it had to do is to fuse this infolding with the sheet on the other side and you then have a mouth, a gut and an anus.

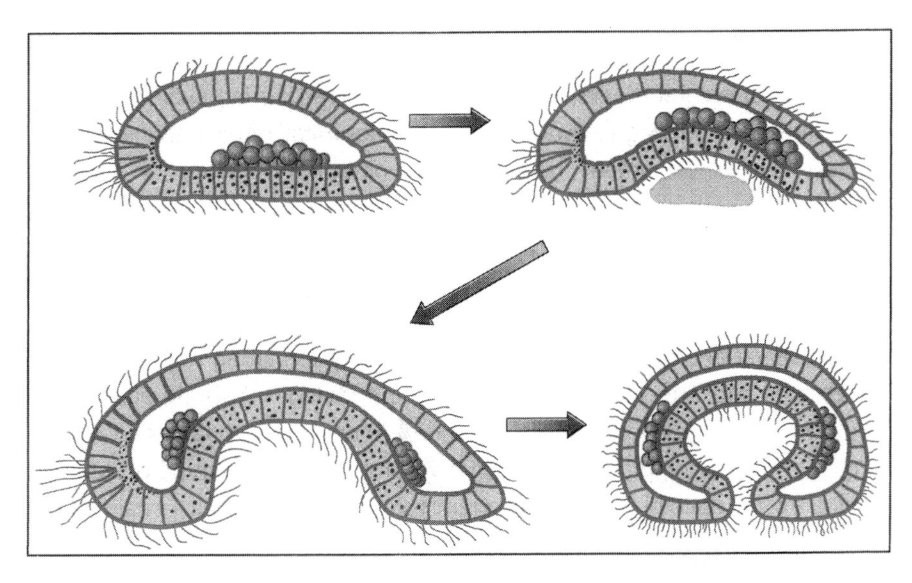

Figure 2.8 A possible scenario for the development of the gastrula. A colonial protozoan in the form of a hollow multicellular sphere could have settled on the sea bottom and developed a gut-like invagination to aid feeding.

Little had to be invented to reach this stage, everything was there in the cell, and in a way the embryo is really much less complicated than individual cells. The complexity of the developmental biology does not lie in the embryo, but lies in the individual cells. If you look at the signals between cells there are less than 10 grand families of signalling molecules between cells and this is trivial by comparison with what goes on inside cells.

Selection pressures

One of the things that needs to be thought about in relation to the evolution of the embryo is the selection pressure on the embryo itself. Now, I like to liken the embryo to medical students at university. Medical students play around, not come to lectures, spend their parents' money, get drunk. Only one thing matters: they have to pass the final exam. It is the same with embryos. They do not have to look for a home, they do not have to mate and energy expenditure is trivial. The only thing they have to do is to reliably give rise to the adult. About 25% of the cell's ATP (adenosine triphosphate – an important molecule used as energy) goes on keeping sodium out of the cell. So just being alive is expensive. Making one more gene or another movement is trivial from an evolutionary perspective. What matters is reliability. There may be little selection for a more efficient way of, for example, gastrulation.

How did cells in the colony evolve different identities and so establish well-defined patterns of cell activities? There are really only two major mechanisms by which cells acquire identity, one is by asymmetric cell division together with cytoplasmic localization. The other relies on interactions. An important mechanism in the evolution of the spatial patterning of the embryo could have involved the Baldwin effect, which I learnt about from one of my teachers, Conrad Waddington. An environmental stimulus, such as contact with the substratum, brings about a change in the cell so that it becomes different from its neighbours. Then having established that machinery it becomes intrinsic. For example there are mutations which lead to the thickening of the soles of the feet due to pressure, then randomly mutated genes that cause this thickening can become selected and make it occur only in the feet. I think it is a nice idea, because I think to get thickening autonomously localized to the feet from the very beginning would be much more difficult. And in the same way, one could think of the cell like this. For example, just having contact with a substratum could cause that region to secrete a protein and then this could have been used by neighbouring cells, and so perhaps morphogen gradients and positional information could have evolved. Then later on this could become genetically determined.

The development of early embryos was probably rather messy, they were not very reliable nor canalized to limit the effects of variations in unrelated genes (Finch and Kirkwood, 2000). It did not matter, they had time – hundreds of millions of years. They could play around, just so long as some of them passed the exam and gave a good phenotype. Reliability in development has not received the attention that it deserves. Isogenic (genetically identical) nematode worms grown under exactly the same conditions all die after about 15 days, but there is considerable variation in the time of death. The idea that clones are identical should be treated with suspicion. Examination of the hypothalamus of human identical twins shows that the number of cells in the hypothalamus can vary by as much as 20%. Development is reliable, yes, it passes the exam but sometimes there are variations in the degree of success.

Acknowledgement

The figures in this chapter are from Wolpert, L. (2002) *Principles of Development*, 2nd edn. Oxford University Press and are reprinted by permission of Oxford University Press.

References

Carroll, S.B., Grenier, J.K. and Weatherbee, S.D. (2001) *From DNA to Diversity*. Blackwell.
Davidson, E. (2001) *Genomic Regulatory Systems: Development and Evolution*. Academic Press.
Finch, C.E. and Kirkwood, T.B.L. (2000) *Chance, Development and Aging*. Oxford University Press.
Gurdon, J.B. and Bourillot, P.Y. (2001) Morphogen gradient interpretation. *Nature*, **413**, 797–803.
Jaegerstern, G. (1972) The evolution of the metazoan life cycle. Academic Press, London.
Kerszberg, M. (1999) Morphogen propagation and action: toward molecular models. *Seminars Cell Devel. Biol.*, **10**, 297.
Kerszberg, M. and Wolpert, L. (1998) The origin of metazoa and the egg: a role for cell death. *Journal of Theoretical Biology*, **193**, 535–537.
Szathmary, E. and Wolpert, L. (2003) *The transition from single cells to multicellularity*, in *Genetic and Cultural Evolution of Cooperation*, pp. 271–290. MIT Press.
Wolpert, L., Beddington, R., Jessell, T. *et al.* (2002) *Principles of Development*. Oxford University Press.
Wolpert, L. and Szathmary, E. (2002) Multicellulosity: Evolution and the Egg. *Nature*, **420**, 745.

The principles of cell signalling

3

JOHN T HANCOCK

3.1 Introduction

Cell signalling is undoubtedly one of the most important areas of modern biology. It is central to our understanding of the workings of all organisms. Not only does the unravelling of cellular signalling enable us to comprehend more fully the biochemical mechanisms of life, it also opens the way for a diverse range of biologically based technologies, from the future manipulation of agricultural crops to the development of new drug therapies.

Every organism, from the simplest single-celled prokaryote, through to the most complex plant or mammal, relies on cell signalling for its very survival. The same, in fact, can be said for the individual cells within a complex multicellular organism, where decisions to survive, multiply, or to commit cellular suicide, are all the result of a vastly complex interplay of cellular signals.

Cells are awash with information to which they may need to respond, or perhaps which they need to ignore. Environmental conditions, such as temperature, forms of radiation such as light, the presence of invading pathogens, or signals sent from other organisms or indeed from other tissues within the same organism, all impact on how the cell needs to behave, for its own benefit, or that of the whole organism. The study of cell signalling encompasses such perception and response to a cell's environment, and it is this that makes the area of cell signalling so crucial to the understanding of modern biology.

This chapter aims to unravel some of the mechanisms and principles used by cells in their cell signalling processes.

3.2 An overview of signalling

One of the most common scenarios in cell signalling is the arrival of a signal at a cell's surface. This is often a molecule to which the cell has to respond. Two basic events are then crucial for the cell: first, the cell has to recognize the signal; secondly, it has to respond to it.

Signals are usually perceived by proteins known as receptors, which are most likely located on the cell's surface, facing out, where they are ready and waiting for the signal to arrive. The receptors first recognize the presence of the signal and then relay the message into the cell.

 ISBN 0-12-428765-4

Commonly, a chain of signalling commands ensues across the interior of the cell, a process commonly referred to as signal transduction. An analogy would be the arrival of a radio message on a submarine. The signal is received by a radio operator (receptor), who tells the captain. The captain, tells the next person, who relays the message to the next person, and so on, until the member of the ship's crew who has to action the signal has been told, perhaps the weapons engineer.

In a cellular situation, the final destination of the signal may be the cytoplasm, where rates of metabolism may be altered, or the mechanisms of muscle contraction are initiated (Figure 3.1). These would be examples of short-term responses. However, the signal may also carry information to the nucleus, leading to a longer term response. Here, the expression of genes might be modified. The DNA sequence comprising a gene carries the information for the manufacture of a protein, so it is the expression of specific genes which lead to the production of specific proteins. It is the proteins which comprise the majority of the cell's machinery, so cell signalling events will dictate which proteins are made, and therefore the types of activity in which the cell might engage. Both the amount and type of protein found in an individual cell are commonly controlled in such a way. Although some proteins are produced all the time, referred to as the products of the expression of 'housekeeping' genes, many genes have their expression regulated, either increased or decreased, by signalling pathways that have been initiated by the recognition of a signal on the cell surface (Figure 3.1).

As the complement of the proteins in a cell are under such close control, cell signalling events are in control of the activity of the cell and can alter the future potential activity of the cell, too. As such, the manipulation of the future of a cell's activity and function is often crucial for

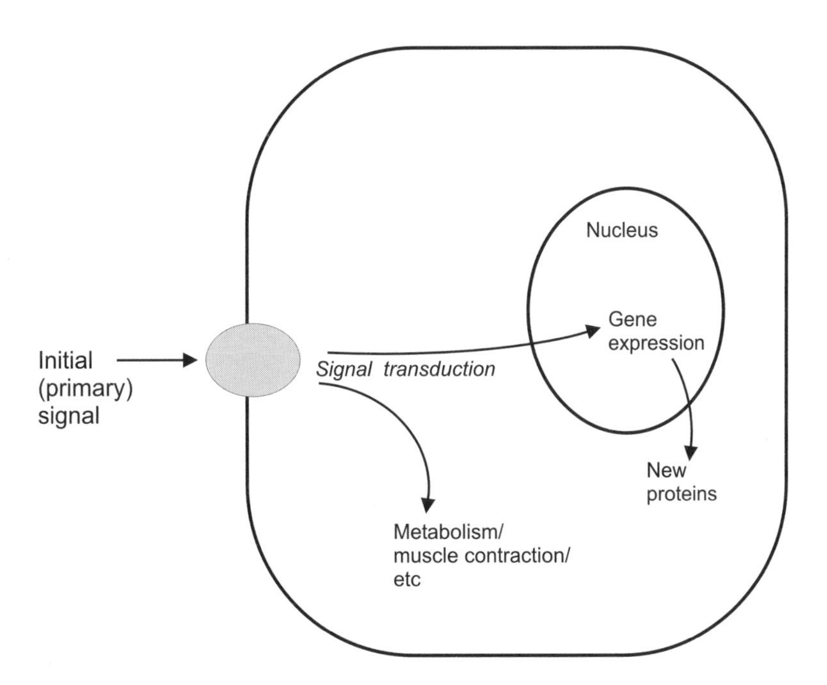

Figure 3.1 The likely routes of cell signalling messages. On the receipt of a signal at the cell surface, the message is transmitted into the cell. The possible location for the cellular functions which are to be controlled by such signals include the cytoplasm where muscle contraction may be initiated, or metabolic rates can be altered, or the signal might alter events in the nucleus. Here, the expression of genes might be regulated, perhaps resulting in new proteins in the cell.

survival of the cell, which may need, for example, to adapt to a new environment, and is vital for the development of cells within multicellular organisms, as will be discussed below.

Traditionally, the signal to which a cell is responding and which is perceived at its surface is termed the primary messenger. Signals which are produced inside cells in response to this signal have been called secondary messengers, but this is a phrase which is now used less, as it is apparent that in many cases a plethora of events are initiated once the primary signal has been recognized. The challenge for the modern biologist is to understand the pattern of signals which are produced and the complex interplay which we can observe taking place between them.

A reading of the literature will quickly reveal that cell signalling research has been comprised of many separately defined areas of study. Research laboratories will have been interested in how a cell responds to a particular stimulus, a hormone perhaps. But cells do not have the luxury of living in such a simple world. The hormones do not arrive all neatly one at a time. The environmental stresses do not happen in isolation. A plant might be short of water, be too hot and be invaded by bacteria all at the same time. Therefore, cells are recognizing signals constantly, responding to them continually and, in the majority of cases, surviving. However, as we shall discuss below, there is only a limited amount of cellular signals that a cell has in its repertoire, and it is far from fully understood how cells untangle the exact message that is being transmitted into its interior to control its internal workings.

3.3 What makes a good signal?

The priorities for any signal are that it should be generated quickly, relay a specific message efficiently and then be removed when no longer needed. For these reasons, most biological signals are relatively small chemicals that are able to be moved efficiently, or can diffuse rapidly, to their site of action.

Relaying of a specific message

The mechanism of relaying the signal from its place of origin to the place of perception depends on the chemical nature of the signal and the distances needed to be travelled. On one extreme, pheromones, which are signals sent between organisms, are spread by the wind, from one individual to the next. Within a single organism, hormones are released by cells to be carried rapidly by the vascular system, for example the blood stream in humans, or perhaps the phloem in plants. In such cases, a cell would release the signalling molecule, which would be free to move in the extracellular environment. On reaching the target cell, it would be perceived and initiate a response (see Figure 3.2A).

Quite often, if the producing cell and target cell are in very close proximity, simple diffusion might be sufficient but it would not be necessary to release the signal to the extracellular medium. Connecting pores exist between cells through which chemicals, such as molecular signals, may pass (Figure 3.2B). In animals, neighbouring cells collude to produce tubular structures called gap junctions, which can allow the passage of signals, piping them from cell to cell. Plants too have similar intercellular connections, called plasmodesmata.

In other cases, if cells are physically neighbours, proteins on the surface of one cell might physically signal to other proteins on the surface of the second cell (Figure 3.2C). In tissue culture, human cells will normally only grow as a monolayer, and once the cells have covered the surface of the dish they stop growing. They can 'feel' the presence of their neighbours and are signalled such. Signalling of this kind is also instrumental in the development of neighbouring cells in organisms.

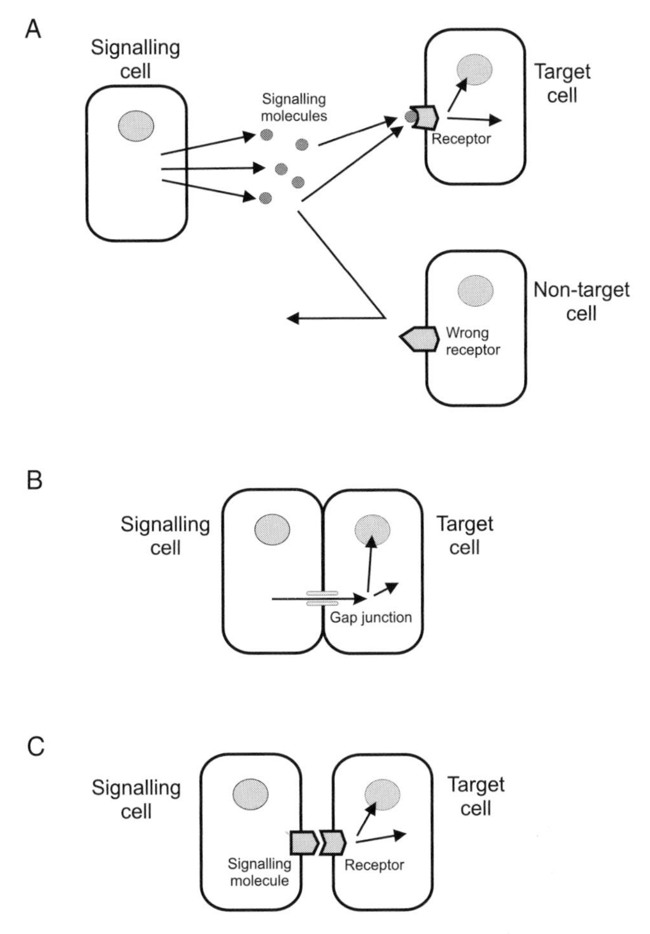

Figure 3.2 Methods for the transfer of signals between cells. **A.** Signalling molecules may be released by one cell, to be perceived by target cells, where a response is initiated. If the cell does not contain the correct receptors, the cell will remain unaware of the signal and not respond. **B.** Signals might be transferred from cell to cell through specialist connections, such as gap junctions. **C.** Proteins on the surface of one cell might be recognized by receptor proteins on neighbouring cells.

As far as intracellular signalling is concerned, diffusion seems to be the key. However, it is apparent that diffusion is not uniform within the cell. If probes are used to study calcium signalling for example, hotspots of high calcium can be seen, while certain parts of the cell have no rise in calcium concentration at all. However, it is not yet clear how such limited diffusion is controlled by cells.

Signals are produced quickly

It is imperative that, in the majority of cases, signals are generated quickly and efficiently. It is little use to the organism if the signal is produced after the event that it is hoping to avoid or respond to. For example, in humans, adrenaline is made as a 'fight and flight' response hormone. It obviously needs to be relaying its signal as the fight or flight is required.

There are two main strategies used by cells to produce a rapid signal. First, the signal may be made, but it is sequestered away so as not to be recognized by the appropriate cellular machinery until such time as it is required. Such a mechanism is shown in Figure 3.3A. Here, signalling molecules, perhaps a hormone such as insulin, are produced by the cell, but kept within lipid vesicles inside the cell. Therefore the potentially receptive cell has no knowledge of them. At the appropriate time, usually in response to other cellular signals, the vesicles are moved rapidly to the cell surface and their contents are released to the outside (Figure 3.3A). They are then free to diffuse or be carried to the target cell where the appropriate response is initiated. It is not only signals which are released from the cell that are controlled in this way. A similar mechanism is also used to control the levels of intracellular calcium, itself a ubiquitous and extremely important cellular signal.

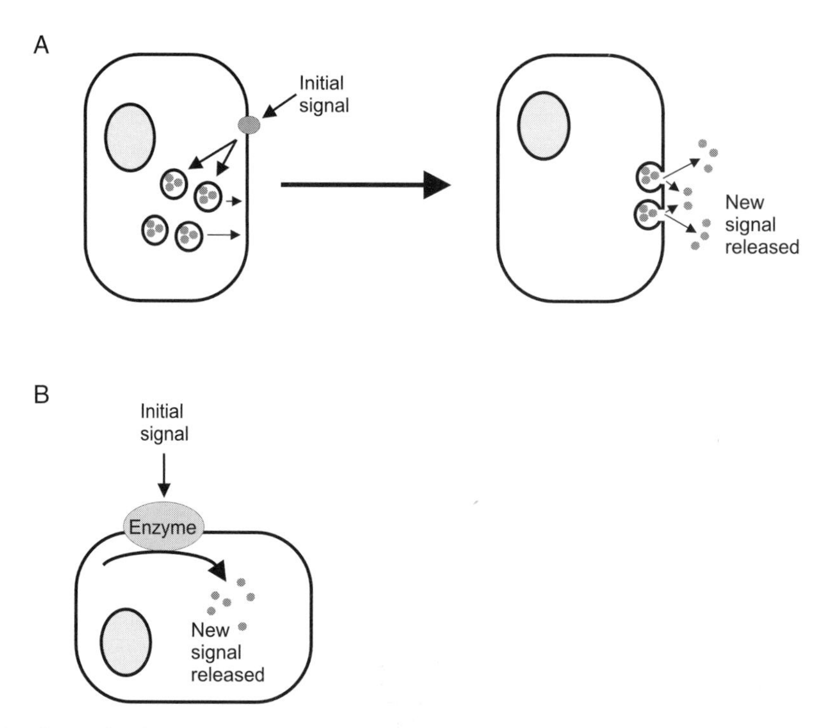

Figure 3.3 Strategies for the generation of signals. **A**. Signalling molecules might pre-exist in the cell and be released as a functional signal at the appropriate time. **B**. Signalling molecules might be made at the appropriate time, usually by enzymes controlled by other signalling events.

Secondly, a signalling molecule may be made on demand. This mechanism appears to be favoured by intracellular signalling pathways. Many small signalling molecules are produced at specific places, at specific times, by specific proteins, or enzymes (Figure 3.3B). Such enzymes commonly are able to catalyse the conversion of one or more substrates to a set of products, at least one of which is the signal. A classical example here is the enzyme adenylyl cyclase (also called adenylate cyclase). This enzyme takes adenosine triphosphate (ATP), a ubiquitous compound found in all cells, and turns it into a new compound by a relatively simple reaction, involving the loss of two phosphate groups to make cyclic adenosine monophosphate (cAMP) (schematically drawn in Figure 3.4). cAMP is again a common signalling molecule, often dubbed the 'hunger signal' as it is often involved in the control of metabolic pathways which result in energy production in the cell. In mammals, for example, it helps to control the breakdown of glycogen to glucose, a crucial cellular food.

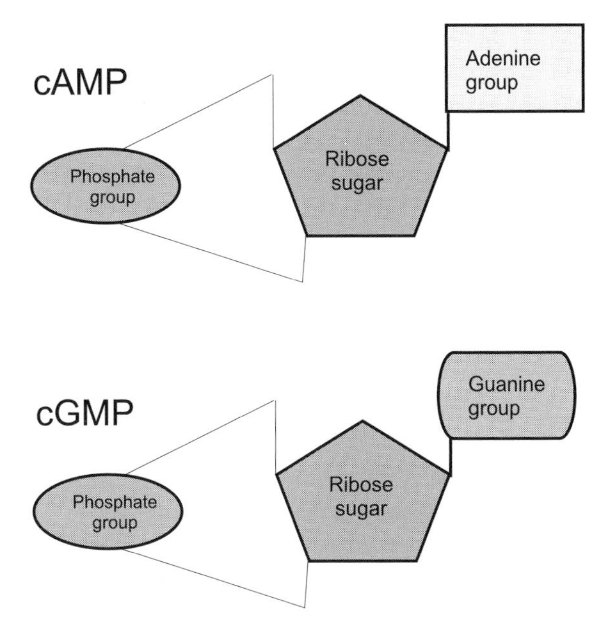

Figure 3.4 Schematic representations of the structures of cAMP and cGMP. These two signalling molecules are extremely similar in their structures, as highlighted when drawn in schematic form. However, they are involved in the control of very different events within the cell.

Removal or reversal of signals

All signals need to be stopped or reversed. Many people will recall that feeling of being suddenly filled with panic, perhaps when they realize that they are late for a train. A rush of adrenaline enables the body to respond rapidly, leading to a production of glucose and hence a source of cellular energy for that final rush down the platform. However, once on the train, one needs to relax, and not continue in the adrenaline 'high'. Cells therefore need to recognize when the signal is no longer valid. A second example can be found in the mechanism of the mammalian eye. Light arrives, there is a response and the organism sees an image. However, the eye has to revert very quickly to the non-stimulated state, otherwise no further light could be perceived. One would end up with a single image, which would never go away.

Several mechanisms are found which enable this return to ground state in cell signalling pathways. Receptors on the cell surface are often internalized into the cell, where they are no longer able to be in the position to perceive incoming signals. These receptors may ultimately be destroyed, or recycled back for another round of signalling.

The physical movement of signalling components may be a strategy used for intracellular signals. The signals may be re-sequestered, such as seen with calcium ions. They are either rapidly removed from the cell or pumped into intracellular compartments (organelles such as mitochondria or the endoplasmic reticulum), where they can remain until the next signal is needed. As such, the calcium can be recycled *ad infinitum*.

Alternatively, and more commonly, an enzyme will be involved in rapidly removing the signal, usually by breaking it down. The intracellular signal cAMP, for example, is broken down by an enzyme called phosphodiesterase to produce AMP, which can then be recycled.

However, there are two considerations here which need to be borne in mind. First, signals are, in many cases, being made and destroyed at the same time and it is the relative rates of production and removal that will determine the overall steady state concentration, or strength, of the signal and, indeed, whether it constitutes a signal at all. Low levels of signals are not always responded to and a threshold level may need to be reached before the cell realizes that it is being triggered. Secondly, over the years, many signals have been thought of as simply being destroyed, with the concomitant production of a second molecule, which had no relevance, but was merely a means to an end. However, it is now recognized that these secondary molecules may themselves be important signals and therefore the removal of one signal may generate another, which will not necessarily have the same effect as the first.

But there is still an important question that we have not yet answered: what are signals composed of?

Extracellular (or intercellular) signals

Most extracellular signals are small water-soluble compounds. Many signals are transmitted between cells by the vascular system of the organisms and therefore need to be able to be dissolved and carried. Even if signals need to be transmitted to cells within a close locality, water is the medium through which they must usually travel.

Extracellular signals commonly are small organic molecules, such as the sex hormone testosterone, or small proteins such as insulin, which is involved in the control of glucose levels in the blood. Some extracellular signals do not appear to be easily grouped with others and might be thought of as almost bizarre, such as the relatively reactive organic compound ethylene, which is involved, among other effects, in the ripening of fruit. Some inorganic molecules have also been identified as important intercellular signals, such as hydrogen peroxide or nitric oxide. Nitric oxide, a reactive gaseous free radical, is important for the control of blood flow in mammals, in particular to the heart and during penile erection.

In humans, the endocrine system uses hormones which are released into the blood, through which they are carried to their site of action. A typical example is the hormone insulin, which is a small protein composed of two polypeptide chains, held together by covalent bonds. More local cell-to-cell signalling is referred to as paracrine, and is typified by the group known as the cytokines. Again, typically small proteins, or peptides, this is an extremely important group of signals, including among their number, tumour necrosis factor and interferon. These signals help to orchestrate the immune response in humans and are often implicated in the control or maintenance of disease states.

Although most extracellular signals are water soluble, not all of them are. Direct cell-to-cell communication as depicted in Figure 3.2C relies on proteins which contain lipid-soluble regions and, furthermore, are not released from the cell. Not even all the released signals are inherently water soluble. Steroids are lipid-soluble compounds which can be carried in the blood and carry information between cells. While in the blood, their solubility is usually bestowed upon them by their association with inherently soluble molecules, such as proteins, which act almost like taxis to carry them to their destination. However, as we shall see below, the perception of these steroids by cells differs from the strategy used by the water soluble signals.

The signals found inside cells

Intracellular signals come in all sorts of shapes and sizes, but the over-arching factor is that the cell can recognize each and everyone of them as unique and can respond swiftly and accurately to their presence.

Among the smallest is the simple ion of calcium, Ca^{2+}, which can neither be created or destroyed by cells, but is moved around creating high concentrations in desired places at the correct times. In contrast, other signals may be large lipids, such as some of the phospholipids comprising the cellular membrane. Alternatively, some of these large lipids can be cleaved, creating smaller lipid-soluble signals and perhaps some water-soluble signals. A frequently occurring example here is the breakdown of the lipid phosphatidylinositol bisphosphate (PIP_2) at the cell membrane into two very important signals, inositol trisphosphate ($InsP_3$) and diacylglycerol (DAG). The former moves into the cell and controls calcium ion signalling, whereas the latter stays associated with the membrane and instigates a second signalling system.

Molecular signals may also have extremely similar molecular structures, but have very different functions. For example, cAMP and cGMP (cyclic adenosine monophosphate and cyclic guanosine monophosphate respectively) look amazingly similar in schematic form, as shown in Figure 3.4. However, they are made by different enzymes, control different signalling events and are removed by different enzymes.

It can be seen, then, that signals can be virtually any molecular structure, as long as the cell has the means to make it or control it and the means to recognize it as being present and it is unique enough not to be confused with other molecular signals.

Electrical signalling

A footnote should be added on electrical signals, used by many organisms, including humans to carry the message extremely quickly over large distances. The electrical charge is propagated along the membrane of the cells and in the case of nerve cells, such a system can carry the message along a defined pathway for a matter of metres, often controlling muscle movements for example. It is thought that even plants have a system of using electrical signalling.

3.4 Perception of signals: why do cells respond?

The perception of a signal, whether it is an extracellular or intracellular signal, is a critical event. If the signal is not recognized, then the message dies and has had no use. Cells must have a way of knowing that the signal is there and have a way of transmitting the message on to the final destination where the desired event takes place. Cells must also respond only to the signals directed to it and not to all signals. If adrenaline is released into the blood stream, not all cells of the body will need to be aware of its presence, or do anything while it is in the circulation. So how do cells decide to respond or not? The key here is the presence of receptors on the cell surface (or sometimes within the cell). With adrenaline, only the cells which contain adrenaline receptors will be able to recognize its presence in the blood.

Receptors are commonly proteins which span the cell membrane, as depicted in Figure 3.5. They contain a region, or domain, which lies outside the cell, poised to detect the signalling molecule. Each receptor type will have a molecular structure which is shaped in three dimensions appropriately for the signalling molecule (often referred to as a ligand) to attach, a process known as ligand binding. Without the unique 3D structure, the signal will not be recognized and go unnoticed.

The receptors usually have a region which spans the membrane, the trans-membrane domain, as well as a region which lies inside the cell, or intracellular domain. On binding the extracellular ligand, the receptor undergoes a change in its 3D structure, a change which also alters the structure of the region inside the cell, transmitting the signal from the outside of the membrane to a region inside the cell, where further events can ensue.

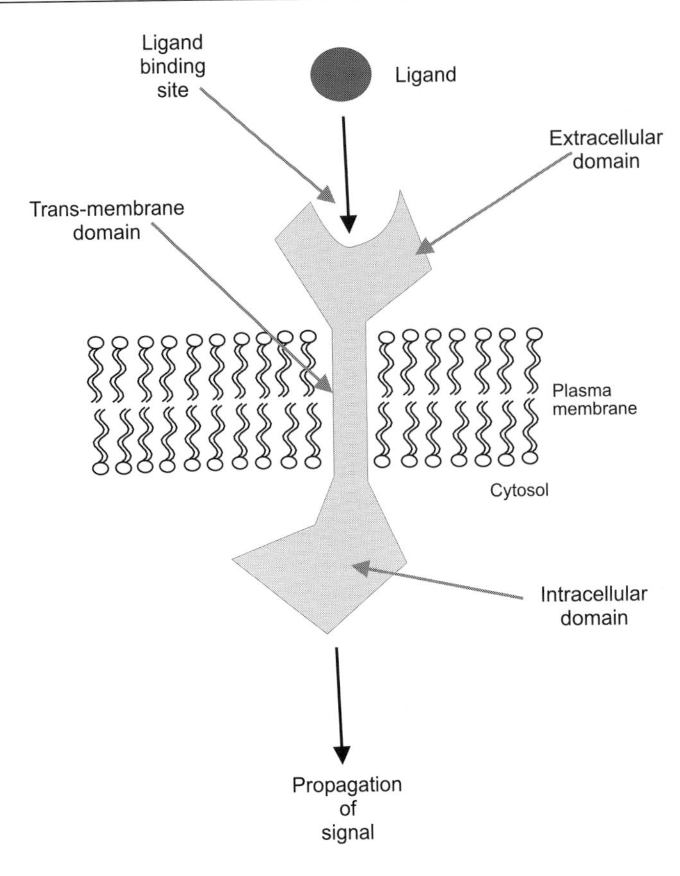

Figure 3.5 The schematic representation of a protein receptor. Protein receptors usually span across membranes. On the outer side, they contain an area of specific three-dimensional shape which recognizes and binds to the initial signalling molecule (the ligand). On the inner side is an area which propagates the signal to the next component in the signalling pathway.

The type of domain that the receptor contains on the inside of the membrane will dictate the type of pathway that the signal will follow. These types have been used by biochemists to categorize the receptors into classes that include the following:

- Those which contain an intrinsic enzyme activity. These proteins will start to produce a product, a new signalling molecule on ligand binding. cGMP can be made by such a system.
- Receptors which associate with G proteins. G proteins are a class of signalling proteins commonly found in nature, which relay the message on to the next part of the signalling mechanism.
- Receptors which recruit enzymes on activation. Structural changes to the receptor allow enzymes to become associated with them, the enzymes themselves then become active and turn on the next part of the signalling system.
- Receptors which are also ion channels. On activation, these proteins undergo a change in shape which opens or closes a channel through which ions can pass. The movement of these ions then causes a further effect in the cell.

An interesting class of receptors which are very important, but do not sit neatly with the others listed above, are those which recognize steroid signals. As the steroids are lipid soluble and can readily enter cells through the lipid membrane there is little need to perceive them at the cell surface. Steroid receptors are therefore found inside the cell, often in the cytoplasm. On binding the steroid, they also will undergo profound structural changes which relay their message. Often, the receptor itself will be moved, in its activated state, to the nucleus where changes in gene expression are initiated.

3.4.1 Structural change is a key signalling mechanism

Most often when signals are perceived, or in many cases when a signal is propagated within the cell, proteins are involved. These proteins may be enzymes that produce a further signalling molecule, or the protein itself is responsible for the relaying of the message. In both cases, the protein may undergo a change in shape, known as a conformational change, similar to that discussed above in relation to receptor–ligand binding.

Proteins are simple chains of amino acids, which have adopted a three-dimensional structure, rather like scrunched-up lengths of string. It is their shape which determines which amino acids are close to each other and are therefore able to interact in three dimensions. For example, the catalytic activity of enzyme proteins is reliant on the right amino acids being in the correct positions to facilitate catalysis. This means that a relatively simple shape change can have a dramatic effect on the protein.

Often proteins are found together in complexes, where two, three or more proteins are closely and intimately associated, such associations being reliant on the shapes of the proteins and on the 'lock and key' fit between them. A change in shape of one of these proteins will disrupt the way the proteins fit together, with two probable results: either the other proteins adopt a new conformation to match, or the association collapses and the proteins separate.

Let us consider what happens in the first situation. Assume a complex is formed of proteins A and B, where A is the protein that perceives the signal and B is the protein which passes the signal to the next protein in the signal transduction cascade, in this case protein C. If protein A changes its shape, this will induce a shape change in protein B, or else the complex will no longer fit together. Therefore protein B is now different and can be recognized by protein C. Protein C would have a shape such that it would not be able to recognize the original form of B, but can recognize the new conformation. Therefore, new associations might now be possible. Protein B will bind to protein C and in doing so will relay the 'information' on to the new protein. The baton has been passed on, so to speak. The formation of such new protein associations is imperative in many of the signal cascades induced by, for example, growth hormones, cytokines and the hormone insulin.

In the second case, where the original conformational change causes the protein complex to dissociate, the released proteins will be free to move and carry the message through the cell, to the next protein or signalling molecule in the cell. Such a system is used by an immensely important class of signalling proteins, called G proteins. These proteins are able to bind a group of molecules called the guanine nucleotides, hence the name G proteins. There are two classes of G proteins but the ones to be considered here are called the trimeric G proteins. In their inactive state, they exist as an association of three proteins, called α, β and γ, and as a complex are commonly further associated with a cell surface receptor. A conformational change caused in the receptor by binding to its signal (as discussed above), causes a conformational change in the G protein, which falls apart into two sections, an independent α protein, and a dimer of β and γ proteins. Both these sections of the original protein then have the potential to move in the cell and trigger further signalling events, as depicted in Figure 3.6.

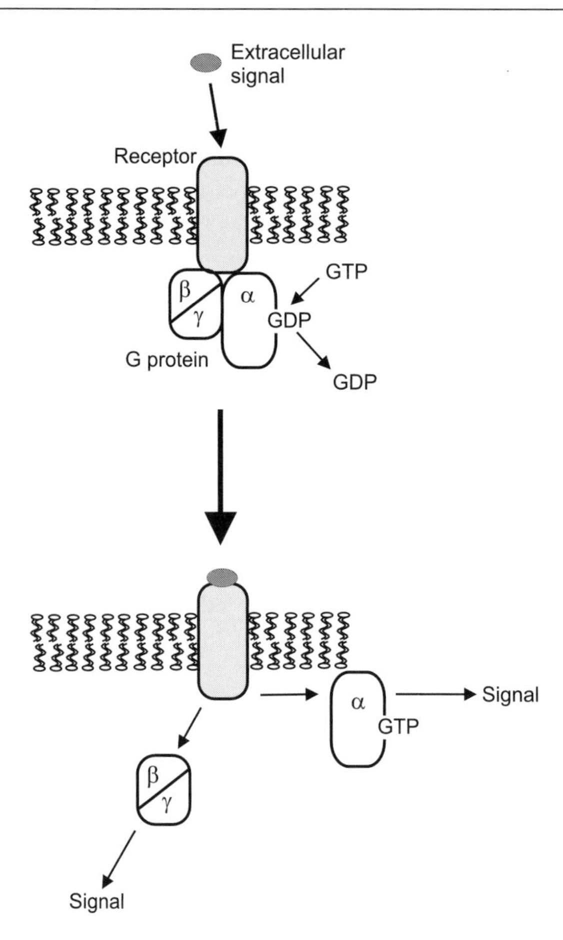

Figure 3.6 Signalling may involve the dissociation of proteins. Often signalling involves the breakdown, or dissociation, of protein complexes. The binding of the ligand to the receptor causes a change in shape of that receptor, causing the protein complex on the other side of the membrane to dissociate. Here, the G protein α subunit moves off, leading to one signal, while the β and γ subunits stay together but might instigate a second signalling event. This dissociate of G proteins is also accompanied by the release of GDP and the binding of GTP by the α subunit.

3.4.2 Phosphorylation

One of the most central events in cell signalling is the process known as phosphorylation, that is the addition of a phosphate group to a protein, as shown in Figure 3.7. The phosphate is donated by the molecule adenosine triphosphate (ATP), which is itself ubiquitous in cells, being produced primarily as a reserve of energy. The phosphate is added to proteins usually on specific amino acids depending on their chemical structure and the molecular environment in which they sit, usually exposed to the outside of the protein. The addition of the phosphate commonly induces a conformational change on the protein, altering its activity if it is an enzyme, or perhaps altering its ability to associate with other proteins. The molecular structure of phosphorylase is shown in Figure 3.8 with and without the phosphate group, and clearly a conformational change has taken place, which has profound effects on its activity.

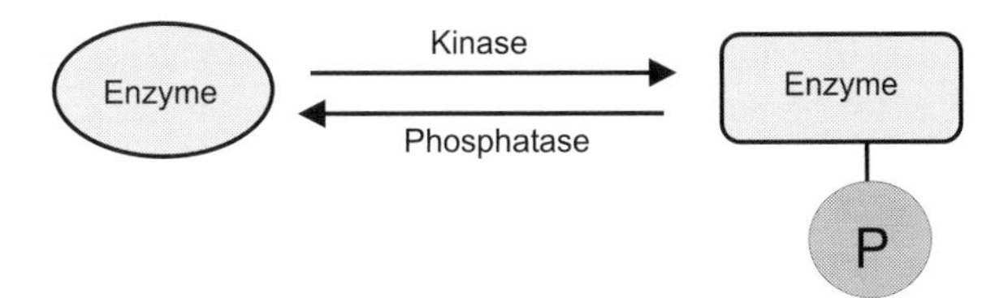

Figure 3.7 Phosphorylation is a key event in cell signalling. Many proteins in cell signalling have their activity altered by changes in their shape, instigated by the addition or removal of a phosphate group, a process called phosphorylation. The phosphate is added by enzymes called kinases, and can be removed by enzymes called phosphatases.

The phosphate is added to proteins by a class of enzymes known as kinases, but like all cell signalling events it needs to be reversible. In this case, the phosphate is removed by a group of enzymes known as phosphatases, which release the protein-bound phosphate as inorganic phosphate and allow the protein to revert to its original shape. Such a system is depicted in the scheme in Figure 3.7.

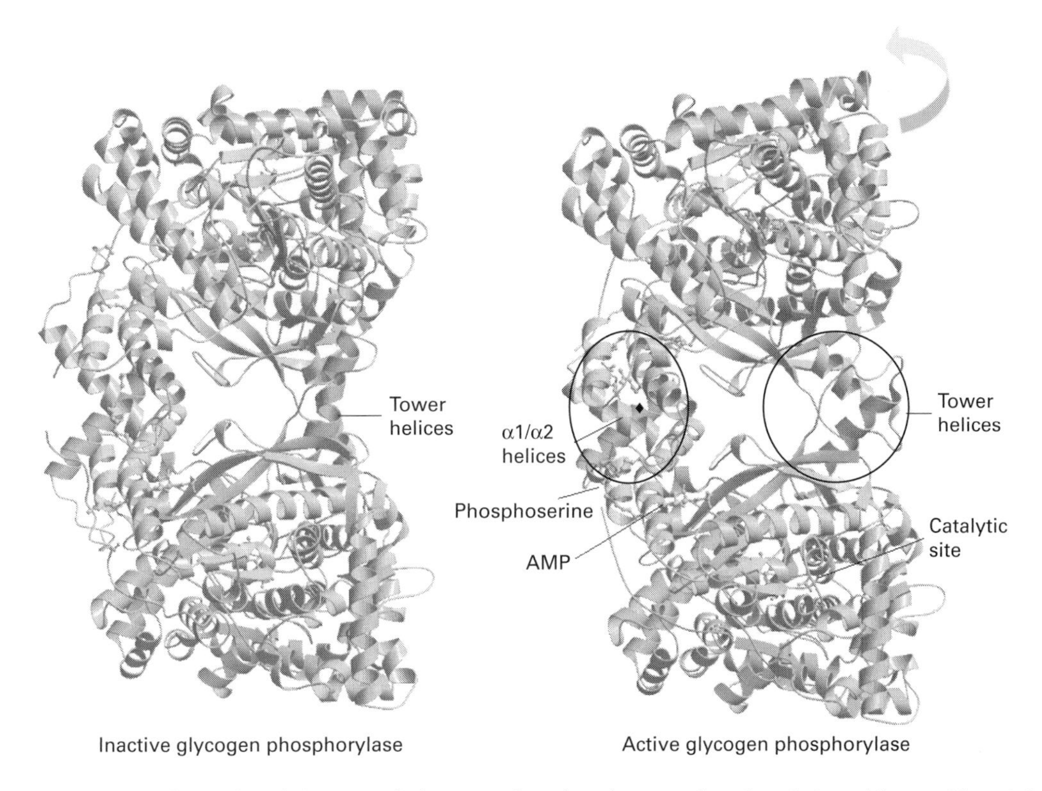

Figure 3.8 Conformational changes of glycogen phosphorylase on phosphorylation. Also see Plate 3.1.

Colour plate 3.1 and Figure 3.8 show the conformational changes of glycogen phosphorylase on phosphorylation. The conversion from the inactive state to the active state induced by phosphorylation of one amino acid (serine-14) on each monomer: the enzyme is a dimer of two identical subunits. Phosphorylation results in new inter-subunit interactions at the $\alpha 1/\alpha 2$ interface that induces a change in the relative orientation of the two subunits. This change of orientation corresponds to a 10° rotation of the upper subunit relative to the lower subunit about an axis perpendicular to the page, located within the $\alpha 1/\alpha 2$ interface. As a consequence, the two tower helices change their relative packing angle from almost anti-parallel (inactive conformation) to perpendicular (active conformation) and this causes a change of structure at the catalytic site located at one end of the tower helix. Areas of most significance are ringed for clarity.

3.4.3 Signalling can be viewed as a pathway

Often signalling is viewed as a pathway of events, known as a signal cascade. While this is a very simplistic view of the events useful pathways can be drawn. Often these start with an extra-cellular signal arriving at the cell, but of course a plethora of signalling events might precede this, controlling the production and the release of these signals from the 'signalling' cell. Once arrived at the 'host' or target cell, however, the signal is perceived by a receptor, which relays the message through the outer cell membrane and into the cell, instigating a chain of events which culminates with the desired response.

The specific example of this, shown in Figure 3.9, has been chosen because it has been well known for some time and is found in many cells. Here, the hormone adrenaline arrives at the cell and binds to its receptor, activating a G protein. This G protein dissociates into sections (as discussed above) and one section, the α protein, moves along the membrane and activates the enzyme adenylyl cyclase. The enzyme rapidly produces cAMP, which activates the kinase cAMP-dependent protein kinase (PKA). This kinase adds a phosphate group to (phos-

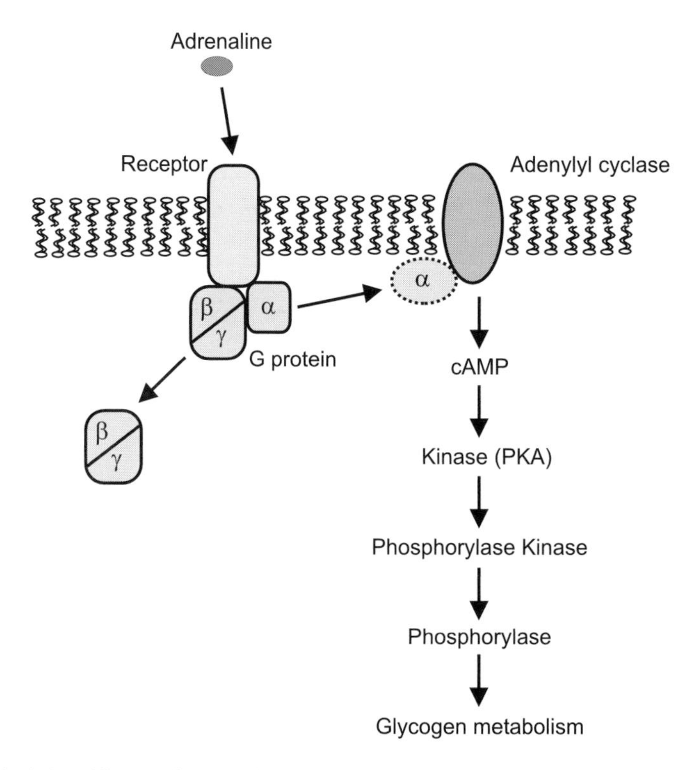

Figure 3.9 A typical signalling pathway. Simple, linear pathways can be drawn. Here a pathway which transmits the signal from adrenaline for the control of the rates of glycogen metabolism is depicted.

phorylates) another kinase, phosphorylase kinase, which itself phosphorylates an enzyme, phosphorylase, activating it (the structures of which are shown in Figure 3.8). This enzyme then breaks down glycogen, a store of material for energy production, which was the desired effect for the original production of adrenaline.

A defined pathway can be drawn, therefore, between the original signal and the cellular event which is altered. Many such pathways can be depicted in this way, some with similar components, and others which are completely different. What they all have in common is that they contain signals which alter the concentration or conformation of the next signalling component in the chain, so propagating the signal on to its goal.

Why does the cell have pathways? Would it not be easier for a single signalling molecule to move straight to the desired end effect? One of the main reasons is that a pathway has the ability to amplify a signal. In our example, one receptor can probably activate several hundreds of G proteins, which can activate many enzymes. These might produce thousands of individual small molecular signals, each of which can activate their target kinases. With each new component in a pathway, there is great potential to increase the strength of the original signal, allowing profound cellular effects to ensue from the recognition of a small amount of initial stimuli. Secondly, each element of a pathway also allows for the potential influence from other pathways, allowing signalling to be a complex interplay of events, something that is discussed in more detail below. In our example, adrenaline might be 'telling' the cell one thing, but several other hormones might be signalling to the cell that that is not such a good idea for the overall survival of the cell or organism.

3.4.4 Signalling is a complex interwoven mesh of messages

As hinted at above, cell signalling should not be viewed simply as a set of straightforward parallel pathways, all independently leading from the outside to the inside of the cell. Even the simple example given above (Figure 3.9), is far more complicated than depicted. For instance, the enzyme PKA has multiple targets, one of which has the effect of inhibiting the storage of glycogen. This makes sense as the pathway which leads to the breakdown of an energy store also stops more being made at the same time, which would be a total waste and would be referred to as a futile cycle. Secondly, each element of the pathway must have a way of having its effect reversed, so the more stages there are in a pathway, the more components are needed to be involved to ensure reversal.

It is important to appreciate that many, perhaps nearly all, cell signalling components have more than one effect and can be influenced by many other components. This means that cell signalling can no longer be seen as a series of independent pathways, but rather as a complex arrangement of pathways that can intermingle and influence each other. Such interplay between pathways is referred to as cross-talk.

A relatively simple example of diverging and converging pathways is shown in Figure 3.10. Here, the extracellular signal binds to a receptor, which activates a G protein, leading to the activation of an enzyme, in a similar manner to the example previously described, but here the enzyme is phospholipase C. Two very different signalling products emerge: diacylglycerol (DAG) and inositol trisphosphate ($InsP_3$). DAG leads the activation of a kinase, which itself has multiple targets. $InsP_3$ leads to the release of calcium ions into the cell's cytoplasm, again with multiple effects possible. The G protein may also have other effects, as do potentially all the other components in this system.

A single extracellular signal may initiate what appears to be a baffling array of events in the cell.

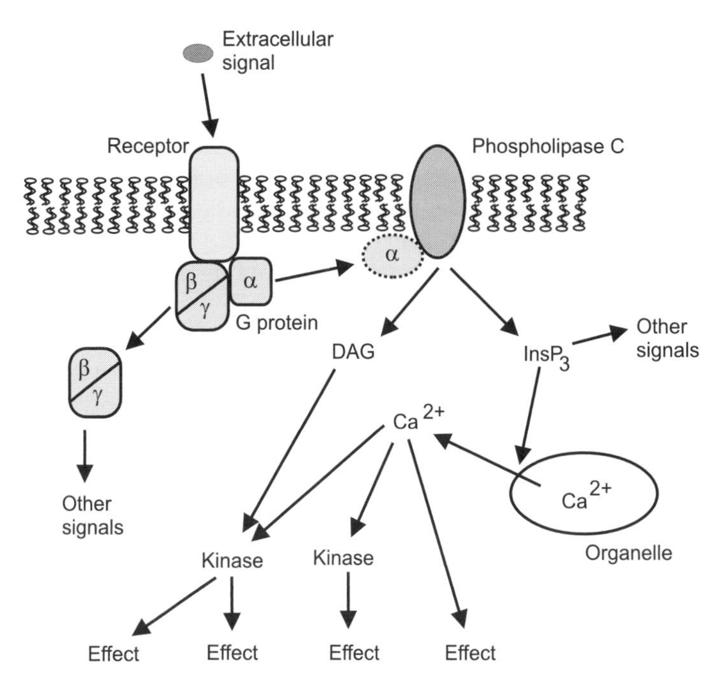

Figure 3.10 A typical complex signalling pathway. Pathways are often not linear, but branching and diverging. Here a pathway involving the enzyme phospholipase C is shown. Although induced by the presence of an extracellular signal, signalling components are produced simultaneously, and a complex web of events ensues. DAG: diacylglycerol; $InsP_3$: inositol trisphosphate.

3.4.5 Neural networks

The studying of isolated pathways in cells, or the study of individual components greatly aids our understanding of how cells respond to stimuli, often with a complex display of intracellular events. However, cells are bombarded with multiple signals and stimuli, any of which could be contradictory. Different pathways might also use common components, for example, several might lead to the alteration of calcium concentrations in cells. One of the challenges today is to unravel the complexities of these systems, especially as the majority of the components have been identified and characterized. To aid in this, computers are being employed, in particular neural networks. In such a system, designated inputs can be used to predict outcomes. Such outcomes can be compared to real biological examples and the computers can be trained so that the predictions are more accurate. It is thought that such computer systems will be able to emulate cellular events, like a virtual cell, so that results of new and multiple signals can be predicted, perhaps helping in therapeutic strategies which impinge on cell signalling mechanisms.

3.5 Signalling and development

Many cell signalling pathways influence cellular and organism development. In a multicellular organism, the individual starts life as a single cell, but by maturity contains millions of cells, of thousands of different types. In the beginning the single cell contains the genetic blueprint, the genome, for building the organism. The single cell divides and many cells are formed, initially all the same. But it is through the expression of particular genes in that genome at the appropriate times, brought about by cell signalling, that individual cells produce their individual complements of proteins, a process known as differentiation. It is the array of proteins found within the cell that enables them to carry out their specific functions and so, for example, the types and amount of each protein will be different in the cells of the heart, compared to those of the liver. It is also these proteins, in the form of receptors and signalling components, which dictate if and how a cell can respond to incoming signals and environmental stimuli.

It is the cell signalling events taking place in the cells as they develop that determine which genes are expressed and at which times they are used and so determine how the cell develops. During the development of the organism, therefore, each cell is picking up multiple signals from its neighbouring cells and from cells in developing tissues and the pattern of signals to which it is exposed and to which it can respond determines its fate in the growing individual.

Signalling also determines if cells live or die. Throughout our lives our cells are dying and being replaced: very little of us is actually our chronological age! If cells did not die as we develop, tissues would not grow and shape properly. For example, in the worm, *Caenorhabditis elegans* normal development is reliant on the death of a set of 131 cells. Even after reaching adulthood, we still have cells dying at a high rate, allowing new and better functioning cells to take over the challenge of making us survive. In plants too, cellular death is a common feature allowing the survival of the organism as a whole. This is clearly seen during the attack of leaves by pathogens. The cell being invaded sends signals to neighbouring cells, inducing them to die and so limiting the spread of the pathogen. These areas of dead, necrotic tissue are seen as spots on the leaves.

So what is it that makes our cells die? Several signals have now been identified which lead to cell signalling cascades resulting in the death of the cell, a process known as programmed cell death, or often referred to as apoptosis (after the Greek for falling leaves). Often this death is associated with defined cellular changes, including breakdown of the DNA comprising the genome in the cell. It is a complex cell signalling mechanism which controls whether a cell lives or dies.

3.5.1 Dramatic effects when things go wrong

Cell signalling events are instrumental in the control of the rates of metabolism, in the rates of gene expression, in the differentiation of cells, in the regulation of their rates of proliferation and in whether they live or die. So what happens when it goes wrong?

Insulin is a major signalling hormone in mammals and when the signalling associated with it goes awry the usual result is diabetes. There are different forms of the disease, but the causes are generally related to the lack of production of the hormone by the signalling cells, the lack of perception of the hormone by the target cells, or a defect in the signalling cascades which insulin invokes. The exact defect will enable an informed strategy for therapeutic manipulation for the individual concerned.

The discovery of the so-called oncogenes that lead to cancer was heralded as one of the major steps forward in biochemistry in the last century. However, it soon became apparent that such genes were simple derivations of normal genes found within all of us and, as such, these genes were labelled proto-oncogenes. Characterization of the proteins encoded for by these genes enabled their normal functions to be determined and, importantly, their role in aberrant biochemical events which lead to the apparent over-proliferation of cells, resulting in tumour formation. It comes as no surprise to discover that the majority of these proteins are involved in cell signalling and that it is defects in signalling cascades which often lead to tumours forming.

One of the best examples of this is a G protein called Ras. It is normally involved in a variety of signalling cascades, often pathways initiated by growth hormones which control proliferation. To turn off the G protein and so switch off the signal that the growth hormone invoked, breakdown of GTP to form GDP is required, a process carried out by the catalytic action of the G protein. However, in the oncogene version of the Ras gene, a mutation has taken place so the amino acid sequence of the protein is altered in the exact place in the protein which is used for this catalysis. The result is that the G protein can no longer break down the GTP, can no longer be turned off and so the 'keep growing' signal initiated by the growth hormone will continue unabated, even if the hormone is removed. The cells continue to proliferate and a tumour results.

Many other examples of oncogenes encoding key signalling components are now known and future therapeutic strategies will no doubt target the functioning of such aberrant proteins.

3.6 Summary

Cell signalling events are in control of the activities of all cells, whether they are unicellular organisms or part of multicellular higher organisms, plants or animals. They are instrumental in the regulation of all major functions, including metabolism, movement and the expression of genes. Signalling determines if cells differentiate, proliferate, live or die.

Biological signals are usually molecular structures that appear to abide by a general set of rules. The signal that they convey can be generated quickly; they can be moved efficiently to relay the message; and importantly, they can be removed rapidly for the cessation of the signal. Such signals are diverse in their structure, from simple ions to complex lipids and proteins, but all are uniquely recognized by the machinery of the cell and can pass on their vital information at the correct time at the right place.

Although many components of both inter- and intracellular signalling are now known and characterized, it is clear that signalling of cells involves a complex and, in most cases, little understood interplay of many converging and diverging pathways. It is also clear that a plethora of diseases, including cancer involve the dysfunctioning of cell signalling mechanisms.

The importance of cell signalling to the understanding of biological sciences in the 21st century is highlighted by the steady and unabated increase in research in the area, with journals now dedicated solely to the subject. Future research will undoubtedly reveal new signalling cascades, along with new signalling molecules and it is the manipulation of the concentrations and activities of such biological signals that will open the way for the enhancement of agricultural outputs and for the development of new therapeutic strategies.

Acknowledgements

I would like to thank Professor David Barford, Institute of Cancer Research, London, for supplying Figure 3.8 (the structures of phosphorylase).

Further reading

Alberts, B., Johnson, A., Lewis J. *et al.* (2002) *Molecular Biology of the Cell* 4th edn. Garland Science. (Chapter 15 in particular.)

Barford, D. and Johnson, L.N. (1989) The allosteric transition of glycogen phosphorylase. *Nature,* **340**, 609–616.

Barford, D., Hu, S.-H. and Johnson, L.N. (1991) Structural mechanism for glycogen phosphorylase control by phosphorylation and AMP. *J. Mol. Biol.,* **218**, 233–260.

Hancock, J.T. (1997) *Cell Signalling*, Addison Wesley. (A single authored text dedicated to the subject, especially concentrating on the components that might comprise signalling pathways.)

Heldin, C.-H. and Purton, M. (eds) (1996), *Signal Transduction.* Nelson Thornes. (A collection of essays on topics in the field.)

Helmreich, E.J.M. (2001) *The Biochemistry of Cell Signalling*. Oxford University Press. (A good text with an emphasis on structure and mechanism.)

Lodish, H., Berk, A., Zipursky, S.L. *et al.* (1999) *Molecular Cell Biology* 4th edn. W.H. Freeman. (Chapter 20 in particular.)

Martinez Arias, A. and Stewart, A. (2002) *Molecular Principles of Animal Development*. Oxford University Press. (An excellent up-to-date book on development.)

Plant Cell (May 2002) **14** supplement, S1–S417. (An issue dedicated to signalling in plant cells, including aspects of plant development.)

Science (May 2002) **296**, 1632–1657. (An excellent collection of articles on a range of aspects of cell signalling.)

From genotype to phenotype: looking into the black box

4

JESSICA A BOLKER

The strategic need to black-box embryology at a certain point in time does not imply that ontogenetic development really is irrelevant to evolutionary understanding.

(R. Amundson, personal communication)

4.1 Introduction

T.H. Morgan's conceptual separation between transmission and developmental genetics in 1926 made the Modern Synthesis possible in the absence of a detailed understanding of development. For the purposes of the Synthesis, genes mattered only as vehicles of heredity, not as participants in the generation of phenotypes. Embryology was simply 'black-boxed' and subsequently largely ignored (Gilbert, 1978; Amundson, 2001). In the last 30 years, however, there has been increasing recognition among both evolutionary and developmental biologists that re-opening the box may help address important issues in both fields. A central question of modern evolutionary developmental biology ('evo-devo'; Hall, 2000) is how development may help explain evolutionary processes and outcomes. Another is how an evolutionary perspective can further understanding of developmental processes, with respect to their functions, their evolutionary origins and their phylogenetic patterns of conservation and change. These questions are being addressed with ever more depth and sophistication, particularly as increasingly powerful molecular tools are applied to a broadening range of species (see Raff, 1996; Gerhart and Kirschner, 1997; Hall, 1998; Carroll *et al.*, 2001 for reviews).

There is a parallel in the history of *in silico* models of evolution, which have often neglected the level of organization between transmitted (i.e. genetic) information and the use of that information to specify the next generation of phenotypes. If the generative algorithms bridging these levels serve a merely computational function, they represent biological development in a drastically oversimplified (and thus potentially misleading) way: most real developmental processes are not invariant, isomorphic mappings of genotype to phenotype. In the biological world, the developmental functions that integrate genetic and environmental information to generate phenotypes are highly structured and themselves evolve. Which is to say: what is in the black box matters. Any model that ignores its structure, and the potential of that structure to evolve, may be severely limited in its ability to represent evolutionary processes or predict their

 ISBN 0-12-428765-4

outcomes – let alone to harness the power of selection (natural or artificial) to evolve solutions to a given functional or computational problem.

The general importance of development to evolution is widely acknowledged: as Amundson (2001) points out, the oft-repeated 'causal completeness' argument, though non-trivial, is 'useful primarily for preaching to the converted'. The harder question is not whether development matters, but exactly how and why (and, ultimately, how to assess its significance relative to other factors) (Amundson, 2001). In this chapter I will describe three aspects of development that may be of particular importance to modelling the evolutionary process: modularity, environmental responsiveness and the capacity of ontogenies (rather than just the phenotypes they generate) to evolve. This is not meant to be an exhaustive list; rather, these three features exemplify ways in which the structure of development may have far-reaching effects on the evolution of both adult phenotypes, and ontogenies themselves.

4.2 Developmental modularity and its consequences

Biological development is highly complex, extraordinarily diverse (despite remarkable conservation at the genetic level), varyingly responsive to environmental factors, contingent on both individual and phylogenetic history, and at once dependent on and decoupled from the genotype. But despite this array of complications, development is also highly ordered, surprisingly robust to a range of perturbations and economical of information: increasing phenotypic complexity need not require a larger genome. One central characteristic of development that may help to explain how these various other attributes have evolved is its modularity.

4.2.1 Definitions of modularity

Biological (including developmental) modules have been defined in a variety of ways (e.g. Atchley and Hall, 1991; Raff, 1996; Arnone and Davidson, 1997; Bolker, 2000); one reason for the diversity is that no single definition is likely to suit all purposes, since assessments of modularity are necessarily context-dependent (Raff and Raff, 2000; Bolker, 2000). Evolutionary and developmental biologists both rely on the concept of modularity, but have historically approached it from opposite directions, resulting in converging but fundamentally disparate definitions (Gass and Bolker, 2003).

For the purposes of the following discussion, I will use a general description (Bolker, 2000): a module is a biological structure, process, or pathway characterized by more internal than external integration; such modules are biological individuals (Hull, 1980; Roth, 1991) with distinct, emergent properties; and individual modules interact in some way with other entities (i.e. they have external as well as internal connectivity; von Dassow and Munro, 1999). Additional attributes specific to developmental modules (Raff, 1996) include discrete genetic specification, hierarchical organization and the ability to undergo transformation on both developmental and evolutionary time scales.

4.2.2 Types of modules

The most intuitively obvious modules are those of structure: elements of the phenotype, or particular sequences of nucleotides in the genome, that behave as individuals. Examples include the vertebrate limb (famous for its evolved morphological and functional diversity) and the set of gene sequences that constitute the paralogue groups of the Hox cluster (variously duplicated in different animal taxa). Structural modules are the kind most often referred to in the context of character evolution. The central diagnostic feature of an evolutionary module is its ability to evolve relatively independently of other characters; indeed, an implicit assumption of modularity underlies the idea that phenotypes are divisible into (more or less) discrete elements or characters in the first place.

Developmentalists' modules sometimes coincide with those perceived by evolutionists (the limb is a good example), but are more likely to be defined with reference to generative processes rather than as elements of adult phenotypes (Gass and Bolker, 2003). Thus, to developmental biologists structural modules may be real but transient (Gass and Bolker, 2003), and the definition of modularity extends beyond structure to encompass processes as well (Gilbert and Bolker, 2001). Process modules, such as signal transduction pathways, serve as critical links (in time and space) between structural modules; their redeployment over evolutionary time-scales is the basis for shifts in connectivity between developmental modules of all types. Many examples of such redeployed signalling modules are described by Carroll *et al.* (2001), Raff (1996) and Gerhart and Kirschner (1997).

4.2.3 Patterns of change in modular systems

Modularity is important not just as a descriptor of phenotypes, but also as an influence on their evolution. A system built of interconnected modules is both economical of information (in whatever form: genetic or software code) and capable of particular kinds of change based on changes in the number, connectivity and context-dependent function of its parts. Raff (1996) describes dissociation, duplication and divergence, and co-option as evolutionary processes that apply to developmental modules, and that generate 'nonrandom variation within the existing modules that can lead to new internal patterns of order' (Raff, 1996, p. 325). The following discussion closely follows his.

4.2.4 Changes in the integration of modules: dissociation, duplication and divergence

Modules are characterized by, and their function depends on, their external connectivity as well as their internal integration. The simplest change in that connectivity is a shift in the strength (or robustness, or reliability) of links between different modules. The significance or function of a module can change dramatically if its external connections change, even if its internal structure remains constant.

Dissociation refers to the weakening (or disappearance) of coupling between modules, such that structures or processes that were once constrained to a common, or at least correlated, evolutionary path are freed to evolve independently. (This term corresponds closely, though not exactly, to Wagner's (1996) 'parcellation'.) Developmental modules may become dissociated from one another with respect to space (leading to heterotopy), time (leading to heterochrony), or connectivity (leading to a different informational and/or structural architecture of the ontogeny as a whole).

Duplication of modules, followed by their divergence, can facilitate dissociation. After duplication, one copy of a module (for example, one copy of a duplicated gene) may retain its original location, timing of expression and association with other modules, while the duplicate evolves a different pattern of external connectivity. The latter copy may take on dramatically different functions as a result of its new context, even if the module itself remains unchanged. (For a distinction between process, such as might be embodied in a particular module, and function, related to the module's effects rather than to its composition, see Gilbert and Bolker, 2001.) If the duplicate modules are completely dissociated from one another, the result may be an entirely new, or newly-independent, selectable character. In some cases, however, both duplicate modules will remain subject to many or all of the original regulatory controls, so that they retain a degree of linkage. If ancestral regulatory controls are retained, but one duplicate module acquires additional external connections (particularly if it comes to regulate new downstream targets), then these newly associated modules may also be drawn into the existing linkage system.

4.2.5 Changes in the deployment of modules: co-option, heterochrony

Modules need not be duplicated to take on new functions (though duplication presumably makes it easier to maintain the original function at the same time as a new one is acquired). Either duplicate or original modules can be redeployed in new places or at new times during development, and thus come to play novel roles and potentially give rise to novel structures or functions. The vertebrate jaw is a modified gill arch; the diverse proteins used to form the eye lens were borrowed from a variety of other contexts; and examples of co-opted process modules, especially signal transduction pathways such as those involving the *wnt* gene family, abound (Raff, 1996; Gerhart and Kirschner, 1997; Carroll *et al.*, 2001; Gilbert and Bolker, 2001).

Since development unfolds over time and both structures and processes are often modular, it naturally follows that modules of structure and of process can be arranged – and, thanks to their dissociability, rearranged – along a time axis. Heterochrony refers to a shift in timing of one module relative to another (for a recent and critical review of the topic, see Zelditch, 2001) and hypotheses of heterochrony are generally based on an evolutionary change in the relative times of appearance of different structural modules (e.g. Richardson, 1995). The concept clearly applies to process as well as structural modules, and structural heterochrony may emerge directly from a lineage-specific shift in the timing of a conserved developmental pathway responsible for the emergence of a particular structure. However, the underlying generative basis for homologous structures can evolve (de Beer, 1971), so ascribing changes in the timing of appearance of structures directly to heterochrony of a conserved generative process is unreliable: an observed structural heterochrony could also arise by the substitution of a different generative process.

4.2.6 Evolutionary consequences of developmental modularity

Duplication and divergence are a ready source of meristic changes in morphology (changes in numbers of serially repeated body segments or other structures), which are relatively common and very probably of adaptive significance (Raff, 1996). These processes also occur at the genetic level, where they can have particularly far-reaching consequences. Repeated duplication events followed by sequence divergence gave rise to the multitude of regulatory gene families essential to metazoan development (for examples see Raff, 1996; Gerhart and Kirschner, 1997; Hall, 1998; Carroll *et al.*, 2001); Raff has suggested that the metazoan radiation itself 'was arguably the consequence of the most encompassing example of duplication and divergence of regulatory genes' (Raff, 1996, p. 338).

Co-option of a pre-existing module can be a source of a new/additional structure or process, eliminating the need to build up the novelty through a fortuitous series of discrete changes that become integrated along the way. Once you have got a wheel, you can invent a bicycle even if the wheel was first developed as part of an entirely different sort of vehicle. You do not need a new mutation for spokes, and another for the hub, and a third for rims and tyres. In the biological realm, the reuse of a surprisingly small number of fundamental signal transduction pathways supports a vast diversity of different interactions between cells and their environments (including other cells) (Raff, 1996; Gerhart and Kirschner, 1997; Hall, 1998; Carroll *et al*., 2001). The Hox cluster is useful for patterning and performs that function at multiple, nested levels in insect development: the same genes function to assign the location of appendages along the body axis, are involved in specifying the type of appendage, and finally act at a very local level as pigmentation patterns form (reviewed by Carroll *et al*., 2001).

At a more fundamental level, Erwin and Davidson (2002; Davidson, 2001) have suggested that the ancient innovation that eventually permitted the evolution of complex metazoa was the assembly of batteries (i.e. modules) of cell-type-specific genes in an ancestral organism deep in the pre-Cambrian. These differentiation modules were then integrated in various ways and organized into disparate body plans, by later-evolving regulatory genes that functioned to pattern the embryo as a whole and to specify which cell type modules should be activated in different regions.

But co-option can also result in constraint, when the same module becomes integrated into the ontogeny of two (or more) otherwise independent phenotypic characters (Wagner, 1996). Features with no functional relationship, that in an ideal world would be selected independently of one another, may become inextricably linked because they share a developmental module. For instance, almost all mammals have exactly seven cervical vertebrae. This constancy has long baffled morphologists, as it seems unlikely to represent a uniquely adaptive condition among such morphologically and functionally diverse lineages as giraffes, whales and shrews. Galis (1999) has suggested that the explanation may lie instead in the fact that the generation of selectable variation in this feature would require mutation of the Hox genes responsible for axial patterning, and such mutations probably expose individuals to a greatly increased risk of early-onset cancers simply because the same genes are involved in regulating cell proliferation. (Interestingly, the few mammalian lineages – as well as other groups of vertebrates – that do show variation in cervical number seem to have a much lower tendency to develop cancers.)

4.3 Why the environment matters

The environment is crucial to the generation of phenotypes, not just as a venue or filter for their selection. The direct role of the environment in the process and structure of development has been reasserted lately by proponents of 'eco-devo' (Gilbert, 2001; Dusheck, 2002a, b; Gilbert and Bolker, 2003). The significance of the environmental input to the black box generating phenotypes was never in doubt to ecologists (e.g. Tollrian and Harvell, 1999) and others concerned primarily with the fitness of phenotypic outcomes. In contrast, modern developmental biology has largely avoided the topic, viewing environmentally-induced developmental variation as noise rather than data (this bias has become a self-fulfilling prophecy as research efforts have centred mainly on a handful of especially 'hard-wired' model species; Bolker, 1995).

The input to the black box of development (out of which phenotypes are to emerge) comprises genotype, environment and also a critical interactive term reflecting the context-dependent function of specific genetic information. Heredity as understood in the post-Synthesis sense of strictly genetic information is not enough to go on to construct a phenotype. Rather, we need to consider the 'broad heredity' defined (or revived) by Amundson (2001), which includes environmental influences in the widest sense of the word. Once we escape the conceptual confines of narrow- (post-Synthesis-) sense heredity, it becomes possible to focus on the significance of environmental variation, and to recognize strategies – including evolved, genetically-based solutions – that organisms use to maximize their fitness in the context of particular environments.

4.3.1 Phenotypic plasticity and developmental reaction norms

Phenotypic plasticity refers to the ability of a single genotype to generate a range of phenotypes based on environmental information incorporated into the process of development (Schlichting and Pigliucci, 1998; Gilbert, 2001; and references therein). Although the term could apply to developmental malformations due to environmental influences, it is most often used to describe a spectrum of adaptive responses to different environments. Some of the best-known examples include the elaboration of protective morphologies in the presence of predators (or, more precisely, of cues indicative of predation risk) (Kruger and Dodson, 1981; Hebert and Grewe, 1985; Agrawal *et al.*, 1999; Tollrian and Dodson, 1999); seasonal variations in colour pattern that enhance crypsis in a changing vegetative environment (Brakefield *et al.*, 1996); and responses (defensive or accommodating, respectively) to microbial predators and symbionts (McFall-Ngai, 2002).

The developmental reaction norm (DRN; Schlichting and Pigliucci, 1998) describes the set of phenotypes generated by a given genotype in response to environmental cues. The DRN is an inherited capacity of the developmental system: it is a genetically-based and therefore selectable character. The underlying mechanistic basis of DRNs, or of plasticity in general, is in most cases poorly understood (though see Nijhout, 1999, Brakefield *et al.*, 1996, and Abouheif and Wray, 2002 for exceptions). Presumably a variety of physiological and ontogenetic processes are involved in recognizing appropriate environmental cues and then transducing that information into ontogenetic changes that are proximately responsible for generating specific adaptive phenotypes. Schlichting and Pigliucci (1998) have argued (from general principles, supported by the handful of available examples) that genetic mechanisms for plasticity are likely to fall into two categories: allelic variation, such as temperature-sensitive variation in enzyme activity, and regulatory variation, in which different environmental conditions operate discrete genetic switches, triggering the activation of qualitatively different downstream cascades. The combination of increasingly sophisticated molecular techniques, genomic/comparative approaches and renewed interest on the part of mechanistically-oriented developmental biologists is likely to foster greatly improved understanding of the basis of plasticity (e.g. Brakefield *et al.*, 1996); such knowledge will elucidate an important aspect of both development and evolution.

4.3.2 Developmental significance of environmental variation

Selective and developmental environments change over evolutionary time; ontogenies must therefore do so as well, in order to continue generating well-adapted phenotypes. Paradoxically, the most efficient way to have a stable phenotypic outcome may sometimes be to have a flexible ontogeny that produces the same, canalized output regardless of genetic and/or environmental perturbations. Conversely, the most efficient way to have an adaptively variable outcome may be to have a fixed ontogeny whose ultimate result directly and predictably reflects (rather than buffers) external environmental variation. In this second case, the 'fixed' element of the ontogeny is its DRN: its capacity to generate a range of environment-appropriate phenotypic outcomes.

Selection for environmental sensitivity as an ontogenetic trait

The degree to which development is flexible (environmentally responsive or plastic) or rigid (buffered or canalized) can be indirectly selected based on the net fitness, integrated across environments, of the sets of phenotypes generated by more or less flexible ontogenetic variants.

The ontogenetic trait 'environmental sensitivity' – called canalization when it has a low value and phenotypic plasticity when it has a high value – itself evolves. Environments characterized by particular types (especially time-scales) of variation, in which reliable cues or predictors of season, predation risk, etc., are available, are likely to favour plastic ontogenies that can make adaptive use of such environmental information (Tollrian and Dodson, 1999).

The ultimate evolutionary significance of plasticity may transcend its immediate usefulness in particular kinds of environments. Like modularity, it provides a means to generate variable yet well-integrated and fully functional phenotypes. It partially decouples genetic from phenotypic variation, which may limit the efficacy of natural selection; but it may also be a critical source of functional, integrated adaptive variations subsequently filtered by selection (West-Eberhard, 1989, 2003; Schlichting and Pigliucci, 1998).

4.4 Evolving development

That development itself evolves is obvious. Whether and how the evolution of development may be significant to evolution more generally is a subtler matter that calls for both detailed explication and a great deal more empirical evidence than is currently available (Amundson, 2001). I will attempt neither a full explication nor a review of the data here (interested readers are referred to Raff, 1996; Gerhart and Kirschner, 1997; Hall, 1998; Carroll *et al.*, 2001), but will briefly address two related issues: the nature of selection on ontogenies and some general consequences of that selection. (Consideration of the implications of developmental evolution for evolutionary algorithms is deferred to the following section.)

4.4.1 How ontogenies evolve and why it matters

The statement 'ontogenies are subject to selection' is not literally true, if one adheres to a narrow definition of selection: selection acts directly only on their phenotypic products, which reflect the integration of developmental environment and genetically-encoded mechanisms. However, this hardly constitutes an argument against the selectability of ontogenies: genes are also screened from direct selection (in fact they are screened largely by the structure of ontogeny – the intervening black box), but there is no debate about whether they are selected. The key distinction (Sober, 1987) is that direct selection *for* phenotypes (i.e. selection based directly on fitness differences whose proximate basis is phenotypic variation) entails indirect selection *of* the genes and ontogenetic features involved in generating particular phenotypes.

If selection of ontogenies is only indirect, how or why might it matter? First, features or characteristics of ontogenies that enhance their (and their resultant phenotypes') evolvability will be selectively favoured and are thus likely to emerge across a wide phylogenetic range of ontogenies. Modularity is one frequently-cited example; its ability to confer a balance of flexibility and robustness, as well as its potential to generate novelty, offer clear evolutionary advantages (Raff, 1996). Plasticity in response to the environment is another: ontogenies that include the capacity to incorporate environmental information in a useful and functional way may thereby gain access to a wide range of potential phenotypes that work – and that are automatically well-fitted to the environments in which they develop.

Second, like anything else that evolves, ontogenies are cobbled together from available (i.e. pre-existing, ancestral) parts and change incrementally by tinkering (Jacob, 1977) rather than wholesale by redesign. This has a variety of consequences. One is the existence of homologous elements in different ontogenies; comparing the form, function and connectivity of homologous modules can yield useful information about both phylogeny and ontogeny itself. Second, certain constraints exist: not everything is possible, and the potential trajectory of future change is biased by history. Third, one might expect to see organizational features indicative of (selected) evolvability: though this expectation can be surprisingly difficult to test. Finally, since ontogenies are selected only indirectly, selection on them depends on the correlation between a particular ontogeny and an especially fit phenotype – itself the resultant of the combination of genotype, environment and their interaction. In the end, ontogeny is the function that integrates all of those sources of information and translates the result into a selectable phenotype.

4.4.2 Implications for evolutionary models

The process of natural selection is relatively straightforward to model (assign fitness values to a range of phenotypes, choose those with the highest values as the source of 'genetic' information for the subsequent generation, construct the next set of phenotypes based on that information – modified to some extent by processes such as mutation, recombination, etc. – and start over). Exactly how each step is represented has profound consequences for how the model works and, in particular, how likely it is to approximate or reproduce real evolutionary processes and outcomes. So the design of each element of the model should ideally draw on what we know about the wet (that is, biological) versions of each of these things.

In parallel with ongoing detailed mechanistic studies of specific processes, developmental biologists are now (re)turning their attention to the larger-scale organization of ontogenies and considering both the developmental mechanisms and the evolutionary ramifications of that architecture. Such structural considerations are likely to have implications for models as well. For example, both the content and the connectivity of developmental modules can evolve, but they need not do so in a correlated way. Modular organization and environmental responsiveness during development are both indirectly selectable features *per se*, and influence the range of phenotypic variation visible to direct selection.

The adaptive importance of environmentally-responsive plasticity suggests that assessment of the environment should factor into algorithms representing ontogeny as well as into those assigning relative fitness values to adult phenotypes. Allowing for the possibility, and particularly the evolution, of plasticity is especially important in models that allow selection criteria to shift in foreseeable ways: such conditions put a premium on the ability to incorporate environmental information into production of the phenotype.

Understanding how development is structured can make evolutionary patterns more explicable (Raff, 1996; Carroll *et al.*, 2001); it may ultimately also make evolutionary processes easier to simulate, and help to bring the structure, function and outcomes of those simulations closer to biological reality.

References

Abouheif, E. and Wray, G.A. (2002) Evolution of the gene network underlying wing polyphenism in ants. *Science*, **297**, 249–252.

Agrawal, A.A., Laforsch, C. and Tollrian, R. (1999) Transgenerational induction of defenses in animals and plants. *Nature*, **401**, 60–63.

Amundson, R. (2001) Adaptation and development: on the lack of common ground. In Orzack, S.H. and Sober, E. eds. *Adaptationism and Optimality*, pp. 303–334. Cambridge Studies in Philosophy and Biology, Cambridge University Press.

Arnone, M.I. and Davidson, E.H. (1997) The hardwiring of development: organization and function of genomic regulatory systems. *Development*, **124**,1851–1864.

Atchley, W.R. and Hall, B.K. (1991) A model for development and evolution of complex morphological structures. *Biol. Rev. Camb. Philos. Soc.*, **66**, 101–157.

Bolker, J.A. (1995) Model systems in developmental biology. *BioEssays*, **17**, 451–455.

Bolker, J.A. (2000) Modularity in development and why it matters to evo-devo. *Amer. Zool.*, **40**, 770–776.

Brakefield, P.M., Gates, J., Keys, D. *et al.* (1996) Development, plasticity and evolution of butterfly eyespot patterns. *Nature*, **384**, 236–242.

Carroll, S.B., Grenier, J.K. and Weatherbee, S.D. (2001) *From DNA to Diversity: Molecular Genetics and the Evolution of Animal Design*. Blackwell Science.

Davidson, E.H. (2001) *Genomic Regulatory Systems: Development and Evolution*. Academic Press.

de Beer, G.R (1971) *Homology: an Unsolved Problem*. Oxford Biol. Readers 11. Oxford University Press.

Dusheck, J. (2002a) It's the ecology, stupid! *Nature*, **418**, 578–579.

Dusheck, J. (2002b) The interpretation of genes. *Natural History*, **111**, 52–59.

Erwin, D.H. and Davidson, E.H. (2002) The last common bilaterian ancestor. *Development*, **129**, 3021–3032.

Galis, F. (1999) Why do almost all mammals have seven cervical vertebrae? Developmental constraints, Hox genes, and cancer. *Journal of Experimental Zoology (Molecular and Developmental Evolution)*, **285**, 19–26.

Gass, G.L. and Bolker, J.A. (2003) Modularity. In Hall, B.K. and Olson, W.M. eds. *Key Concepts and Approaches in Evolutionary Developmental Biology*, pp. 260–267. Harvard University Press.

Gerhart, J. and Kirschner, M. (1997) *Cells, Embryos, and Evolution: Toward a Cellular and Developmental Understanding of Phenotypic Variation and Evolutionary Adaptability*. Blackwell Science, Inc.

Gilbert, S.F. (1978) The embryological origins of the gene theory. *J. Hist. Biol.*, **11**, 307–351.

Gilbert, S.F. (2001) Ecological developmental biology: developmental biology meets the real world. *Devel. Biol.*, **233**, 1–12.

Gilbert, S.F. and Bolker, J.A. (2001) Homologies of process and modular elements of embryonic construction. In Wagner, G.P. ed. *The Character Concept in Evolutionary Biology*. Yale University Press.

Gilbert, S.F. and Bolker, J.A. (2003) Ecological development biology: preface to the symposium. *Evolution and Development*, **5**, 3–8.

Gilbert, S.F., Opitz, J. and Raff, R.A. (1996) Resynthesizing evolutionary and developmental biology. *Dev. Biol.*, **173**, 357–372.

Hall, B.K. (1998) *Evolutionary Developmental Biology* (2nd edn). Kluwer Academic Publishers.

Hall, B.K. (2000) Evo-devo or devo-evo – does it matter? *Evolution and Development*, **2**, 177–178.

Hebert, P.D.N. and Grewe, P.M. (1985) *Chaoborus*-induced shifts in the morphology of *Daphnia ambigua. Limnol. Oceanogr.*, **30**, 1291–1297.

Hull, D.L. (1980) Individuality and selection. *Ann. Rev. Ecol. Syst.*, **11**, 311–332.

Jacob, F. (1977) Evolution and tinkering. *Science*, **196**, 1161–1166.

Kruger, D.A. and Dodson, S.I. (1981) Embryological induction and predation ecology in *Daphnia pulex. Limnol. Oceanogr.*, **26**, 219–223.

McFall-Ngai, M.J. (2002) Unseen forces: the influence of bacteria on animal development. *Dev. Biol.*, **242**, 1–14.

Nijhout, H.F. (1999) Control mechanisms of polyphenic development in insects. *BioScience*, **49**, 181–192.

Raff, E.C. and Raff, R.A. (2000) Dissociability, modularity, evolvability. *Evolution and Development*, **2**, 235–237.

Raff, R.A. (1996) *The Shape of Life*. Chicago University Press.

Richardson, M.K. (1995) Heterochrony and the phylotypic period. *Devel. Biol.*, **172**, 412–421.

Roth, V.L. (1991) Homology and hierarchies: problems solved and unresolved. *J. Evol. Biol.*, **4**, 167–194.

Schlichting, C.D. and Pigliucci, M. (1998) *Phenotypic Evolution: a Reaction Norm Approach*. Sinauer Associates.

Sober, E. (1987) *The Nature of Selection: Evolutionary Theory in Philosophical Focus*. Bradford Books, MIT Press.

Tollrian, R. and Dodson, S. I. (1999) Inducible defenses in cladocera: constraints, costs, and multipredator environments. In Tollrian, R. and Harvell, C. D. eds. *The Ecology and Evolution of Inducible Defenses*, pp. 177–202. Princeton University Press.

Tollrian, R. and Harvell, C.D. eds. (1999) *The Ecology and Evolution of Inducible Defenses*. Princeton University Press.

von Dassow, G. and Munro, E. (1999) Modularity in animal development and evolution: elements of a conceptual framework for EvoDevo. *Journal of Experimental Zoology (Molecular and Developmental Evolution)*, **285**, 307–325.

Wagner, G.P. (1996) Homologues, natural kinds and the evolution of modularity. *Am. Zool.*, **36**, 36–43.

West-Eberhard, M.J. (1989) Phenotypic plasticity and the origins of diversity. *Ann. Rev. Ecol. Syst.*, **20**, 249–278.

West-Eberhard, M.J. (2003) *Development Plasticity and Evolution*. Oxford University Press.

Zelditch, M.L. (ed.) (2001) *Beyond Heterochrony: the Evolution of Development*. John Wiley and Sons Inc.

Plasticity and reprogramming of differentiated cells in amphibian regeneration

5

JEREMY P BROCKES and **ANOOP KUMAR**

5.1 Introduction

Regeneration means the regrowth of lost or destroyed parts or limbs. The only adult vertebrates that can regenerate their limbs are the urodele amphibians, of the order *Caudata*. Regeneration of the salamander limb, together with several other examples of urodele regeneration, was first reported by Spallanzani in 1768. The ability to regenerate large sections of the body plan is widespread in metazoan phylogeny and the discovery of this ability was an important aspect of the emergence of experimental biology in the eighteenth century (Dinsmore, 1991). It provoked intense public discussion on the nature of generation and the basis of individual identity, which are issues that continue to be explored in present-day debates about reproductive cloning. As a problem in molecular cell biology, urodele regeneration provides crucial information about the reversal and plasticity of the differentiated state (Brockes, 1997; Brockes *et al.*, 2001). In addition, limb regeneration is a key system in which to study how positional identity in cells is established (Maden, 1982; Nardi and Stocum, 1983; Pecorino *et al.*, 1996; Torok *et al.*, 1998). Although much work on development and evolution serves to underline the similarities and continuity between different phylogenetic contexts (Raff, 1996; Carroll, 2001), it is an all-embracing concern of research on regeneration to understand the relevant distinctions between species that regenerate and those that do not (Alvarado 2000; Brockes *et al.*, 2001). In the case of urodeles and mammals, this takes on a biomedical imperative in view of the current interest in regenerative medicine.

From our mammalian perspective, the ability of an adult newt to regenerate its limbs (colour plate 5.1a, b) might seem exceptional or even exotic, but it is unlikely that regeneration arose independently in different phylogenetic contexts (Goss, 1969; Alvarado, 2000; Brockes *et al.*, 2001). In only six phyla are there no examples of adult regeneration, yet in all contexts in which it is found, there seem to be examples of closely related species that have marked differences in regenerative ability. The hypothesis from phylogeny is that regeneration is a primordial attribute of metazoans (of the kingdom Animalia (animals), which comprises ~35 phyla of multicellular organisms) that has been lost subsequently for reasons that are not yet understood. In analysing the cellular and molecular mechanisms that underlie the regenerative responses in urodeles, it has been informative to compare them with the mammalian case. This helps to pinpoint the crucial differences between urodeles and mammals and will shed light ultimately on the evolutionary basis of regeneration.

5.2 Regenerative responses in urodeles

An adult newt can regenerate its jaws (Ghosh *et al.*, 1994), lens (Reyer, 1954), retina (Mitashov, 1996) and large sections of the heart (Oberpriller and Oberpriller, 1974), as well as its limbs and tail, in response to molecular events that signal tissue damage or removal. Although the processes of tissue restoration unfold in a different way in the heart, limbs and tail of an adult newt, in all cases, the outcome seems to depend on the plasticity of differentiated cells (i.e. their ability to build new tissue) that remain after tissue removal.

5.2.1 Plasticity in the heart

After removal of the apical region of the newt ventricle, the heart seals by contraction around the clot (Figure 5.1a). The adult cardiomyocytes (heart muscle cells) re-enter the cell cycle and divide in a zone that surrounds the clot (Oberpriller and Oberpriller, 1974; Oberpriller *et al.*, 1995). If the animal is injected with tritiated thymidine to identify those cells that are in S phase (the phase of the eukaryotic cell cycle in which DNA is synthesized) $\sim$10% of the cardiomyocytes in this region are labelled in a one-day period. In comparable experiments with the adult mammalian heart, very few cells label after injury (Soonpaa and Field, 1998). Thymidine labelling carried out in conjunction with ultrastructural studies has identified cardiac myofibrils (skeletal-muscle fibres that consist of single long multinucleate cells) as a marker of cell identity.

5.2.2 Plasticity in the iris

After removal of the newt lens (lentectomy), the reactive population comprises pigmented epithelial cells, which are invariably located at the dorsal pupillary margin of the iris (Figure 5.1b). These cells re-enter the cell cycle, lose their pigment granules and convert into lens cells, a process that is referred to as transdifferentiation (Eguchi *et al.*, 1974; Eguchi, 1988; Okada, 1991). Elegant clonal cultures of newt pigment-epithelial cells have established conclusively that these cells transdifferentiate into lens (Eguchi and Okada, 1973; Eguchi *et al.*, 1974).

Transdifferentiation

Transdifferentiation was originally defined by Okada (1991) as the conversion of one differentiated cell type into another. The original, and still the most studied, example is the conversion of pigmented epithelial cells from the iris into lens cells, which occurs during lens regeneration in newts and in cultures of pigment cells from various species (Figure 5.1b). This can still occur to some extent if cell division is blocked, although it normally occurs in conjunction with cell-cycle re-entry.

In urodele limb regeneration, transdifferentiation could occur in a switch between the chondrogenic (cartilage forming) connective tissue lineage and the myogenic (muscle forming) lineage. If cartilage is selectively labelled and introduced into a limb blastula, the occurrence of labelled nuclei in muscle is negligible.

Myotubes are structures comprising multiple nuclei, formed by the fusion of proliferating myoblasts (undifferentiated precursors to muscle cells) and characterized by the presence of certain muscle-specific markers. If cultured myotubes are labelled by microinjection of

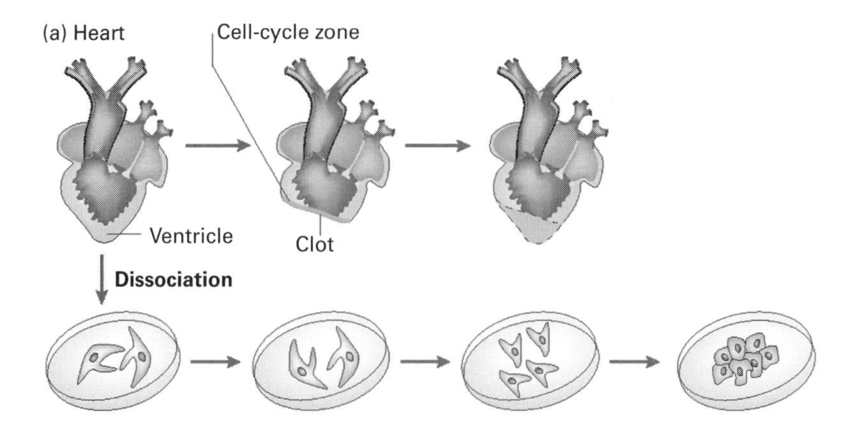

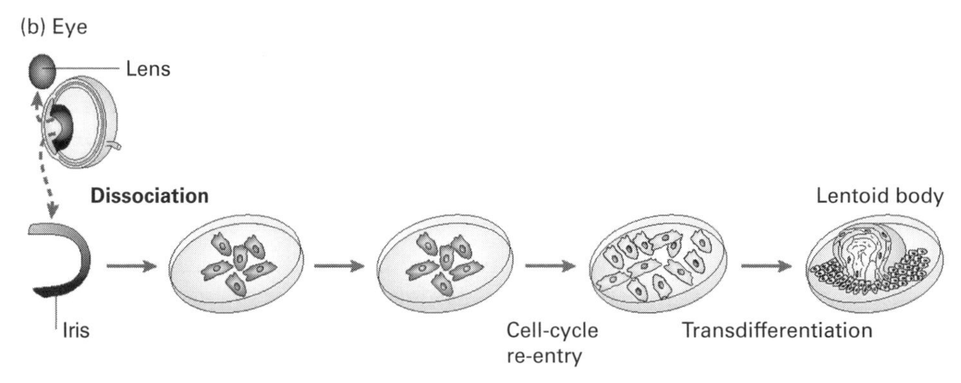

Figure 5.1 Plasticity of differentiated cells during regeneration of newt heart or lens. a. After removal of the apex of the ventricle, the heart is sealed by a clot, and the adult cardiomyocytes return to the cell cycle in a zone around the clot, proliferate and regenerate the ventricle. If cardiomyocytes are dissociated into a culture dish, they can divide and form groups of beating cells. It is striking that some cardiomyocytes complete the cell cycle and then resume beating. b. After removal of the lens, the pigmented cells will re-enter the cell cycle, de-pigment and transdifferentiate into lentoid bodies (a small cluster of lens cells) that express lens markers.

rhodamine-dextran (Lo *et al.*, 1993) or with an integrated retrovirus (Kumar *et al.*, 2000), small clones of labelled mononucleate cells are detected occasionally in the cartilage of the regenerate.

There are several cases of transdifferentiation in mammals (Tosh and Slack, 2002). One example is the conversion of pancreatic exocrine cells (that make up part of the exocrine gland, which discharges its secretion through a duct) to hepatocytes (the parenchymal cells of the liver that are responsible for the synthesis, degradation and storage of a wide range of substances) *in vivo* under conditions of copper deficiency, as well as *in vitro*. Other examples include the transition between smooth and skeletal muscle in the developing oesophagus and the conversion of myoblasts to adipocytes (which forms fat or adipose tissue).

There has been much recent interest in transdifferentiation of stem cells; for example, the ability of haematopoietic stem cells or mesenchymal stem cells (which form immature connective tissue) to give rise to neural and other epithelial derivatives after transplantation. The interpretation of these results is controversial at present (Tsai *et al.*, 2002), in view of the possibility of heterogeneity in the starting stem cell populations and the contribution of cell fusion between different stem cells. However, transdifferentiation remains an important area for understanding cell plasticity.

5.2.3 Plasticity in the limb

After amputation of the limb at any point along its proximodistal axis (shoulder to fingertip), the wound surface is covered rapidly by epithelial cells, which form the wound epidermis at the end of the stump. In contrast to lens regeneration, which depends on the plasticity of epithelial cells of the iris, in limb regeneration, it is the mesenchymal cells in a zone underlying the wound epidermis that re-enter the cell cycle. The cartilage, connective tissue and muscle cells lose their differentiated characteristics and become blastemal cells – the progenitor cells of the regenerate. These cells divide to form the blastema, a mesenchymal growth zone that undergoes proliferation, differentiation and morphogenesis to regenerate the limb (Tsonis, 1996) (see colour plate 5b).

Extensive evidence indicates that multinucleate newt myotubes and myofibres re-enter the cell cycle and undergo conversion to mononucleate cells. Such evidence comes from the results of experiments on muscle cells in culture (Tanaka *et al.*, 1997, 1999; Velloso, 2001), cells implanted from culture into a limb blastema and observations of live myofibres in a regenerating tail (Echeverri *et al.*, 2001), which are all discussed later. In addition, earlier observations of implantation of labelled cartilage also provide strong evidence for reversal of this cell type (Steen, 1968).

In each of these three cases of regeneration, the initial mobilization of differentiated cells occurs within a zone of $\sim$100 μm from the site of tissue removal. It is obviously crucial that such changes do not propagate into the main body of the tissues that are concerned and the signalling events that link tissue removal with plasticity clearly accomplish this spatial restriction. Although these three cases involve re-entry into the cell cycle, the cardiomyocyte is the only example to retain differentiated function, a feature that is also observed in culture. Although in one sense cell division is necessary to generate the extra cells that are needed for the regenerate, it is less clear how important it is for the events of plasticity and reversal and this issue needs to be addressed experimentally in each context.

The urodele strategy for regeneration of most structures is therefore the respecification of differentiated cells to a local progenitor cell, rather than a pluripotent cell. So, if iris epithelial cells are transplanted to the limb blastema, they give rise to a lens (Reyer *et al.*, 1973; Ito *et al.*, 1999) and limb blastemas always give rise to a limb after transplantation, even after relocation to the anterior chamber of the eye (Kim and Stocum, 1986). This is in contrast to recent findings of the extensive plasticity of stem cells after transplantation in mammals (Clarke *et al.*, 2000). At present, there is no evidence that adult stem cells can contribute to urodele limb regeneration, although it is not possible to exclude this potentiality completely. One advantage of the urodele mechanism might be that it allows the progenitor cells to derive local cues from their differentiated parental cells.

5.3 Plasticity of differentiated cells in mammals?

Plasticity of cellular differentiation provides a convenient cellular assay to compare a differentiated urodele cell with its mammalian counterpart. It is important to recognize that there are examples of regeneration in mammals that do involve plasticity. For example, liver regeneration seems comparable to cardiac regeneration in newts in that the hepatocytes divide without loss of differentiated function (Michalopoulos and DeFrances, 1997). The regeneration of myelinated peripheral nerves requires that Schwann cells divide and lose expression of myelin before they redifferentiate in conjunction with the regenerating axons (Aguayo *et al.*, 1976; Weinberg and Spencer, 1976). It therefore seems to be more closely related than liver regeneration to the general case of the urodele mechanism. It seems unlikely that the regulation of the differentiated state is fundamentally different in urodeles and mammals. To study this further, we have compared urodele multinucleate skeletal muscle cells with their mouse counterparts. To what extent are the two intrinsically different in terms of regulation of the differentiated state or responsiveness to injury-related signals, or are there distinctive signals in the urodele context that mouse cells never encounter?

5.4 Reversal of muscle differentiation

Multinucleate skeletal myotubes are formed by the fusion of mononucleate precursor cells. The myotube enters a state of post-mitotic (post cell division) arrest in which it is entirely refractory (resistant) to the growth factors that stimulate division of its precursors (Olwin and Hauschka, 1988). The change in cell structure from mononucleate to multinucleate together with the post-mitotic arrest provides two indices for the reversal of the myogenic phenotype (de-differentiation). Newt A1 cells, which were originally derived from limb mesenchyme (Ferretti and Brockes, 1988), fuse in culture to form multinucleate myotubes, which express markers of late muscle differentiation, such as myosin heavy chain, and are also refractory to protein growth factors (Ferretti and Brockes, 1988; Tanaka *et al.*, 1999).

5.4.1 Assaying for de-differentiation in the blastema

A1 cells can be stably infected with retroviruses that express genes for alkaline phosphatase or green fluorescent protein (GFP), so that after differentiation, labelled myotubes can be readily distinguished and purified from any labelled mononucleate cells (Kumar *et al.*, 2000). Other methods of labelling involve microinjecting the myotubes with a lineage tracer, such as rhodamine dextran (Lo *et al.*, 1993; Velloso *et al.*, 2000), or labelling them with a lipophilic cell tracker dye (Kumar *et al.*, 2000).

Implantation of labelled myotubes under the wound epidermis of an early limb blastema showed that many were converted to mononucleate cells, which divided and contributed to the regenerate – a process that is referred to here as cellularization (Lo *et al.*, 1993; Kumar *et al.*, 2000; Velloso *et al.*, 2000) (Figure 5.2, see colour plate 5.2). If the myotubes were double labelled in their nuclei and cytoplasm before implantation, the mononucleate progeny derived both markers, which shows that both compartments of the multinucleate muscle cells were conserved after cellularization (Lo *et al.*, 1993; Velloso *et al.*, 2000).

A second index for the reversal of differentiation is re-entry to S phase by the nuclei in a myotube. After implantation of myotubes into a newt blastema and injection of the animals with 5-bromodeoxyuridine (BrdU) to label nuclei in S phase, residual myotubes had several positively stained nuclei. In some cases, all of the nuclei in a myotube were labelled (Kumar *et al.*, 2000). It is not clear if such myotubes undergo subsequent cellularization after re-entry, but it is clear that the environment of the blastema leads to reversal of both aspects of myogenic differentiation.

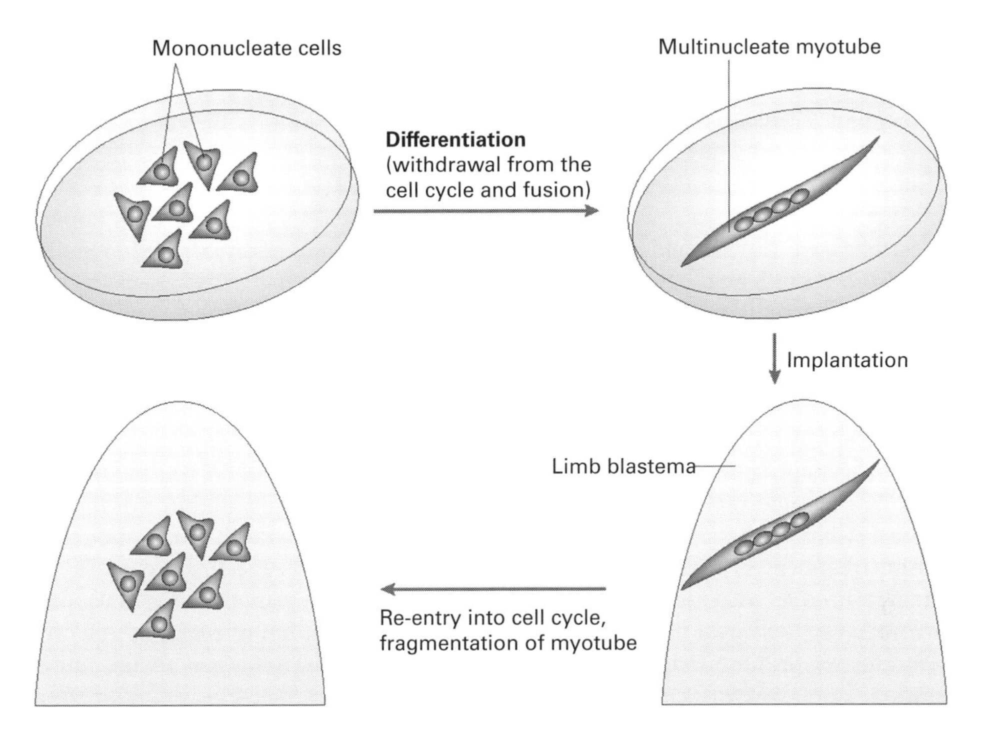

Figure 5.2 Plasticity of urodele myogenesis. Mononucleate cells fuse to form multinucleate myotubes, but this process can be reversed. After implantation into a limb blastula, the nuclei within myotubes re-enter S phase and the myotubes are fragmented into viable mononucleate cells, which divide and contribute to the regenerate. Reproduced with permission from Brockes *et al.* (2001) © 2001 Cambridge University Press.

Implantation is an extremely useful approach because it enables prior transfection (infection of a cell with purified viral nucleic acid, resulting in subsequent replication of the virus in the cell), labelling and purification of the myotubes in culture (see colour plate 5.2). Importantly, both aspects of reversal that are discussed above have also been observed in endogenous multinucleate muscle fibres after amputation. First, the nuclei within muscle fibres at the site of limb amputation were labelled after BrdU injection into adult newts (Kumar *et al.*, 2000). The identity of the labelled nuclei was verified after serial sectioning, which showed that they were present in muscle fibres. Second, the regenerating tail of the larval axolotl was visualized under Nomarski optics (which forms images of high contrast and resolution in unstained cells using birefringent prisms and polarized light) to observe the behaviour of single fibres after microinjection of a lineage tracer. Such fibres fragmented synchronously into mononucleate cells and underwent rapid proliferation. The activation of this process apparently required both 'clipping' at the end of the fibre as well as tissue injury in the vicinity (Echeverri *et al.*, 2001).

These experiments have established that the response of myotubes that are implanted into the limb blastema is similar to that of resident myofibres and underline the importance of a more detailed analysis of both re-entry to the cell cycle and cellularization. Furthermore, it is noteworthy that the muscle fibres make an important cellular contribution to the blastema (Echeverri *et al.*, 2001). The authors estimate that a 4-day old blastema contains ~900 cells, of which approximately 150 come from the cellularization of muscle fibres.

5.5 Cell-cycle re-entry in newt myotubes

Newt myotubes clearly differ from their vertebrate counterparts in that they enter and traverse S phase after serum stimulation in culture (Tanaka *et al.*, 1997, 1999). The nuclei in the myotube double their DNA content and are stably arrested in G2. The response to serum is not observed for other vertebrate myotubes, with the exception of mouse cells that lack both copies of the retinoblastoma gene *Rb* (Schneider, 1994). pRb has a familiar and essential role in regulating the transition from G1 to S phase. Several lines of evidence (Novitch *et al.*, 1996; Zacksenhaus *et al.*, 1996) indicate that it is important for maintenance of the differentiated state in vertebrate myotubes, not only for stable arrest from the cell cycle (Schneider, 1994), but also for transcription from certain muscle promoters that depend on activation of members of the myocyte enhancer factor 2 (MEF2) family of transcription factors (Novitch *et al.*, 1999). Newt myotubes express pRb, but serum stimulates a pathway that leads to its inactivation by phosphorylation and hence triggers progression from G1 into S phase (Tanaka *et al.*, 1997). If the myotubes are injected with a plasmid that encodes mammalian $p16^{INK4}$ – a specific inhibitor of the cyclin-D–CDK4 protein kinase that inactivates Rb – they are effectively blocked from entering S phase after serum treatment (Tanaka *et al.*, 1997). These data provide the first clear evidence that differentiated urodele cells are intrinsically different from their mammalian counterparts. The newt myotube can activate a pathway that leads to phosphorylation of pRb and re-entry to the cell cycle and we propose that the regulation of this pathway is pivotal to the initiation of regeneration.

5.5.1 Regulation of cell-cycle re-entry by thrombin

The activity of vertebrate serum on newt myotubes is not due to the presence of typical protein growth factors, such as platelet derived growth factor (PDGF) or epidermal growth factor (EGF); these factors act on mononucleate A1 cells but not on the myotubes (Tanaka *et al.*, 1997, 1999) (Figure 5.3). The post-mitotic arrest, as defined in mouse myotubes, is therefore comparable to that in newt myotubes and this is consistent with the pivotal role of pRb in both

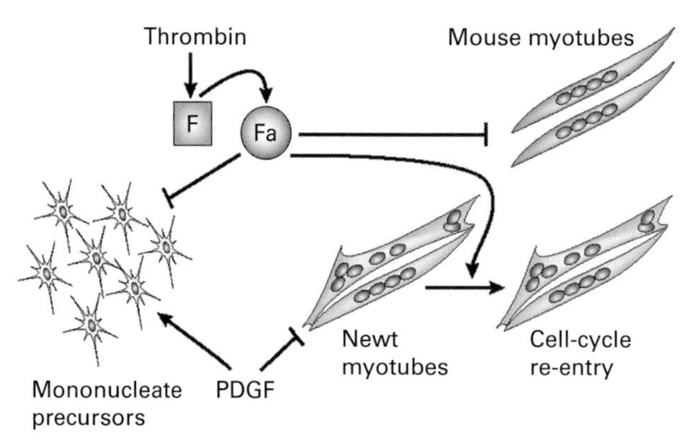

Figure 5.3 Thrombin activity counteracts post-mitotic arrest in newt myotubes. Thrombin acts indirectly by cleaving a substrate (F) that is present in vertebrate serum or plasma (Tanaka *et al.*, 1999) to induce cell-cycle re-entry. Interestingly, the resulting activity, referred to as Fa, does not affect mononucleate cells, whereas growth factors such as platelet-derived growth factor (PDGF) have the reverse specificity. Mouse myotubes are completely refractory to the thrombin-generated activity.

cases. The distinction is that the urodele myotubes are responsive to an activity in serum that has no effect on the mouse cells. Serum is the soluble fraction of clotted blood and results from the activation of *prothrombin* to generate the serine protease thrombin. Thrombin activates the clotting cascade and various other events that mediate the response to injury. It is possible to pre-incubate sub-threshold concentrations of serum with thrombin and then inactivate all of the residual protease activity. Such digests can generate considerable activity on the newt myotubes (Tanaka *et al.*, 1999). When crude prothrombin is activated *in vitro*, the resulting thrombin preparations contain a distinct activity that acts directly on newt myotubes in serum-free medium (Tanaka *et al.*, 1999). We hypothesize that this ligand is generated downstream of prothrombin activation both in regeneration and in culture, and it is this factor that acts on the myotubes and other differentiated cells to promote re-entry (Figure 5.3).

The activation of vascular prothrombin after injury occurs in relation to a protease complex known as tissue factor, which is assembled on the cell surface. Thrombin formation is subject to strict spatial and temporal regulation, as it is essential that clot formation is restricted to the wound area and does not spread. It is therefore tempting to speculate that urodele regeneration – for example, in the heart, limb or eye – is linked to the acute events of injury or tissue removal by the local activity of thrombin. Thrombin activity is increased locally in the early mesenchymal blastema of the limb (Tanaka *et al.*, 1999), but more striking are recent findings that prothrombin is selectively activated on the dorsal margin of the iris after lentectomy and that inhibition of thrombin in this context blocks cell-cycle re-entry on the dorsal margin (Imokawa and Brockes, 2003). We propose that activation of prothrombin and generation of an activity referred to as Fa (Figure 5.3) are important for the initiation of urodele regeneration in several contexts. It will be important to determine whether Fa is active on various differentiated cell types.

5.5.2 Cell-cycle re-entry by mouse myonuclei

Although myotubes of the mouse C2 cell line are not normally responsive to serum and the thrombin-based manipulations (Tanaka *et al.*, 1997), their nuclei can be induced to re-enter the cell cycle by this pathway under some experimental conditions. This involves fusing mononucleate newt A1 and mouse C2 cells to obtain viable hybrid interspecies myotubes (heterokaryons) (Velloso *et al.*, 2001). The mouse nuclei can be identified by a species-specific antibody to nuclear lamin B. These cells grow at 33 °C under conditions that support serum-induced cell-cycle re-entry of A1 myotubes, as well as the post-mitotic arrest. If the heterokaryons are stimulated with serum or thrombin, mouse nuclei frequently enter S phase – sometimes in conjunction with newt nuclei and sometimes alone (Velloso *et al.*, 2001). So, mouse myotubes have apparently lost responsiveness to the thrombin-derived factor (Tanaka *et al.*, 1997, 1999; Velloso *et al.*, 2001), but their nuclei remain responsive to the intracellular consequences of the pathway, such as the phosphorylation of pRb and, in this respect, are quite comparable to newt nuclei.

5.6 Cellularization

An important mechanistic issue is the relationship between the two aspects of plasticity – re-entry to the cell cycle and cellularization (Kumar *et al.*, 2000). One possibility is that nuclei proceed into mitosis and the myotube is fragmented by cytokinesis (the process of cytoplasmic division). Alternatively, myotube nuclei could bud off the main axis of the myotube in the presence or absence of cell-cycle re-entry. A1 newt myotubes can be blocked from cell-cycle re-entry either by X-irradiation or by microinjection of a plasmid that encodes mammalian p16^{INK4}. Such arrested myotubes, when implanted into a limb blastema together with a differentially labelled control population, were effectively converted to mononucleate cells (Velloso *et al.*, 2000) (Figure 5.4). So, although re-entry and cellularization occur in parallel after implantation, they are not linked mechanistically. One other example of cellularization that is not dependent on mitosis and which has some parallels with the process in newt myotubes, is that of the avian osteoclast-like multinucleated giant cells (Solari *et al.*, 1995). These cells are formed by fusion of mononuclear monocytes and, in culture they bud off mononuclear osteoclasts from the apical surface.

These findings on the independence of re-entry and cellularization have been reinforced by observations on striated muscle fibres, which can be dissociated from the limbs of larval salamanders and maintained in culture as adherent cells. These cells are apparently activated by the

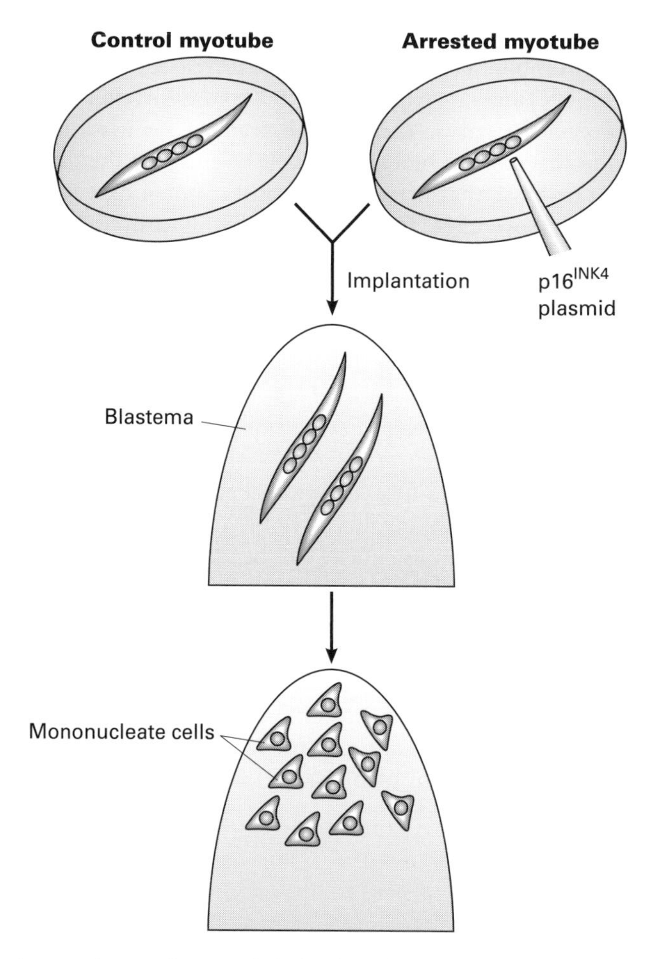

Figure 5.4 Arrested myotubes undergo cellularization after implantation into the newt blastema. The figure shows a cultured newt myotube injected with a plasmid that expresses the cyclin-dependent-kinase inhibitor p16^{INK4} to arrest cell-cycle progression along with a control myotube that is labelled with the lineage tracer rhodamine-dextran. After implantation, both give rise to mononucleate cells in the blastema (Velloso *et al.*, 2000).

dissociation process to undergo several aspects of morphological plasticity, which can be analysed by time-lapse studies and by microinjection of a lineage tracer into the fibres. Some of them break into smaller fibres, or discharge viable multinucleate buds that subsequently adhere to the substrate. The most striking observation is that nuclei can aggregate in some fibres to form a lobulated, 'cauliflower' structure, which gives rise to a colony of dividing mononucleate cells. These events all occur in the absence of S-phase entry in the muscle-fibre nuclei (A. Kumar *et al.*, unpublished observations). This accessible system offers the potential for a more detailed exploration of the mechanisms of cellularization.

5.6.1 Cellularization of mammalian myotubes

An important impetus for the study of cellularization has come from the recent discovery of two methods that induce this process in mouse myotubes. In one approach, mononucleate C2 cells were stably transfected with the homeobox gene *Msx-1* linked to a conditional promoter (Odelberg *et al.*, 2000). Several studies have previously indicated that *Msx* genes promote cell proliferation and that their expression is inversely correlated with differentiation (Song *et al.*, 1992; Woloshin *et al.*, 1995; Hu *et al.*, 2001). After fusion of the stably transfected C2 cells, expression of *Msx-1* in myotubes was induced and this led to a decrease in the expression of myogenic regulatory genes. About 5% of the myotubes were induced to cleave into viable fragments – as described above for the larval myofibres – and another 5% fragmented into mononucleate cells, which proliferated. In some cases, the clonal progeny of a single myotube was isolated, propagated and shown to be capable of either chondrogenic, adipogenic or myogenic differentiation, depending on the culture conditions (Odelberg *et al.*, 2000). It has been proposed (Koshiba *et al.*, 1998) that Msx-1 is a master regulator of the programme for cellularization that is expressed in urodele regeneration and that it can also induce this programme in mammalian myotubes.

A second impetus to the analysis of these issues has come from the application of 'chemical genetics'. A large combinatorial library of trisubstituted purines was screened to identify a compound that would induce mammalian myotubes to undergo the regenerative response of newt myotubes (Rosania *et al.*, 2000). One compound, with methoxybenzyl and isopropyl substituents, effectively fragments the myotubes within 24 hours to yield viable mononucleate cells that can divide and also fuse again to re-form myotubes. This compound – called myoseverin – seems to have two activities in mammalian myotubes. First, it depolymerizes microtubules. Second, it induces changes in the expression of a specific complement of genes that is involved in repair, wound healing and regeneration (Figure 5.5a). Although there are other agents that can fragment microtubules, they tend not to yield viable mononucleate cells that can fuse into myotubes again (Perez *et al.*, 2002). Therefore, it is possible that both activities – i.e. microtubule depolymerization and changes in gene expression – are important for the generation of viable mononucleate cells.

One possibility is that the compound that is responsible for fragmenting myotubes might also, by chance, regulate genes that are implicated in tissue remodelling and repair. The alternative and more attractive hypothesis, as proposed by Rosania *et al.* (2000), is that myoseverin can activate the expression of a programme that mediates cellularization and other functions that are relevant to regeneration. DNA microarray analysis has identified ~90 genes so far, the expression of which was up or downregulated at least twofold by myoseverin (Rosania *et al.*, 2000). Interestingly, many of these genes belong to categories that are regulated in fibroblasts in response to serum (Iyer *et al.*, 1999) (Figure 5.5b). A crucial question, which is now open to investigation, is whether the mechanism of myoseverin

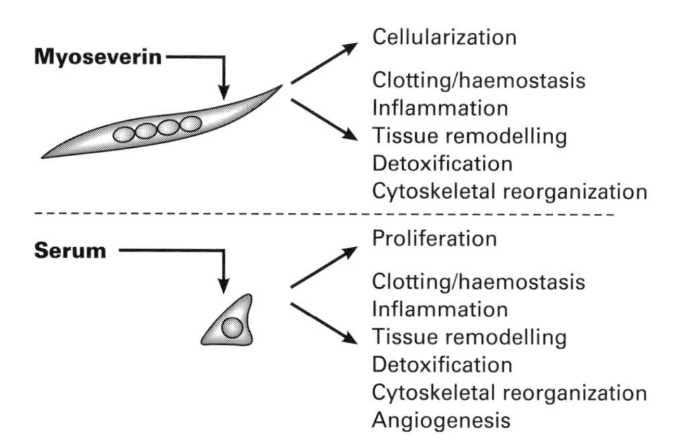

Figure 5.5 Categories of genes that are regulated after myoseverin treatment of myotubes or serum stimulation of fibroblasts. In addition to the overall response of proliferation or cellularization, both myoseverin treatment (upper) and serum stimulation (lower) alter the expression of categories of genes that are implicated in tissue repair and remodelling and wound healing (Iyer *et al.*, 1999; Rosania *et al.*, 2000). The interpretation has been that fibroblasts encounter serum in the context of tissue injury and that the changes in expression level of these genes orchestrate the behaviour and activities of this cell type. For myoseverin, it is plausible that as cellularization occurs during urodele regeneration, the programme of gene expression might also occur normally in this context.

action overlaps with the endogenous programme of cellularization in urodeles, as well as with the action of Msx-1 that has been outlined above.

The results with myoseverin raise the question of the identity of the endogenous signal that activates this response in the urodele muscle fibres during the initial phase of regeneration. A recent paper (McGann *et al.*, 2001) has provided evidence that ligands that are present in crude extracts of the early regenerating newt limb can induce both A1 newt myotubes and C2 mouse myotubes to undergo cell-cycle re-entry and cellularization. The cellularization event indicates the presence of myoseverin-like activity on both newt and mouse cells. However, the S-phase re-entry of mouse myotubes is somewhat surprising in view of the evidence discussed earlier that there is a clear difference in responsiveness to serum and thrombin between newt and mouse myotubes. Further analysis of the factors that are present in the crude extracts should provide more information.

5.7 Perspective

In view of the evolutionary and biomedical issues, it is interesting to compare the relevant examples of plasticity in myotubes and pigment epithelial cells from urodeles and mammals. Although mammalian myotubes are thought not to re-enter the cell cycle or undergo conversion to mononucleate cells under normal circumstances, the intracellular pathways that mediate these responses are apparently intact, although it remains to be determined if the myoseverin or Msx-1 programme(s) does overlap with the urodele programme. Recent results (McGann *et al.*, 2001) have raised the possibility that there are one or more distinctive signals for regeneration in the blastema. Nevertheless, the thrombin-generated activity is apparently a ubiquitous signal to which mammalian cells have lost responsiveness (Tanaka *et al.*, 1999). Although newts are the only adult vertebrates that can regenerate their lens, pigmented epithelial cells from several vertebrates can be converted to lens cells in culture (Agata *et al.*, 1993). It will be necessary to identify the blastemal signals (Tanaka *et al.*, 1999; McGann *et al.*, 2001) that trigger these responses of differentiated cells, as well as the molecular basis of any differences between the differentiated cells of the two species. It is already clear that this approach to regeneration – i.e. the investigation of the plasticity of differentiated cells – is a productive and informative one, particularly in the absence of a comprehensive genetic analysis.

In view of the overlap, or even correspondence, in pathways between urodeles and mammals, it could be predicted that regenerative responses would be subject to genetic variability. The most striking example at present is the MRL mouse, a strain that has a markedly enhanced ability to restore tissue after wounds such as an ear punch (Clark *et al.*, 1998; Heber-Katz, 1999). After cryogenic infarction, this mouse can restore significant cardiac function and this response is associated with cell-cycle progression of cardiomyocytes in the vicinity of the lesion (Leferovich *et al.*, 2001). The analysis of the various loci that are associated with these aspects of the MRL phenotype should be informative (McBrearty *et al.*, 1998).

At present, the main approach to mammalian regenerative medicine is the isolation of appropriate stem cells, followed by manipulations that are aimed at directing their differentiation towards the morphogenesis of complex structures (Bianco and Robey, 2001). Although this attracts considerable interest at present, many of the applications are sufficiently problematic to warrant consideration of alternative and complementary approaches. The urodele strategy – the limited respecification of residual differentiated cells – is so successful that it would be surprising if it were not eventually tried as a therapeutic approach in some contexts of mammalian regeneration. The example of myoseverin (Rosania *et al.*, 2000; Perez *et al.*, 2002) shows that the responses that are discussed in this review are a potential target for therapeutics that are directed at regulating the stability of the differentiated state.

Acknowledgement

This chapter is modified and reprinted by permission from Nature Reviews Neuroscience from Brockes, J.P. and Kumar, A. (2002) *Nature Reviews in Molecular Cell Biology*, **3**, 566–574, Macmillan Magazines Ltd.

References

Agata, K., Kobayashi, H., Itoh, Y. Mochii, M., Sawada, K. and Eguchi, G. (1993) Genetic characterization of the multipotent dedifferentiated state of pigmented epithelial cells *in vitro*. *Development*, **118**, 1025–1030.

Aguayo, A.J., Epps, J., Charron, L. and Bray, G.M. (1976) Multipotentiality of Schwann cells in cross-anastomosed and grafted myelinated and unmyelinated nerves: quantitative microscopy and radioautography. *Brain Res.*, **104**, 1–20.

Alvarado, A.S. (2000) Regeneration in the metazoans: why does it happen? *Bioessays* **22**, 578–590. *This reviews the evolutionary questions about regeneration and its origins, such as why some animals regenerate but others apparently do not.*

Bianco, P. and Robey, P.G. (2001) Stem cells in tissue engineering. *Nature*, **414**, 118–121.

Brockes, J.P. (1997) Amphibian limb regeneration: rebuilding a complex structure. *Science*, **276**, 81–87. *An overview of* Urodele *regeneration that focuses on plasticity and positional identity.*

Brockes, J.P., Kumar, A. and Velloso, C.P. (2001) Regeneration as an evolutionary variable. *J. Anat.*, **199**, 3–11.

Carroll, S.B. (2001) Chance and necessity: the evolution of morphological complexity and diversity. *Nature*, **409**, 1102–1109.

Clark, L.D., Clark, R.K. and Heber-Katz, E. (1998) A new murine model for mammalian wound repair and regeneration. *Clin. Immunol. Immunopathol.*, **88**, 35–45.

Clarke, D.L. *et al.* (2000) Generalized potential of adult neural stem cells. *Science*, **288**, 1660–1663.

Dinsmore, C.E. (1991) *A History of Regeneration Research*. Cambridge University Press.

Echeverri, K., Clarke, J.D. and Tanaka, E.M. (2001) *In vivo* imaging indicates muscle fiber dedifferentiation is a major contributor to the regenerating tail blastema. *Dev. Biol.*, **236**, 151–164.

Eguchi, G. (1988) Cellular and molecular background of wolffian lens regeneration. *Cell Differ. Dev.*, **25**, S147–S158.

Eguchi, G. and Okada, T.S. (1973) Differentiation of lens tissue from the progeny of chick retinal pigment cells cultured *in vitro*: a demonstration of a switch of cell types in clonal cell culture. *Proc. Natl Acad. Sci. USA*, **70**, 1495–1499.

Eguchi, G., Abe, S.I. and Watanabe, K. (1974) Differentiation of lens-like structures from newt iris epithelial cells *in vitro*. *Proc. Natl Acad. Sci. USA*, **71**, 5052–5056.

Ferretti, P. and Brockes, J.P. (1988) Culture of newt cells from different tissues and their expression of a regeneration associated antigen. *J. Exp. Zool.*, **247**, 77–91.

Ghosh, S., Thorogood, P. and Ferretti, P. (1994) Regenerative capability of upper and lower jaws in the newt. *Int. J. Dev. Biol.*, **38**, 479–490.

Goss, R.J. (1969) *Principles of Regeneration*. Academic Press.

Heber-Katz, E. (1999) The regenerating mouse ear. *Semin. Cell Dev. Biol.*, **10**, 415–419.

Hu, G., Lee, H., Price, S.M. *et al.* (2001) Msx homeobox genes inhibit differentiation through upregulation of cyclin D1. *Development*, **128**, 2373–2384.

Hughes, R.N. (1989) *A Functional Biology of Clonal Animals*. Chapman and Hall.

Imokawa, Y. and Brockes, J.P. (2003) *Selective acturation of thrombin is a critical determinant for vertebrate lens regeneration*. Curr. Biol. **13**, 877–881.

Ito, M., Hayashi, T., Kuroiwa, A. and Okamoto, M. (1999) Lens formation by pigmented epithelial cell reaggregate from dorsal iris implanted into limb blastema in the adult newt. *Dev. Growth Differ.*, **41**, 429–440.

Iyer, V. R. *et al.* (1999) The transcriptional program in the response of human fibroblasts to serum. *Science*, **283**, 83–87.

Johnson, S.L. and Weston, J.A. (1995) Temperature-sensitive mutations that cause stage-specific defects in Zebra fish fin regeneration. *Genetics*, **141**, 1583–1595.

Kim, W.S. and Stocum, D.L. (1986) Retinoic acid modifies positional memory in the anteroposterior axis of regenerating axolotl limbs. *Dev. Biol.*, **114**, 170–179.

Koshiba, K., Kuroiwa, A., Yamamoto, H., Tamura, K. and Ide, H. (1998) Expression of *Msx* genes in regenerating and developing limbs of axolotl. *J. Exp. Zool.*, **282**, 703–714.

Kumar, A., Velloso, C.P., Imokawa, Y. and Brockes, J.P. (2000) Plasticity of retrovirus-labelled myotubes in the newt limb regeneration blastema. *Dev. Biol.*, **218**, 125–136.

Leferovich, J.M. *et al.* (2001) Heart regeneration in adult MRL mice. *Proc. Natl Acad. Sci. USA*, **98**, 9830–9835.

Lo, D.C., Allen, F. and Brockes, J.P. (1993) Reversal of muscle differentiation during urodele limb regeneration. *Proc. Natl Acad. Sci. USA*, **90**, 7230–7234. *This study found that implanted labelled myotubes underwent reversal of muscle differentiation in the limb blastema.*

Maden, M. (1982) Vitamin A and pattern formation in the regenerating limb. *Nature*, **295**, 672–675.

McBrearty, B.A., Clark, L.D., Zhang, X.M., Blankenhorn, E.P. and Heber-Katz, E. (1998) Genetic analysis of a mammalian wound-healing trait. *Proc. Natl Acad. Sci. USA*, **95**, 11792–11797.

McGann, C.J., Odelberg, S.J. and Keating, M.T. (2001) Mammalian myotube dedifferentiation induced by newt regeneration extract. *Proc. Natl Acad. Sci. USA*, **98**, 13699–13704.

Michalopoulos, G.K. and DeFrances, M.C. (1997) Liver regeneration. *Science*, **276**, 60–66.

Mitashov, V.I. (1996) Mechanisms of retina regeneration in *Urodeles*. *Int. J. Dev. Biol.*, **40**, 833–844.

Nardi, J.B. and Stocum, D.L. (1983) Surface properties of regenerating limb cells: evidence for gradation along the proximodistal axis. *Differentiation*, **25**, 27–31.

Novitch, B.G., Mulligan, G.J., Jacks, T. and Lassar, A.B. (1996) Skeletal muscle cells lacking the retinoblastoma protein display defects in muscle gene expression and accumulate in S and G2 phases of the cell cycle. *J. Cell Biol.*, **135**, 441–456.

Oberpriller, J.O. and Oberpriller, J.C. (1974) Response of the adult newt ventricle to injury. *J. Exp. Zool.*, **187**, 249–253.

Oberpriller, J.O., Oberpriller, J.C., Matz, D.G. and Soonpaa, M.H. (1995) Stimulation of proliferative events in the adult amphibian cardiac myocyte. *Ann. NY Acad. Sci.*, **752**, 30–46.

Odelberg, S.J., Kollhoff, A. and Keating, M.T. (2000) Dedifferentiation of mammalian myotubes induced by *msx1*. *Cell*, **103**, 1099–1109. *This study showed that the expression of* Msx-1 *in mouse myotubes leads to the generation of mononucleate pluripotent progeny.*

Okada, T.S. (1991) *Transdifferentiation*. Clarendon.

Olwin, B.B. and Hauschka, S.D. (1988) Cell surface fibroblast growth factor and epidermal growth factor receptors are permanently lost during skeletal muscle terminal differentiation in culture. *J. Cell Biol.*, **107**, 761–769.

Pecorino, L.T., Entwistle, A. and Brockes, J.P. (1996) Activation of a single retinoic acid receptor isoform mediates proximodistal respecification. *Curr. Biol.*, **6**, 563–569.

Perez, O.D., Chang, Y.T., Rosania, G., Sutherlin, D. and Schultz, P.G. (2002) Inhibition and reversal of myogenic differentiation by purine-based microtubule assembly inhibitors. *Chem. Biol.*, **9**, 475–483.

Raff, R.A. (1996) *The Shape of Life*. University of Chicago.

Reyer, R.W. (1954) Regeneration of the lens in the amphibian eye. *Quart. Rev. Biol.*, **29**, 1–46.

Reyer, R.W., Woolfitt, R.A. and Withersty, L.T. (1973) Stimulation of lens regeneration from the newt dorsal iris when implanted into the blastema of the regenerating limb. *Dev. Biol.*, **32**, 258–281.

Rosania, G.R. *et al.* (2000) Myoseverin, a microtubule-binding molecule with novel cellular effects. *Nature Biotechnol.*, **18**, 304–308. *This study isolated a trisubstituted purine from a combinatorial library that induces cellularization of mouse myotubes.*

Schneider, J.W., Gu, W., Zhu, L., Mahdavi, V. and Nadal-Ginard, B. (1994). Reversal of terminal differentiation mediated by p107 in *Rb–/–* muscle cells. *Science*, **264**, 1467–1471.

Solari, F. *et al.* (1995) Multinucleated cells can continuously generate mononucleated cells in the absence of mitosis: a study of cells of the avian osteoclast lineage. *J. Cell Sci.*, **108**, 3233–3241.

Song, K., Wang, Y. and Sassoon, D. (1992) Expression of Hox-7.1 in myoblasts inhibits terminal differentiation and induces cell transformation. *Nature*, **360**, 477–481.

Soonpaa, M.H. and Field, L.J. (1998) Survey of studies examining mammalian cardiomyocyte DNA synthesis. *Circ. Res.*, **83**, 15–26.

Steen, T.P. (1968) Stability of chondrocyte differentiation and contribution of muscle to cartilage during limb regeneration in the axolotl (*Siredon mexicanum*). *J. Exp. Zool.*, **167**, 49–78.

Tanaka, E.M., Gann, A.A., Gates, P.B. and Brockes, J. P. (1997) Newt myotubes re-enter the cell cycle by phosphorylation of the retinoblastoma protein. *J. Cell Biol.*, **136**, 155–165.

Tanaka, E.M., Drechsel, D.N. and Brockes, J. P. (1999) Thrombin regulates S-phase re-entry by cultured newt myotubes. *Curr. Biol.*, **9**, 792–799. *This reference provides the first clear evidence that a* Urodele *differentiated cell – the skeletal myotube – is intrinsically different from its mammalian counterpart.*

Torok, M.A., Gardiner, D.M., Shubin, N.H. and Bryant, S.V. (1998) Expression of *HoxD* genes in developing and regenerating axolotl limbs. *Dev. Biol.*, **200**, 225–233.

Tosh, D. and Slack, J.M. (2002) How cells change their phenotype. *Nature Rev. Mol. Cell Biol.*, **3**, 187–194.

Tsai, R.Y., Kittappa, R. and McKay, R.D. (2002) Plasticity, niches, and the use of stem cells. *Dev. Cell*, **2**, 707–712.

Tsonis, P.A. (1996) *Limb Regeneration*. Cambridge University Press.

Velloso, C.P., Simon, A. and Brockes, J.P. (2001) Mammalian postmitotic nuclei re-enter the cell cycle after serum stimulation in newt/mouse hybrid myotubes. *Curr. Biol.*, **11**, 855–858.

Velloso, C.P., Kumar, A., Tanaka, E.M. and Brockes, J.P. (2000) Generation of mononucleate cells from post-mitotic myotubes proceeds in the absence of cell cycle progression. *Differentiation*, **66**, 239–246.

Weinberg, H.J. and Spencer, P.S. (1976) Studies on the control of myelinogenesis. II. Evidence for neuronal regulation of myelin production. *Brain Res.*, **113**, 363–378.
Woloshin, P., Song, K., Degnin, C., Killary, A.M., Goldhamer, D.J., Sassoon, D. and Thayer, M.J. (1985) MSX1 inhibits myoD expression in fibroblast x 10T1/2 cell hybrids. *Cell*, **82**, 611–620.
Zacksenhaus, E. *et al.* (1996) pRb controls proliferation, differentiation, and death of skeletal muscle cells and other lineages during embryogenesis. *Genes Dev.*, **10**, 3051–3064.

Section 2

Analytical Models of Developmental Biology

Qualitative modelling and simulation of developmental regulatory networks

6

HIDDE DE JONG, JOHANNES GEISELMANN
and **DENIS THIEFFRY**

6.1 Introduction

The analysis of genetic regulatory networks responsible for cell differentiation and development in prokaryotes and eukaryotes will greatly benefit from the recent upscaling to the genomic level of experimental methods in molecular biology. Hardly imaginable only 20 years ago, the sequencing of complete genomes has become a routine job, highly automated and realized in a quasi-industrial environment[1]. The miniaturization of techniques for the hybridization of labelled nucleic acids in solution to DNA molecules attached to a surface has given rise to DNA chips, tools for measuring the level of gene expression in a massively parallel way (Lockhart and Winzeler, 2000). The development of proteomic methods based on two-dimensional electrophoresis, mass spectrometry and the double-hybrid system, allows the identification of proteins and their interactions at the genomic scale (Pandey and Mann, 2000).

In addition to high-throughput experimental methods, mathematical and bioinformatical approaches are indispensable for the analysis of genetic regulatory networks. Given the large number of components of most networks of biological interest, connected by positive and negative feedback loops inside and between cells, an intuitive comprehension of the spatio-temporal evolution of a developmental system is often difficult, if not impossible to obtain. Mathematical modelling, supported by computer tools, can contribute to the analysis of a regulatory network by allowing the biologist to focus on a restricted number of plausible hypotheses. The formulation of a mathematical model requires an explicit and non-ambiguous description of the hypotheses being made on the regulatory mechanisms under study. Furthermore, simulation using the model yields predictions on the behaviour of the cell or embryo that can be verified experimentally.

A variety of methods for the modelling and simulation of genetic regulatory networks have been proposed in the literature (McAdams and Arkin, 1998; Smolen *et al.*, 2000; Hasty *et al.*, 2001; de Jong, 2002). The use of formal methods to study regulatory networks is currently subject to two major constraints. First of all, the biochemical reaction mechanisms underlying the interactions are usually not or incompletely known. This prevents the formulation of

[1] http://igweb.integratedgenomics.com/GOLD/

 ISBN 0-12-428765-4

detailed kinetic models, such as those developed for the genetic switch controlling phage λ growth (McAdams and Shapiro, 1995) or the feedback mechanisms regulating tryptophan synthesis in *E. coli* (Santillán and Mackey, 2001). A second constraint arises from the general absence of quantitative information on most kinetic parameters and molecular concentrations. As a consequence, traditional methods for numerical analysis are difficult to apply.

Few of the modelling and simulation methods that have been developed so far are capable of handling these constraints. In this chapter, we review two related methods that form an exception to the rule. On the one hand, we present the qualitative simulation of genetic regulatory networks described by piecewise-linear (PL) differential equations. On the other hand, we provide an overview of the analysis of genetic regulatory networks by means of asynchronous, multivalued logic. Both methods are based on *coarse-grained* models that, while abstracting from the precise molecular mechanisms involved, capture essential aspects of gene regulation. Moreover, these methods allow a *qualitative* analysis of the dynamics of the genetic regulatory systems to be carried out. Although the methods are based on different formalisms, differential and logical equations, they share important biological intuitions, in particular the description of gene activation in terms of on/off switches.

Both methods are supported by computer tools that allow the user to enter a model of a genetic regulatory network, simulate or analyse its qualitative behaviour and interpret the results in biological terms. We will illustrate the methods and the tools by their application to two model systems for development: the choice between vegetative growth and sporulation in *B. subtilis* and the genetic control of segmentation in the early *Drosophila* embryo. These examples show that, in order to understand the functioning of an organism in terms of the interactions in regulatory networks, it is not always necessary to model the process down to individual biochemical reactions. In fact, when a global understanding of the evolution of spatiotemporal patterns of gene expression is sought, we suggest that it might be profitable to employ coarse-grained and qualitative models of the type discussed in this chapter.

6.2 Dynamical modelling of genetic regulatory networks

Gene expression is a complex process regulated at several stages in the synthesis of proteins (Lewin, 1999). Apart from the regulation of DNA transcription, the best-studied form of regulation, the expression of a gene, may be controlled during RNA processing and transport (in eukaryotes), RNA translation and post-translational modification of proteins. The degradation of proteins and intermediate RNA products can also be regulated in the cell. The proteins fulfilling the above regulatory functions are themselves produced from other genes. This gives rise to *genetic regulatory networks* consisting of interactions between DNA, RNA, proteins and small molecules. Figure 6.1 shows an example of a simple genetic regulatory network. Genes *a*

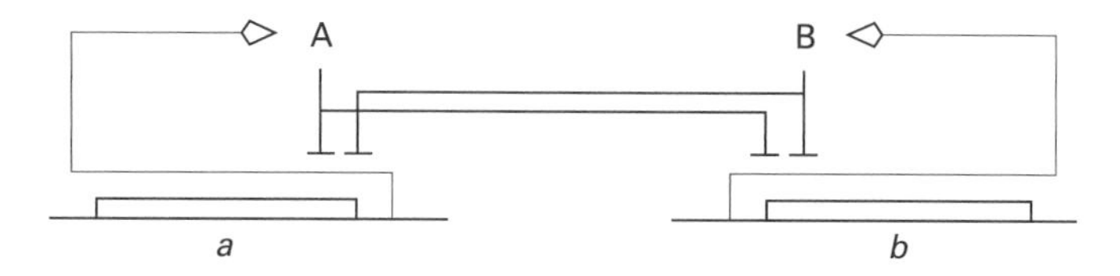

Figure 6.1 Example of a simple regulatory network, consisting of the genes *a* and *b*, proteins A and B, and their mutual interactions. The notation follows, in a somewhat simplified form, the graphical conventions proposed by Kohn (2001).

and b, transcribed from separate promoters, encode proteins A and B, each of which independently controls the expression of both genes. More specifically, proteins A and B repress gene a as well as gene b at different concentrations. Repression of the genes is achieved by binding of the proteins to regulatory sites overlapping with the promoters.

6.2.1 Graph models

Probably the most straightforward way to model the genetic regulatory network in Figure 6.1 is to view it as a *directed graph*. That is, we define a tuple (V, E), with V a set of vertices and E a set of edges. A directed edge is a tuple (i, j) of vertices, where i denotes the head and j the tail of the edge. The vertices of a directed graph correspond to genes or other components of the regulatory network, while the edges denote interactions among the genes. By labelling the vertices and edges, we can express further information on the genes and their interactions. For instance, defining a directed edge as (i, j, s), with s equal to $+$ or $-$, allows one to indicate whether i is activated or inhibited by j. In the example at hand, we would obtain the model shown in Figure 6.2.

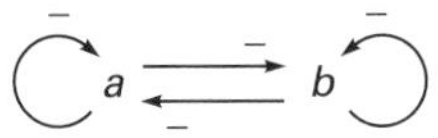

$V = \{a, b\}$
$E = \{\langle a, b, - \rangle, \langle b, a, - \rangle, \langle a, a, - \rangle, \langle b, b, - \rangle\}$

Figure 6.2 Directed graph representing the genetic regulatory network in Figure 6.1.

A number of operations on graphs can be carried out to infer biologically relevant information about genetic regulatory networks. A search for the paths linking two genes, for instance, may provide clues about redundancy in the network. Cycles in the network point at feedback relations that are important for homeostasis and differentiation. Global connectivity characteristics, such as the average and the distribution of the number of regulators per gene, give an indication of the complexity of the network. More sophisticated analyses involving the comparison of sets of homologous genes in a given organism or across different species may provide interesting insights into the evolution of genetic regulatory networks.

Notwithstanding their intuitiveness and genericity, graph models usually only describe the structure of a regulatory system, leaving the dynamical aspects of gene regulation implicit. This is why most modellers of development have preferred more explicitly dynamical formalisms, although the graph representation remains valuable as a common framework for describing regulatory networks (Kohn, 2001).

6.2.2 Differential equation models

Being arguably the most widespread formalism to model dynamical systems in science and engineering, *ordinary differential equations (ODEs)* (Boyce and DiPrima, 1992; Strogatz, 1994) have been widely used to analyse genetic regulatory systems. The ODE formalism models the concentrations of RNAs, proteins and other molecules by time-dependent variables having non-negative real values. Regulatory interactions take the form of functional and differential relations between the concentration variables. More specifically, gene regulation is modelled by non-linear equations expressing the rate of production or degradation of a component of the system as a function of the concentrations of other components. The equations have the mathematical form:

$$\frac{dx_i}{dt} = f_i(x), \; 1 \leq i \leq n, \qquad (1)$$

where $\boldsymbol{x} = [x_1,\ldots, x_n]^T \geq 0$ is the vector of concentrations of proteins, mRNAs or small molecules and $f_i : \mathbb{R}^n \rightarrow \mathbb{R}$ a usually non-linear function. The rate of synthesis of ith component is seen to be dependent upon the concentrations x, possibly including x_i.

Figure 6.3(a) shows how the regulatory network in Figure 6.1 can be modelled in terms of differential equations. The model consists of four variables denoting the concentration of mRNA and protein for genes a and b. The transcriptional inhibition of these genes is described by means of sigmoidal functions h: $\mathbb{R}^2 \rightarrow [0, 1]$, which is motivated by the usually non-linear, switch-like character of gene regulation. The translation of mRNA and the degradation of mRNA and proteins are assumed to be non-regulated and proportional to the substrate concentration. Due to the non-linearity of f_i, analytical solution of the rate equation (**1**) is not normally possible. In special cases, qualitative properties of the solutions, such as the number

$$\frac{dx_{ra}}{dt} = \kappa_{ra}\, h^{-}(x_{pb}, \theta^1_{pb})\, h^{-}(x_{pa}, \theta^2_{pa}) - \gamma_{ra}\, x_{ra}$$

$$\frac{dx_{pa}}{dt} = \kappa_{pa}\, x_{ra} - \gamma_{pa}\, x_{pa}$$

$$\frac{dx_{rb}}{dt} = \kappa_{rb}\, h^{-}(x_{pa}, \theta^1_{pa})\, h^{-}(x_{pb}, \theta^2_{pb}) - \gamma_{rb}\, x_{rb}$$

$$\frac{dx_{pb}}{dt} = \kappa_{pb}\, x_{rb} - \gamma_{pb}\, x_{pb}$$

(a) $$h^{-}(x, \theta) = \frac{\theta^2}{x^2 + \theta^2}$$

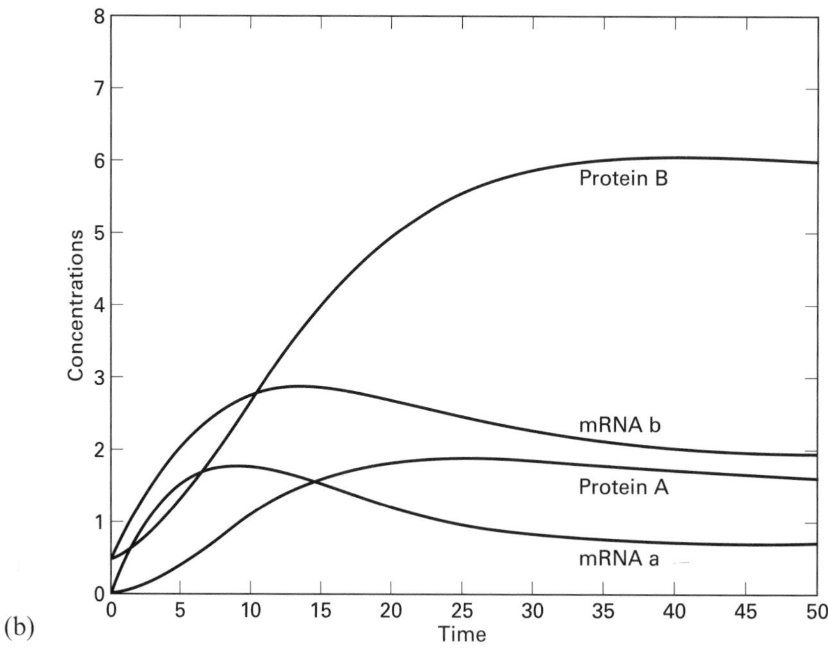

(b)

Figure 6.3 (a) ODE model of the regulatory network in Figure 6.1. The variables x_{pa}, and x_{pb} denote the concentration of protein A and B, the variables x_{ra}, and x_{rb} the concentration of the corresponding mRNA, the parameters κ_{ra}, κ_{pa}, κ_{rb}, and κ_{pb} production rates, the parameters γ_{ra}, γ_{pa}, γ_{rb}, and γ_{pb} degradation rates, and the parameters $\theta^1{}_{pa}$, $\theta^2{}_{pa}$, $\theta^1{}_{pb}$, and $\theta^2{}_{pb}$, threshold concentrations. The variables are non-negative and the parameters positive. (b) Time-concentration plot resulting from a numerical simulation of the system described in (a), for selected values for the parameters.

and the stability of steady states and the occurrence of limit cycles, can be established. Most of the time, however, one has to take recourse to numerical techniques. In Figure 6.3(b) the results of a numerical simulation of the example network are shown. As can be seen, the system reaches a steady state in which protein A is present at a high concentration, whereas protein B is nearly absent. For different initial conditions, but the same parameter values, a steady state may be reached in which the concentrations of A and B are reversed.

Differential equations of the form (**1**) do not take into account the spatial dimension of developmental processes, essential though in multicellular organisms. The equations can be generalized by defining compartments that correspond to cells or nuclei, by introducing concentration variables specific to each compartment and by allowing diffusion between the compartments to take place. In the limit of the number of compartments, the resulting equations can be approximated by *partial differential equations* (PDEs). Partial differential equations are even more difficult to solve analytically than ordinary differential equations and, in almost every situation of practical interest, their use requires numerical techniques (Boyce and DiPrima, 1992).

6.2.3 Stochastic models

An implicit assumption underlying (**1**) and differential equations more generally, is that concentrations of substances vary continuously and deterministically. Both of these assumptions may be questionable in the case of gene regulation, due to the usually small number of molecules of certain components (Ko, 1992; Gibson and Mjolsness, 2001; McAdams and Arkin, 1997, 1999). Instead of taking a continuous and deterministic approach, some authors have proposed to use discrete and stochastic models of gene regulation. Discrete amounts (X) of molecules are taken as state variables, and a joint probability distribution $p(X, t)$ is introduced to express the probability that at time t the cell contains X_1 molecules of the first species, X_2 molecules of the second species, etc. The time evolution of the function $p(X, t)$ can then be specified as follows:

$$p(X, t + \Delta t) = p(X, t)(1 - \sum_{j=1}^{m} \alpha_j \Delta t) + \sum_{j=1}^{m} \beta_j \Delta t, \quad (2)$$

where m is the number of reactions that can occur in the system, $\alpha_j \Delta t$ the probability that reaction j will occur in the interval $[t, t + \Delta t]$ given that the system is in the state X at t, and $\beta_k \Delta t$ the probability that reaction j will bring the system in state X from another state in $[t, t + \Delta t]$ (Gillespie, 1977, 1992). Rearranging (**2**) and taking the limit as $\Delta t \rightarrow 0$, gives the *master equation* (van Kampen, 1997):

$$\frac{\partial}{\partial t} p(X, t) = \sum_{j=1}^{m} (\beta_j - \alpha_j p(X, t)). \quad (3)$$

Compare this equation with the rate equation (**1**) above. Whereas equation (**1**) determines how the state of the system changes with time, equation (**3**) describes how the probability of the system being in a certain state changes with time. Notice that the state variables in the stochastic formulation can be reformulated as concentrations by dividing the number of molecules X_i by a volume factor.

Although the master equation provides an intuitively clear picture of the stochastic processes governing the dynamics of a regulatory system, it is even more difficult to solve by analytical means than the deterministic rate equation. In order to approximate the solution of the master equation, *stochastic simulation methods* have been developed (Gillespie, 1977; Morton-Firth and Bray, 1998). Given a set of possible reactions, the temporal evolution of the state $\boldsymbol{X}$, the number of molecules of each species, is predicted. The evolution of the state is determined by stochastic variables τ and ρ, representing the time interval between two successive reactions and the type of the next reaction, respectively. In each state, a value for τ and ρ is randomly chosen from a set of values whose joint probability density function $p(\tau, \rho)$ has been derived from the same principles as those underlying the master equation (**3**).

In Figure 6.4(a) a few examples of reactions occurring in the network of Figure 6.1 are shown: dimerization of the repressor A, binding of the repressor complex A·A to the promoter region, fixation of DNA polymerase to the promoter in the absence of the repressor complex, transcription of the gene b, etc. Typical results of a stochastic simulation of the example network are shown in Figure 6.4(b). Notice the noisy aspect of the time evolution of the protein and mRNA concentrations. This effect, reflecting the stochastic nature of the initiation of transcrip-

$$
\begin{aligned}
\mathrm{A+A} &\longleftrightarrow \mathrm{A_2} \\
\mathrm{A_2 + DNA}_b &\longleftrightarrow \mathrm{A_2 \cdot DNA}_b \\
\mathrm{RNAP + DNA}_b &\longleftrightarrow \mathrm{RNAP \cdot DNA}_b \\
\mathrm{RNAP \cdot DNA}_b &\longleftrightarrow \mathrm{RNAP + DNA}_b + \mathrm{RNA}_b \\
\mathrm{B+B} &\longleftrightarrow \mathrm{B_2} \\
\mathrm{B_2 + DNA}_b &\longleftrightarrow \mathrm{B_2 \cdot DNA}_b \\
\mathrm{B_2 \cdot DNA}_b + \mathrm{A_2} &\longleftrightarrow \mathrm{A_2 \cdot B_2 \cdot DNA}_b \\
\mathrm{A_2 \cdot DNA}_b + \mathrm{B_2} &\longleftrightarrow \mathrm{A_2 \cdot B_2 \cdot DNA}_b
\end{aligned}
$$

(a)

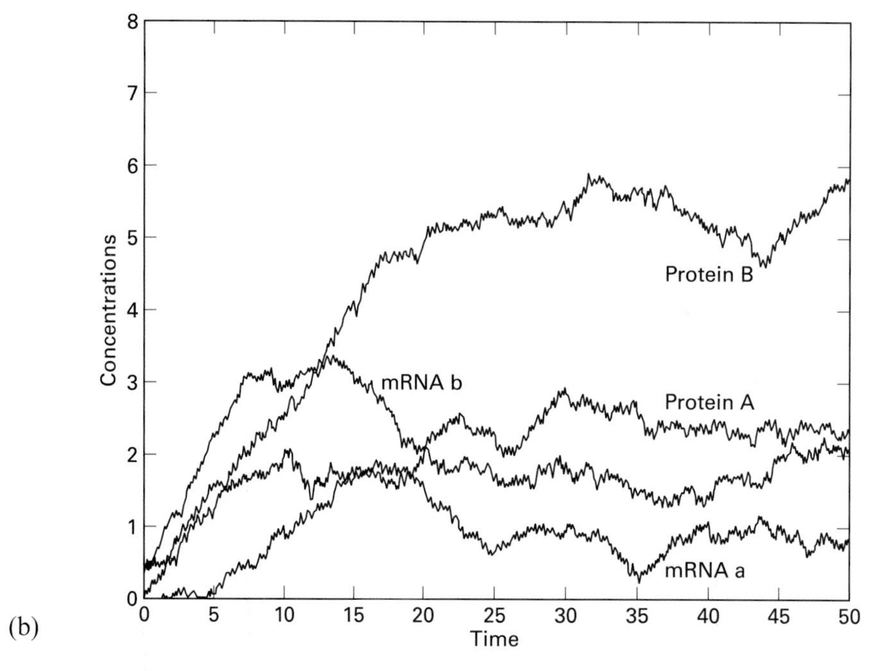

(b)

Figure 6.4 (a) Some of the reactions involved in the expression of gene b in the regulatory network of Figure 6.1. The following abbreviations are used: A and B (proteins A and B), A_2 and B_2 (homodimers of A and B), RNAP (RNA polymerase), DNA_b (promoter region of gene b), and RNA_b (mRNA b). (b) A typical time-concentration plot resulting from stochastic simulation of the reaction system described in (a).

tion and the number of protein molecules produced per transcript, may have important developmental consequences. More particularly, fluctuations in the rate of gene expression may lead to phenotypic variation in an isogenic population (McAdams and Arkin, 1997, 1999). Indeed, starting from the same initial conditions, two different simulations may lead to qualitatively different outcomes. Whereas in one simulation protein A may be ultimately present at a high concentration and B at a low concentration like in Figure 6.4(b), another simulation could lead to the opposite result.

In summary, differential equation and stochastic models provide detailed descriptions of genetic regulatory networks, down to the molecular level. In addition, they can be used to make precise, numerical predictions of the behaviour of regulatory systems. Many excellent examples of the application of these methods to prokaryote and eukaryote development can be found in the literature. McAdams and Shapiro (1995) have simulated the choice between lytic and lysogenetic growth in bacteriophage λ using non-linear differential equations, while Arkin and colleagues have studied the same system by means of a detailed stochastic model (Arkin *et al.*, 1998). In a series of publications, the groups of Novak and Tyson have developed ODE models of the kinetic mechanisms underlying cell cycle regulation in *Xenopus* (Borisuk and Tyson, 1998) and in yeast (Novak *et al.*, 1998) (see Tyson *et al.* (2001) for a review). Differential equation models for the segmention of *Drosophila* have been proposed, focusing on the formation on the expression patterns of the gap, the pair-rule and the segment-polarity gene products in the trunk of the embryo (Reinitz *et al.*, 1995, 1998; von Dassow *et al.*, 2000).

In many situations of biological interest, however, the application of differential equation and stochastic models is seriously hampered. In the first place, the biochemical reaction mechanisms underlying regulatory interactions are usually not known, or incompletely known. This means that it is difficult to specify the rate functions f_i in (**1**) and the reactions j in (**3**). In the second place, quantitative information on kinetic parameters and molecular concentrations is only seldom available, even in the case of well-studied model systems like sporulation in *B. subtilis* and segmentation of the *Drosophila* embryo. As a consequence, the numerical simulation methods mentioned above are often difficult to apply.

Apart from being difficult to attain, the numerical precision and molecular detail provided by the differential equation and stochastic models may be unnecessary or even inappropriate. Assuming that computer technology would develop to the point that whole cells and organisms can be simulated on the molecular level, by tracing hundreds of thousands of reactions occurring in parallel, such brute-force strategies are not guaranteed to yield insight into the functioning of living systems. In fact, it may be far from straightforward to find meaningful patterns in the masses of data generated by these simulations. For many purposes, such as understanding the genetic control of developmental processes, a qualitative understanding of the dynamics of the system on a coarse-grained level of description is sufficient and appropriate.

6.3 Qualitative modelling, analysis and simulation

In this section, we will describe two methods that are based on qualitative and coarse-grained models of genetic regulatory networks. Although the methods have been developed in different formalisms, differential and logic equations, they share fundamental assumptions about the modelling of gene regulation. In the next two sections, these methods are applied in studies of the initiation of sporulation in *B. subtilis* and of the early control of segmentation in the *Drosophila* embryo along the anterior-posterior axis.

6.3.1 Qualitative simulation using piecewise-linear differential equation models

Consider again the differential equation model shown in Figure 6.3(a). This model can be simplified by making two assumptions. First, instead of distinguishing between transcription and translation, the rate of synthesis of the proteins could be directly expressed in terms of the concentration of the regulatory protein. Second, the sigmoid regulation function h could be approximated by a step function $s^- : \mathbb{R}^2 \rightarrow \{0, 1\}$. Figure 6.5 indicates how the sigmoid function approaches the step function as it becomes increasingly steep.

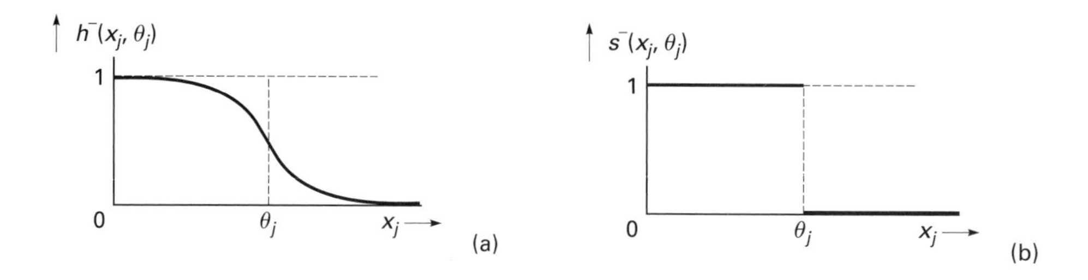

Figure 6.5 Regulation functions: (a) sigmoid function h^- and (b) step function s^-.

The differential equations resulting from these simplifications, shown in Figure 6.6(a), are *piecewise-linear* (PL) and have favourable mathematical properties facilitating their qualitative analysis. More precisely, the dynamics of genetic regulatory networks can be modelled by differential equations of the following general form (Glass and Kauffman, 1973; Snoussi, 1989; Mestl *et al.*, 1995):

$$\dot{x}_i = f_i(\boldsymbol{x}) - g_i(\boldsymbol{x})x_i, \quad 1 \leq i \leq n, \tag{4}$$

where $\boldsymbol{x} = [x_1,..., x_n]^T \geq 0$ is a vector of cellular protein concentrations. The rate of change of each concentration x_i is defined as the difference of the rate of synthesis $f_i(\boldsymbol{x})$ and the rate of degradation $g_i(\boldsymbol{x})\ x_i$ of the protein. The function $f_i : \mathbb{R}^n_{\geq 0} \rightarrow \mathbb{R}_{\geq 0}$ consists of a sum of step function expressions, each weighted by a rate parameter, which expresses the logic of gene regulation (Thomas and d'Ari 1990; Mestl *et al.*, 1995). The function $g_i : \mathbb{R}^n_{\geq 0} \rightarrow \mathbb{R}_{>0}$ is defined analogously. In our example, these functions have a simple form, e.g. $f_a(x_b, x_b) = \kappa_a s^-(x_b, \theta^1{}_b)s^-(x_a, \theta^2{}_a)$ and $g_a = \gamma_a$ in the case of gene a (Figure 6.6(a)). More complex expressions can represent the combined effects of several regulatory proteins.

The dynamical properties of the PL models can be analysed in the n-dimensional phase space box $\Omega = \Omega_1 \times \ldots \times \Omega_n$, where every Ω_i, $1 \leq i \leq n$, is defined as $\Omega_i = \{x_i \in \mathbb{R}_{\geq 0} \mid 0 \leq x_i \leq max_i)$. max_i is a parameter denoting a maximum concentration for the protein. Given that the protein encoded by gene i has p_i threshold concentrations, the $n-1$-dimensional threshold hyperplanes $x_i = \theta^{k_i}$, $1 \leq k_i \leq p_i$, partition Ω into (hyper)rectangular regions that are called *domains* (de Jong *et al.*, 2002b). Figure 6.6(c) shows the subdivision into domains of the two-dimensional phase space box of the example network. We distinguish between domains like D^4 and D^7, which are located on (intersections of) threshold planes, and domains like D^1, which are not. The former domains are called *switching* domains, whereas the latter are called *regulatory* domains. The phase space box in Figure 6.6(c) is partitioned into 9 regulatory and 16 switching domains.

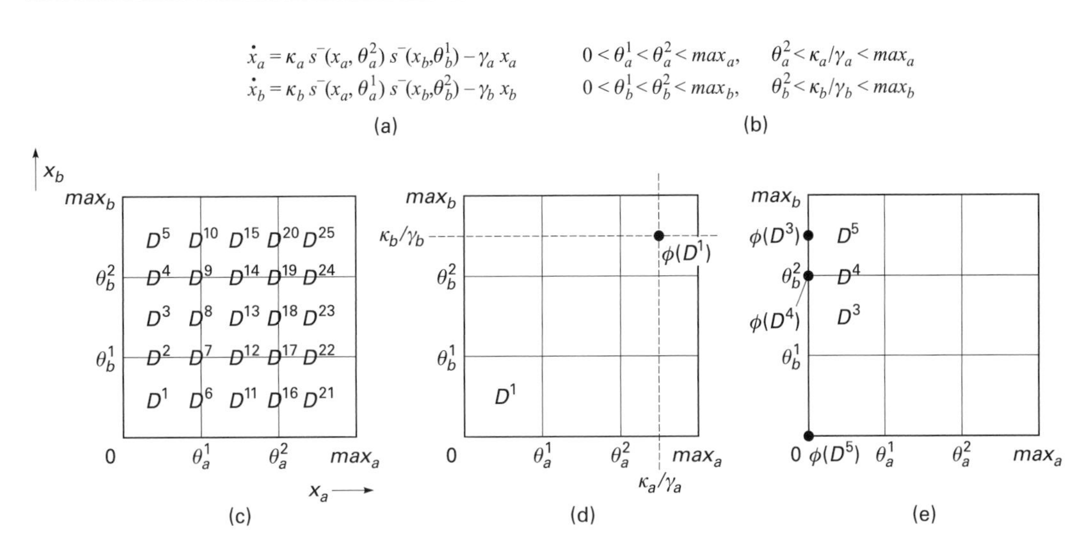

$\dot{x}_a = \kappa_a s^-(x_a, \theta_a^2) s^-(x_b, \theta_b^1) - \gamma_a x_a$
$\dot{x}_b = \kappa_b s^-(x_a, \theta_a^1) s^-(x_b, \theta_b^2) - \gamma_b x_b$
(a)
$0 < \theta_a^1 < \theta_a^2 < max_a, \quad \theta_a^2 < \kappa_a/\gamma_a < max_a$
$0 < \theta_b^1 < \theta_b^2 < max_b, \quad \theta_b^2 < \kappa_b/\gamma_b < max_b$
(b)
(c)
(d)
(e)

Figure 6.6 (a) PLDE model of the regulatory network in Figure 6.1. For the meaning of the variables and parameters see Figure 6.3. (b) Parameter inequalities supplementing the PLDE model. (c) Subdivision of the phase space into regulatory and switching domains. (d), (e) Phase space analysis of the model in regulatory domain D^1 and switching domain D^4. The analysis has been carried out for the parameter inequalities in (b).

When evaluating the step function expressions of **(4)** in a regulatory domain, f_i and g_i reduce to sums of rate constants. More precisely, in a regulatory domain D, f_i reduces to some μ_i^D, and g_i to some ν_i^D. It can be shown that all solution trajectories in D monotonically tend towards a stable equilibrium $\Phi(D) = \{(\mu_1^D/\nu_1^D, \ldots, \mu_n^D/\nu_n^D)\}$, the *target equilibrium* (Glass and Kauffman, 1973; Mestl *et al.*, 1995; Snoussi, 1989). The target equilibrium level μ_i^D/ν_i^D of the protein concentration x_i gives an indication of the strength of gene expression in D. If $\Phi(D) \cap D \neq \{\}$, then all trajectories will remain in D. If not, they will leave D at some point. In regulatory domain D^1 shown in Figure 6.6(d), the trajectories tend towards $\Phi(D^1) = \{(\kappa_a/\gamma_a, \kappa_b/\gamma_b)\}$. Since $\Phi(D_1) \cap D_1 = \{\}$, the trajectories starting in D will leave this domain at some point. Different regulatory domains generally have different target equilibria. For instance, in regulatory domain D^3, the target equilibrium is given by $\{(0, \kappa_b/\gamma_b)\}$ (not shown).

In switching domains, f_i and g_i may not be defined, because some concentrations assume their threshold value. Moreover, f_i and g_i may be discontinuous in switching domains. In order to cope with this problem, the system of differential equations **(4)** is extended into a system of differential inclusions, following an approach widely used in control theory (Gouze and Sari, 2003). Using this generalization, it can be shown that, in the case of a switching domain D, the trajectories either traverse D instantaneously or remain in D for some time, tending towards a target equilibrium set $\Phi(D)$. Here, $\Phi(D)$ is the smallest closed convex set including the target equilibria of regulatory domains having D in their boundary, intersected with the hyperplane containing D (see de Jong *et al.*, 2002b) for technical details). If $\Phi(D) \cap D \neq \{\}$, then the trajectories may remain in D. If not, they will leave D at some point. Following this approach, the target equilibrium set $\Phi(D^4)$ is given by the intersection of the linear segment connecting the points $(0, \kappa_b/\gamma_b)$ and $(0, 0)$, and the threshold hyperplane $x_b = \theta^2_b$ (Figure 6.6(e)). Consequently, $\Phi(D^4)$ equals $\{(0, \theta^2_b)\}$, and the solutions arriving at D^4 from D^3 or D^5 slide along the threshold plane towards $(0, \theta^2_b)$. Because $\Phi(D^4)$ is included in D^4, $(0, \theta^2_b)$ is an equilibrium of the system. Closer analysis reveals that it is stable.

Most of the time, precise numerical values for the threshold and rate parameters in (**4**) are not available. Instead, we will specify qualitative constraints on the parameter values, as explained in de Jong *et al.* (2002b). Having the form of algebraic inequalities, these constraints can usually be inferred from biological data. The first constraint is obtained by ordering the p_i threshold concentrations of gene i, yielding the *threshold inequalities*. Second, the possible target equilibrium levels μ_i^D/ν_i^D of x_i in different regulatory domains $D \subseteq \Omega$ can be ordered with respect to the threshold concentrations, giving rise to the *equilibrium inequalities*. A set of parameter inequalities for the example of Figure 6.6(a) is shown in Figure 6.6(b).

A *quantitative* PL model of a genetic regulatory network consists of state equations (**4**) and numerical parameter values. In a qualitative PL model, on the other hand, the state equations are supplemented by threshold and equilibrium inequalities. Let $\boldsymbol{x}$, defined on some time-interval $[0, \tau]$, be a solution of a quantitative PL model describing a genetic regulatory network. Furthermore, at some time-point t, $0 \leq t \leq \tau$, $\boldsymbol{x}(t) \in D$. A qualitative description of $\boldsymbol{x}$ at t consists of the domain D, supplemented by the relative position of D and $\Phi(D)$. We call this the qualitative state of the system. On $[0, \tau]$ the solution traverses a sequence of domains $D^0, \ldots, D^m$ in Ω. Whenever $\boldsymbol{x}$ enters a new domain, the system makes a transition to a new qualitative state. The sequence of qualitative states corresponding to the sequence of domains is called the *qualitative behaviour* of the system on the time-interval. Given a qualitative PL model and initial conditions in a domain D, the aim of *qualitative simulation* is to determine the possible qualitative behaviours of the system (Kuipers, 1994). More precisely, denoting by X the set of solutions $\boldsymbol{x}(t)$ of all quantitative PL models corresponding to the qualitative model, such that $\boldsymbol{x}(0) = \boldsymbol{x}_0 \in D^0$, the aim of qualitative simulation is to find the set of qualitative behaviours abstracting from some $\boldsymbol{x} \in X$. In de Jong *et al.* (2002), a simulation algorithm is described that generates a set of qualitative behaviours by recursively determining qualitative states and transitions from qualitative states, starting at the qualitative state associated with the initial domain D^0. This results in a *transition graph*, a directed graph of qualitative states (vertices) and transitions between qualitative states (edges). The transition graph contains *qualitative equilibrium states* or *qualitative cycles*. These may correspond to equilibrium points or limit cycles reached by solutions in X, and hence indicate functional modes of the regulatory system.

Figure 6.7(a) shows the transition graph for a qualitative simulation of the example of Figure 6.6(a), starting in the regulatory domain D^1. As can be seen, the simulation results in five qualitative behaviours leading to different qualitative equilibrium states. In QS^{16}, associated with the switching domain D^{16} in Figure 6.6(c), protein A is present at a high concentration ($x_a = \theta^2{}_a$), whereas protein B is present at a low concentration ($0 \leq x_b < \theta^1{}_b$). In QS^4, associated with D^4, protein A is present at a low concentration ($0 \leq x_a < \theta^1{}_a$) and protein B at a high concentration ($x_b = \theta^2{}_b$). In QS^7, associated with D^7, protein A and protein B are present at intermediate concentrations ($x_a = \theta^1{}_a$ and $x_b = \theta^1{}_b$). The qualitative equilibrium states QS^4 and QS^{16} correspond to stable equilibria of the system, whereas QS^7 corresponds to an unstable equilibrium.

A sequence of qualitative states in the transition graph represents a predicted qualitative behaviour of the system. It has been demonstrated that the transition graph generated by the simulation algorithm covers all qualitative behaviours abstracting from some $\boldsymbol{x} \in X$ (de Jong *et al.*, 2002b). That is, whatever the exact numerical values for the parameters are, if these values are consistent with the threshold and equilibrium inequalities specified in the qualitative PL model, the qualitative shape of the solution is described by a sequence of states in the transition graph.

This qualitative simulation method has been implemented in Java 1.3 in the program *Genetic Network Analyser* (GNA) (de Jong *et al.*, 2002). GNA is available for non-profit academic research purposes.[2] The core of the system is formed by the simulator, which generates a

[2] http://www-helix.inrialpes.fr/gna

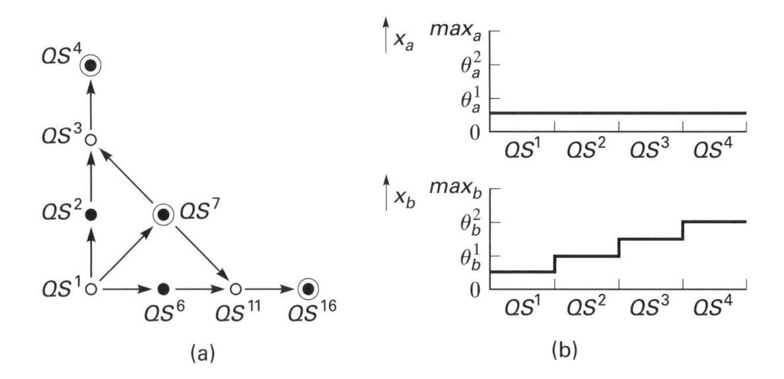

Figure 6.7 (a) Transition graph resulting from a simulation of the example system starting in the domain D^1. Qualitative states associated with regulatory domains and switching domains are indicated by unfilled and filled dots, respectively. Qualitative states associated with domains containing an equilibrium point are circled. (b) Description of the qualitative behaviour (QS^1, QS^2, QS^3, QS^4), analogous to the time plots of Figures 6.3 and 6.4.

transition graph from a qualitative PL model and initial conditions. The input of the simulator is obtained by reading and parsing text files specified by the user. A graphical user interface (GUI), named *VisualGNA,* assists the user in specifying the model of a genetic regulatory network as well as in interpreting the simulation results.

6.3.2 Qualitative analysis using generalized logical models

Since the 1970s, several authors have proposed to simplify the description of regulatory networks even further through the use of Boolean algebra (Kauffman, 1969, 1993; Thomas, 1973, 1991). In this context, a gene product will be considered as *present* (*absent*) when its concentration or activity exceeds (remains below) a certain threshold level. Interactions between genes can then be formalized in terms of logical equations of the form:

$$v_i(t+1) = b_i(\boldsymbol{v}(t)),\ 1 \leq i \leq n, \tag{5}$$

where v_i is a Boolean variable representing the activity of gene i, $b_i : \{0, 1\}^n \rightarrow \{0, 1\}$ a Boolean function, and $\boldsymbol{v} = [v_1,\ldots, v_n]^T$ the variables associated with genes in the network. In most simulations, authors have treated these equations under a synchronicity assumption, i.e. the computation of $v_i(t+1)$ in terms of $\boldsymbol{v}(t)$ is carried out synchronously for all i. This assumption often leads to simulation artifacts which can be avoided by considering specific delays for each change in value (upward or downward) of a variable. In many situations, however, a Boolean representation oversimplifies the system being modelled, leading to the loss of important qualitative information and impeding the generation of biologically meaningful results. This realization led Thomas to generalize the logical approach in order to:

- use multilevel variables whenever needed (i.e. variables taking the values 0, 1, 2,...)
- explicitly take into account threshold values (denoted by s^1, s^2,...)
- define logical parameters to replace the logical operators
- treat value changes under an asynchronicity assumption.

In the context of this *generalized logical approach,* a regulatory system can reach *regular* logical states, i.e. states with integer values, or *singular* logical states, i.e. states located on one or several thresholds (Thomas, 1991). In the network of Figure 6.1, two variables are distinguished, v_a and v_b, which can each take on the values $\{0, s^1, 1, s^2, 2\}$. Examples of regular logical states are 00 and 12, while $0s^1$ and s^1s^2 are instances of singular logical states.

Figure 6.8(a) gives the logical translation of the network of Figure 6.1 in the form of a regulatory matrix. Each box of the matrix corresponds to one interaction with its sign (all signs are negative here as we have only inhibitory interactions). The numbers 1 and 2 order the thresholds associated with the different regulatory effect of a given variable. Thus, −2 in the upper-left box means that the product A (logical variable v_a) inhibits the expression of its own gene (logical function F_a) above its second threshold s^2. Similarly, −1 in the upper-right box means that the product B inhibits the expression of the gene *a* above its first threshold. As in the previous section, we care only about the relative positioning of the different thresholds attached to the variables.

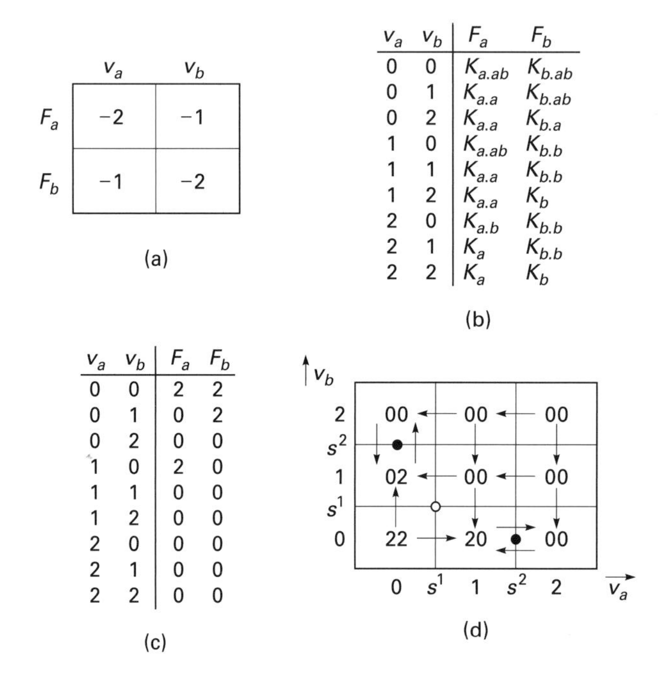

v_a	v_b	F_a	F_b
0	0	$K_{a.ab}$	$K_{b.ab}$
0	1	$K_{a.a}$	$K_{b.ab}$
0	2	$K_{a.a}$	$K_{b.a}$
1	0	$K_{a.ab}$	$K_{b.b}$
1	1	$K_{a.a}$	$K_{b.b}$
1	2	$K_{a.a}$	K_b
2	0	$K_{a.b}$	$K_{b.b}$
2	1	K_a	$K_{b.b}$
2	2	K_a	K_b

(b)

v_a	v_b	F_a	F_b
0	0	2	2
0	1	0	2
0	2	0	0
1	0	2	0
1	1	0	0
1	2	0	0
2	0	0	0
2	1	0	0
2	2	0	0

(c)

Figure 6.8 Generalized logical model for the example network in Figure 6.1. (a) Regulatory matrix for the network. (b) State table defining the rates of gene expression (logical functions F_a and F_b) in terms of logical parameters (Ks). (c) State table for specific values of the logical parameters corresponding to the simulation of Figure 6.7. (d) Phase space-like representation of the logical states and their mutual transitions obtained with the parameters in (c).

In the context of the generalized logical formalism, one can compute a general state table giving the values of the logical functions (rates of gene expression) corresponding to each value combination of the variables (levels of regulatory products), in terms of logical parameters (*K*s). In the notation of these logical parameters, the first subscript identifies the function, whereas the others enumerate all positive influences for the given state, i.e. activator levels above their regulatory thresholds or inhibitor levels below their regulatory thresholds. For example, in

Figure 6.8(b), $K_{a \cdot ab}$ stands for a situation where the regulatory products A and B are both below their inhibitory thresholds with respect to gene *a*.

Depending on the values given to the logical parameters, this table covers different dynamical situations. In Figure 6.8(c), one example of a parameter set is given, corresponding to the situation described in Figure 6.6. Figure 6.8(d) shows all possible transitions in the form of a phase-space representation. Consistent with the situation described in Figure 6.7(a), the system has three steady states, two stable s^20 and $0s^2$, and one unstable s^1s^1.

An important difference between the generalized logical model and the piecewise-linear model presented above resides in the treatment of the combinations of interactions regulating a common target gene. Here, the logical parameters constitute a generic way to represent all possible situations (such as AND, OR, or XOR relationships) in terms of specific combinations of values for a well-defined set of logical parameters. Changing the values of some of the logical parameters before computing the state table would give rise to different dynamics and (possibly) a different attractor configuration. To obtain some of these dynamical behaviours with the piecewise-linear model, the structure of the equations themselves may need to be changed. The generalized logical model is thus more flexible, though it comes at the price of defining a larger number of parameters (which can each take a limited number of logical values).

An important outcome of the generalized logical formalization lies in the possibility of computing specific parameter value constraints for any given feedback circuit (i.e. a closed chain of interactions) to generate specific dynamical properties (Thomas and d'Ari, 1990; Thomas *et al.*, 1995). Indeed, it is possible to show that, for proper parameter values, *positive* circuits (involving an even number of negative interactions) generate multistationarity, whereas *negative* circuits (involving an odd number of negative interactions) may generate homeostasis, possibly in the form of damped or sustained oscillations. For the parameter values selected, the simulation of Figure 6.8 corresponds to a situation where all three circuits (one positive two-element and two negative autoinhibitory circuits) are functional, leading to a combination of multistationarity (there are indeed two stable states and one unstable state) and homeostasis (in each of the two stable states, one of the variables has a threshold value).

More generally, in the context of the generalized logical formalism, it is possible:

- for given parameter values, to check the potential dynamical role (or functionality) of each feedback circuit
- to compute the parameter constraints to render any (combination) of circuit(s) simultaneously functional
- to analytically compute the parameter constraints such that the system has a particular (set of) steady state(s)
- for given parameter constraints, to compute all steady states of a complex multilevel logical model.

From a computational perspective, our logical approach takes the form of a series of Java classes, collectively called *GIN-sim* (Chaouiya *et al.*, 2002). At the present stage, both synchronous and asynchronous simulation tools have been fully implemented. Starting with a set of initial conditions, GIN-sim generates a graph of sequences of logical states (each describing the levels of the different regulatory products), qualitatively representing all permitted state transitions corresponding to the network structure encoded in the original regulatory graph. The initial conditions and the parameter values can be defined by the user or by default, including the number of distinct levels for each regulatory product and the qualitative weights of the different combinations of interactions on each gene.

As of now, our software prototype reads the interaction graph and the user parameter definitions from a plain text file and gives the results of the simulation in terms of a graph of sequence of states written in another text file (Chaouiya *et al.*, 2002). These files are due to evolve soon towards a Graph XML standard. For the moment, we adopt the GML standard (Graph Modelling Language) which is the ancestor of almost all XML-based standards for graphs. These files can be processed by a series of Java graphical libraries allowing the user to visualize, edit and format the two types of graphs.

In addition, we are implementing a series of analytical tools for the isolation, the comparison and the labelling of regulatory modules in both classes of graphs. Some of these tools will implement the feedback circuit analysis mentioned above, as well as a new symbolic computational approach allowing the analytic derivation of all (regular and singular) steady states of a logical model, thereby avoiding the enumeration of all logical states in order to find those which correspond to interesting attractors (this approach was initially developed by V. Devloo, ULB, Belgium).

This logical approach has been applied to various biological regulatory models (e.g. Thomas and d'Ari, 1990; Sánchez *et al.*, 1997; Mendoza *et al.*, 1999; Sanchez and Thieffry, 2001). It will be illustrated below by a model analysis of a cross-regulatory module involved in the segmentation of the trunk during the fly embryogenesis.

6.4 Initiation of sporulation in *B. subtilis*

Under conditions of nutrient deprivation, the Gram-positive soil bacterium *Bacillus subtilis* can abandon vegetative growth and form a dormant, environmentally-resistant spore instead. During vegetative growth, the cell divides symmetrically and generates two identical cells. During sporulation, on the other hand, cell division is asymmetric and results in two different cell types: the smaller cell (the *forespore*) develops into the spore, whereas the larger cell (the *mother cell*) helps to deposit a resistant coat around the spore and then disintegrates (Figure 6.9). The decision to abandon vegetative growth and initiate sporulation involves a radical change in the genetic programme, the pattern of gene expression, of the cell. The switch of genetic programme is controlled by a complex genetic regulatory network integrating various environmental, cell-cycle and metabolic signals. Due to the ease of genetic manipulation of *B. subtilis*, it has been possible to identify and characterize a large number of the genes, proteins and interactions making up this network. Currently, more than 125 genes are known to be involved (Stragier and Losick, 1996; Fawcett *et al.*, 2000).

Sporulation in *B. subtilis* has become a model for understanding the principles that underlie developmental decisions in higher organisms. The differentiation of the forespore and mother cell following division of a sporulating cell can be seen as the simplest instance of development in a multicellular organism. It prefigures the complexity of developmental processes in multicellular eukaryotic organisms, where differentiation and growth have a spatial as well as a temporal dimension. In a developing embryo, the cells each have their own replicate of the genetic regulatory network, while the networks in neighbouring cells are coupled through diffusion or membrane-bound receptor proteins (Section 6.5). The regulatory interactions inside and between the cells give rise to the spatiotemporal evolution of the pattern of gene expression driving the development of a fertilized egg into an adult organism.

The qualitative simulation method based on PL models will be illustrated by analysing the genetic regulatory network underlying the initiation of sporulation in *B. subtilis*. A graphical representation of the regulatory network controlling the initiation of sporulation is shown in Figure 6.10, displaying key genes and their promoters, proteins encoded by the genes and the regulatory action of the proteins. References to the experimental literature used to compile the network are given in de Jong *et al.* (2002a).

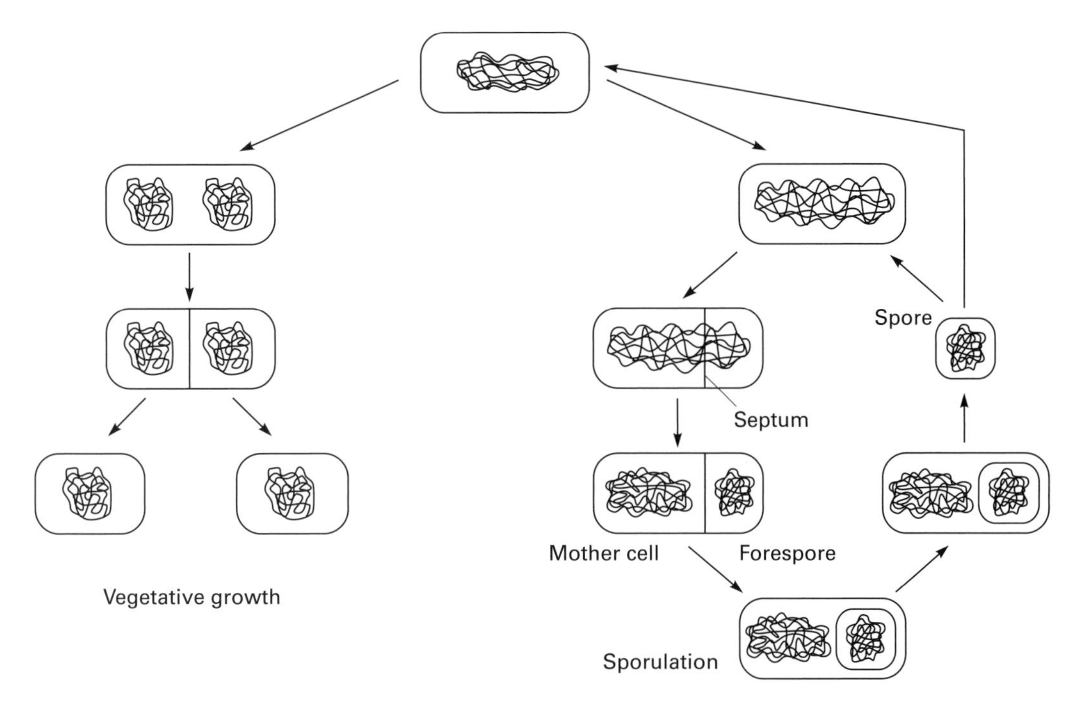

Figure 6.9 Life cycle of *B. subtilis:* decision between vegetative growth and sporulation (adapted from Levin and Grossman, 1998).

The network is centred around a *phosphorelay,* which integrates a variety of environmental, cell-cycle and metabolic signals. Under conditions appropriate for sporulation, the phosphorelay transfers a phosphate to the Spo0A regulator, a process modulated by kinases and phosphatases. The phosphorelay has been simplified in this chapter by ignoring intermediate steps in the transfer of phosphate to Spo0A. However, this simplification does not affect the essential function of the phosphorelay: modulating the phosphate flux as a function of the competing action of kinases and phosphatases (here KinA and Spo0E). Under conditions conducive to sporulation, such as nutrient deprivation or high population density, the concentration of phosphorylated Spo0A (Spo0A~P) may reach a threshold value above which it activates various genes that commit the bacterium to sporulation. The choice between vegetative growth and sporulation in response to adverse environmental conditions is the outcome of competing positive and negative feedback loops, controlling the accumulation of Spo0A~P (Hoch, 1993; Grossman, 1995).

Notwithstanding the enormous amount of work devoted to the elucidation of the network of interactions underlying the sporulation process, very little quantitative data on kinetic parameters and molecular concentrations are available. We have used the qualitative simulation method summarized in Section 6.3.1 to model the sporulation network and to simulate the response of the cell to nutrient deprivation. To this end, the graphical representation of the network has been translated into a PL model supplemented by qualitative constraints on the parameters. The resulting model consists of nine state variables and two input variables. The 49 parameters are constrained by 58 parameter inequalities, the choice of which is largely determined by biological data (de Jong *et al.*, 2002a).

GNA has been used to simulate the response of a wild-type *B. subtilis* cell to nutrient depletion and high population density. Starting from initial conditions representing vegetative growth, the

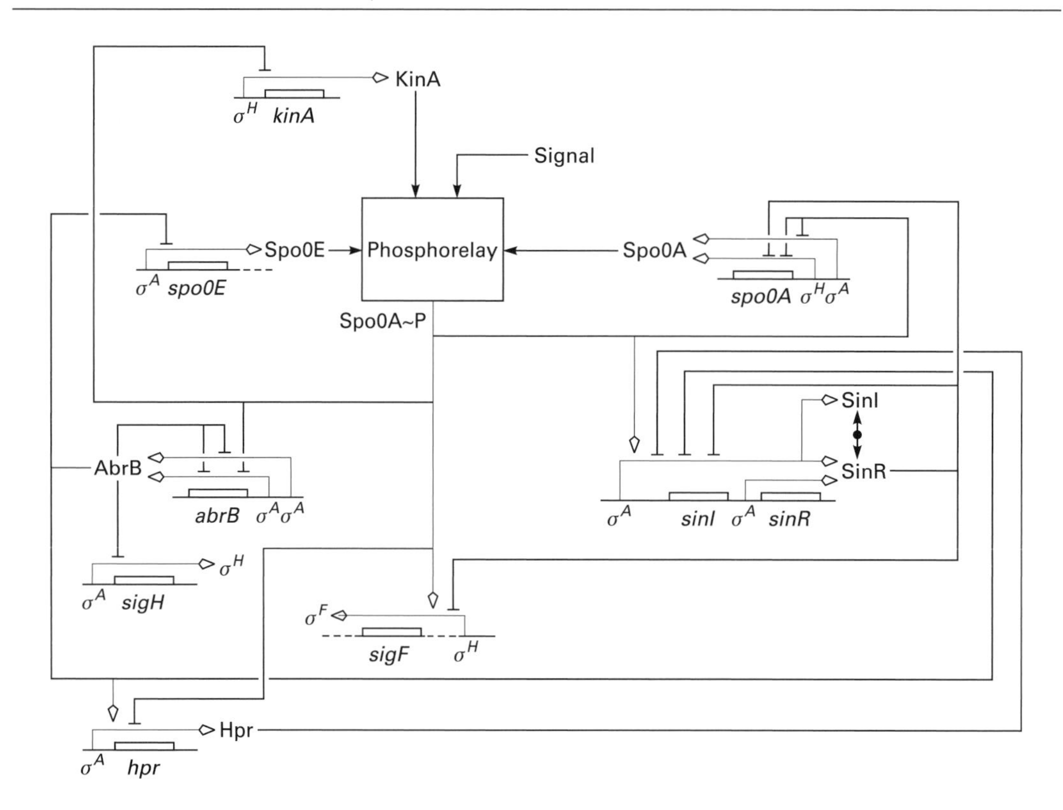

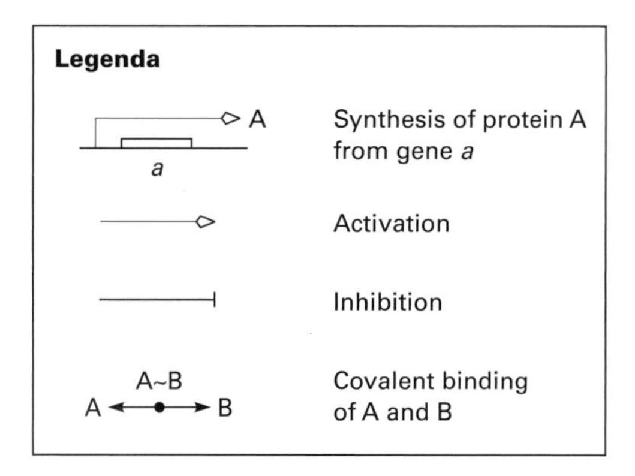

Figure 6.10 Key genes, proteins and regulatory interactions making up the network involved in *B. subtilis* sporulation. In order to improve the legibility of the figure, the control of transcription by the sigma factors σ^A and σ^H has been represented implicitly, by annotating the promoter with the corresponding sigma factor.

system is perturbed by a sporulation signal that causes KinA to autophosphorylate. Simulation of the network takes less than a few seconds to complete on a PC (500 MHz, 128 MB of RAM) and gives rise to a transition graph of 465 qualitative states. Many of these states are associated with switching domains that the system traverses instantaneously. Since the biological relevance of these states is limited, they can be eliminated from the transition graph. This leads to a reduced transition graph with 82 qualitative states, part of which is shown in Figure 6.11.

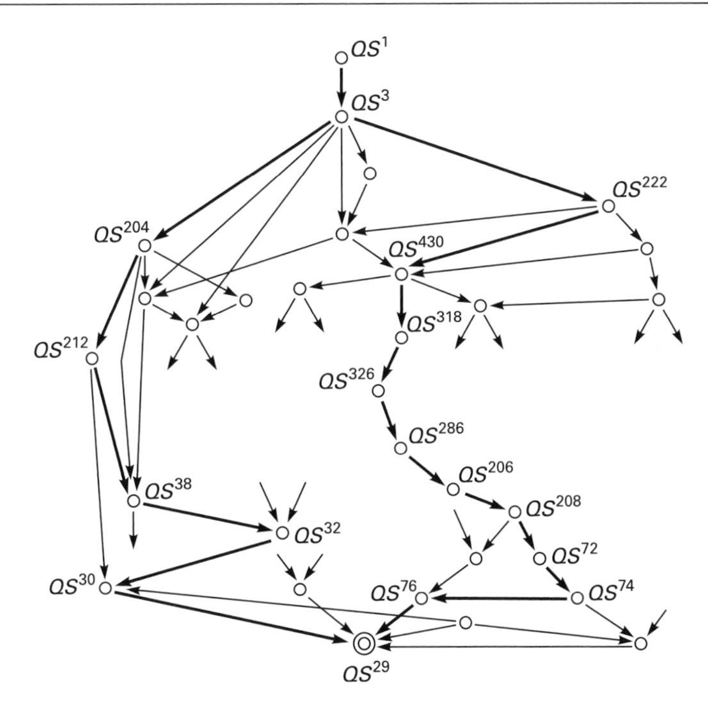

Figure 6.11 Fragment of the state transition graph produced for vegetative growth conditions, when the sporulation signal is present.

The transition graph faithfully represents the two possible responses to nutrient depletion that are observed for *B. subtilis:* either the bacterium continues vegetative growth or it enters sporulation. A typical qualitative behaviour for sporulation as well as for vegetative growth is shown in Figure 6.12. The initiation of sporulation is determined by positive feedback loops acting through Spo0A and KinA and a negative feedback loop involving Spo0E. When the rate of accumulation of the kinase KinA outpaces the rate of accumulation of the phosphatase Spo0E, we observe transient expression of *sigF*, i.e. a spo^+ phenotype (Figure 6.12a). If the kinetics of these processes are inversed, *sigF* is never activated and we observe a spo^- phenotype (Figure 6.12b). Deletion or overexpression of genes in the network of Figure 6.10 may disable a feedback circuit, leading to specific changes in the observed sporulation phenotype. The results of the simulation of a dozen examples of sporulation mutants are discussed in de Jong *et al.* (2002a).

6.5 Early embryogenesis in *D. melanogaster*

Doubtless one of the most extensively studied developmental processes in higher organisms is the early embryogenesis of *Drosophila melanogaster*. Saturated mutagenesis followed by careful screening of mutant phenotypes has led to the identification of the key regulatory genes controlling the formation of segments along the anterior-posterior axis of the embryo (Johnston and Nüsslein-Volhard, 1992; Rivera-Pomar and Jäckle, 1996), prefiguring the specific arrangement of body structures, first in the larva and later in the adult fly. The setting of segmentation involves dozens of genes, expressed either maternally during oogenesis or in the zygotic syncytium. These genes form a hierarchical genetic network with different modules, each responsible for a step in

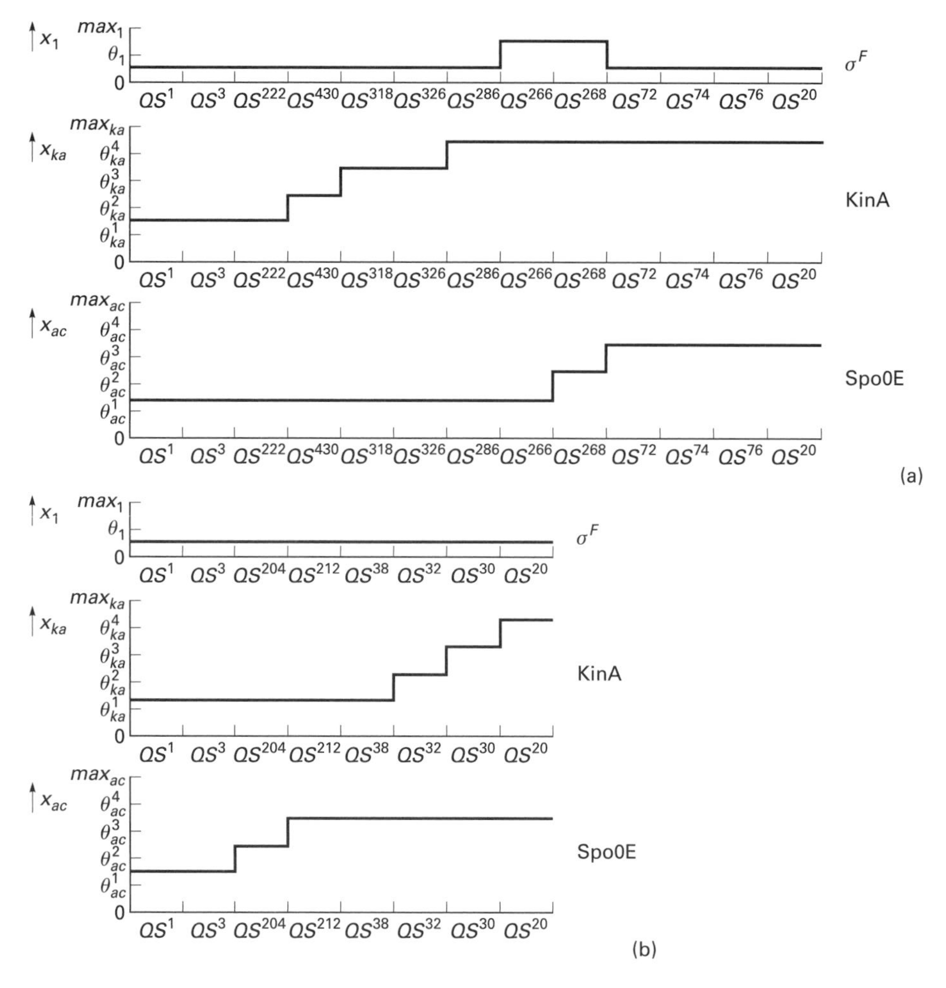

Figure 6.12 (a) Temporal evolution of selected protein concentrations in a typical qualitative behaviour corresponding to the *spo*$^+$ phenotype. (b) Idem, but for a typical qualitative behaviour corresponding to the *spo*$^-$ phenotype.

the processing of the initial gradients of maternal products, ultimately leading to specific and robust stripes of zygotic gene expression. Figure 6.13 shows the main classes of segmentation genes and their regulatory relationships, together with some examples of expression patterns.

For the sake of simplicity, we will focus here only on one, relatively simple regulatory module, defined by the gap genes and their cross-regulatory interactions in the trunk region (Figure 6.14). At the onset of the first zygotic gene expression, three maternal products, Bcd, Hb_{mat}, and Cad are each gradually distributed along the anterior-posterior axis. Together, these maternal products define different functional inputs on the gap genes *gt*, hb_{zyg} and *kni*. The gap genes are consequently differentially activated along the trunk of the embryo. Furthermore, cross-regulations (predominantly cross-inhibitions) among gap genes amplify these initial differences ultimately leading to well-differentiated expression domains. At a later stage, the maternal and gap products act together on the pair-rule genes, leading to a further refinement of the segmented gene expression pattern. Finally, all these genes will define the expression of the segment-polarity genes, which ultimately and stably encode the segmental borders.

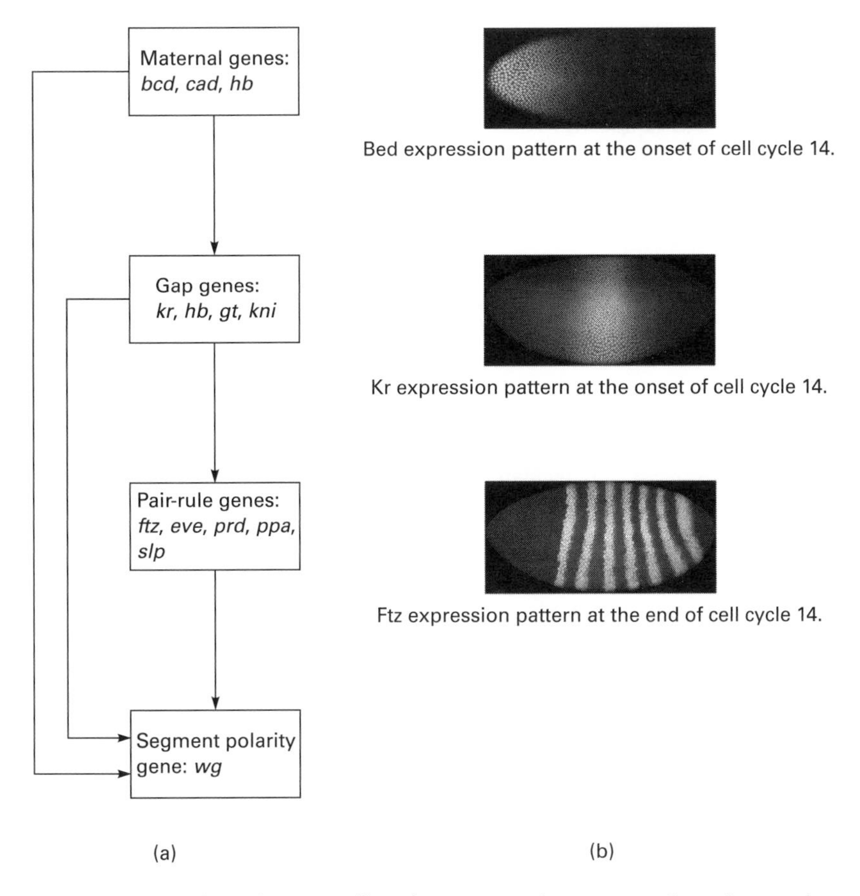

Figure 6.13 The genetic hierarchy controlling the segmentation process along the anterior-posterior axis of the *Drosophila* embryo. (a) The four main classes of genes progressively specifying segmentation. (b) Typical expression patterns for three of the four classes of genes, as revealed by *in situ* antibody labelling (pictures kindly provided by J. Reinitz, SUNY, New York).

The gap cross-regulatory module has been modelled using the generalized logical formalism introduced in Section 3.2 (Sánchez and Thieffry, 2001). The variables in the model correspond to the four gap genes and three maternal inputs in Figure 6.14. The system encompasses seven feedback circuits:

- two positive circuits: *gt-kr*, *gt-kni-kr-hb*
- one conditional autocatalytic circuit: *hb*
- three negative circuits: *gt-kr-hb*, *gt-kni-kr*, *hb-kni-kr*
- one dual (positive/negative) circuit: *hb-kr*.

However, these seven circuits cannot all be simultaneously functional for given parameter values. In fact, our computation suggests that only one of these circuits is playing a key dynamical role, namely the positive circuit involving the cross-inhibitions between *kr* and *gt*.

On the basis of available genetic and molecular data, it is possible to derive a set of parameters giving rise to simulation results consistent with wild-type as well as single loss-of-function phenotypes. Figure 6.15 shows the wild-type simulations. Each column represents a region of the trunk of the embryo, characterized by specific maternal inputs (different values associated

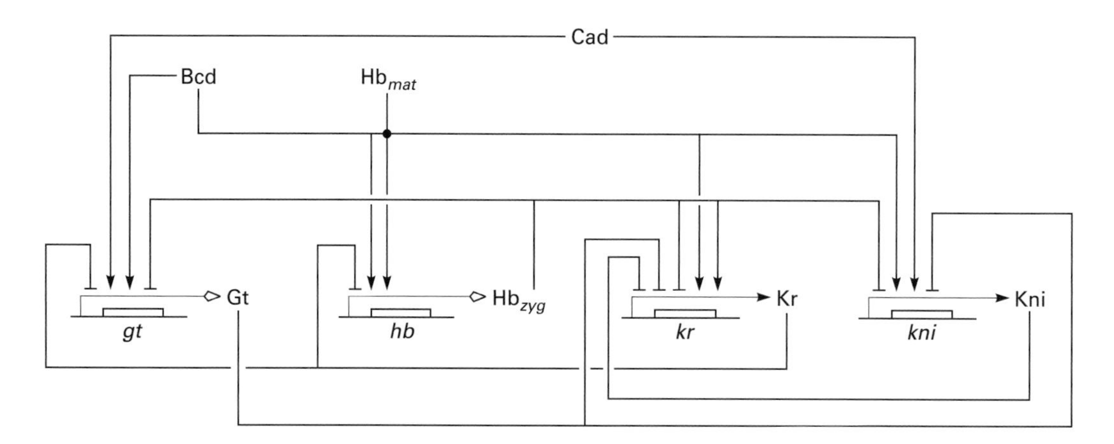

Figure 6.14 Key genes, proteins and regulatory interactions making up the gap module involved in *Drosophila* segmentation. The maternal inputs (Bcd, Hb_{mat}, Cad) are shown in the upper half of the figure and the gap genes (Gt, Hb_{zyg}, Kr, Kni) in the lower half of the figure.

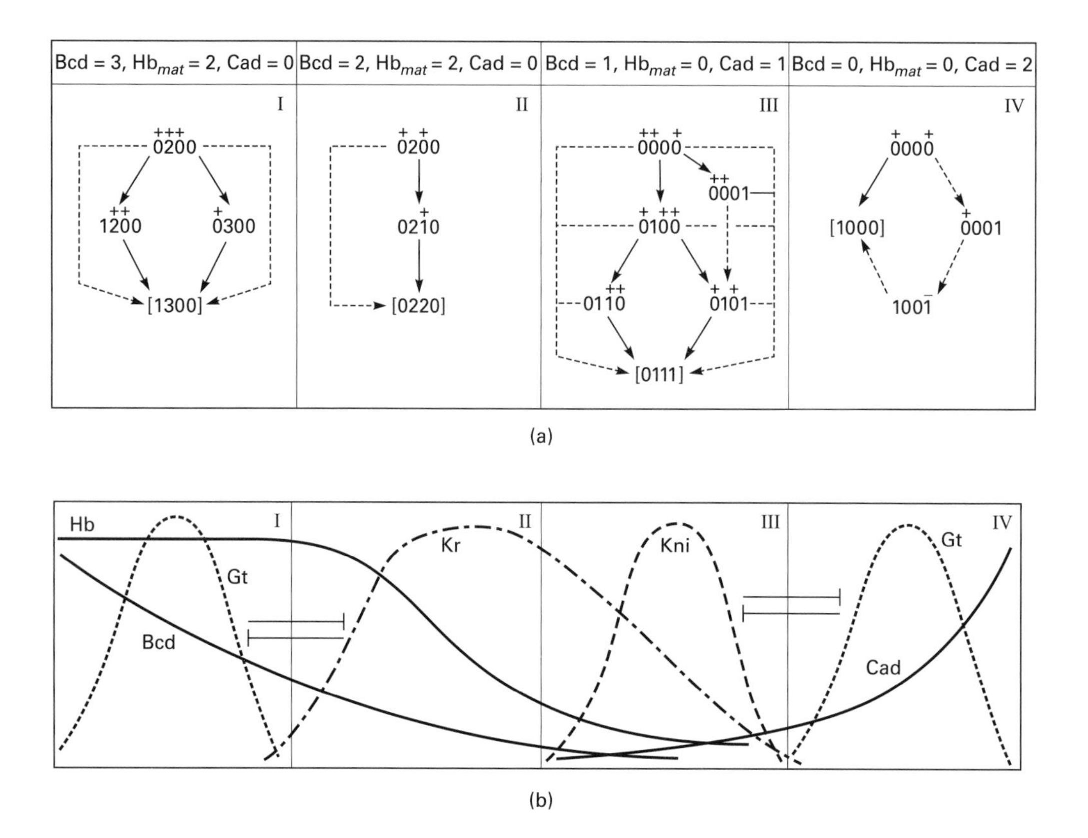

Figure 6.15 (a) Results of the logical simulation of the gap module in four broad regions along the anterior (left, I)-posterior (right, IV) axis of the trunk of the embryo. Solid arrows stand for single transitions, whereas dotted arrows represent (less likely) multiple transitions. The different regions are characterized by different values for the maternal inputs (Bcd, Hb_{mat} and Cad). (b) Matching of the equilibrium states reached in each of the regions with the qualitative profiles of gap regulatory products published in Rivera-Pomar and Jäckle (1996).

Table 6.1 Results of the simulation of maternal and gap loss-of-function mutations, ectopic expressions and cis-regulatory mutations. **The bold values of the logical variables** in the equilibrium states reached under the above conditions indicate the differences with the wild-type equilibrium state (I–IV correspond to the four anterior to posterior regions in Figure 6.15).

Genetic background	*Equilibrium state (Gt, Hb, Kr, Kni)*				*Observations/predictions*
	I	II	III	IV	
Wildtype	1300	0220	0111	1000	
bcd loss-of-function	0001	0001	0001	1000	Loss of Gt in I, loss of Hb in I–III and of Kr in I–II. Kni expands anteriorly into I–II
bcd loss-of-function	0300	0220	0111	0001	Kni expands posteriorly into IV
kr loss-of-function	1300	1200	1100	1000	Gt expands into II–III, loss of Kni in III
gt-kr loss-of-function	0300	0200	0101	0001	Kni expands posteriorly into IV
gt ectopic expression	1300	1210	1110	1000	Lower Kr in II, loss of Kni in III
kni ectopic expression	1301	1201	0111	1001	Loss of Kr and activation of Gt in II
gt-kr ectopic expression	1301	1201	1101	1001	Loss of Kr in II-III
Inactivation of Hb autoregulation	0[1–2]20	0[1–2]20	0111	1000	Activation of Kr and repression of Gt in I
Inactivation of Kr regulation of *gt*	1300	1210	1110	1000	Gt expands into II–III, lower Kr and Kni in II and III, respectively

with the Bcd, Hb and Cad products), the regions from left to right corresponding to the most anterior to more posterior regions of the trunk of the embryo. The asynchronous transition graphs shown in (a) contain several alternative pathways of gene (in)activation (solid arrows indicate the most plausible pathways, whereas the dotted arrows indicate less plausible, multiple transitions). The key point to note is that a single equilibrium state of gene expression is reached in every region. Each of these states is stable when the gap gene module is considered on its own, but has to be transient in the context of the whole segmentation process. In any case, these states qualitatively match the patterns reported for the expression of the gap genes at the onset of cellularization, as shown in (b), adapted from Rivera-Pomar and Jäckle (1996). Notice the mutual exclusion of Gt and Kr, attributable to the cross-inhibitory (positive) circuit involving these genes.

On the basis of this model, it is relatively easy to perform simulations of different categories of mutants or perturbations. To illustrate this point, we give the results of a few simulations of maternal and gap loss-of-function mutations, ectopic expressions and cis-regulatory mutations

(Table 6.1) (Sánchez and Thieffry, 2001). In each of the simulations, not only the final expression pattern of the mutated gene is changed, but indirect effects also occur and affect the expression of other gap genes in the different regions of the embryo. The most important biological conclusions are summarized in the right-hand column. Note that some of these simulations (e.g. the double *gt-kr* loss-of-function mutant) correspond to mutant situations which are still awaiting experimental test and thus constitute *bona fide* predictions. This indicates how the model can be used as a platform to perform *in silico* experiments, enabling the rapid selection of interesting situations to be further tested experimentally.

6.6 Discussion

In living organisms, from bacteria to humans, developmental processes are controlled by genetic regulatory networks consisting of interactions between DNA, RNA, proteins and small molecules. The size and complexity of these networks preclude an intuitive understanding of the dynamics of a developmental system. They require the use of modelling and simulation tools, with a solid foundation in mathematics and computer science, to predict the behaviour of the system in a systematic way. A variety of methods for the computational analysis of genetic regulatory networks have been developed since the 1960s. Classical approaches based on detailed, numerical models of regulatory processes, either deterministic or stochastic, are hampered by two problems: a lack of knowledge on the reaction mechanisms underlying interactions and a general absence of quantitative information on kinetic parameters and molecular concentrations.

In this chapter, we have briefly presented two alternative approaches, describing genetic regulatory networks by means of qualitative and coarse-grained models. Although these methods have been elaborated in different formal contexts, differential and logical equations, they are based on similar abstractions of the biological processes. The application of these methods has been illustrated by means of two examples of prokaryote and eukaryote development: the initiation of sporulation in *B. subtilis* and the setting of segmentation in the *Drosophila* embryo.

The qualitative and coarse-grained models provide a global description of changing patterns of gene expression during the development of an organism. In the sporulation example, we can predict the two qualitatively different responses of *B. subtilis* cells to nutrient deprivation: the initiation of sporulation (spo^+) or the continuation of vegetative growth (spo^-). The spo^+ phenotype arises from the (transient) expression of *sigF*, whereas in the spo^- phenotype *sigF* is not expressed at all (Figure 6.12). The occurrence of one response rather than the other depends on the relative strength of competing positive and negative feedback loops affecting the accumulation of Spo0A$\sim$P. In the case of *Drosophila* segmentation, the application of the logical modelling approach led to the delineation of the most crucial interactions and regulatory circuits acting during the earliest stages of the segmentation process. It sheds new light on the role of gap-gene cross-interactions in the transformation of graded maternal information into discrete gap-gene expression domains. Furthermore, the resulting logical model allows a qualitative reproduction of the patterns of gene expression experimentally characterized, in wild-type as well as in mutant situations.

The predictions obtained through the qualitative methods discussed in this chapter are less precise than their numerical counterparts. For example, we can predict that, under conditions of nutrient deprivation, the KinA and Spo0E concentrations will increase over time. However, we cannot predict the numerical value to which the concentrations will rise. Similarly, the logical model of the gap module leads to the delineation of four qualitatively distinct gap expression domains along the anterior-posterior axis of the trunk of the embryo, but it does not lead to a quantitative localization of the borders of these expression domains, nor does it provide a quantitative prediction of the corresponding mRNA or protein levels. If more precise predictions are needed, the qualitative approaches need to be supplemented by conventional quantitative methods, like those reviewed in Section 6.2. Although the qualitative behaviour predictions lack numerical precision, the examples demonstrate that they do, nevertheless, capture essential features of the dynamics of the regulatory system and provide interesting insights into the underlying regulatory logic.

The genetic regulatory networks discussed in this chapter are sufficiently large and complex to impede reliable predictions of the outcome of physiological and genetic perturbations without the use of computer tools. However, the complete genetic regulatory network underlying sporulation is an order of magnitude larger than the part involved in the decision to initiate spore formation. Similarly, the different stages of pattern formation in the *Drosophila* embryo involve many genes besides those included in the model of Section 6.5. This raises the question whether the qualitative and coarse-grained models are upscalable to larger systems than those treated thus far. It would seem that the limits of size and complexity can be pushed further up through the development of techniques for the efficient computation of qualitative equilibrium states and the automated verification of properties satisfied by qualitative behaviours in transition graphs. Nevertheless, it can be reasonably expected that the analysis of regulatory networks involving several dozens of genes will encounter computational problems related to the exponential growth of the number of possible states with the number of genes in the network.

A conventional engineering strategy for constructing large and complex systems is to decompose the system into loosely-coupled modules. Each of these could be simulated separately, followed by an analysis of the interactions between the modules on a more abstract and hence computationally less demanding level. It seems that this strategy is also valuable when studying biological systems that have evolved under the pressure of natural selection (Hartwell *et al.*, 1999). Formally, cross-regulatory modules can be defined as strongly connected components in the regulatory graph describing all the interactions known to occur between the genes of a given cell or organism (Thieffry and Sánchez, 2003). In such modules, each gene (in)directly regulates all other genes, including itself. These modules range from simple (positive or negative) feedback circuits up to complex sets of intertwined circuits. A proper delineation of the dynamical roles of feedback circuits at the scale of large regulatory networks calls for the development of new qualitative analytical tools.

A good example of modular organization is provided by the genetic regulatory network responsible for *Drosophila* segmentation, consisting of a hierarchical and temporal cascade of three cross-regulatory modules, exactly matching the classes of genes defined by the developmental geneticists (i.e. the gap, pair-rule and segment-polarity genes). The network regulating the response of *B. subtilis* to adverse environmental conditions provides another example of modularity. In this chapter, we have focused on a single module, responsible for the decision to initiate sporulation but, in reality, this module interacts with other modules, regulating the development of genetic competence and motility, the production of antibiotics and other responses.

The regulatory networks responsible for the initiation of sporulation in *B. subtilis* and the segmentation of the *Drosophila* embryo belong to the best-studied model systems of prokaryote and eukaryote development. This has allowed us to compile data from published sources and to reach a reasonably complete model of the core regulatory network controlling these developmental processes. The coherence between the model predictions and the observed biological behaviours constitute a validation of our approaches based on the use of coarse-grained and qualitative models of regulatory networks. The results warrant future application of the methods to biological systems that are understood to a lesser degree.

Acknowledgements

The qualitative simulation method based on PL models has been developed by H. de Jong, J. Geiselmann, J.-L. Gouze (INRIA Sophia-Antipolis), C. Hernandez (INRIA Rhône-Alpes, now at the Swiss Institute for Bioinformatics, Geneva), M. Page (INRIA Rhône-Alpes) and T. Sari (Université de Haute Alsace, Mulhouse). The sporulation network has been modelled by H. de Jong and J. Geiselmann, with important contributions by G. Batt (INRIA Rhône-Alpes) and C. Hernandez. The logical modelling of the *Drosophila* segmentation process is the product of a long-standing and ongoing collaboration between L. Sánchez (CIB, Madrid) and D. Thieffry. C. Chaouiya (LGPD, Marseille) is playing a key role in the development of the logical simulation software suite GIN-sim. The authors acknowledge financial support from the Programme Bioinformatique inter-EPST (Ministère de la Recherche, France).

References

Arkin, A., Ross, J. and McAdams, H.A. (1998) Stochastic kinetic analysis of developmental pathway bifurcation in phage A-infected *Escherichia coli* cells. *Genetics*, **149**, 1633–1648.

Borisuk, M.T. and Tyson, J.J. (1998) Bifurcation analysis of a model of mitotic control in frog eggs. *Journal of Theoretical Biology*, **195**, 69–85.

Boyce, W.E. and DiPrima, R.C. (1992) *Elementary Differential Equations and Boundary Value Problems*. 5th edn. John Wiley & Sons.

Chaouiya, C., Sabatier, C., Verheecke-Mauze, C. *et al.* (2002) GIN-tools Vers une suite logicielle pour l'integration, l'analyse, et la simulation des reseaux genetique. In Nicolas, J. and Thermes, C. (eds). *Recueil des Actes des Journees Ounertes Biologie Informatique Mathematiques, JOBIM 8001*, pp. 17–26. Saint-Malo.

de Jong. H. (2002) Modeling and simulation of genetic regulatory systems: A literature review. *Journal of Computational Biology*, **9**(1), 69–105.

de Jong, H., Geiselmann, J., Batt, G. *et al.* (2002a) Qualitative simulation of the initiation of sporulation in *B. subtilis*. Technical Report RR-4527, INRIA.

de Jong, H., Geiselmann, J., Hernandez, C. and Page, M. (2003) Genetic Network Analyzer: Qualitative simulation of genetic regulatory networks. *Bioinformatics*, **19**(3), 336–344.

de Jong, H. Gouzé, J.-L., Hernandez, C. *et al.* (2002b) Qualitative simulation of genetic regulatory networks using piecewise-linear models. Technical Report RR-4407, INRIA.

Fawcett, P., Eichenberger, P., Losick, R. and Youngman, P. (2000) The transcriptional profile of early to middle sporulation in *Bacillus subtilis*. *Proceedings of the National Academy of Sciences of the USA*, **97**(14), 8063–8068.

Gibson, M.A. and Mjolsness, E. (2001) Modeling the activity of single genes. In Bower, J.M. and Bolouri, H. (eds). *Computational Modeling of Genetic and Biochemical Networks*, pp. 1–48. MIT Press.

Gillespie, D.T. (1977) Exact stochastic simulation of coupled chemical reactions. *Journal of Physical Chemistry*, **81**(25), 2340–2361.

Gillespie, D.T. (1992) A rigorous derivation of the chemical master equation. *Physica D*, **188**, 404–425.

Glass, L. and Kauffman, S.A. (1973) The logical analysis of continuous non-linear biochemical control networks. *Journal of Theoretical Biology*, **39**, 103–129.

Gouzé, J.-L. and Sari, T. (2003) A class of piecewise linear differential equations arising in biological models. *Dynamical Systems*, **17**(4), 299–316.

Grossman, A.D. (1995) Genetic networks controlling the initiation of sporulation and the development of genetic competence in *Bacillus subtilis*. *Annual Review of Genetics*, **29**, 477–508.

Hartwell, L.H., Hopfield, J.J., Leibler, S. and Murray, A.W. (1999) From molecular to modular cell biology. *Nature*, **402**(supplement), C47–C52.

Hasty, J., McMillen, D., Isaacs, F. and Collins, J.J. (2001) Computational studies of gene regulatory networks: *in numero* molecular biology. *Nature Review Genetics*, **2**(4), 268–279.

Hoch, J.A. (1993) Regulation of the phosphorelay and the initiation of sporulation in *Bacillus subtilis*. *Annual Review of Microbiology*, **47**, 441–465.

Johnston, D. St. and Nüsslein-Volhard, C. (1992) The origin of pattern and polarity in the *Drosophila* embryo. *Cell*, **68**(2), 201–220.

Kaufman, S.A. (1969) Metabolic stability and epigenesis in randomly constructed genetic nets. *Journal of Theoretical Biology*, **22**, 437–467.

Kauffman, S.A. (1993) *The Origins of Order: Self-Organization and Selection in Evolution*. Oxford University Press.

Ko, M.S.H. (1992) Induction mechanism of a single gene molecule: Stochastic or deterministic? *BioEssays*, **14**(5), 341–346.

Kohn, K.W. (2001) Molecular interaction maps as information organizers and simulation guides. *Chaos*, **11**(1),1–14.

Kuipers, B. (1994) *Qualitative Reasoning: Modeling and Simulation with Incomplete Knowledge*. MIT Press.

Levin, P.A. and Grossman, A.D. (1998) Cell cycle and sporulation in *Bacillus subtilis*. *Current Opinion in Microbiology*, **1**(6), 630–635.

Lewin, B. (1999) *Genes VII*. Oxford University Press.

Lockhart, D.J. and Winzeler, E.A. (2000) Genomics, gene expression and DNA arrays. *Nature*, **405**, 827–836.

McAdams, H.M. and Arkin, A. (1997) Stochastic mechanisms in gene expression. *Proceedings of the National Academy of Sciences of the USA*, **94**, 814–819.

McAdams, H.H. and Arkin, A. (1998) Simulation of prokaryotic genetic circuits. *Annual Review of Biophysics and Biomolecular Structure*, **27**, 199–224.

McAdams, H.H. and Arkin, A. (1999) It's a noisy business! Genetic regulation at the nanomolar scale. *Trends in Genetics*, **15**(2), 65–69.

McAdams, H.H. and Shapiro, L. (1995) Circuit simulation of genetic networks. *Science*, **269**, 650–656.

Mendoza, L., Thieffry, D. and Alvarez-Buylla, E.R. (1999) Genetic control of flower morphogenesis in *Arabidopsis thaliana*: A logical analysis. *Bioinformatics*, **15**(7–8), 593–606.

Mestl, T., Plahte, E. and Omholt, S.W. (1995) A mathematical framework for describing and analysing gene regulatory networks. *Journal of Theoretical Biology*, **176**, 291–300.

Morton-Firth, C.J. and Bray, D. (1998) Predicting temporal fluctuations in an intracellular signalling pathway. *Journal of Theoretical Biology*, **192**, 117–128.

Novak, B., Csikasz-Nagy, A., Gyorffy, B. *et al.* (1998) Mathematical model of the fission yeast cell cycle with checkpoint controls at the G1/S, G2/M and metaphase/anaphase transitions. *Biophysical Chemistry*, **72**, 185–200.

Pandey, A. and Mann, M. (2000) Proteomics to study genes and genomes. *Nature*, **405**, 837–846.

Reinitz, J., Mjolsness, E. and Sharp, D.H. (1995) Model for cooperative control of positional information in *Drosophila* by bicoid and maternal hunchback. *Journal of Experimental Zoology*, **271**, 47–56.

Reinitz, J., Kosman, D., Vanario-Alonso, C.E. and Sharp, D.H. (1998) Stripe forming architecture of the gap gene system. *Developmental Genetics*, **23**, 11–27.

Rivera-Pomar, R. and Jäckle, H. (1996) From gradients to stripes in *Drosophila* embryogenesis: filling in the gaps. *Trends in Genetics*, **12**(11), 478–483.

Sánchez, L. and Thieffry, D. (2001) A logical analysis of the *Drosophila* gap genes. *Journal of Theoretical Biology*, **281**, 115–141.

Sánchez, L., van Helden, J. and Thieffry, D. (1997) Establishment of the dorso-ventral pattern during embryonic development of *Drosophila melanogaster*. A logical analysis. *Journal of Theoretical Biology*, **189**, 377–389.

Santillán, M. and Mackey, M.C. (2001) Dynamic regulation of the tryptophan operon: a modelling study and comparison with experimental data. *Proceedings of the National Academy of Sciences of the USA*, **98**(4), 1364–1369.

Smolen, P., Baxter, D.A. and Byrne, J.H. (2000) Modeling transcriptional control in gene networks: methods, recent results, and future directions. *Bulletin of Mathematical Biology*, **62**, 247–292.

Snoussi, E.H. (1989) Qualitative dynamics of piecewise-linear differential equations: a discrete mapping approach. *Dynamics and Stability of Systems*, **4**(3–4), 189–207.

Stragier, P. and Losick, R. (1996) Molecular genetics of sporulation in *Bacillus subtilis*. *Annual Review of Genetics*, **30**, 297–341.

Strogatz, S.H. (1994) *Nonlinear Dynamics and Chaos: With Applications to Physics, Biology, Chemistry, and Engineering*. Perseus Books.

Thieffry, D. and Sánchez, L. (2003) Qualitative analysis of gene networks: Towards the delineation of trans-regulatory modules. In Schlosser, G. and Wagner, G. (eds). *Modularity in Development and Evolution*. University of Chicago Press.

Thomas, R. (1973) Boolean formalization of genetic control circuits. *Journal of Theoretical Biology*, **42**, 563–585.

Thomas, R. (1991) Regulatory networks seen as asynchronous automata: a logical description. *Journal of Theoretical Biology*, **153**, 1–23.

Thomas, R. and d'Ari, R. (1990) *Biological Feedback*. CRC Press.

Thomas, R., Thieffry, D. and Kaufman, M. (1995) Dynamical behaviour of biological regulatory networks: I. biological role of feedback loops and practical use of the concept of the loop-characteristic state. *Bulletin of Mathematical Biology*, **57**(2), 247–276.

Tyson, J.J., Chen, K. and Novak, B. (2001) Network dynamics and cell physiology. *Nature Reviews Molecular Cell Biology*, **2**(12), 908–916.

van Kampen, N.G. (1997) *Stochastic Processes in Physics and Chemistry*. Elsevier.

von Dassow, G., Meir, E., Munro, E.M. and Odell, G.M. (2000). The segment polarity network is a robust developmental module. *Nature*, **406**, 188–192.

Models for pattern formation and the position-specific activation of genes

7

HANS MEINHARDT

7.1 The problem of pattern formation

A higher organism consists of many different cell types that have precise spatial relationships with each other. Starting with a single cell, the fertilized egg, this complex structure is reformed in each life cycle. The similarity of identical twins indicates how precisely this process is controlled by the genes. However, reference to the genes does not provide an explanation for the generation of spatial structures *per se*, since in each cell division both daughter cells obtain the same genetic information. This leads to the question of how different parts of the developing organism can become different from each other in a reproducible way.

Before it was possible to isolate particular molecules involved in the control of development, important insights could only be obtained from the perturbation of normal development and the observation of the subsequent development. Tissue fragments were removed or transplanted either to ectopic positions or to organisms that were in a different developmental stage. Genetic material was changed by mutagenesis etc. These experiments revealed that development is a very robust process and often normal development is possible even after severe perturbation. More recently, the new tools of molecular biology have opened a new inroad. It may be tempting to assume that a complete understanding can be achieved just by measuring the distributions of the relevant molecules at all stages. As a rule, however, this only shifts the problem. If the local concentration of a particular substance elicits a particular structure we need insights on why this concentration maximum has been formed at precisely this position. This is especially true if patterns emerge in cells that were initially more or less equivalent. Models and theories provide the necessary step from observation to paradigm regardless of whether the regulatory behaviour or the distributions of molecules have been observed.

Since models for biological pattern formation are expected to contain non-linear reactions with strong feedback terms, our intuition about the properties of the hypothetical interactions is often insufficient. Only a mathematical formulation combined with computer simulations allows a reliable comparison between theory and experiment. In this way, models can be checked and if necessary, modified. We have developed several models for basic steps as well as for their linkage to account for the generation of more complex systems. In this chapter we will focus on models that account for elementary steps in primary pattern formation and for the position-dependent activation of genes.

 ISBN 0-12-428765-4

7.2 Organizing regions

Although many eggs have some overall structure, the final pattern cannot already be present within the egg in a hidden form. At an early stage, many embryos can be fragmented into two parts and each part forms a complete organism. The sea urchin embryo or the mouse embryo at an 8-cell stage are examples, indicating that in these cases the embryo is not a mosaic-like arrangement of differently determined cells that have a fixed further pathway. Instead, communication must exist between different parts such that the removal of some parts becomes detected and the missing parts replaced.

Some small specialized regions play a decisive role for the overall organization of the developing organism. Such organizing regions direct pattern formation in the surrounding tissue. The gastric opening of the freshwater polyp hydra (Browne, 1909; see also Figure 7.6) and the Spemann-Organizer in amphibians are examples of such regions. Upon transplantation they are able to instruct the surrounding cells. Small regions with an organizing influence on the surrounding tissue are also found in the generation of substructures. For instance, at the posterior margin of a chick wing bud a small nest of cells exists that organizes the anteroposterior pattern of the limb (Tickle *et al.*, 1975). Transplantation of these cells into a more anterior position causes limb duplications. Very few transplanted cells may be sufficient to establish a new organizing region. To account for the long range effect of small specialized regions and for the spatial continuity observed after many experimental interferences, Wolpert (1969) introduced the concept of 'positional information': a local source region produces a signalling substance. By diffusion and decay, a graded concentration profile is generated. The cells are assumed to interpret this positional information by a concentration-dependent response. This concept, however, leaves open the question of how a local source region is generated in initially equivalent cells.

7.2.1 Local self-enhancement and long-range antagonistic effects as the driving force of pattern formation

We have proposed that primary pattern formation proceeds by local self-enhancement coupled to a long-range antagonistic effect (Gierer and Meinhardt, 1972; Gierer, 1981; Meinhardt, 1982). Patterns are formed since small deviations from a homogeneous distribution have a strong positive feedback such that the deviations grow further. The long-ranging antagonistic effect keeps the emergent elevation localized and suppresses the onset of a similar patterning at larger distances.

Pattern formation starting from almost homogeneous initial conditions is very common also in non-animated systems; sand dunes, rivers, clouds, electric discharges during lightning are examples (Figure 7.1). It is easy to see that these pattern formations are based on the same principle. A sand dune provides a windshield that accelerates the deposition of more sand. Erosion proceeds faster at the site of an initial injury since more water collects there. Since the total amount of water or sand is limited, a local accumulation must be accompanied by an overall decrease elsewhere.

7.2.2 The activator–inhibitor scheme

A biochemical realization of this general principle requires the interaction of at least two types of molecules (Figure 7.2, equation **1**). For instance, a short-ranging substance – the activator – promotes its own production (autocatalysis) as well as that of its rapidly diffusing antagonist, the

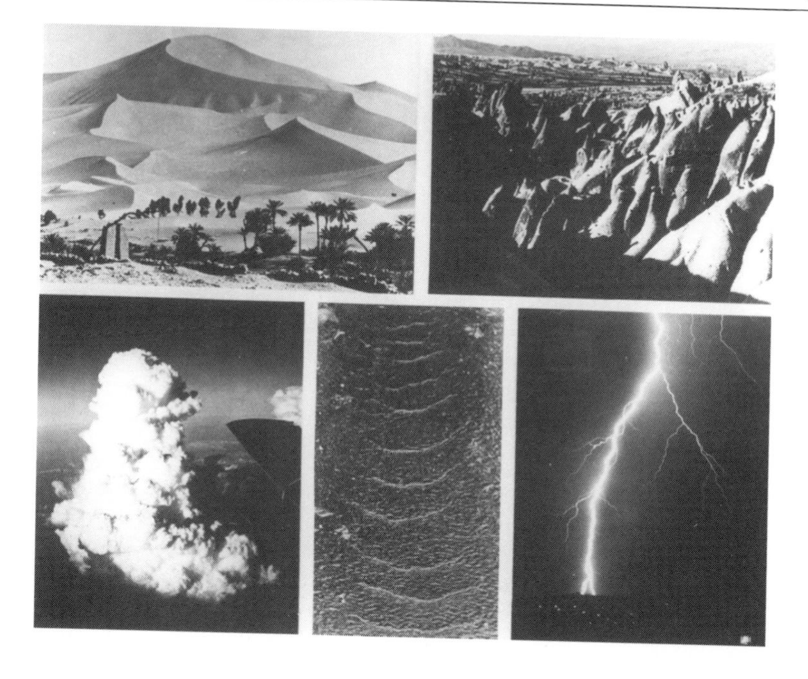

Figure 7.1 Pattern formation in non-living nature. Local self-enhancement acts as driving force. *Sand dunes* may arise from a sand deposit behind a small wind shelter; this deposit increases the wind shelter and thereby accelerates further deposition, and so on. *Erosion* proceeds faster at the site of an injury. More water collects in the incipient valleys, accelerating the erosion there. *Thunderstorm clouds*: warmed-up air expands, becomes lighter and moves upwards. There, the surrounding air is even colder and this accelerates the upstream. *Waves in layers of downstreaming water* are formed despite a uniform rainfall. The speed of the water depends on friction with the ground. A thicker layer has less friction and is therefore faster. Any local increase leads to a local acceleration and the water then catches up with water already further down, amplifying the local increase and the speed further. *Lightning* with sharp contours is formed despite the fact that charged clouds are diffuse. Under the influence of the voltage-difference ions are accelerated, producing more ions by collision with other atoms, leading to an avalanche effect. In all these situations, an effect of lateral inhibition is involved too. If the sand or water accumulates at a particular location, it is depleted at others. The upstream at some location necessitates downstreams at others, etc. (from Meinhardt, 1982).

inhibitor. The concentrations of both substances can be in a steady state. A general increase of the activator concentration would be compensated by a corresponding increase in the inhibitor concentration that brings the activator back to the initial concentration. However, such an equilibrium is locally unstable. Any local increase of the activator will increase further due to the autocatalysis since the surplus of the inhibitor diffuses rapidly into the surroundings of this local increase. It inhibits the activator production there while the local activator elevation grows further (Figure 7.2). Even minute random fluctuations are sufficient for pattern initiation. The following set of partial differential equations describes a possible interaction. It relates the change per time unit of the activator a and the inhibitor h as functions of the actual concentrations.

$$\frac{\partial a}{\partial t} = \frac{\rho a^2}{h} - \mu_a a + D_a \frac{\partial^2 a}{\partial x^2} + \rho_a \tag{1a}$$

$$\frac{\partial h}{\partial t} = \rho a^2 - \mu_h h + D_h \frac{\partial^2 h}{\partial x^2} + \rho_h \tag{1b}$$

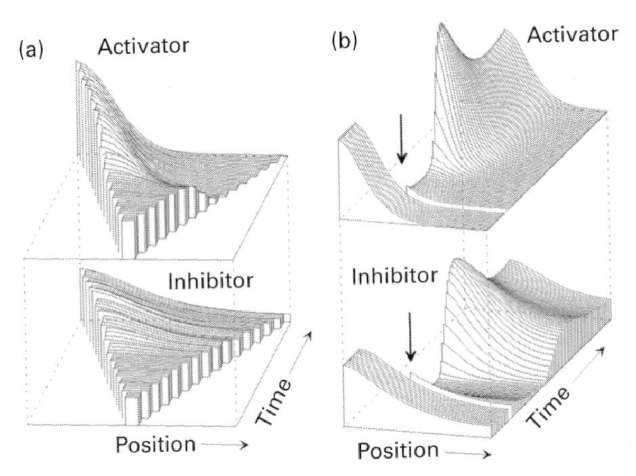

Figure 7.2 Pattern formation by autocatalysis and long-range inhibition. We assume a molecule (termed activator) that catalyses its own production and that of its highly diffusing antagonist, the inhibitor (equation **1**). Concentrations in a linear array of cells are plotted as a function of time. (A) In a growing field, small random fluctuations in the source density ρ are sufficient to initiate pattern formation whenever a critical size is exceeded. A maximum appears on one side of the field. Thus, the model explains how in initially equivalent cells a polar pattern emerges; one side becomes different from the other. (B) Regeneration: after removal of the activated region (the time of interference is indicated by the arrow) a new activator maximum can be formed after the decay of the remnant inhibitor. A crucial condition is that the remaining fragment is not too small (for the regeneration of small fragments see Figure 7.5).

where t is time, x is the spatial coordinate, D_a and D_h are the diffusion coefficients, μ_a and μ_h the decay rates of a and h. The source density ρ describes the ability of the cells to perform the autocatalysis. A small activator-independent activator production ρ_a can initiate the system at low activator concentrations.

For computer simulation, these equations have been approximated by difference equations for discrete cells. For instance, (**1a**) is approximated by

$$\mathrm{da}_{i,t} = \rho\, \mathrm{a}_{i,t} / \mathrm{h}_{i,t} - \mu \mathrm{a}_{i,t} + D_a (\mathrm{a}_{i-1,t} + \mathrm{a}_{i+1,t} - 2\, \mathrm{a}_{i,t}) + \rho_a \qquad \textbf{(1c)}$$

where $i = 1, 2 \ldots$ is the number of the cell in a linear array of cells. This allows the calculation of the activator and inhibitor change in each cell in a given time unit. Adding these changes to the actual concentrations leads to the new profiles at the later moment; $\mathrm{a}_{i,t+1} = \mathrm{a}_{i,t} + \mathrm{da}_{i,t}$. The total time course is calculated from a repetition of many such iterations. Computer programs that allow such simulations are given elsewhere (Meinhardt, 2003). In all simulations shown the boundaries are assumed to be impermeable.

A condition for the formation of stable patterns is that the diffusion of the activator (or any other mode of spread) is much slower than that of the inhibitor, i.e. $D_a << D_h$ must be satisfied. Further, the activator has to have a longer time constant than the inhibitor, $\mu_a < \mu_h$, otherwise oscillations would occur. As discussed below in some detail, the patterns that can be generated in this way correspond to many patterns observed in living systems. In small or growing fields, monotonic gradients are formed that are appropriate to supply positional information (Figures 7.2, 7.3A). In fields large compared with the range of both substances, spatial periodic patterns will result (Figure 7.3B). The mechanism has regulatory properties that allow, for instance, re-formation of a graded concentration profile after an interference.

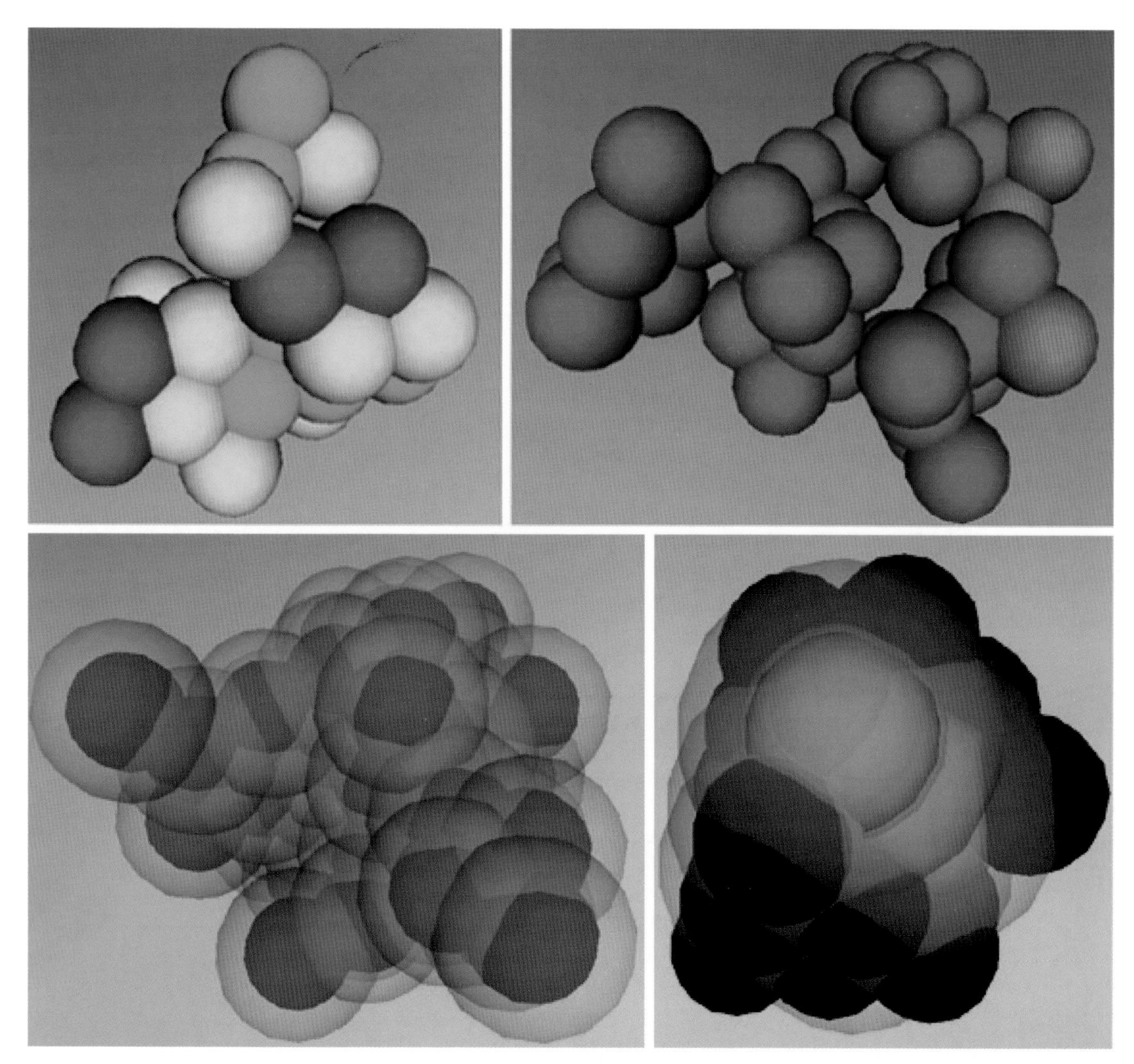

Plate 1.1 (Chapter 1). Four embryos with differentiated cells developed by the Evolutionary Developmental System. An example of differentiation with three cell types depicted in green, yellow and red (top left). Embryo with two cell types and internal cavity caused by apoptosis (top right). Embryo with diffusing proteins visible (bottom left). Note that all cells have differentiated to a single cell type and emit the same protein. Beginnings of cube development with proteins visible (bottom right). There are two cell types: internal cells (which are emitting several proteins) and external cells (black).

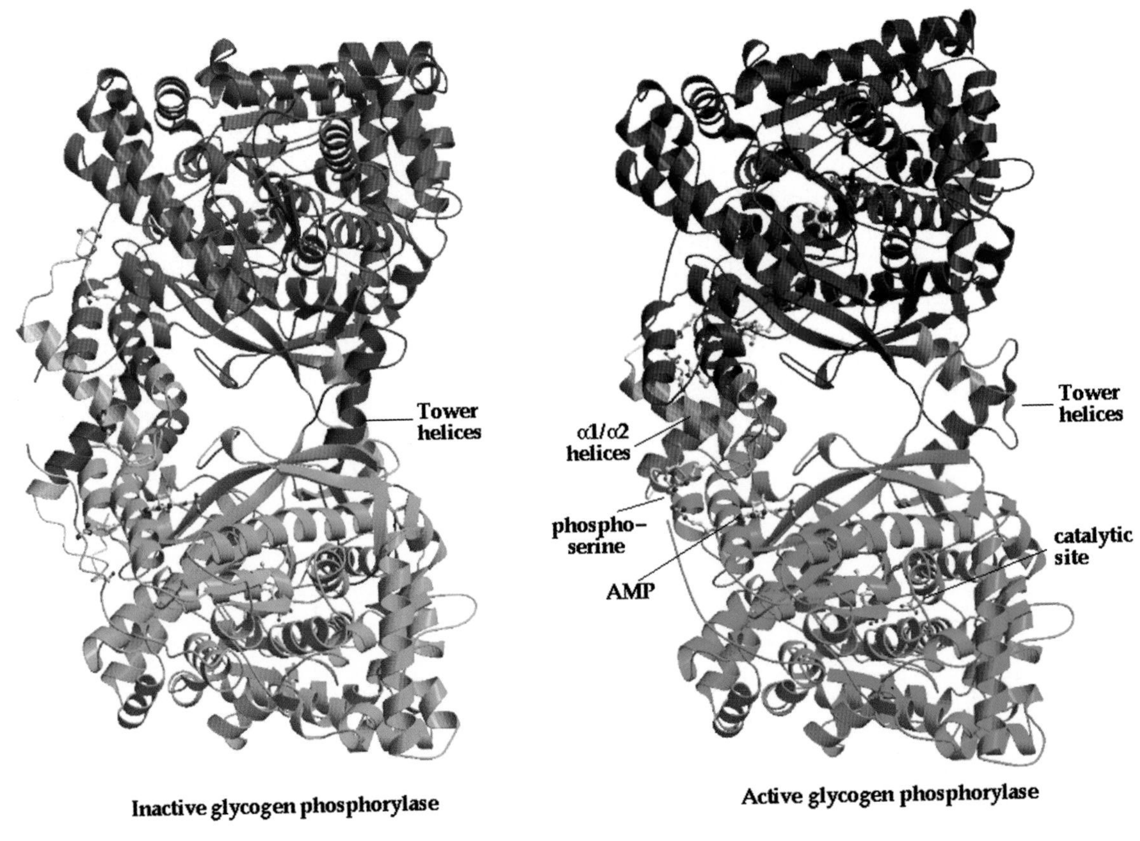

Plate 3.1 (Chapter 3). The conformational changes of glycogen phosphorylase on phosphorylation.

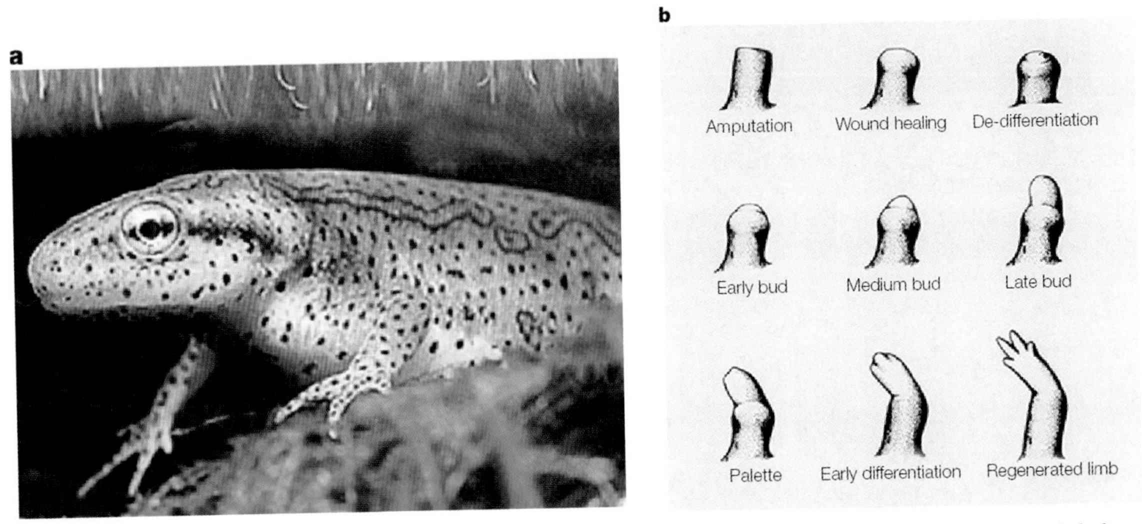

Plate 5.1 (Chapter 5). *Urodele* limb regeneration. A. The North American red spotted newt, *Notophthalmus viridescens*. B. Stages of limb regeneration in an adult newt. Reproduced with permission from (Iten & Bryant, 1973) (c) 1973 Springer-Verlag.

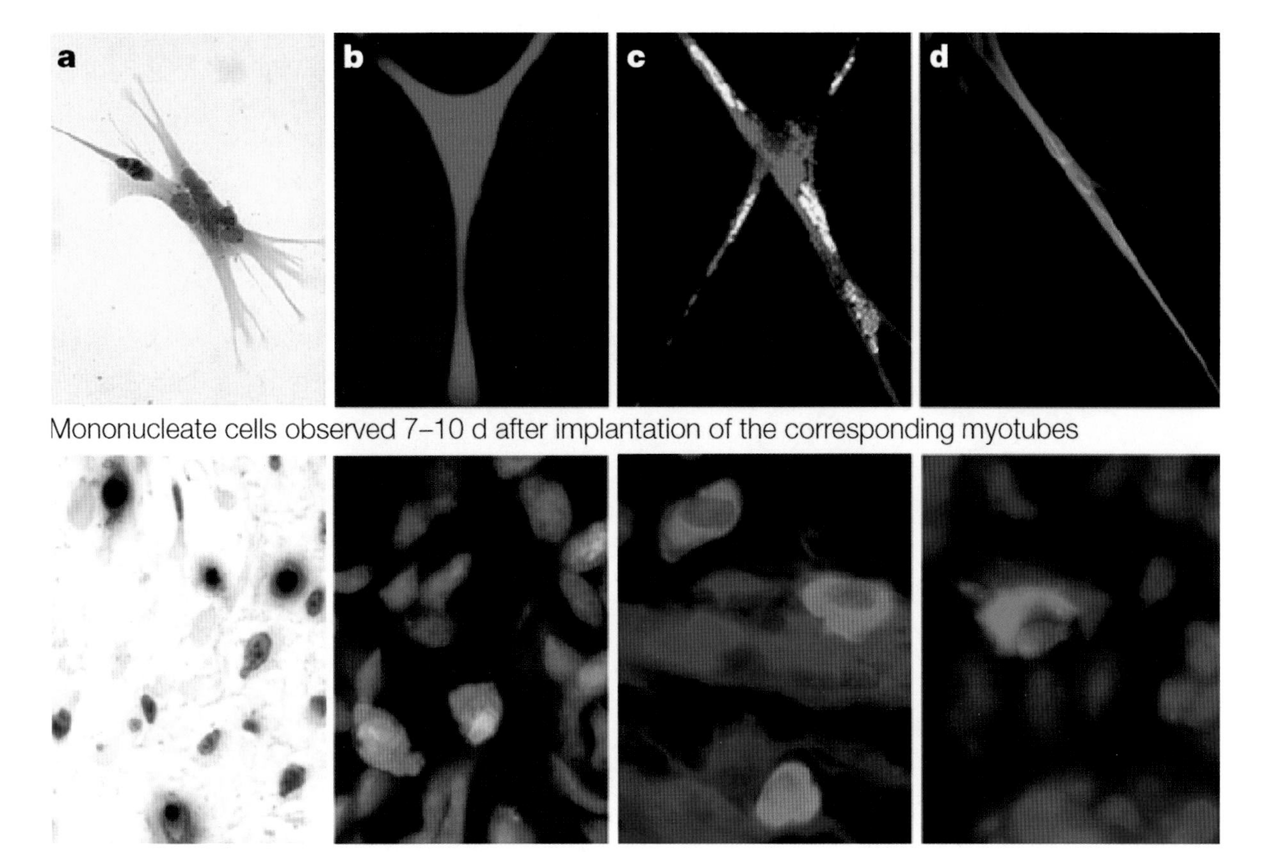

Plate 5.2 (Chapter 5). Plasticity of urodele myogenesis. Examples of implantation assays with newt myotubes that are labelled with an integrated provirus that expresses (a) the marker enzyme human placental alkaline phosphatase, (b) the lineage tracer rhodamine-dextran, (c) the lipophilic cell tracker dye PKH-26, or (d) green fluorescent protein. Bottom panels show mononucleate cells observed 7–10 days after implantation of the corresponding myotubes. Nuclei in (b), (c) and (d) are stained with Hoecht dye, whereas those in (a) are stained with haematoxylin. Reproduced with permission from (Brockes *et al.*, 2001) © 2001 Cambridge University Press.

Plate 7.1 (Chapter 7). Formation of oblique branching lines on shells of a tropical snail *Oliva porphyria*. The background shows a simulation.

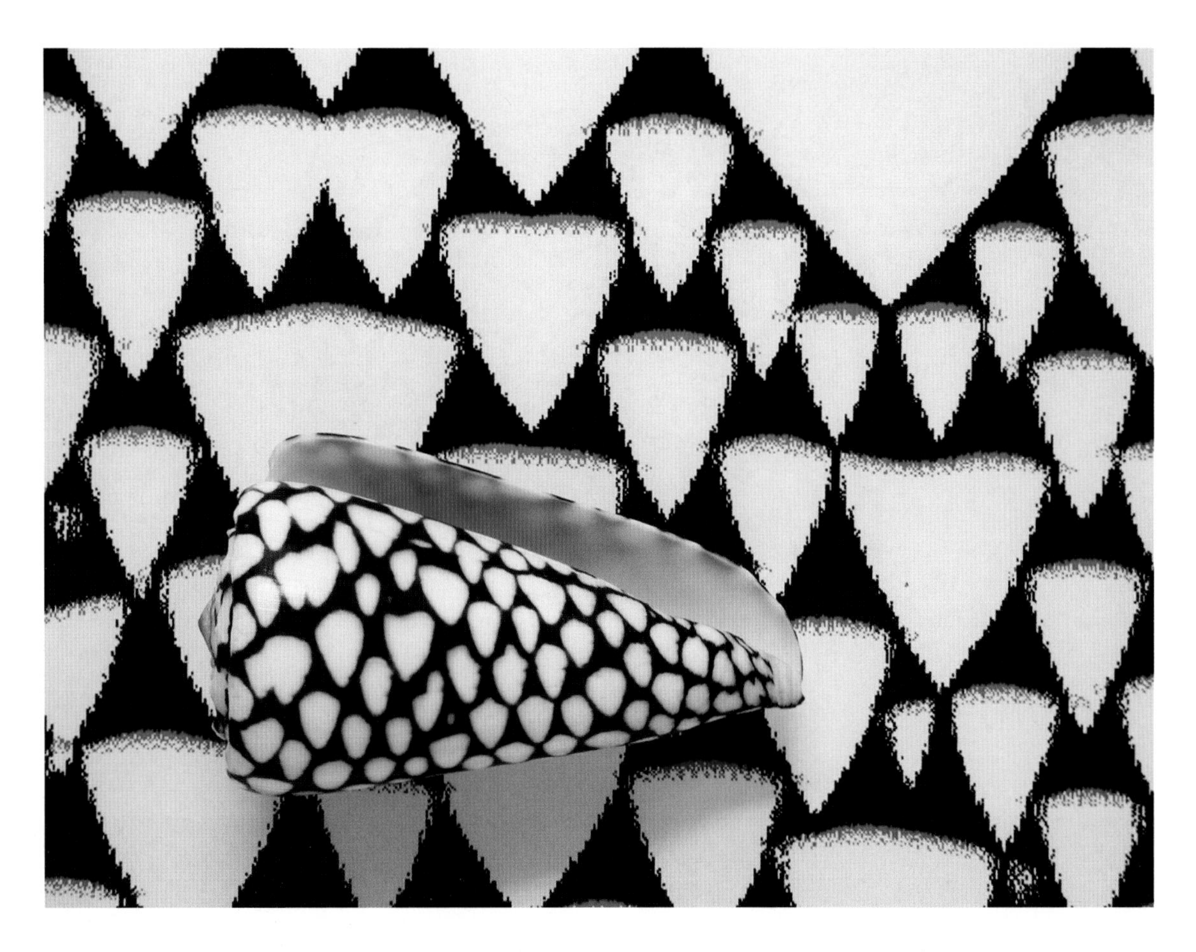

Plate 7.2 (Chapter 7). Shell patterning by superimposition of two time-dependent patterns. Shell of *Conus marmoreus*. A model is shown in the background.

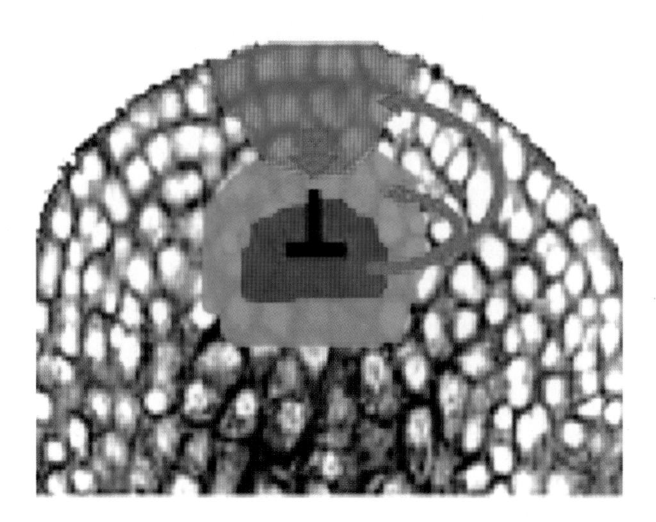

Plate 8.1 (Chapter 8). Illustration of suggested expression patterns of the CLAVATA1,3 and WUSCHEL genes in the SAM. Blue corresponds to WUS, green to CLV1 and red to CLV3. The arrows represent the interactions suggested by experiments (Fletcher *et al.*, 1999, Brand *et al.*, 2000, Schoof *et al.*, 2000, Bowman and Eshed, 2000).

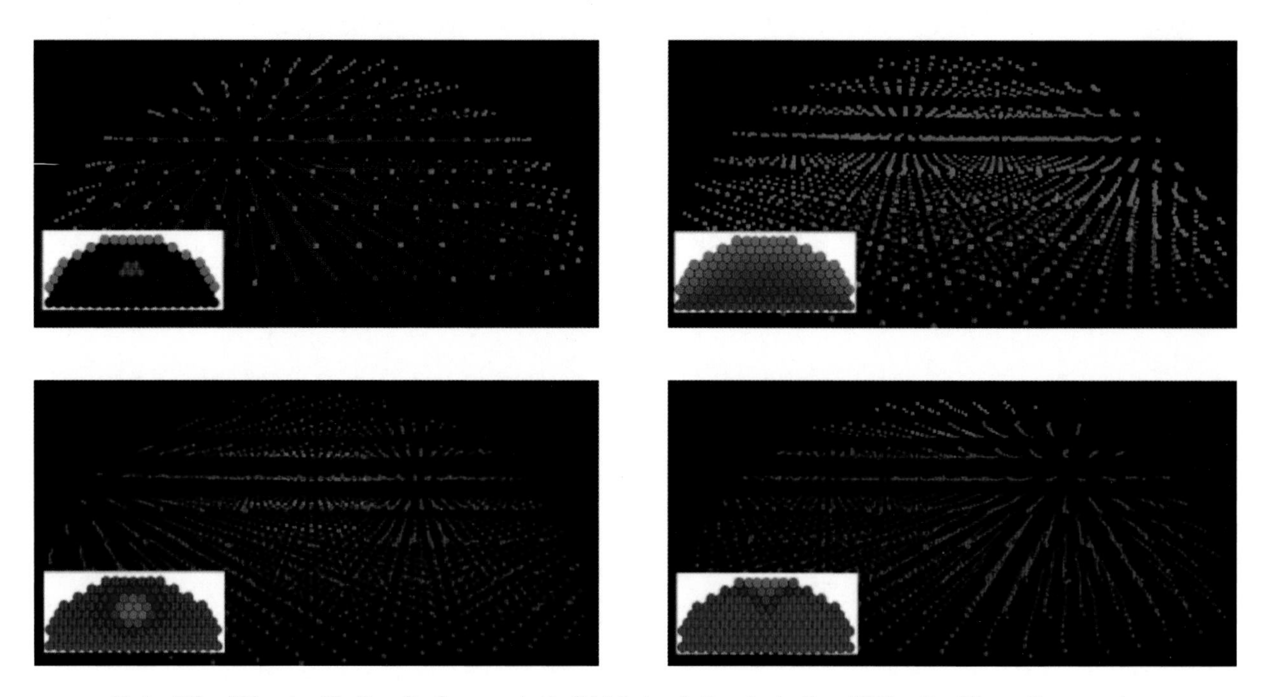

Plate 8.2 (Chapter 8). Results from a static SAM simulation including 1765 cells. The cells are colour coded by protein concentration values. The WUSCHEL region is initiated and the system is simulated until the expression patterns are stabilized in all cells. (a) WUS (blue) and a L1-specific gene (red). (b) The L1 diffusive signal Y (red is high protein concentration, blue is low). (c) CLV1 (colours as in b). (d) CLV3 (colours as in b).

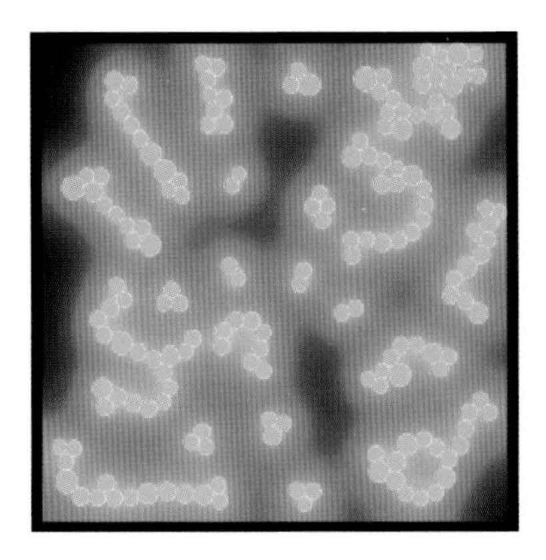

Plate 12.1 (Chapter 12)
Size regulation.

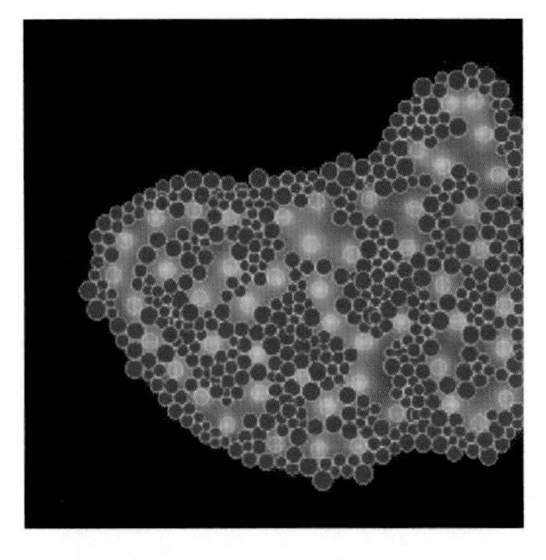

Plate 12.2 (Chapter 12)
Differentiation and shape formation.

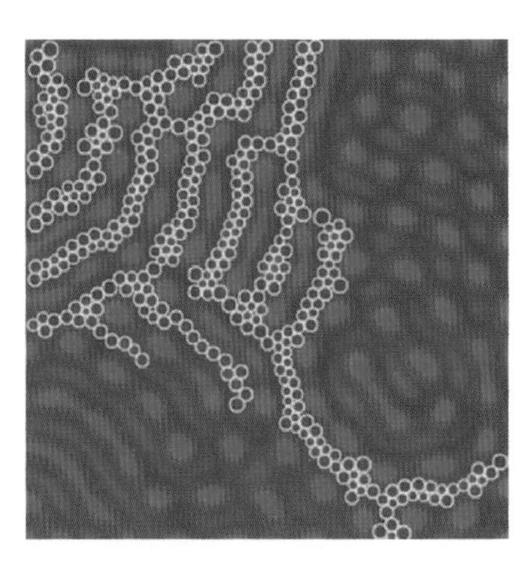

Plate 12.3 (Chapter 12)
Cells on RD patterns.

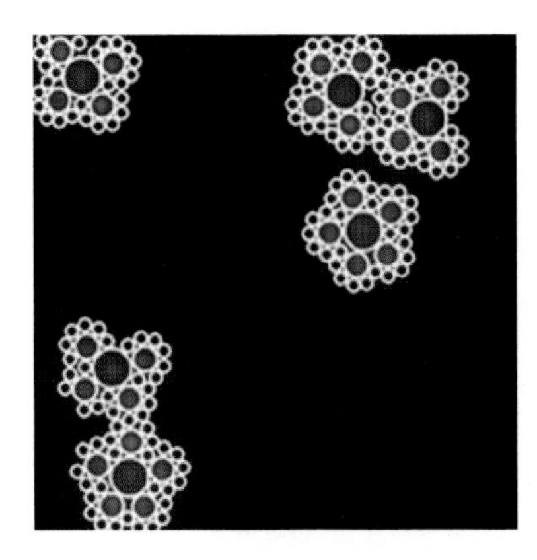

Plate 12.4 (Chapter 12)
Hierarchical structures.

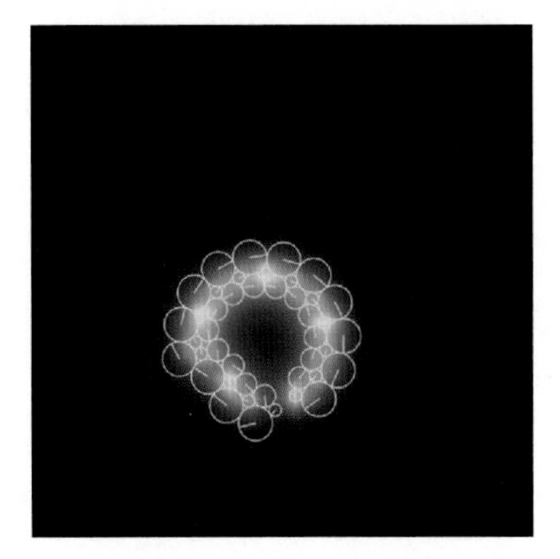

Plate 12.5 (Chapter 12)
Curling, segmented chain.

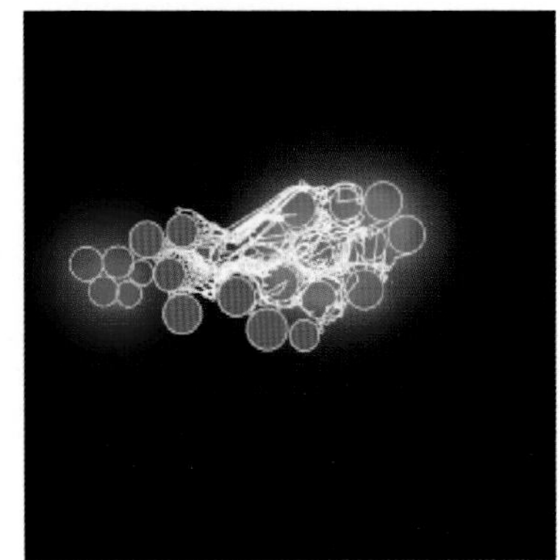

Plate 12.6 (Chapter 12)
One-to-one network.

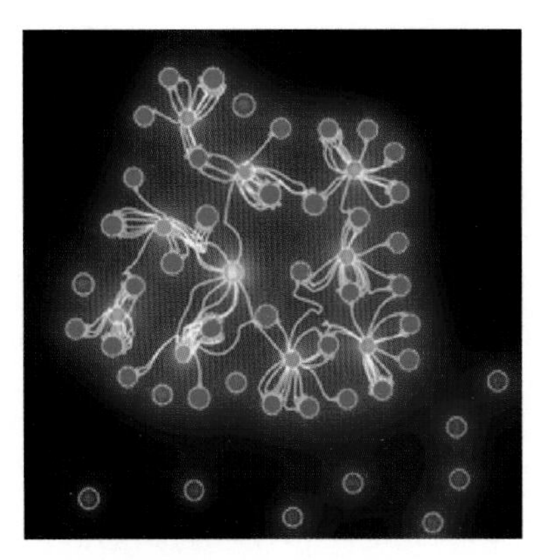

Plate 12.7 (Chapter 12)
Two-level hierarchical network.

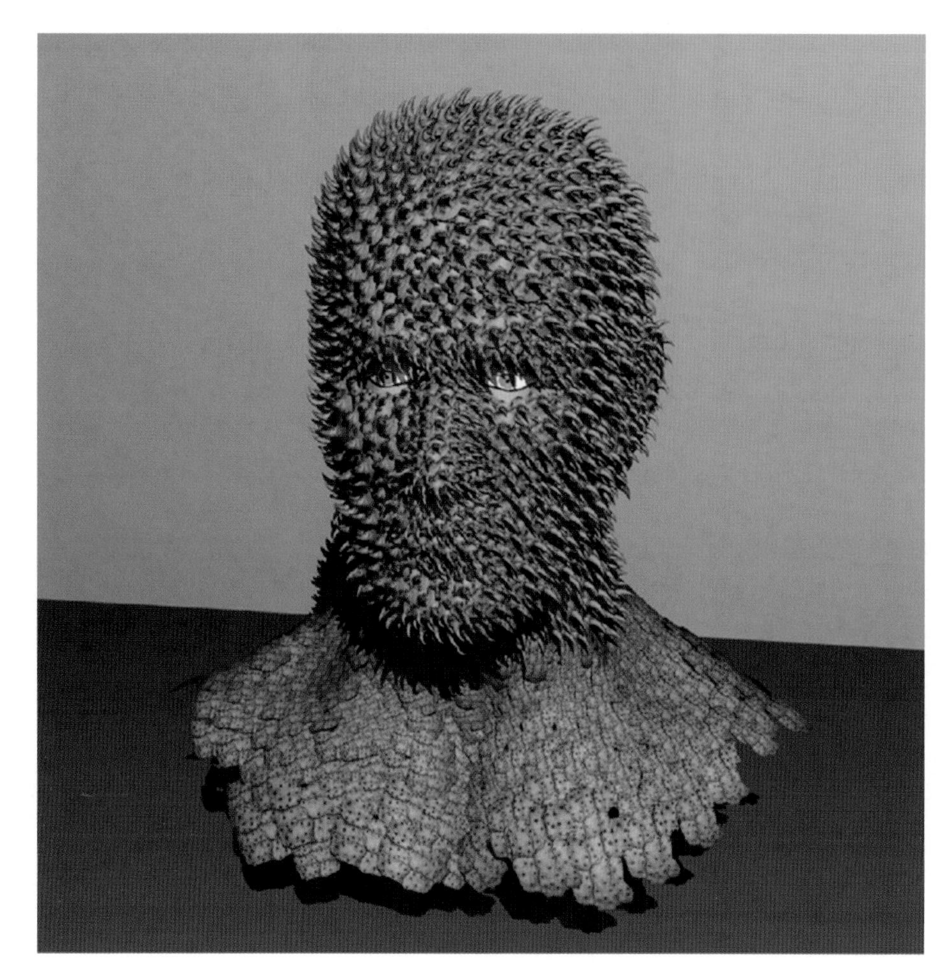

Plate 12.8 (Chapter 12). Developmental monster generation.

Plate 12.9 (Chapter 12). Developmentally-motivated equations determine the patterns on these spheres.

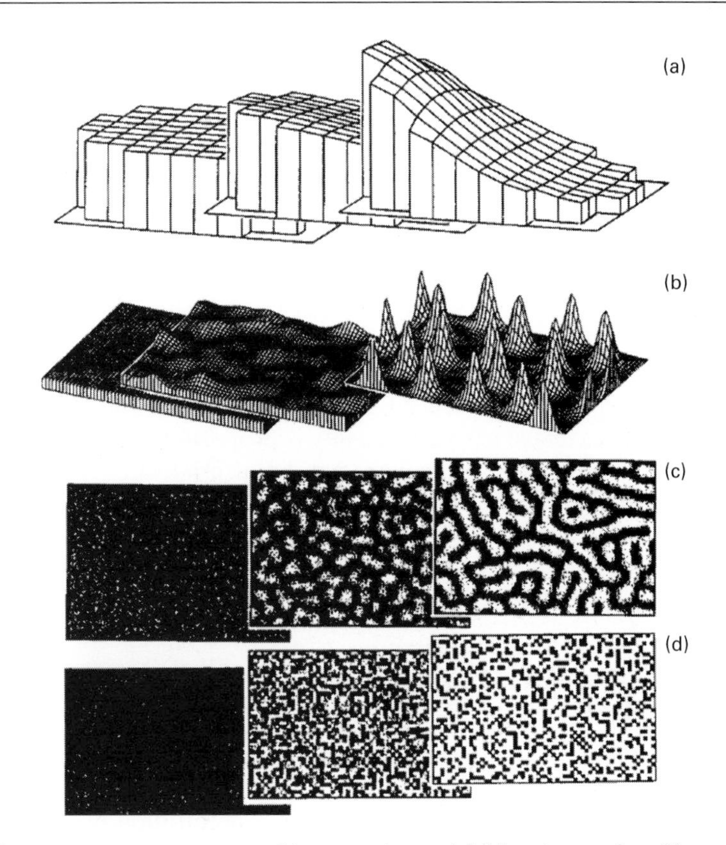

Figure 7.3 Elementary patterns generated by an activator-inhibitor interaction. Shown are the initial, an intermediate and the final activator distributions. (A) Monotonic gradients are formed if the range of the activator is comparable to the size of the field (Figure 7.2). The pattern orients itself along the longest extension of the field. This is appropriate to initiate the formation of an embryonic axis. (B) A more or less regular arrangement of peaks results in fields that are large compared to the range of the inhibitor. (C) Stripe-like distributions result if the autocatalysis saturates (Meinhardt, 1989). (D) The same parameters, but without activator diffusion, lead to a segregation into two cell types, as required, for instance, in the early prestalk-prespore patterning of the slime mould *Dictyostelium discoideum*.

In contrast, if , $D_a > D_h$ and $\mu_a > \mu_h$ oscillations and travelling waves can result. The pigmentation patterns on shells of molluscs will be discussed later as an example of this mode of pattern formation.

7.2.3 Comparison with Turing's mechanism

The possibility of generating a pattern by the reaction of two substances with different diffusion rates was discovered by Turing (1952). Turing exemplified the mechanism he proposed by the following set of equations (Turing, 1952, p. 42):

$$\frac{\partial x}{\partial t} = 5x - 6y + 1 \quad \textbf{(2a)}$$

$$\frac{\partial y}{\partial t} = 6x - 7y + 1 \quad [+Diffusion] \quad \textbf{(2b)}$$

Both equations look very similar. It is not immediately obvious why such interaction leads to a pattern. However, it is easy to see that x is autocatalytic and that y acts as an antagonist since it lowers the x concentration. Therefore, Turing's mechanism is also based on local self-enhancement and long-ranging inhibition and leads to the same types of patterns (Bard and Lauder, 1974; Lacalli and Harrison, 1978).

The mechanism proposed by Turing has an essential drawback: its molecular basis is not reasonable. According to equation **2a**, the number of x molecules disappearing per time unit is assumed to be proportional to the number of y molecules but independent of the number of x molecules. In other words, x molecules can disappear although x molecules are no longer present. This can lead to negative concentrations. Turing had seen this problem and proposed to ignore negative concentrations. A reasonable removal term would require that the number of molecules disappearing per time unit depend on the number of molecules present (like the number of people dying on average per year in a city is proportional of the number of inhabitants). Such a destruction term, however, requires non-linear production terms such as given in equation (**1**). Non-linear equations, however, are more difficult to treat mathematically. Knowing that a balance of local self-enhancement and long-ranging inhibition is crucial simplifies considerably the finding and understanding of appropriate interactions.

7.3 Formation of graded concentration profiles

The generation of the primary body axes is a very important step in early embryogenesis. One side of the developing organism must become different from the other. It is a property of the mechanism outlined above that in growing fields, polar concentration profiles will be formed whenever a critical size is exceeded (Figure 7.2A). In a growing field a high concentration on one side and a low concentration at maximum distance is the first pattern that emerges since a non-marginal maximum requires space for two slopes, a space that is not available if the critical size is just exceeded. If the size is smaller, any incipient activator pattern will be smoothed out due to its rapid re-distribution within the small field. Thus, the critical size depends on the range of the activator molecules. The range depends on the diffusion rate and the lifetime of the molecule. A molecule with a longer half-life can move further away from its origin. The resulting polar pattern can be maintained during further growth (Figure 7.5).

Since a homogeneous activator-inhibitor distribution represents an unstable situation, any inhomogeneity can initiate pattern formation. The stimulus can be very unspecific. Any asymmetry in oxygen supply, in the pH or in temperature could be sufficient. Local disadvantages would shift the maximum towards an opposite region. Pattern initiation can result from fluctuations of the activator concentration or from an inhomogeneity in the source density, i.e. the ability of the cells to perform the pattern forming reaction (ρ in equation **1**). According to the model, such stimulus only orients the pattern. The pattern itself is fairly independent of the initial stimulus. Therefore, no precision is required for the initiating signal. A stronger initiating asymmetry has the advantage that the pattern reaches the final steady state much faster since no time-consuming competition is required between opposite sides of the field. In addition, there is less danger that in a somewhat larger field a symmetric instead of a polar pattern is formed.

Eggs with strong as well as no initial internal asymmetries are known. An example for the latter is the brown alga *Fucus* (Jaffe, 1968). Any external asymmetry (for instance gradients in pH, in temperature, in illumination, in the streaming of the surrounding water or the proximity of other *Fucus* eggs) is able to orient the outgrowth, underlining the instability of the system. In contrast, in amphibians, the position of the sperm entry polarizes the egg. The point of invagination during gastrulation, the classical organizer of Spemann and Mangold (1924) is formed opposite the sperm entry (see Harland and Gerhart, 1997). However, after removal of this intrinsic asymmetry by dissociation and re-aggregation of cells, the pattern can still be formed (Nieuwkoop, 1973). Thus, the initiating asymmetry is used, but it is not necessary for pattern formation.

Several molecular systems are known that are close to the proposed scheme. For instance, *Nodal* is required for mesoderm formation and for the left-right patterning in vertebrates. It has an autoregulatory feedback on its own production and is regulated by *lefty-2*, a secreted factor of longer range (Schier and Shen, 2000). In organizer formation, molecules of higher organisms (the *Wnt/ß-catenin* pathway) play an essential role. This pathway is well preserved: *ß-catenin* RNA isolated from the little freshwater polyp hydra, injected into an early *Xenopus* embryo, can trigger a second embryonic axis (Hobmayer *et al.*, 2000). Evidence for autoregulation exists there too, but the long-ranging component is yet unknown. Also for hydra, very elegant transplantation experiments have revealed the ranges of the activating inhibitory influences (Technau *et al.*, 2000). The blue-green alga *Anabaena* forms long chains of cells. About every seventh cell becomes a nitrogen-fixating cell, a heterocyst. An autocatalytic and an inhibitory component have been observed (Yoon and Golden, 1998; Buikema and Haselkorn, 2001). Likewise, in *Drosophila*, self-enhancement and long-range inhibition is involved in the patterning of dorsal appendages of the egg (Wasserman and Freeman, 1998) and in the selection of sensory mother cells (see Culi and Modolell, 1998; Sun *et al.*, 1998).

7.3.1 Pattern regeneration

Many developing systems show pattern regulation. As mentioned, the separation of an early sea urchin embryo in two parts can lead to two complete embryos. After removal of the so-called head of a hydra, a new head regenerates.

An activator-inhibitor system shows substantial pattern regulation. With the removal of the site of high activator concentration, the site of inhibitor production is removed too. The remnant inhibitor decays until a new activator maximum is triggered in the remaining cells from low-level activator production ρ_a in equation (**1**). The pattern is restored in a self-regulatory way (Figure 7.2B). A condition for pattern regeneration in a small fragment is that the remaining fragment is larger than the minimum field size required for pattern formation.

7.3.2 Elementary pattern types

The actual pattern that is generated by an activator-inhibitor mechanism depends on the parameter. Figure 7.3 shows a collection of several elementary patterns. As mentioned, if the range of the activator is of the order of field size, only one marginal maximum can emerge, convenient to generate a polar structure along one axis of the field (Figure 7.3A). If, however, the range of the inhibitor is much smaller than the field size, multiple maxima will emerge. The resulting pattern may not be strictly regular but a minimum and maximum distance between the maxima is maintained (Figure 7.3B). More regular periodic patterns are formed as the patterning already works during growth. Stripe-like concentration profiles result if the activator saturates at high concentration. Saturation results in a reaction according to equation (**1**) by the employment of an autocatalytic term of the type $a^2/(1+\kappa a^2)$ instead of a^2 only (Figure 7.3C). Stripe formation occurs since the saturation of the activator autocatalysis leads also to a limitation of the inhibitor production. More cells become activated until sufficient inhibitor is produced, although at a lower level. Thus, the activated regions have the tendency to enlarge. However, in order to become activated, a close neighbourhood to the non-activated cells is required in order for the inhibitor to escape. The requirement of large activated patches on the one hand and of a direct neighbourhood of non-activated cells on the other seems contradictory. However, in a stripe-like activation pattern, both conditions are satisfied: each activated cell has an activated neighbour and non-activated cells are close by. When initiated by random fluctuations and without orienting cues, the stripes may have a random orientation too and may bifurcate. Transitions between patch- and stripe-like patterns can be frequently seen in the skin pattern of tropical fishes. Kondo and Asai (1995) have observed the dynamic regulation of these stripes on growing fishes and have shown that this can be reproduced by models employing a saturating self-enhancement. In the absence of other constraints, if initiated by random fluctuations the stripes have random orientations too.

It is a property of such stripe-forming mechanisms that the stripes and the interstripes have a similar width. The formation of a single narrow straight stripe-like pattern is crucial for the bilateral body plan, acting as the reference level for the mediolateral position (i.e. for the back-to-belly axis). This requires a coupling between a stripe-forming and a patch-forming system whereby the patch-forming system makes sure that only a single stripe is formed. The Spemann-Organizer can be regarded as a patch-like device that locally initiates and elongates the midline (like an airplane that locally elongates a vapour trail). The stripe system can have the restricted lateral inhibition to avoid a decay into isolated patches while as a result of the strong lateral inhibition of the organizer, only a single stripe can be formed (Meinhardt, 2001, 2002).

In addition to saturation, stripe formation requires a moderate diffusion of the activator in order that the activated regions appear in coherent regions. In the absence of activator diffusion, activated cells appear in a salt-and-pepper-like fashion, whereby the number of activated cells are a certain proportion of the total number of cells (Figure 7.3D). The regulation and initial patterning of prestalk-prespore cells seems of this type (Maeda and Maeda, 1974; Meinhardt, 1983c).

7.3.3 Formation of periodic patterns during growth

If a tissue grows by cell divisions at random positions, the distances between existing maxima increase. New maxima can emerge. This can happen in two different ways. One possibility is that at large distances to existing maxima the inhibitor concentration can become so low that autocatalysis is no longer repressed and new maxima are inserted (Figure 7.4A). The insertion of new bristles after each mould in the bug *Rhodnius* (Wigglesworth,1940) presumably has this cause. If the autocatalysis saturates, a second mode is possible – the splitting and shifting of existing maxima (Figure 7.4B). Saturation leads, as mentioned above, to an enlargement of maxima. The centre of such an enlarged maximum may become deactivated since the cells there have more problems removing the inhibitor compared to the cells located at the flank.

7.3.4 The activator-depleted substrate interaction

As mentioned, in many inorganic pattern-forming systems the antagonistic reaction results from the depletion of a necessary ingredient (see Figure 7.1). Likewise, the antagonistic effect in a pattern-forming reaction can also result from the depletion of a substrate or co-factors consumed during the autocatalytic activator production.

$$\frac{\partial a}{\partial t} = \rho\, s\, a^2 - \mu_a a + D_a \frac{\partial^2 a}{\partial x^2} + \rho_0 \quad \textbf{(3a)}$$

$$\frac{\partial s}{\partial t} = \delta - \rho\, s\, a^2 - \mu_s s + D_s \frac{\partial^2 s}{\partial x^2} \quad \textbf{(3b)}$$

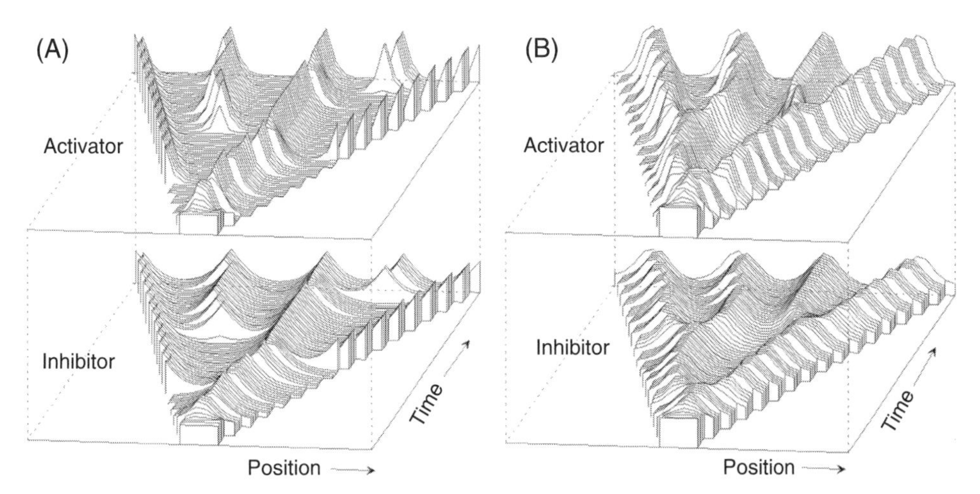

Figure 7.4 Formation of periodic patterns during growth. (A) Insertion of new maxima: whenever the distance to an existing maximum becomes large, the inhibition may become too low to suppress the onset of new activator production. (B) If the autocatalysis saturates at high levels of the self-enhancing loop, the maximum broadens. The spatial extension of the maximum increases as more space becomes available into which the inhibitor can escape. If a critical extension of an activator maximum is reached, the centre of the widening maximum becomes confronted with problems to get rid of the inhibitor and becomes deactivated. A split of a maximum occurs.

This reaction has similarities with the so-called Brusselator reaction (Prigogine and Lefever, 1968; Lefever, 1968) but is somewhat simpler. It has properties more similar to an activator-inhibitor reaction with saturation. This is because the substrate concentration cannot become less than zero. Thus, the depletion mechanism has an inherent upper boundary level for the activator production, as it is obtained by the explicit saturation discussed further above.

7.3.5 Other modes of realizations

There are several other interactions that satisfy our condition of pattern formation. For instance, the autocatalysis may be realized by two substances that inhibit each other. Assume two substances, a and b. If both substances inhibit each other, an increase of a would lead to a stronger inhibition of b and the reduced b level leads to a further increase of a, in the same way as if a would be autocatalytic.

An interesting form of lateral inhibition can be found in the interactions on which segmentation is based. Using classical observations we predicted that segmentation is achieved by the interaction of several feedback loops that locally exclude each other but on long range help each other (Meinhardt and Gierer, 1980). Thus, lateral inhibition of the same feedback loop is replaced by a lateral activation of a competing feedback loop. For instance, if loop a and loop b locally exclude each other but activate each other on longer range, this enforces the correct neighbourhood between a cells and b cells. A stripe-like pattern is especially stable since the long common border between an a-stripe and a b-stripe allows an efficient mutual stabilization.

The unravelling of the genetic network on which the segmental compartmentalization of *Drosophila* is based has provided full support for the predicted interactions (reviewed in Ingham and Martinez-Arias, 1992; DiNardo *et al.*, 1994; Perrimon, 1994). The specification of the posterior compartment is accomplished by the *engrailed (en)* gene, which is autoregulatory (Heemskerk *et al.*, 1991). Likewise, the specification of the anterior compartment is based on a feedback loop employing *wingless (wg)* and other components (Li and Noll, 1993; Manoukian *et al.*, 1995; Lee and Frasch, 2000). The cell states depend on each other through reciprocal signalling: the *wingless* (*wg*) signal is necessary to maintain the autoregulation of *en* in posterior cells. The *hedgehog* (*hh*) signal produced under *en* control upregulates *cubitus interruptus (ci)* in anterior cells, a gene that, in turn, activates *wg* (Alexandre *et al.*, 1996). The repression of *ci* by *en* in the posterior compartment results in the suppression of *wg* at that location (Eaton and Kornberg, 1990). By theoretical explorations of more complex genetic networks, von Dassow *et al.* (2000) found a direct positive feedback loop for *engrailed* to be indispensable for a correct segmentation of *Drosophila*, confirming the prediction of our minimum model proposed a long time ago.

7.3.6 The wavelength problem: maintenance of a graded concentration profile during growth

Usually the size of morphogenetic fields increases during the growth of the embryo. In a reaction as given by equation **1**, a graded concentration profile can be maintained only over a range of about a factor of two. With increasing field size, a tendency exists to change from a monotonic into a symmetric and ultimately into a periodic distribution (see Figure 7.4). This is inappropriate if the graded concentration should be used as positional information for the determination of the primary body axes in the growing embryo.

Several possibilities exist to circumvent this problem. The conversion of the dynamically regulated pattern into a stable pattern by a concentration-dependent gene activation will be discussed further below (see Figures 7.8 and 7.9). Alternatively, the appearance of secondary maxima can be suppressed. If the inhibitor has a small activator-independent production term (ρ_h in equation **1b**), the nominator in equation **1a** remains finite even at very low activator concentrations. This leads to a second stable state at low activator concentrations. This can inhibit the onset of autocatalysis at low activator concentrations. Thus, the field can grow without the secondary maxima appearing. However, this mechanism is inappropriate to model systems that show pattern regulation over a large range of sizes since, as a result of the baseline inhibitor level, an activator maximum would be unable to regenerate after its complete removal.

7.3.7 Maintenance of a graded distribution by a feedback of the pattern onto the competence

Pattern regulation and growth are not mutually exclusive. For instance, the fresh water polyp hydra maintains its polar structure over substantial growth but, nevertheless, a fragment 1/10 of the normal body size is still able to regenerate. The hydra has presumably the longest history in the study of regeneration (Trembly, 1744). Head regeneration of a fragment always occurs at the side pointing towards the original head. Morgan (1904) interpreted similar observations with *Tubularia* by assuming that a systematic change in the ability for head regeneration exists and that a competition takes place. The tissue originally closer to the head has a head start and wins the competition. The relative position is decisive.

The range of dominance of the activated over the non-activated region, the apical dominance, can be increased by an order of magnitude if a feedback of the pattern onto the ability to generate the pattern exists (Meinhardt and Gierer, 1974). This ability we have called source density and can be described as competence of the cell to perform the pattern-forming reaction.

So far in the simulations, apart from small random fluctuations, the sources of the activator and inhibitor synthesis are assumed to be homogeneously distributed. If, however, an increased activator or inhibitor concentration leads to an increase of the source density, the cells distant from the maximum lose the ability to perform the pattern. Consequently, it is less likely that the inhibition emanating from the existing maximum can be overcome, such that a secondary maximum can appear. Thus, the dominance of an existing maximum is enlarged and a polar pattern can be maintained although substantial growth occurs (see Figure 7.7). Source density also shows a graded distribution. Possible additions to equation **1** for the change of the source density ρ are given in (**4a**) and (**4b**):

$$\frac{\partial \rho}{\partial t} = c\, a - \mu_\rho \rho + D_\rho \frac{\partial^2 a}{\partial x^2} \quad \textbf{(4a)}$$

or

$$\frac{\partial \rho}{\partial t} = c \frac{h}{\rho} - \mu_\rho \rho \quad \textbf{(4b)}$$

The feedback of the pattern onto the source can be accomplished either by the activator (**4a**) or by the inhibitor (**4b**). If the activator is employed, a diffusible intermediate must be involved since the competence has to be graded over the total field. If the long-ranging inhibitor is involved, the long range of the inhibitor leads anyway to a smoothly graded distribution, but the increase of ρ has to be slowed down by ρ in the denominator (**4b**) otherwise an overall explosion would occur. In an interaction according to equations (**1**) with (**4b**) the inhibitor has a dual role (Figure 7.5): it inhibits the activator autocatalysis and increases the capability to perform the autocatalysis. Why do both effects not just cancel each other? The inhibitor must have a high diffusion rate and a short time constant to enable a stable pattern formation. In contrast, the source density must be non-diffusible and must have a long time constant. Therefore, after removal of an activated region, the remnant inhibitor declines rapidly while the source density remains essentially unchanged until a new activator maximum is triggered. The asymmetry of the source density orients the regeneration, i.e. the polarity is maintained (Figure 7.5).

A somewhat anthropomorphic analogy may provide a better intuition for the mechanisms envisaged: a king, prime minister, president or the like usually has a strong tendency to suppress others from taking over power – a long-range inhibition. On the other hand, he promotes individuals in his surroundings to obtain different levels in competence, to become ministers etc. In this way, the centre of power generates a hierarchy. Only those in the highest rank of the hierarchy are usually able to take over. Only these have to be hindered to do so. If the centre of power were to become void, due to this non-uniformity, it is usually clear who will win the subsequent competition.

7.3.8 Evidence for the patterning by multiple systems of different wavelength

If hydras or fragments of hydras are dissociated into individual cells and these cells are allowed to re-aggregate, within 4 days new head and foot structures are formed (Gierer *et al.*, 1972). This is a striking example for the formation of a pattern from an initially uniform situation. By searching for molecules that are known to play a decisive role in organizer formation in higher organisms, homologues of the *ß-catenin*, *wingless* and *Tcf* (Hobmayer *et al.*, 2000) as well as

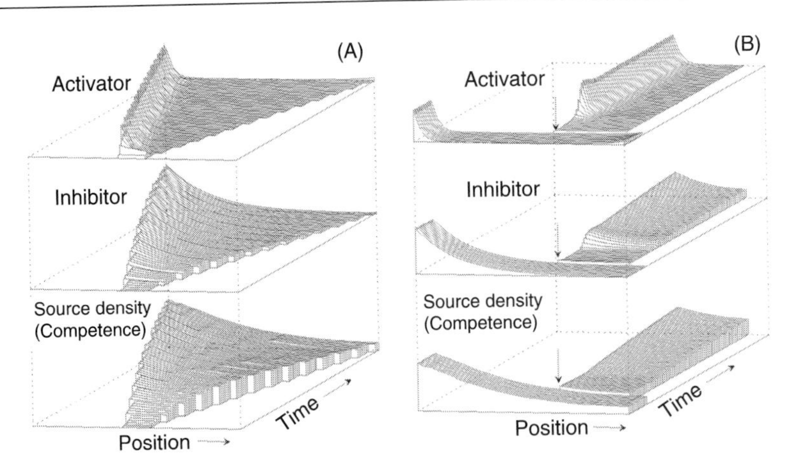

Figure 7.5 Maintenance of a polar distribution in a growing field. (A) The formation of additional maxima (organizing regions) during growth can be suppressed by a feedback of the pattern on the ability of the cell to synthesize the activator, i.e. by a feedback on the source density ρ in an interaction according to equations **1** and **4** (compare with Figure 7.4). The source density (competence, lowest profile in (A) and (B)) reaches a graded distribution. In regions of low source density, the initiation of secondary maxima is unlikely. The first maximum generated at an early stage becomes dominant. (B) Nevertheless, a small fragment can regenerate even when derived from a region of low source density. During regeneration, the orientation of the gradient is maintained as a result of the graded competence.

Brachyury (Technau and Bode, 1999) have been found. All these genes are expressed in the head region (Figure 7.6). Upon regeneration these molecules behave as expected from our theory: the activity reappears in the region where the regenerating head is expected, a process that requires about 2 hours. Ectopic expression of *Hyβ-catenin* RNA in *Xenopus* can induce a secondary embryonic axis, indicating that the crucial pathways for organizer formation are well preserved in evolution. Although all these genes are expressed in the head region, there are differences in the expression pattern: Hy*β-catenin* and *Hy-tcf* have a somewhat wider distribution. During bud formation, these are distributed over the entire bud in a graded fashion. Later, the high levels become restricted to the head region. In contrast, from the beginning, *Hy-wnt* is more confined to the surrounding of the opening of the gastric column. Thus, both systems have different wavelengths. These different expression patterns suggest that these molecules are not just part of the same autoregulatory loop.

Both *β-cat*/*Tcf* on the one hand and *Hy-Wnt* on the other behave differently during re-aggregation of cells (Figure 7.6). Taking cells from the gastric column that do not express these genes, *Hyβ-cat*/*Tcf* first become active in a homogeneous way. Later, isolated local concentrations emerge that eventually give rise to head formation. In contrast, no such initial homogeneous distribution has been found in *Hy-Wnt* (Figure 7.6 C–E; Hobmayer *et al.*, 2000). The activity emerges directly in very sharp spots that later form the oral opening. This suggests that at first relatively smooth *Hyβ-cat*/*Tcf* distributions are formed by a genuine patterning process. The highest levels of these peaks are then used to trigger the *Hy-Wnt* activity. This behaviour can be reproduced in simulations (Figure 7.6I).

This sequence of events suggests a somewhat different mechanism of using different wavelengths as suggested above: that a pattern with a longer wavelength is formed first and used to localize a second pattern with a shorter wavelength. However, in the adult organism the system with the longer wavelength is still confined to the head region, while experiments indicate that a graded competence exists that spans the entire animal. The molecular nature of this gradient is

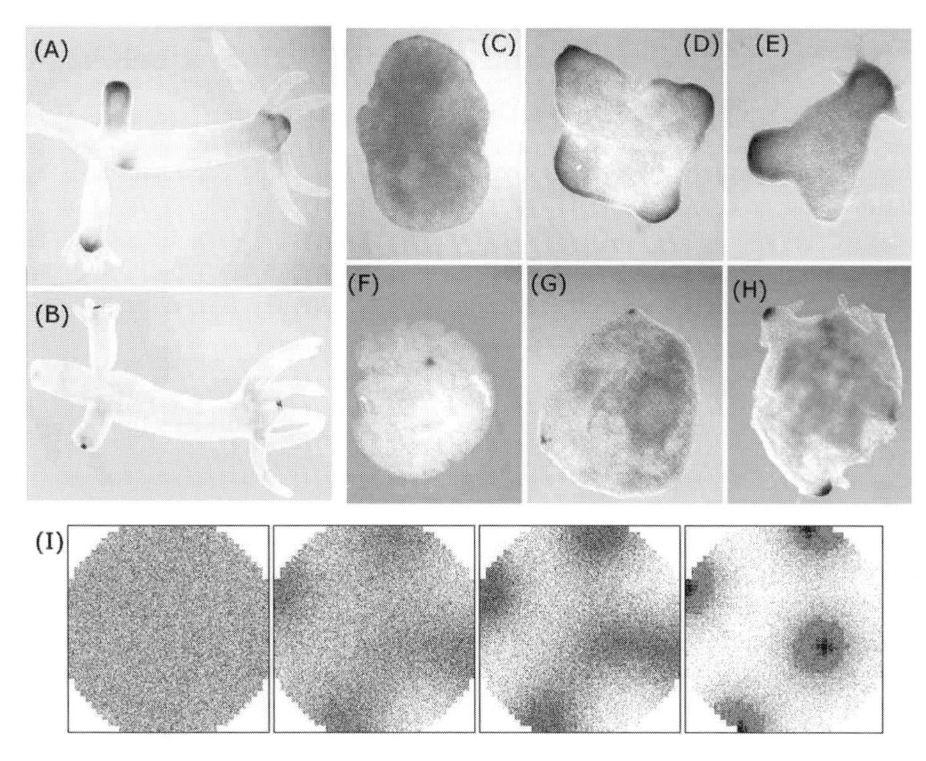

Figure 7.6 Patterning with different wavelengths in Hydra. (A, B) While *Hy-ßcat* (A) and *Hy-Tcf* expression are more smoothly distributed in the head region, the *Hy-Wnt* expression is sharply confined to the oral opening (B). During pattern formation in re-aggregated cells, (C–E) *Hy-ßcat* and *Hy-Tcf* first appear uniformly distributed and become subsequently more restricted to regions that eventually form the heads. (F–H) In contrast, *Hy-Wnt* appears directly in sharp spots that form the future oral opening (Hobmayer *et al*., 2000). (I) The nested pattern formation can be accounted for by the assumption that *Hy-ßcat*/*Hy-Tcf* is part of the primary system since it proceeds, as theoretically expected, first through a uniform distribution (grey). High *Hy-ßcat*/*Hy-Tcf* concentration is the precondition to trigger *Hy-Wnt* (black) that becomes more localized than the initiating *Hy-ßcat*/*Hy-Tcf* pattern (photographs courtesy of B. Hobmayer, T. Holstein and colleagues; see Hobmayer *et al*., 2000).

not yet clear. However, it is known how this gradient can be modified. Using diacyl-glycerol (DAG) the tissue of the whole animal can reach near-head properties (Müller, 1990). After foot removal, for instance, a second head regenerates instead of a foot. This suggests that at least three systems are involved in the formation of the head region: the primary system that sharpens later on (*Hyß-cat*/*Tcf*), the system that is triggered by high levels of the primary system to provide a sharp signal for the oral opening (*Hy-Wnt*) and a very long-ranging system that can be influenced by DAG. While arguments can be given that *Wnt* has an autoregulatory element, no molecule has been found that has the characteristic properties of the inhibitor. Since about 2 hours are sufficient for the re-establishment of the head signal, the communication over the entire axis, i.e. the spread of the antagonist, must be a rapid process. The mechanism that allows such a rapid communication between such distant regions is unknown. In contrast, since it takes about 2 days to reverse the polarity of hydra tissue (by transplanting a head to the foot region and *vice versa*) the polarity-determining gradient must have a much longer time constant.

7.4 Biochemical switches and gene activation: a cell has to remember what it has learned

In higher organisms, the pattern generated by a reaction-diffusion mechanism is necessarily transient since, due to growth, the polar pattern cannot be maintained over the whole expansion of the growing organism. This requires that at an appropriate stage the cells make use of position-specific signals, i.e. they become determined for a particular pathway by activating particular genes. Afterwards the cells maintain this determination whether or not the evoking signal is still present.

The simplest system with a long-term memory consists of a gene whose gene product feeds back on the activation of its own gene in a non-linear way. The production must saturate at high concentrations to avoid unlimited growth. The interaction described by the following equation provides an example for a reaction that leads to threshold behaviour.

$$\frac{\partial g}{\partial t} = \frac{\rho_g g^2}{(1 + \kappa g^2)} - \mu_g g + m \tag{5}$$

At low g concentration, the linear decay rate dominates and the g concentration decreases further to zero. Above a threshold, the non-linear production term dominates. The concentration of g increases further until a steady state at high concentration is reached due to the saturation. However, if a signal m is above a certain threshold, only the high state is stable (Figure 7.7A). Once reached, the system remains at this state even if m becomes small later on (Meinhardt, 1976). Figure 7.7B shows the switching from low to high g concentration in a portion of a field under the influence of a graded m concentration.

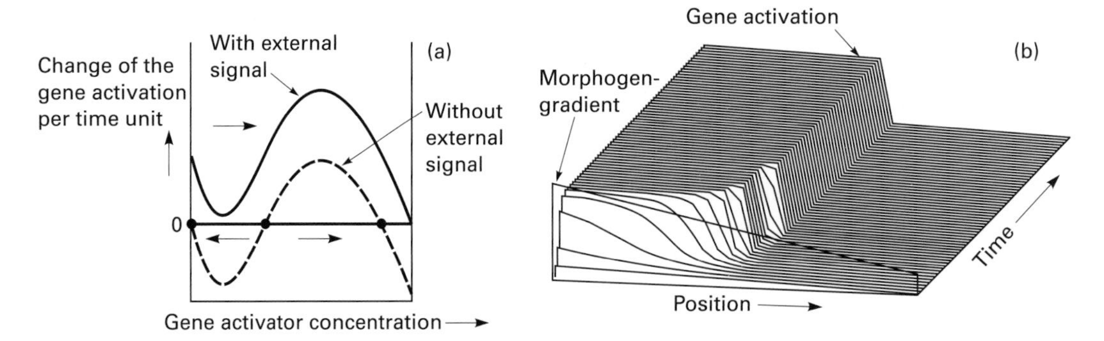

Figure 7.7 Switching behaviour of a substance with non-linear, saturating feedback on its own production rate. (A) In a reaction according to equation **5**, in the absence of a morphogen m, two stable states exist where the change of gene activation is zero: one at a low and one at a high activation level. At higher m concentration, the low gene activation becomes unstable and a switch occurs to the high activation. (B) If a chain of cells is exposed to a graded m distribution, all cells exposed at least to a certain m concentration switch irreversibly into the high state. Regions of high and low g concentrations appear. No zone of transition exists, although the controlling signal has a smoothly graded distribution. This gene activation is maintained even after the signal is no longer available.

7.4.1 Gene activation – pattern formation among alternative genes

For several developmental systems, evidence exists that a single morphogen gradient controls the activation of several genes in a concentration-dependent manner, leading to a region-specific gene activation. To see which type of mechanisms can be involved, it is helpful to realize the formal similarities of the activation of a particular gene and the formation of a pattern. In pattern formation, a particular substance should be produced at a particular location while this production is suppressed at other locations. Correspondingly, cell determination requires the activation of a particular gene out of a set of alternative genes while the remaining genes of the set should be suppressed. Gene activation may thus be regarded as a pattern formation in the gene space.

For a position-dependent gene activation under the influence of a graded signal, a particular concentration range has to lead to the activation of a particular gene. An analysis of experiments with early insect development has suggested that measuring the concentration is not a single and instantaneous event, but instead proceeds step by step until a state is achieved that corresponds to the local morphogen concentration (Meinhardt, 1978, 1982). The analogy given in Figure 7.8A provides some intuition as to how unidirectional promotion is assumed to work.

Starting with a default activation of gene 1, gene 2 becomes activated in the region in which the morphogen concentration surpasses a particular threshold level. In this process, gene 1 can

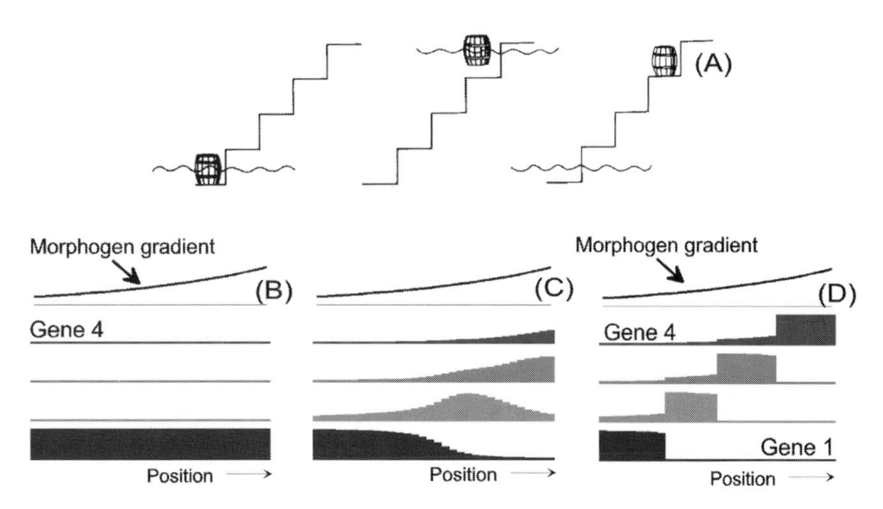

Figure 7.8 Space-dependent gene activation by a step-wise promotion under control of a morphogen gradient. (A) An analogy to illustrate the promotion: a flood can lift a barrel located at the base of a staircase up to a higher level. It will remain even when the flood recedes. The position of the barrel is a measure for the highest level the flood has attained. A higher flood at later times can lift the barrel to an even higher level. In contrast, a second lower flood has no influence. The barrel remains where it came to rest before. (B–D) Simulation of gene activation: a set of genes (gene 1, gene 2 ...) is assumed whose gene products feed back on the activation of the corresponding genes. In addition, all genes compete with each other for activity. This has the consequence that only one of the genes can be active within one cell. The signal (morphogen) causes a promotion from gene 1 to gene 2 etc. until the local morphogen concentration is insufficient for a further step. Sharply confined regions of gene activities emerge despite the fact that the initiating signal is smoothly graded. Once obtained, the gene activation is maintained even if the signal declines later on (Meinhardt, 1978, 1982). A requirement from the model is that each subsequent gene is less sensitive for the signal but is better in the autoregulation. Thus, at a sufficient level of the signal, a switch occurs although this gene activation is less sensitive towards the signal. The system is a biochemical realization of an analogue-to-digital converter.

be switched off due to mutual competition. In a still smaller region, gene 3 will become active, and so on. Each of these steps is essentially irreversible. Although only small concentration differences exist in neighbouring cells, sharp borders are formed in which either one or another gene becomes active (Figure 7.8).

In analogy to the activator-inhibitor system for pattern formation, one can formulate the following set of equations for the activation of particular genes via their gene products g_i (Meinhardt, 1978, 1982):

$$\frac{\partial g_i}{\partial t} = \frac{c_i g_i^2 + m\delta_i g_{i-1}}{\kappa_1 g_1^2 + \kappa_2 g_2^2 + \kappa_3 g_3^2 \ldots} - \mu_i g_i$$

Each gene product g_i is autocatalytic. Instead of a spatial long-ranging inhibition, a mutual repression of the alternative gene products take place. Due to the resulting competition, in a particular cell only one gene of the set can be active (Figure 7.8). Several possibilities exist for a coupling between the gene switching system and the morphogen gradient *m*. In the example given above, the cells switch from one activated gene g_{i-1} to the next, g_i. The number of steps depends on the morphogen concentration. Since each gene feeds back on its own activation, a gene remains active, independent of whether the signal is present or not. The result of this 'interpretation' of positional information is that in groups of neighbouring cells a particular gene is active. An abrupt transition from one activated gene to the subsequent one takes place between neighbouring cells despite the smooth distribution of the morphogen. The result is an ordered sequence of gene activities in space (Figure 7.8).

Meanwhile several systems with this behaviour have been described. In cells from the blastula stage of *Xenopus* (an amphibian), a low concentration of *activin* (a member of a family of growth factors) causes an activation of the *Xbra* gene while for the gene *Xgsc* a high concentration is required (Gurdon *et al.*, 1994). These genes remain activated after removal of *activin*. However, applying *activin* first in low and later in high concentrations leads to a reprogramming of cells from *Xbra* to *Xgsc*. Gurdon *et al.* called this a ratchet-like switching, in full agreement with the mechanism I proposed some years ago. A similar regulation supporting the promotion scheme has been found in the organization of the hindbrain (Gould *et al.*, 1997; Grapin-Botton *et al.*, 1998).

7.5 Patterns on shells of tropical molluscs: a natural picture book to study time-dependent pattern formation

The mechanisms discussed so far lead to patterns that are stable in time. However, in many systems pattern formation remains dynamic all the time. An interesting example are the pigment patterns on the shells of tropical molluscs. A mollusc can enlarge its shell only at the growing edge. As the rule, pigment is only produced at this edge. Thus, the mollusc provides us with a time record of a highly dynamic pattern – a very unique feature. This opens the possibility to decode the patterning process (see colour plates 7.1 and 7.2). These patterns are presumably without functional significance. Therefore, nature was able to play without endangering the species, leading to an incredible diversification of patterns in the different species.

The basic patterns consist of lines that are perpendicular, parallel or oblique to the edge. Keeping in mind that these patterns are time records of one-dimensional patterning it is clear that these patterns result from a stable pattern, either from synchronous oscillations or from travelling waves respectively. These patterns can be simulated with similar reactions as described above. Activated cells are assumed to produce pigment. Travelling waves, for instance, arise if a slowly spreading self-enhancing process is coupled with a non-spreading antagonistic reaction. The spread of an epidemic is a good example of such a process. Neighbouring regions become infected (e.g. by the virus), while the antagonistic reaction (e.g. the immune reaction) brings the system back to the original situation.

A large class of shell patterns require an extension of the simple two-component system in that they require one self-enhancing reaction and *two* antagonists. One antagonist has (as discussed above) a long range and a short-time constant and is responsible for a pattern in space. A second one has a long-time constant and a short range. The local accumulation of this antagonist in the course of time enforces changes in the activation pattern. Either maximum becomes permanently shifted into neighbouring positions, a process that leads to travelling waves. Alternatively, a local activation may become suppressed and reappear at a different location. A surprising diversity of patterns can be generated in this way.

Travelling waves in excitable media usually annihilate each other upon collision. The shell given in Figure 7.9, however, shows crossings of oblique lines. Keeping the space-time character of this pattern in mind, these crossings indicate the penetration of waves – a behaviour that is very unusual for waves in excitable media. For instance, two fronts of a forest fire cannot penetrate each other. Wave penetrations are, however, known from the so-called solitons in physics. The two-antagonist model allows an explanation. In the correct parameter range, the first (essential non-diffusible) antagonist keeps a cell, once activated, in a steady state. The diffusible inhibitor extinguishes the activation of the preceding cell. Due to the activator diffusion, the activation can spread. Therefore, apparently normal travelling waves result. However, if two waves collide, the situation is different. No newly activated cell is available that extinguishes the activation at the point of collision. These regions would remain in a temporary

Figure 7.9 Pigment pattern on the shell of *Tapes literatus* in front of a computer simulation. Shell patterns are time records of a one-dimensional patterning process that occurs at the growing edge. Oblique lines are records of travelling waves. Crossings indicate that two waves do not annihilate each other but penetrate each other upon collision – a unusual behaviour for waves in excitable media. Such patterns arise if a pattern, generated by local autocatalysis and lateral inhibition becomes destabilized by a second long-lasting and local-acting inhibitor (for details see Meinhardt and Klingler, 1987; Meinhardt, 2003).

steady state until the refractory period of the neighbouring region is over. These cells will become re-infected. The newly activated cells extinguish via the diffusible inhibitor the steady state activation of the cells at the point of collision.

This mode of travelling wave formation is a remarkable reversion of the mechanism of stabilization of a single activator maximum discussed above (see Figure 7.5). For the penetration of travelling waves, a local *self-destabilization* has to be assumed, that leads to a permanent displacement into a neighbouring region and thus, in the time record, to travelling waves. In contrast, to reinforce the dominance of a single maximum a positive feedback, i.e. a *self-stabilization* is required.

This two-antagonist mechanism has turned out to be appropriate for the explanation of pattern formation in very different systems, e.g. of how the signals for leaf initiation jump around in the leaf forming zone next to the apical meristem of a shoot to produce the alternate or helically arranged leaves (Meinhardt *et al.*, 1998). A similar mechanism can generate temporary signals on the surface of a single cell that leads to local protrusions. Minute external signals are sufficient to determine on which side these protrusions occur. This mechanism allows an oriented cell movement under the influence of minute signals (Meinhardt, 1999). A related process leads to polar out-of-phase oscillation in *E. coli* that allows a reliable initiation of the cell division apparatus in the centre of the cell (Meinhardt and de Boer, 2001).

Colour plate 7.1 illustrates the formation of oblique branching lines on shells of a tropical snail *Oliva porphyria*. The oblique lines are time records of travelling waves of pigment production along the growing edge of the shell. Branches indicate the sudden formation of backwards waves. The background shows a simulation. Travelling waves are assumed to result from an activator-inhibitor system. The activator has a longer diffusion range but a shorter time constant than the inhibitor, the inhibitor does not diffuse. Branching occurs by a temporary transition from an oscillatory to a steady state activator production whenever the number of travelling waves drops below a critical level. The controlling agent is a hormone-like substance (not shown in the simulation) that spreads rapidly within the organism and which is produced at a rate proportional to the local activator concentration. The hormone thus provides a measure of how many travelling waves are present. The hormone stabilizes the inhibitor. Below a certain hormone concentration the inhibitor life time becomes so short that the just activated cells switch from the oscillatory into a steady state mode of activator production. Groups of cells remain activated. Backwards waves are initiated by re-infection after the refractory period of the neighbouring, still oscillating, cells is over (after Meinhardt and Klingler, 1987; Meinhardt, 1998).

Colour plate 7.2 illustrates shell patterning of *Conus marmoreus* by superposition of two time-dependent patterns. Model in the background: a first reaction is tuned to a steady state activation and leads to the spread of pigment production. A second reaction (plotted in red, not visible in the natural pattern) extinguishes the first. It is triggered whenever cells have been activated for a certain period. Due to its high diffusion rate, pigment production stops almost simultaneously over larger regions, leading to the upper end of the white drop-like pattern elements. The white areas becomes black again in the course of time by the spread from the regions in which the activation survived (after Meinhardt and Klingler, 1987; Meinhardt 1997, 1998).

7.6 Conclusion and outlook

Our models have demonstrated that essential steps in pattern formation during development can be explained by relatively simple coupled biochemical interactions. An explanation of many phenomena can be given without the addition of unreasonable assumptions. All the ingredients used, mutual activation and inhibition of biochemical reactions and diffusion are known to exist. The computer simulations show the consistency of the models. Many elements postulated by the models have now found direct support in the new molecular-genetic techniques. The investigations on the molecular level have revealed that the actual realizations are even more complex. One of the major simplifications in the models is the assumption of a simple diffusion. However, in order that a signal can be transmitted from one cell to the next, ligands have to be secreted by the signal-sending cells, the receiving cells have to express receptors at their surfaces and a cascade has to transmit the signal from the cell surface to the nucleus. However, to understand the logic behind the process, the assumption of a simple exchange by diffusion is a reasonable first approximation.

Models have been developed for many other situations that are beyond the scope of the present chapter: for the initiation of legs and wings in insects and vertebrates (Meinhardt, 1983a, b), for the generation of net-like structures (Meinhardt, 1976, 1982), for the orientation of main body axes in a developing organism (Meinhardt, 2001, 2002). These models show that a theoretical approach can be an essential supplement to the experimental observations in order to find an understanding of a most fundamental process in biology: how a complex organism can arise from a single fertilized egg. These models are far from being complete. Movements and rearrangement of cells and the control of growth are two aspects not yet integrated. Much is to be done to integrate the plethora of recent data into more refined models.

Acknowledgements

I wish to express my sincere thanks to Professor Alfred Gierer. Much of the basic work described in here emerged from a fruitful collaboration over many years. Some of the simulations are available in an animated form at http://www.eb.tuebingen.mpg.de/abt.4/meinhardt/theory.html.

References

Alexandre, C., Jacinto, A. and Ingham, P.W. (1996) Transcriptional activation of hedgehog target genes in Drosophila is mediated directly by the cubitus interruptus protein, a member of the gli family of zinc-finger DNA-binding proteins. *Genes Dev.*, **10**, 2003–2013.

Bard, J.B. and Lauder, I. (1974) How well does Turing's theory of morphogenesis work. *J. Theor. Biol.*, **45**, 501–531.

Browne, E.N. (1909) The production of new hydrants in Hydra by insertion of small grafts. *J. Exp. Zool.* **7**, 1–23.

Buikema, W.J. and Haselkorn, R. (2001) Expression of the Anabaena hetR gene from a copper-regulated promoter leads to heterocyst differentiation under repressing conditions. *PNAS*, **98**, 2729–2734.

Culi, J. and Modolell, J. (1998) Proneural gene self-stimulation in neural precursors – an essential mechanism for sense organ development that is regulated by Notch signaling. *Genes Dev.* **12**, 2036–2047.

DiNardo, S., Heemskerk, J., Dougan, S. and O'Farrell, P.H. (1994) The making of a maggot; patterning the Drosophila embryonic epidermis. *Current Opinion in Genetics and Development*, **4**, 529–534.

Eaton, S. and Kornberg, T. (1990) Repression of cubitus interruptus dominant expression in the posterior compartment by engrailed. *Genes Dev.*, **4**, 1074–1083.

Gierer, A. (1981) Generation of biological patterns and form: some physical, mathematical, and logical aspects. *Prog. Biophys. molec. Biol.*, **37**, 1–47.

Gierer, A., Berking, S., Bode, H. *et al.* (1972) Regeneration of hydra from reaggregated cells. *Nature New Biology* **239**, 98–101.

Gierer, A. and Meinhardt, H. (1972) A theory of biological pattern formation. *Kybernetik* **12**, 30–39.

Gould, A., Morrison, A., Sproad, G. *et al.* (1997) Positive cross-regulation and enhancer sharing: two mechanisms for specifying overlapping Hox expression patterns. *Genes Dev.* **11**, 900, 913.

Grapin-Botton, A., Bonnin, M.A., Sieweke, M. and Le Douarin, N.M. (1998) Defined concentrations of a posteriorizing signal are critical for MadB/Kreisler segmental expression in the hindbrain. *Development* **125**, 1173–1181.

Gurdon, J.B., Harger, P., Mitchell, A. and Lemaire, P. (1994) Activin signalling and response to a morphogen gradient. *Nature*, **371**, 487–492.

Harland, R. and Gerhart, J. (1997) Formation and function of Spemann's organizer. *Annual Review Cell and Dev. Biol.*, **13**, 611–667.

Heemskerk, J., DiNardo, S., Kostriken, R. and O'Farrell, P.H. (1991) Multiple modes of engrailed regulation in the progression towards cell fate determination. *Nature*, **352**, 404–410.

Hobmayer, B., Rentzsch, F., Kuhn, K. *et al.* (2000) Wnt signalling molecules act in axis formation in the diploblastic metazoan hydra. *Nature*, **407**, 186–189.

Ingham, P.W. and Martinez-Arias, A. (1992) Boundary and fields in early embryos. *Cell*, **68**, 221–235.

Jaffe, F. (1968) Localization in the developing Fucus egg and the general role of localizing currents. *Adv. Morphogenesis*, **7**, 295–328.

Kondo, S. and Asai, R. (1995) A viable reaction-diffusion wave on the skin of Pomacanthus, the marine Angelfish. *Nature*, **376**, 765–768.

Lacalli, T.C. and Harrison, L.G. (1978) The regulatory capacity of Turing's model for morphogenesis, with application to slime moulds. *J. Theor. Biol.*, **70**, 273–295.

Lee, H.H. and Frasch, M. (2000) Wingless effects mesoderm patterning and ectoderm segmentation events via induction of its downstream target sloppy paired. *Development*, **127**, 5497–5508

Lefever, R. (1968) Dissipative structures in chemical systems. *J.Chem.Phys.*, **49**, 4977–4978.

Li, X. and Noll, M. (1993) Role of the gooseberry gene in Drosophila embryos: maintenance of the wingless expression by a wingless-goosberry autoregulatory loop. *EMBO J*, **12**, 4499–4509.

Maeda, Y. and Maeda, M. (1974) Heterogeneity of the cell population of the cellular slime mold *Dictyostelium discoideum* before aggregation and its relation to subsequent locations of the cells. *Exp. Cell Res.*, **84**, 88–94.

Manoukian, A.S., Yoffe, K.B., Wilder, E.L. and Perrimon, E.L. (1995) The porcupine gene is required for wingless autoregulation in Drosophila. *Development*, **121**, 4037–4044.

Meinhardt, H. (1976) Morphogenesis of lines and nets. *Differentiation*, **6**, 117–123.

Meinhardt, H. (1978) Space-dependent cell determination under the control of a morphogen gradient. *J. Theor. Biol.*, **74**, 307–321.

Meinhardt, H. (1982) *Models of Biological Pattern Formation*. Academic Press. (an electoric remake can be downloaded from http://www.eb.tuebingen.mpg.de/abt.4/meinhardt/theory.html

Meinhardt, H. (1983a) Cell determination boundaries as organizing regions for secondary embryonic fields. *Dev. Biol.*, **96**, 375–385.

Meinhardt, H. (1983b) A boundary model for pattern formation in vertebrate limbs. *J. Embryol. exp. Morphol.*, **76**, 115–137.

Meinhardt, H. (1983c) A model for the prestalk/prespore patterning in the slug of the slime mold *Dictyostelium discoideum*. *Differentiation*, **24**, 191–202.

Meinhardt, H. (1989) Models for positional signalling with application to the dorsoventral patterning of insects and segregation into different cell types. *Development*, (Supplement 1989), 169–180.

Meinhardt, H. (1997) Biological pattern-formation as a complex dynamic phenomenon. *International J. Bifurcation and Chaos*, **7**, 1–26.

Meinhardt, H. (2003) *The Algorithmic Beauty of Sea Shells*. 3rd enlarged edn. Springer.

Meinhardt, H. (1999) Orientation of chemotactic cells and growth cones: models and mechanisms. *J. Cell Sci.*, **112**, 2867–2874.

Meinhardt, H. (2001) Organizer and axes formation as a self-organizing process. *Int. J. Dev. Biol.*, **45**, 177–188.

Meinhardt, H. (2002) The radial-symmetric hydra and the evolution of the bilateral body plan: an old body became a young brain. *Bioessays*, **24**, 185–191.

Meinhardt, H. and de Boer, P.A.J. (2001) Pattern formation in *E. coli*: a model for the pole-to-pole oscillations of Min proteins and the localization of the division site. *PNAS*, **98**, 14202–14207.

Meinhardt, H. and Gierer, A. (1974) Applications of a theory of biological pattern formation based on lateral inhibition. *J. Cell Sci.*, **15**, 321–346.

Meinhardt, H. and Gierer, A. (1980) Generation and regeneration of sequences of structures during morphogenesis. *J. Theor. Biol.*, **85**, 429–450.

Meinhardt, H. and Klingler, M. (1987) A model for pattern formation on the shells of molluscs. *J. Theor. Biol*, **126**, 63–69.

Meinhardt, H., Koch, A.J. and Bernasconi, G. (1998) Models of pattern formation applied to plant development. In: Barabe, D. and Jean, R.V. (eds). *Symmetry in Plants*, pp. 723–758. World Scientific Publishing.

Morgan, T.H. (1904) An attempt to analyse the phenomena of polarity in tubularia. *J. Exp. Zool.*, **1**, 587–591.

Müller, W. (1990) Ectopic head and foot formation in Hydra: diacylglycerol-induced increase in positional values and assistance of the head in foot formation. *Differentiation*, **42**, 131–143.

Nieuwkoop, P. (1973) The 'organization centre' of the amphibian embryo, its origin, spatial organization and morphogenetic action. *Adv. Morph.*, **10**, 1–39.

Perrimon, N. (1994) The genetic-basis of patterned baldness in Drosophila. *Cell*, **76**, 781–784.

Prigogine, I. and Lefever, R. (1968) Symmetry breaking instabilities in dissipative systems. II. *J. Chem. Phys.*, **48**, 1695–1700.

Schier, A.F. and Shen, M.M. (2000) Nodal signalling in vertebrate development. *Nature*, **4**, 385–389.

Spemann, H. and Mangold, H. (1924) Über Induktion von Embryonalanlagen durch Implantation artfremder Organisatoren. *Wilhelm Roux' Arch. Entw. mech. Org.*, **100**, 599–638.

Sun, Y., Jan, L.Y. and Jan, Y.N. (1998) Transcriptional regulation of atonal during development of Drosophila peripheral nervous system. *Development*, **125**, 3731–3740.

Technau, U. and Bode, H.R. (1999) Hybra1, a brachyury homologue, acts during head formation in hydra. *Development*, **126**, 999–1010.

Technau, U., Cramer von Laune, C., Rentzsch, F., Luft, S. *et al.* (2000) Parameters of self-organization in hydra aggregates. *Proceedings of the National Academy of Sciences of the United States of America*, **97**, 12127–12131.

Tickle, C., Summerbell, D. and Wolpert, L. (1975) Positional signalling and specification of digits in chick limb morphogenesis. *Nature*, **254**, 199–202.

Turing, A. (1952) The chemical basis of morphogenesis. *Phil. Trans. B*, **237**, 37–72.

von Dassow, G., Meir, E., Munro, E.M. and Odell, G.M. (2000) The segment polarity network is a robust development module. *Nature*, **406**, 188–192.

Wasserman, J.D. and Freeman, M. (1998) An autoregulatory cascade of egf receptor signaling patterns in the Drosophila egg. *Cell*, **95**, 355–364.

Wigglesworth, V.B. (1940) Local and general factors in the development of 'pattern' in *Rhodnius prolixus*. *J. Exp. Biol.*, **17**, 180–200.

Wolpert, L. (1969) Positional information and the spatial pattern of cellular differentiation. *J. Theor. Biol.*, **25**, 1–47.

Yoon, H.S. and Golden, J.W. (1998) Heterocyst pattern-formation controlled by a diffusible peptide. *Science*, **282**, 935–938.

8 Signalling in multicellular models of plant development

HENRIK JÖNSSON, BRUCE E SHAPIRO, ELLIOT M MEYEROWITZ and **ERIC MJOLSNESS**

8.1 Introduction

The shoot apical meristem (SAM) is the source of the complete part of a plant above ground. *Arabidopsis thaliana* has become a model system for dicot plants (Meinke *et al.*, 1998; The-Arabidopsis-Genome-Initiative, 2000), and it has a SAM of about 10^3 cells. It retains this size and its almost half-spherical shape throughout the post-embryonic life of the plant. The SAM can be divided into cytologically defined zones where the central zone is at the very apex, the peripheral zone is on the sides, and the rib meristem is in the central parts of the meristem (Steeves and Sussex, 1989; Meyerowitz, 1997).

The central zone is the stem cell domain, the peripheral zone is where new leaf and flower primordia originate and the rib meristem cells are differentiating into cells contributing to the stem and inner parts of the plant. An interesting property of the SAM is that it is self-organizing. For example, if the shoot is divided into two parts, both parts can reorganize into new functional SAMs (Steeves and Sussex, 1989).

In recent years, analysis of expression patterns of important genes have refined the regions of different cell types in the SAM (see for example Bowman and Eshed, 2000). The mutant phenotypes and expression patterns in plants are used to explore the roles of different genes in the development of the SAM. These experiments are far from exhaustive but parts of a genetic network, controlling the development of the SAM, have been identified (Fletcher *et al.*, 1999; Brand *et al.*, 2000; Schoof *et al.*, 2000).

The modest size and the amount of experimental data of the SAM makes it an appropriate candidate for developmental modelling. The focus of this chapter will be on interactions between the CLAVATA1 (CLV1), CLAVATA3 (CLV3), and WUSCHEL (WUS) genes (Fletcher *et al.*, 1999; Brand *et al.*, 2000; Schoof *et al.*, 2000), which is widely discussed in the recent literature. Since these genes are expressed in non-overlapping spatial domains of the SAM, a model incorporating intercellular interactions and signalling is required. We here extend a multicellular model framework (Shapiro and Mjolsness, 2001; Mjolsness *et al.*, 2002) to incorporate a more thorough description of intercellular protein signalling. This allows for simulations of the WUSCHEL/CLAVATA network in the SAM which, *in-silico*, is able to reproduce gene expression domains from experiments on wild type plants.

 ISBN 0-12-428765-4

We argue and show that a model framework with interaction and signalling between cells, such as the one we describe, is a powerful tool to investigate the behaviour of various parts of a developmental system. The model can be used to falsify or discard hypotheses and also to introduce and test new ones.

8.2 The shoot apical meristem

All cells of the above ground part of a plant originate from a group of stem cells found at the growing tip of the shoot called the shoot apical meristem (SAM). The SAM forms during plant embryogenesis and, after seed germination, it remains a collection of undifferentiated cells approximately uniform in shape and size. At the same time it provides at its flanks the cells that will become lateral structures (leaves with attendant second-order meristems and flowers), and at its base the cells that will make the stem, including pith and vasculature. Thus, through the life of the plant the addition of cells to the meristem by cell division and the departure of cells to form differentiated structures, must be closely balanced. Furthermore, the pattern of cell divisions must be highly controlled, to maintain the uniform meristematic shape and to provide flanking structures in appropriate positions (e.g. spiral phyllotaxis).

The interactions between the genes CLV1, CLV3 and WUS are proposed as a possible explanation of how plants control the size of the stem cell region in the shoot (Bowman and Eshed, 2000; Weigel and Jurgens, 2002). The expression domains and interactions of the CLV1, CLV3 and WUS genes are suggested from a number of experiments (Fletcher *et al.*, 1999; Brand *et al.*, 2000; Schoof *et al.*, 2000) and are illustrated in colour plate 8.1. The WUSCHEL protein, which is a homeodomain transcription factor, upregulates the expressions of the CLV1 and CLV3 genes. On the other hand, CLV3 encodes a small, putative extracellular protein and CLV1 a receptor kinase protein which both act in a network downregulating the WUS expression. This feedback network is proposed to regulate the size of the stem cell region of plants. An increased WUSCHEL domain will generate larger CLV1/3 regions which, in turn, will repress the WUS region. As can be seen in colour plate 8.1 the expression domains are not overlapping spatially. The stem cell marking CLV3 is expressed at the very apex, while CLV1 and WUS are expressed in a central region below, sometimes referred to as an 'organizing centre' (Bowman and Eshed, 2000).

8.3 Modelling the SAM

Although the interaction network described in Section 8.2 is widely discussed in the literature, some questions that arise with its assumptions are not. For example, if WUS is a positive regulator of CLV3, why is it that it only upregulates CLV3 in a region above its own expression domain? Why not within it, or symmetrically around it? These kinds of questions are particularly addressable using a modelling framework. In *in-silico* models it is possible to include known data in an expanded network to create expression patterns of the known genes.

8.3.1 The generic model

The central part of the model are the proteins and the regulation of their production. The proteins are implemented by concentration values in each cell and the dynamics are described by a neural network inspired genetic regulatory network (GRN) model (Mjolsness *et al.*, 1991; Marnellos and Mjolsness, 1998).

$$\tau_a \dot{v}_a^i = g(u_a^{(i)} + h_a) - \lambda_a v_a^{(i)} + \gamma_a v_a^{(i)} + \gamma_a v_{a,ext}^{(i)} \tag{1}$$

where

$$u_a^{(i)} = \sum_b T_{ab} v_b^{(i)} + \sum_j \Lambda_{ij} \left(\hat{T}_{ab} v_b^{(j)} + \sum_{bc} \tilde{T}_{ac}^{(1)} \tilde{T}_{cb}^{(2)} v_b^{(j)} v_c^{(i)} \right) \tag{2}$$

$$g(x) = \frac{1}{2}\left(1 + \frac{x}{\sqrt{1+x^2}}\right). \tag{3}$$

In equations (**1–3**), the vs are the protein concentrations, a,b are indices for the proteins and i, j are indices for the cells. The model describes intracellular interactions encoded by the T matrix, as well as intercellular interactions between proteins of neighbouring cells. The intercellular part contains a direct interaction ($\hat{T}$) and also a ligand-receptor type of interaction ($\tilde{T}^{(1)}$, $\tilde{T}^{(2)}$). In our simulation, only the direct interaction term is used, but the ligand-receptor interaction would be suitable for the CLV1/3 signal, omitted here. The λ term is a degradation term, τ is a time parameter, and h is a parameter regulating the basal expression level for a gene. Λ_{ij} describes the connection between cells. The addition of the γ-term represents contributions from other kinds of processes. The simulation presented in this chapter uses diffusion as a model for transferring a molecular signal between cells, which defines the γ-term in this case.

The cells are approximated to have spherical shapes. A mechanical interaction between cells is described by a spring potential between each pair of neighbouring cells with a relaxing distance equal to the sum of the radii. The interaction is softly truncated for larger distances, such that there is no interaction between cells that are not neighbours.

Although the simulation presented here only represents a non-growing model, the framework does include cell growth and proliferation (Shapiro and Mjolsness, 2001). Growth is implemented as being radially symmetric and the rate may be a function of both the mass (size) of the cell and the protein concentrations within the cell. Different models of the cell-cycle are implemented, where the simplest implementation uses only the cell size and the cells divide when the mass reaches a threshold value. A more biologically interesting implementation uses the Goldbeter model (Goldbeter, 1991) and the period can be tuned by a binding of proteins to the cyclin protein as proposed in Gardner *et al.* (1998). Using this model, growth is decoupled from the cell proliferation. At a cell division the total mass is conserved. In the first step of the division, two smaller spheres are created and placed partly on top of each other and then the action of the spring force moves the new cells apart, in a direction dependent on interactions with neighbouring cells.

8.3.2 Simulation

The model framework is implemented both in C/C++ and in Cellerator (Shapiro and Mjolsness, 2001; Shapiro *et al.*, 2002), which is a Mathematica package. The simulation presented is from the C/C++ implementation of a non-growing SAM. The static solution is found by integrating the time differential equations until a stable configuration is found. The differential equation solver is a 5th order Runge-Kutta using adaptive step sizes, based on the *odeint* function from Numerical Recipes (Press *et al.*, 1992). The simulations have been performed on a PC running Linux.

In the simulation there are many parameters, the most important of which are the parameters in the GRN-equations (**1–2**). Experiments only suggest whether interactions are up- or down-regulating and no rates are given. Hence, the parameters are tuned by hand. More than one set of parameters result in the behaviour shown in this chapter.

8.3.3 Results

Figure 8.1 shows a hypothesized network that includes the CLV1,3 and WUS genes, as well as yet unidentified genes. Marked in the red box is the network implied by experiments (see colour plate 8.1). The added gene *X* is not yet identified, but experiments indicates that the WUSCHEL protein does not move between cells (Gross-Hart *et al.*, 2002), requiring some other protein/molecule transferring the information. The *X* in the network could actually represent a larger subnetwork of genes that in the end upregulates both CLV1 and CLV3. The important property for the model to succeed is that a 'signal' originates from the WUSCHEL region, that it decreases with distance from the region and that it upregulates CLV1 and CLV3. In the proposed network we have implemented *X* as a protein which can move between cells by diffusion. For CLV1, *X* is the only input and CLV1 is expressed if and only if the concentration of *X* is above a threshold value within the cell.

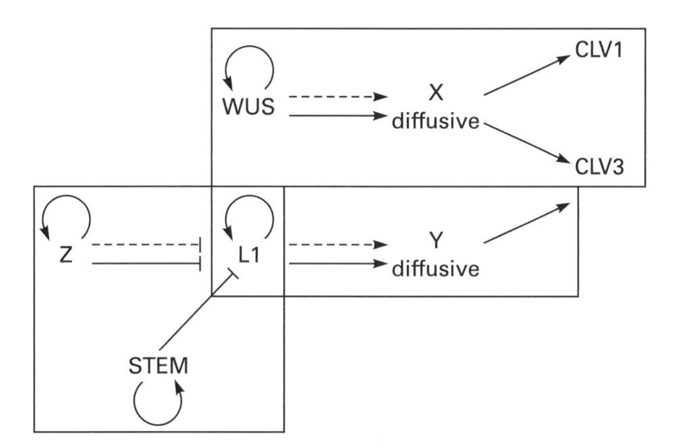

Figure 8.1 Hypothesized network for developmental control in the plant shoot. Solid line arrows represent intracellular interactions and dashed represent intercellular.

The main hypothesis introduced in the network model is shown in the middle box in Figure 8.1. It suggests how WUS can be involved in promoting the CLV3 domain. The idea is that there is a second gene, *L*1, involved which is expressed only in the surface layer (L1) of cells in the SAM. The protein *Y* represents a signal which originates from the *L*1-expressing cells and diffuses into the SAM. The CLV3 gene is turned on only if the total signal (*X* and *Y*) exceeds a threshold value in the cell.

A second hypothesis, also included in the network, is marked with a box in Figure 8.1 (bottom left). This part of the network is used to create the L1-specific expression pattern for the *L*1 gene, regardless of the initial expression patterns of the included genes. *Z* is a gene which is expressed all over the SAM.

Since only the cells of the SAM are included in the simulations, the lowest layer of cells represents the start of the stem in the plant. The boundary is implemented as a gene expressed only in these lowest layer cells (*STEM*). Note also that the suggested repression from CLV1,3 on WUS is not yet included.

In colour plate 8.2, the results of a non-growth simulation of the SAM are presented. The final (stable) expression pattern for some of the genes in the network is shown.

The simulation starts with an initial concentration of WUSCHEL protein in the cells shown in colour plate 8.2a, analogous to the known WUS domain. There is no external input to the WUS gene and the expression stays on in these cells throughout the simulation. The stem indicating protein is initiated with a non-zero protein concentration in the bottom cell layer. This gene also stays on in initiated cells only throughout the simulation. All other proteins are initiated with zero initial concentration in all cells and the expression patterns shown in colour plate 8.2 are the stable self-organizing expression domains for the model network.

Colour plate 8.2a shows the final expression of the *L*1 gene in red. The expression is restricted to the cells of the surface layer. From the *L*1 expressing cells there is a diffusive signal sent into the inner regions of the SAM which is shown in the concentration value of the *Y* protein in colour plate 8.2b.

Colour plate 8.2c and d show the resulting expression patterns for the CLV1 and CLV3 genes. Both show close resemblance with the domains found in real SAMs.

8.4 Discussion

It is indeed encouraging that the proposed network actually produces a stable configuration where the CLV1 and CLV3 expression domains agree with experimental data. The model shows a self-organization of the expression domains with the only requirement that the WUSCHEL gene is expressed in its known region.

A large portion of the network consists of hypothesized components. Some of these may correspond to real genes. There are for example an L1-sharp expressing gene (ATML1, (Lu *et al.*, 1996)) as well as a diffusely L1-peaked gene (ACR4, (Tanaka *et al.*, 2002)), which might be analogous to the *L*1 and *Y* genes respectively. The homeodomain transcription factor, ATML1, is also thought to bind to its own promoter. This is an essential component in the submodel corresponding to our second hypothesis; the network that automatically generates an L1-specific gene. ATML1 binds to the L1-box motif (Abe *et al.*, 2001), which appears in a number of L1-specific gene promoters, including its own. Hence, it is a good target for mutation experiments, testing our hypotheses and we have initiated a discussion to create experiments for this.

The proposed network can explain experimental data, but it remains speculative. The more important contribution is to show how the introduced model framework is valuable as a tool when evaluating a developmental system. We have used it to suggest possible explanations of experimental data in the SAM, including hypotheses open for verification. In the model framework, a gene is only an extra variable and an interaction is a parameter in the differential equations. The simulations presented here need only a couple of minutes to finish on a typical personal computer (1GHz AMD processor). Hence, the framework provides a tractable method to refine new hypotheses *in silico* for future verification *in vivo*.

Acknowledgements

Thanks to Barabara Wold, Marcus Heisler and Venugopala Reddy, for inspiring discussions. The research described in this chapter was carried out, in part, by the Jet Propulsion

Laboratory, California Institute of Technology, under contract with the National Aeronautics and Space Administration. Further support came from the California Institute of Technology President's Fund. HJ was in part supported by Knut and Alice Wallenberg Foundation through the Swegene consortium.

References

Abe, M., Takahashi, T. and Komeda, Y. (2001) Identification of a cis-regulatory element for l1 layer-specific gene expression, which is targeted by an l1-specific homeodomain protein. *The Plant Journal*, **26**(5), 487–494.

Bowman, J.L. and Eshed, Y. (2000) Formation and maintenance of the shoot apical meristem. *Trends in Plant Science*, **5**(3), 110–115.

Brand, U., Fletcher, J.C., Hobe, M. *et al.* (2000) Dependence of stem cell fate in arabidopsis on a feedback loop regulated by clv3 activity. *Science*, **289**, 617–619.

Fletcher, J.C., Brand, U., Running, M.P. *et al.* (1999) Signalling of cell fate decisions by clavata3 in *Arabidopsis* shoot meristems. *Science*, **283**, 1911–1914.

Gardner, T.S., Dolnik, M. and Collins, J.J. (1998) A theory for controlling cell cycle dynamics using a reversibly binding inhibitor. *Proc. Natl. Acad. Sci. USA*, **95**, 14190–14195.

Goldbeter, A. (1991) A minimal cascade model for the mitotic oscillator involving cycline and cdc2 kinase. *Proc. Natl. Acad. Sci. USA*, **88**, 9107–9111.

Gross-Hardt, R., Lenhard, M. and Laux, T. (2002) WUSCHEL signalling functions in interregional communication during Arabidopsis ovule development. *Gene Dev*, **16**(9), 1129–1138, May 1 2002.

Lu, P., Porat, R., Nadeau, J.A. and O'Neill, S.D. (1996) Identification of a meristem l1 layer-specific gene in arabidopsis that is expressed during embryonic pattern formation and defines a new class of homeobox genes. *The Plant Cell*, **8**, 2155–2168.

Marnellos, G. and Mjolsness, E.D. (1998) A gene network approach to modelling early neurogenesis in Drosophila. In *Pacific Symposium on Biocomputing '98*, pp. 30–41. World Scientific.

Meinke, D.W., Cherry, J.M., Dean, C. *et al.* (1998) *Arabidopsis thaliana:* a model plant for genome analysis. *Science*, **282**, 678–682.

Meyerowitz, E.M. (1997) Genetic control in cell division patterns in developing plants. *Cell*, **88**, 299–308.

Mjolsness, E., Sharp, D.H. and Reinitz, J. (1991) A connectionist model of development. *Journal of Theoretical Biology*, **152**, 429–454.

Mjolsness, E.D., Jonsson, H. and Shapiro, B.E. (2002) Modelling plant development with gene regulation networks including signalling and cell division. In *Proceedings of the Third International Conference on Bioinformatics of Gene Regulation and Structure*, Vol. 2, pp. 128–129. Russian Academy of Sciences.

Press, W.H., Teukolsky, S.A., Vetterling, W.T. and Flannery, B.P. (1992) *Numerical Recipes in The Art of Scientific Computing*. Cambridge University Press.

Schoof, H., Lenhard, M., Haecker, A. *et al.* (2000) The stem cell population of arabidopsis shoot meristems is maintained by a regulatory loop between the clavata and wuschel genes. *Cell*, **100**, 635–644.

Shapiro, B.E. and Mjolsness, E. D. (2001) Developmental simulation with cellerator. In *Proceedings of the Second International Conference on Systems Biology*, pp. 342–351. Omnipress.

Shapiro, B.E., Levchenko, A., Meyerowitz, E., Wold, B. and Mjolsness, E.D. (2002) Cellerator: extending a computer algebra system to include biochemical arrows for signal transduction simulations. *Bioinformatics*, **19**(5), 677–678.

Steeves, T. and Sussex, I. (1989) *Patterns in Plant Development*. Cambridge University Press.

Tanaka, H., Watanabe, M., Watanabe, D. *et al.* (2002) Acr4, a putative receptor kinase gene of arabidopsis thaliana, that is expressed in the outer cell layers of embryos and plants, is involved in proper embryogenesis. *Plant Cell Physiology*, **43**(4), 419–428.

The-Arabidopsis-Genome-Initiative (2000) Analysis of the genome sequence of the flowering plant *Arabidopsis thaliana*. *Nature*, **408**, 796–815.

Weigel, D. and Jurgens, G. (2002) Stem cells that make stems. *Nature*, **415**, 751–754.

Computing an organism: on the interface between informatic and dynamic processes

9

PAULIEN HOGEWEG

9.1 Introduction

How do you compute an organism? It is certainly open to dispute in what sense this question makes any sense. An argument against this possibility is that because of percolation of micro-scale physical processes, 'computing' an organism is impossible (Conrad, 1995, 1996). Why then this provocative title? It is because indeed the interaction of processes at different space and time scales is my main theme. Such interactions are a hallmark of biotic systems. Therefore, they should be explored and exploited in our attempts to model and understand them. In the model formulation described in this chapter, I try to minimize the explicit definition of inter-level interactions, while enabling the possibility for such interactions to emerge. The Glazier and Graner (1993) 'cellular Potts model' is at the core of our modelling efforts; long-range interactions in this model are a side effect of volume conservation of fluid-like cells, whose dynamics are modelled at the sub-cellular level.

A second reason for the title of this chapter is Lee Segel's (Segel, 2001) commentary 'Computing an organism' about our model on *Dictyostelium* development, which is based on the interplay between two well-known and well characterized 'physical' processes, i.e. excitable media as a model for long-range cAMP signalling, and cellular adhesion. In Section 9.4, I will briefly discuss how it is indeed the interplay between these processes that (literally) 'shapes' the organism in that model. Nevertheless, 'informatic' processes, in this case gene regulation, are (almost) fully absent in the model.

In Section 9.5, I explore the interplay between gene regulation and cell adhesion and the resulting morphogenetic processes. To this end, I will use, and indeed need, an evolutionary process in order to focus on generic features of the (evolved) mechanism of morphogenesis that rely on the interplay between 'informatic' and 'physical' processes.

Finally, I will study how evolutionary processes on the one hand exploit, and on the other hand may tend to reduce the interplay between 'uncontrolled' physical and evolved informatic processes.

 ISBN 0-12-428765-4

9.2 Classical models of development: pattern formation and morphogenesis

In this section, I will briefly discuss two different classical approaches for studying morphogenesis in order to set the stage for models attempting to incorporate the interaction between different levels of description.

9.2.1 Pattern formation in reaction diffusion systems

Allegedly Turing said, when praised for his well known reaction-diffusion mechanism for pattern formation (the so-called Turing patterns (Turing, 1952)), 'well, the stripes are easy, but what about the horse part?' Nevertheless, morphogenesis is still commonly modelled in terms of 'Turing patterns' and related reaction-diffusion systems.

The appeal of such models is in their apparent simplicity (and therewith tractability) and their power to generate complex patterns (which do sometimes resemble observed patterns), from almost nothing. Interestingly, the well-studied system of stripe formation in the early embryogenesis of *Drosophila* is a striking counter example of this kind of simplicity in biological systems: each of the stripes is generated by a different regulatory mechanism. In hindsight: why not use the differences which are there anyway, especially since the stripes while being initially similar, should later on differentiate? For a recent chapter exploring the (evolution of) hierarchical control versus uniform control see Salazar-Ciudad *et al.* (2001).

Moreover, a point to keep in mind is that, although 'diffusion' appears to be a simple process, in a cellular system in which the diffusing substance has to enter the cell via receptors it is not so simple, as Kerszberg and Wolpert (1998) have shown. Nevertheless, both at intra- and intercellular scales, pattern formation by 'short-range activation and long-range inhibition' appears to be a powerful mechanism to enhance or to (re)generate spatial differentiation (for a recent review see Meinhardt and Gierer, 2000). Similarly, excitation waves (of e.g. calcium, cAMP), can be modelled profitably by reaction diffusion systems. In my opinion, however, models of development and morphogenesis should preferably go a step beyond such a 'uniform' explanation, in which only the formation of a prepattern is addressed independently of the (always inhomogeneous) medium in which it takes place and independently of the actual 'morphogenesis' that it is supposed to trigger. One reason is that, although the 'generic' patterns formed by these processes certainly do occur in biotic systems, an elephant is not such a generic pattern.

9.2.2 Graftal patterns in ancestor based systems

Lindenmayer (1968a, b, 1975) introduced parallel grammar-like rewriting systems (known as L-systems) to describe (plant) development. In contrast to reaction-diffusion systems, this formalism focuses on discrete cells that are in discrete states representing differences in gene expression. The next state of a cell depends on its current state and the state of the neighbours. The next state rules are applied to all cells simultaneously. In addition to state changes cell divisions occur, giving rise to a new cell, which is subsequently subjected to the same rewriting rule. Thus, growth is an inherent feature of the system. The neighbourhood relation in these systems is entirely 'ancestry based': the notion of space is not included in the original formulation. This feature provides the tractability of these actually 'developing' systems.

The structures to which these systems can give rise were termed 'graftals' by Smith (1984), who considered them to be a generalization of 'fractals'. When represented as 3D structures they are strikingly 'life-like' (Figure 9.1, Hogeweg and Hesper, 1974; Smith, 1984; Prusinkiewicz and Lindenmayer, 1990). One should, however, keep in mind that such a 3D representation is entirely a post-processing step. The impact and flexibility of the post-processing step is illustrated in Figure 9.1b: the entire series of pictures does not represent 'development' in the sense of L-systems – they all represent the same stage! The differences are due to the varying length of the branches and the angle of the branch relative to its lower order branch: while these decrease from lower to higher-order branches in the left-most patterns, they increase in the right-most patterns.

Figure 9.1 3D representation of a 2-state bracketed L-system by post processing. Upper panel: by manipulating, e.g. thickness of branches, a very life-like pattern is obtained (image courtesy of A.R. Smith). Lower panel: manipulation of branch length of one stage of the development suggests a developmental sequence.

We conclude that in both these classical approaches to modelling 'development' there is a strong separation between 'development' as modelled and the actual form that arises. In contrast, in the following sections we will develop a modelling approach that allows us to combine a spatial embedding with a growing cell-based system. Such a system allows us to study gene regulation, cell differentiation, intercellular signalling, cell growth, cell division and cell death and the actual morphogenesis that is the result of the interaction between all these processes.

9.3 Modelling approach: two scales and beyond

The core of the models I use in the next two sections is the Glazier and Graner (1993) model for cell sorting. They developed the model as an extension of the Large Q Potts model. In the model, a biological cell (hereafter called cell) is represented by a number of cellular automaton cells (here called sites) with the same state. Interaction between the cells is based on an 'energy' (H) minimization rule that operates at the scale of the sites but includes variables and parameters that are defined for cells. Most notably, the minimization term includes a cell volume conservation term and the energy bonds (Jij) are defined at the level of the cells as well. At each time step, N times a site is selected randomly, and so is one of its neighbours. _H is calculated if the site would copy the state of the neighbouring site and the probability of update is determined using the Boltzmann equation. Thus cells expand and retract 'pseudopodia', can be deformed, grow and shrink, and can therefore, squeeze along each other: these are all features that are very hard or impossible to represent in models defined on a single scale.

Apart from its intrinsic versatility, the beauty of this model is in the ease with which it can be interfaced with other processes in such a way that the interaction between the processes gives rise to novel phenomena. In the next two sections we present two such extensions. In Section 9.4 we study the interaction between excitable media and this cell sorting model. We show that large parts of the development of cellular slime moulds can be modelled from just these two ingredients, once the model parameters are set. Very little 'interference' by genes is necessary to form a functioning multicellular organism from a random assemblage of single cells. In contrast, in Section 9.5 we study the interaction of gene regulation and cellular adhesion.

We do not do so by trying to mimic an existing organism, but we evolve 'critters' and study the morphogenetic mechanisms that ensue, as well as the evolutionary process itself. We show that it is indeed the interaction of gene regulation and cell adhesion that shape both morphogenesis and evolution.

9.4 Computing an organism: 'from single cells to slender stalk'

Dictyostelium discoideum (Dd) is an organism with a curious life-cycle, which is often used as paradigm for studying minimal conditions that can generate functional multicellularity. They live and feed as unicellular amoeboids; when food becomes scarce they aggregate into a mount, the mount topples over and crawls away as a slug, guided by light and temperature gradients. At a suitable spot it settles once more and develops into a fruiting body that has the form of a slender stalk with a spore mass on top. Cell sorting in the mount and the slug stage results in spore cells in the back/bottom of the cell mass, and prestalk cells in the front/tip. The formation of the fruiting body thus involves complex cell movements that lead to a reversal of this positioning.

An overview of the results is given in Figure 9.2. The model incorporates some experimentally well-established properties of Dd., i.e. differential cell adhesion and long range cell signalling.

The latter takes the form of an excitable medium: cAMP release is triggered by cAMP. It is modelled by a generic PDE model for excitable media and we use a discretization corresponding to the sites of the above described GG model of cell adhesion. Chemotaxis of the amoeboids up the cAMP gradient is modelled on the scale of the sites in the GG model: extension of a

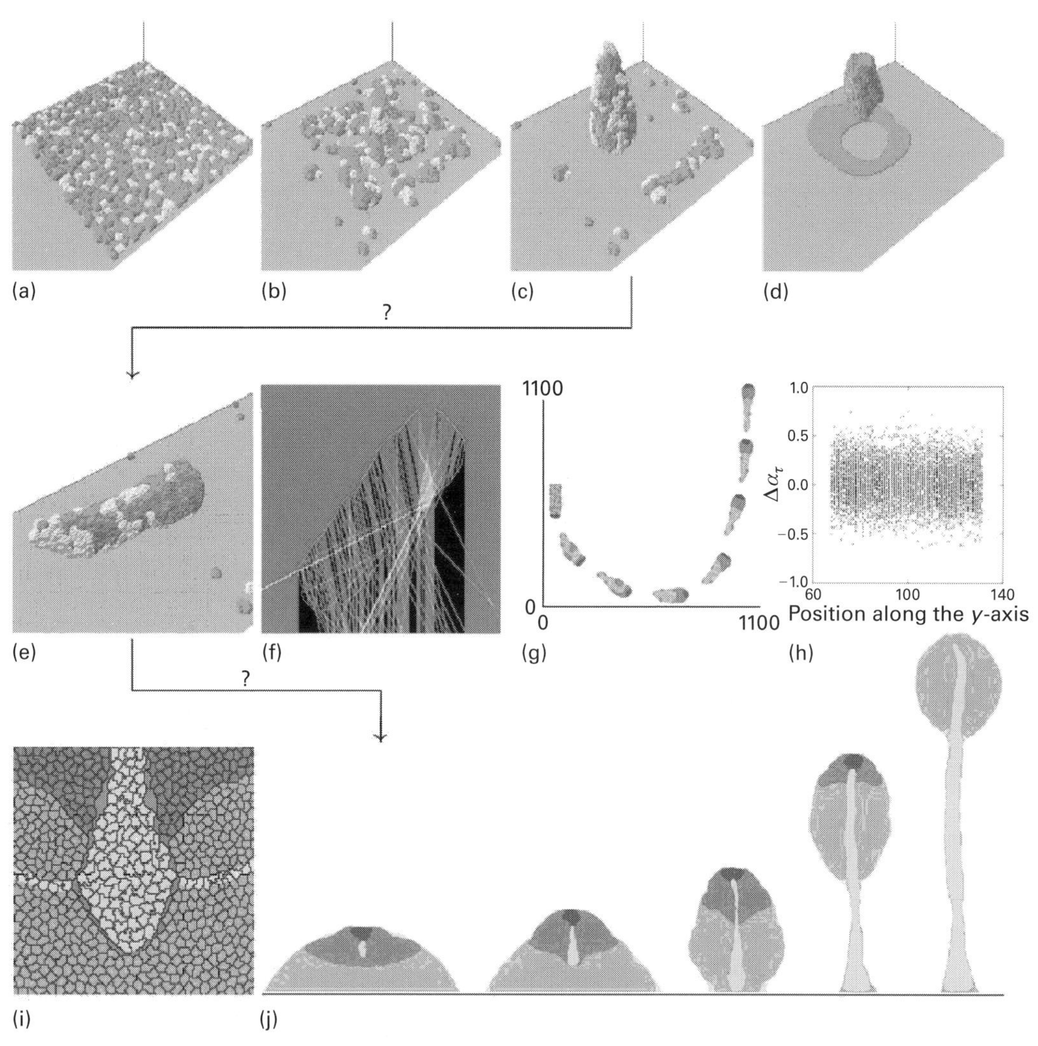

Figure 9.2 The development of *Dictyostelium* in a model combining cell adhesion and cell signalling, (a)–(e) from single cells to crawling slug; (f) the 'lens effect' focusing the light; (g) orientation of the slug towards the light; (h) Dd will show thermotaxis even in such a noisy gradient; (i) the stalk cells are squeezed down by the pressure waves generated by the periodically upward motion of the (pre)spore cells; (j) the formation of a slender stalk with a spore mass on top. Within the same modelling framework we have been able to produce (almost) the entire development of Dd (Savill and Hogeweg, 1997; Marée *et al.*, 1999a, b; Marée and Hogeweg, 2001).

phillopodia in the direction of the gradient is (periodically) more likely. The initial condition is a random distribution of single cells of two types (prestalk and prespore cells) which differ in the surface energy parameters only. A subset of the prestalk cells produces cAMP autonomously – the other cells relay the signal and all cells respond chemotactically to the signal. From these ingredients the aggregation (Figure 9.2 upper panel), orientation (Figure 9.2 middle panel) and culmination (Figure 9.2 lower panel) into a fruiting body unfold. We here comment on the features which emerge due to interaction of the excitable medium and cell adhesion (for more details see the above mentioned publications):

- Aggregating amoeboids form streams (Figure 9.2b). As a result of the cell adhesion, stream formation increases the speed of the aggregation. Single cells only move up the gradient during a short period. The 'pushing' and 'pulling' of other cells (caused by the adhesion and volume conservation) means that the period of directed motion is extended in the streams.
- In the mount and slug stage the prestalk cells sort out to the top/front and prespore cells at the bottom/back of the cell mass (Figure 9.2a–e). The question whether this is caused by chemotaxis or adhesion is hotly discussed in the literature (see the recent chapter of Clow *et al.* (2000)). Our model shows that this 'either/or' question is ill-conceived: the two 'causes' enhance each other. No difference in chemotaxis strength is needed, but chemotactic motion is needed to accomplish the cell sorting in the correct time frame.
- Orientation of the slug towards light is caused by an intricate interplay between the shape of the excitation waves, the chemotaxis and momentum caused by cell crowding. Light is focused at the opposite site of the cell mass, because of the different breaking index of the cells and the 'outside' (Figure 9.2f). Increased light causes the production of ammonia (modelled by another PDE); NH_3 decreases the sensitivity of the cells toward cAMP, the waves become skewed and the slug as a whole orients toward the light source, even if it starts out in the opposite direction (Figure 9.2g).
- The orientation is very robust. In the case of a temperature gradient (also mediated by ammonia production) we studied the influence of noise on a very shallow gradient. Even in the case shown in Figure 9.2h almost all slugs do turn into the direction of the (barely visible) gradient. This robustness is caused by the information integration in the waves (which remain straight) and again the crowding of the cells.
- During the culmination phase the prestalk cells differentiate into stalk cells, which move towards the bottom and anchor. The prespore cells move upwards to form the spore mass. The upward movement of the spore cells can easily be understood in terms of, once again, the chemotaxis toward cAMP that is still being signalled by auto-cycling prestalk cells at the top. We show in our model that, indirectly the downward movement is caused by this signal as well. Stalk cells do not relay and do not respond to cAMP, and produce a stiff extracellular matrix. They are squeezed down as a result of the periodic pressure waves caused by the upward movement of the prespore cells (see Figure 9.2i) provided that a small group of stalk cells at the tip is not surrounded by the stiff slime. The combination of upward and downward movement produces the so-called 'reverse fountain' that transforms a heap of cells into a slender stalk with a spore mass on top (Figure 9.2j).
- Also the stalk formation is very resistant to noise. The stalk will go straight down, correcting any initial misalignment through different arrival times of the pressure waves (data not shown).

Note the two question marks in Figure 9.2. The seemingly 'easy' step of toppling over of the mount is not yet understood in our model (or in experiments: it is not simply caused by gravity). The transition from slug to the culminant is also not yet modelled. Note also the relative minor role of changes in gene expression in the model: most of the process unfolds without assuming changes in the behaviour or parameters of the individual cells. Only during the culmination phase do we model a simple 'maturation' process: auto-cycling prestalk cells become stalk cells, and relaying prestalk cells become auto-cycling prestalk cells; both these maturation steps are triggered by adjacency to these cell types.

We conclude that a functional multicellular critter and its repertoire of observations and reactions of the environment can be the result of the interaction between two well-studied simpler 'modules', i.e. the cell adhesion model and cell signalling as excitable media. In modelling both these processes we ignored much biological detail but, nevertheless, we uncovered the importance of entanglement of various processes. In the next section, we will focus on the interaction between gene regulation and cell adhesion, i.e. we will study how genes can assume a more 'active' role than setting of the parameters of an otherwise (almost) autonomously unfolding process, as is the case in the model presented in this section.

9.5 Evolving morphogenesis: the interplay between cell adhesion and cell differentiation

Whereas, in the previous section we modelled an existing organism using processes known from experiments, in this section, we do not try to mimic any particular organism. Instead we try to uncover some generic features of evolving and developing systems and the interplay between inherited information (genes and gene expression) and the physical dynamics of the system. We do this by studying these issues in critters which we evolve using a very general fitness criterion: the studied features are the side effects of the evolution rather than its 'goal' (see Hogeweg, 1998 for a discussion of this strategy to uncover 'generic/non-generic' features of complex systems).

9.5.1 The model

An overview of the model is shown in Figure 9.3. It combines the GG model with a Boolean gene regulation network in each cell and an evolutionary process to focus on networks which generate morphogenesis. (For a detailed description of the model see Hogeweg (2000a).) Development starts with one large cell that undergoes a number of prescheduled cleavages; during the first two cleavages 'maternal factors' may cause cell differentiation. The rules for subsequent cell growth and division are based on the experimental observation that stretch can indeed lead to cell growth and division and squeezing of cells leads to apoptosis (Chen *et al.*, 1997; Ruoslahti, 1997). Note that both these processes involve an elaborate sequence of changes in gene expression; we here implement only the effect. Cell division occurs in the model when a cell has reached a size that is twice some reference size and cell division is through the middle and perpendicular to the longest axis. This seems indeed to be the default situation and modelling of the spindle formation has shown that it will self-organize accordingly (Dogterom *et al.*, 1995). However, *in vivo* many mechanisms are used to alter the division plane. We will examine below their effect on developmental robustness.

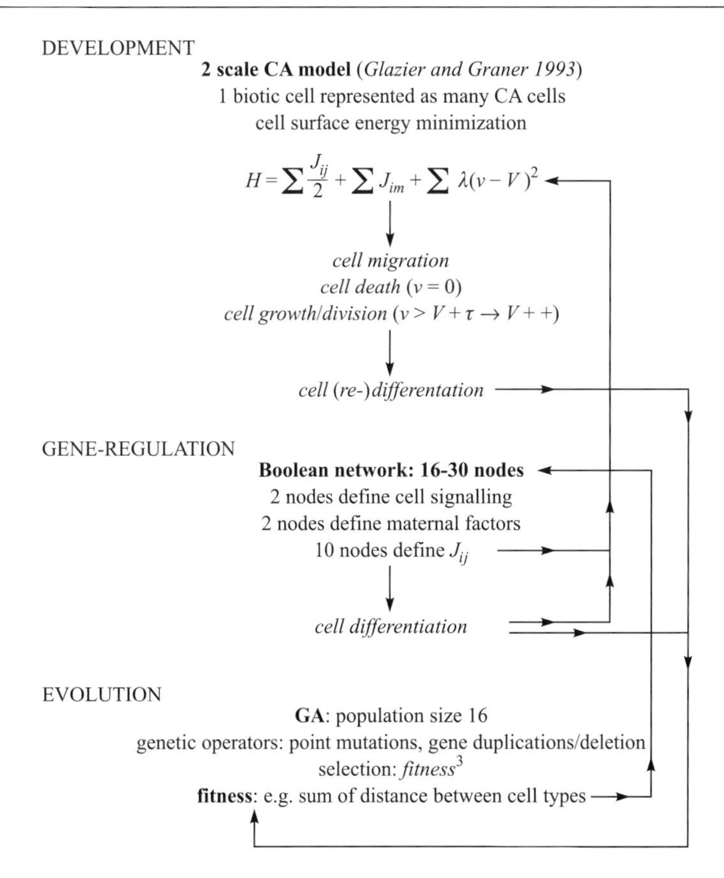

Figure 9.3 Overview of the model: entanglement between gene regulation, development and evolution.

In earlier work (Hogeweg, 2000a, b, 2002) we have shown:

- Differentiation into cell types varies between (a) stable differentiation, i.e. differentiation is maintained independent of intercellular signalling, (b) history dependent differentiation, in which the cell type is dependent on a sequence of neighbourhood conditions, and (c) differentiation which is fully specified by the current neighbourhood only. Each of these types of differentiation is associated with different types of morphogenesis.
- Morphogenesis results as 'sustained transient' from surface energy minimization and 'intrinsic conflict', which is maintained by cell differentiation, cell growth and cell death. Without continued 'interference', the initial, high energy state would change, through shape changes to, at the end a 'blob'-like low-energy shape.
- These intrinsic conflicts lead to automatic orchestration of adhesion, migration, differentiation, cell growth/division and death. It results in 'pseudo-isomorphic outgrowth'. Although, the shapes do change during 'maturation', a 'critter' preserves its general appearance.
- Many different morphemes result from combinations of a few mechanisms:
  - An important mechanism is meristematic growth, a layer of dividing cells that differentiate into non-dividing (or rarely dividing) cells of several types. The zone is maintained because cells redifferentiate if 'out of line': cells differentiation that is fully dependent on neighbourhood is 'used'.

  - A related mechanism we dubbed elongation by 'budding': a small group of differentiated cells is pushed outwards, because an other cell type on the one hand tends to engulf them, but on the other sticks together more firmly than to the 'bud'. Again, the situation is maintained, because cells which do, nevertheless, engulf, differentiate into bud-type cells. Like the previous one, this mechanism depends on neighbourhood dependent cell differentiation. The elongation shown in Figure 9.4 is an example of this.
  - Another often occurring mechanism is convergence extension, which occurs due to maximization of the contact line between stably differentiated cell types and often involves redifferentiation of subtypes of these cell types when the contact zone increases.
  - Elongation can also result from intercalation of stably diverged cell types and their subsequent growth and division.
  - Finally engulfing is an intrinsic mechanism of differential cell adhesion. In our model it often induces neighbourhood dependent cell differentiation.
- The evolutionary dynamics show many of the features known to result from a non-linear genotype-phenotype mapping, i.e. neutral paths and punctuated equilibria at the phenotypic level, although the shape of the quasispecies distribution differs from the simpler examples studied before (e.g. Huynen *et al.*, 1996; van Nimwegen *et al.*, 1999). We will examine this below in more detail.
- Moreover, the evolutionary dynamics give rise to interesting mosaic-like variation at the phenotypic level, i.e. repeated reinvention of similar morphotypes occurs in one evolutionary history.

All these features involve the interaction of the intra- and intercellular levels, which generate long-range correlations. In the next two sections we will examine the relationship between the 'informatic' processes, i.e. gene regulation, and the dynamic processes, i.e. cell behaviour due to differential adhesion, by focusing on the morphological variation that occurs among critters with identical genomes. In other words, we will study how the genetic and inheritable information can on the one hand exploit and on the other hand 'tame' the dynamics of the system.

9.5.2 Developmental variability and robustness: the role of asymmetric cleavage

Developmental robustness seems to be an important feature for organisms, at least for all properties contributing to its fitness. In our model, gene expression is deterministic but dependent on signals of neighbouring cells. In contrast, cell adhesion and the resulting changes in shape of the cells and cell movements, growth and cell death are stochastic. Through cell neighbourhood they can change gene expression, which in its turn may affect cell adhesion parameters. In this way, small stochastic differences can be blown up and quite different critters may develop from identical genomes. The positive feedback is in fact required to obtain and maintain interesting shapes despite the circular initial conditions and the tendency of the minimization rule to produce 'blobs'. The mechanisms of morphogenesis described above all depend on it. Here we ask the question: in what way can and/or does the inheritable information harness the physical dynamics in such a way that the positive feedback leads to similar critters which, nevertheless, do not simply correspond to the attractor of the 'physical' dynamics?

Figure 9.4 shows an example of a genome that can lead to critters with quite different shapes and also with different cell types. The upper three rows show the development of three of these and other examples of 'mature' critters are shown in the lower part of the figure. From the

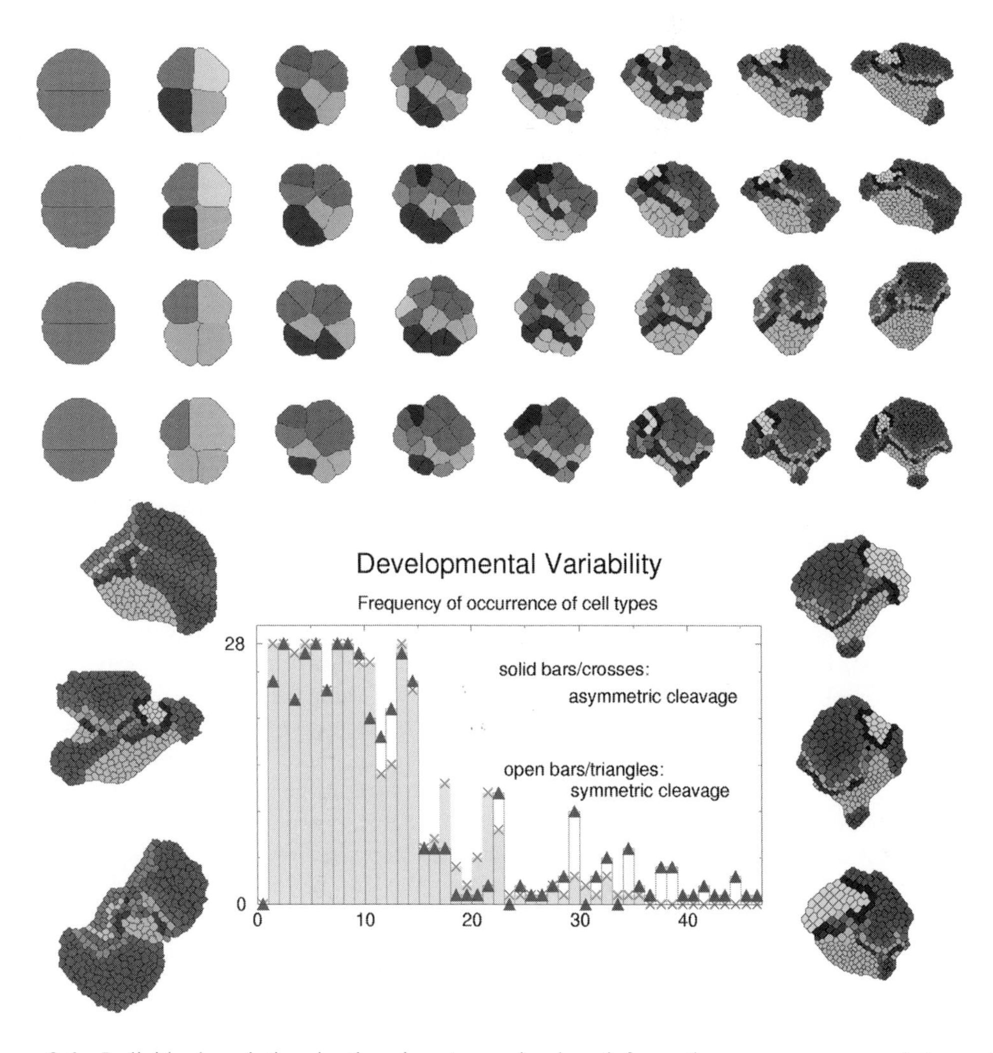

Figure 9.4 Individual variation in the phenotypes developed from the same genome and its partial stabilization by asymmetric early cleavage. Upper panels: developmental sequence. Stages shown just after cleavage and cell redifferentation (3 upper panels show symmetric division, 4th panel shows asymmetric division). Left hand morphemes: variants developed under symmetric cleavage. Right-hand morphemes: variants developed under symmetric cleavage. Histogram shows number of occurrences of cell types as indicated.

developmental sequences we see that a source of variation in the final critter is slight variation in the division planes in early (or somewhat later) cleavage division. This leads to variation in cell contacts and therewith in gene expression and therewith in the adhesion forces and resulting cell shape, which will lead to further differentiation in cell contacts. Note, however, that in many respects convergence of the critters and cell types also occurs. Note for example the occurrence of green and grey cell types at different stages of the development.

It is well known that the location and direction of the division plane is precisely regulated in early cleavages (for a recent review see Nishida *et al.*, 1999). Successive cleavages are usually perpendicular and asymmetric division leading to macromeres and micromeres is seen in many species. The tight regulation supposedly serves to capture cytoplasmic localized factors in one daughter cell only. Indeed, regulation of localization of maternal factors and of the division plane seems to be coupled directly. A second 'purpose' seems to be the determination of cell contacts. Interestingly, several types of regulation of the division plane operate in different organisms and even in different stages of the same organism. This suggests that it evolved repeatedly and suggests important selection pressures to do so.

Here we study how asymmetric division alters the variability observed in the critter displayed in Figure 9.4. In our model, maternal factors are assumed to be captured perfectly in any case. We limit asymmetry, however, to the first two cleavages, i.e. the ones in which cell differentiation due to maternal factors occurs in our model. Thus we investigate the role of determining cell contacts in the first divisions on the long-term development.

The fourth row of Figure 9.4, displays the development of the critter with asymmetric division. The asymmetric division limits cell contacts after the second division to correspond to those in the third row which occurs in 50% of the cases in the symmetric case. However, influence of asymmetric division appears to extend beyond this stage. It is interesting to see the similarity of the cell types in the third stage: the same cell types occur in rows 1, 2, 4, while row 3 differs. In later stages, as a result of the size differences of the cells, the cell deformation and movement is more constrained than it is in the default, symmetric case.

In the histogram we show that not only shape differences are reduced because of the asymmetric division, but also variation in cell types is reduced. More cell types always occur for development with asymmetric first and second cleavages than for symmetric first cleavages. Moreover, many more cell types occur once or a few times in the set generated by symmetric cleavage than that generated by asymmetric cleavage. We conclude that asymmetric cell division in early stages indeed seems to be an efficient strategy to exploit the potential of cell interactions for early differentiation, while limiting the amount of developmental variation.

In the next section, we demonstrate the tendency of evolutionary dynamics to minimize developmental variation and mutational sensitivity. We do this in the standard model with symmetric cell division.

9.5.3 Evolution of developmental and mutational robustness

In Figure 9.5, upper panel, the fitness of an evolving population is plotted. The dots represent the fitness of all members of the population, the filled circles are the common ancestors of the final population. In this case, we use a pseudo-morphological criterion for fitness, instead of our usual fitness criterion on number and differentiation of cell types. The criterion is 'total path length' of identical cell types, i.e. selection favours long layers of cells of the same type (maximum of two layers per cell type are counted), as well as many cell types. In fact during the entire period shown in the figure the 'strategy' for maximizing this criterion remains similar: concentric cell layers develop using engulfing, cell type induction and maximization of the interface between layers.

The evolutionary dynamics show the following interesting features:

- Two types of stepwise improvement occur:
  - At time = 479, a new phenotype is discovered that has much higher fitness than the previous ones. The phenotypic transition is shown in the two pictures on the left. The new genotype takes over the population very rapidly: the two vertical lines show that the entire population on the right is ascended from one-and-the-same individual on the left line.

  - Despite the rapid take-over, there is only a very temporary increase in the median fitness of the population and a slight increase in the average fitness. At around time = 900, however, a stepwise increase in median (but not average) fitness occurs. The previously discovered phenotype is stabilized both developmentally and mutationally. Nevertheless, aberrant development still does occur. The common ancestor at time = 1308 is an example: it has a fitness of 46 and only three cell types. Nevertheless, its genotype is identical to its ancestor, which has six cell types and fitness of 83. The corresponding phenotypes are displayed on the right.
- In the period that neither the maximum fitness, the fitness of the common ancestors, the median or the average fitness changes, the shape of the quasi-species still does change. In the later period, there are more individuals with very high fitness and very low fitness, as shown in the histogram of Figure 9.5. Still aberrant development does occur, however, their fitness remains fairly high.
- The shape of the gene regulation network changes drastically. The left network is the one at time = 479, the right one is time = 2145; the corresponding critters are displayed just above the networks. The expression of more genes is unregulated in later stages of the evolution and the network has become much shorter. Thus mutational changes have less effect.

We conclude that the simple evolutionary dynamic implemented does more than simply select fit individuals. Developmental and mutational stability are strongly selected as well. Selection to robustness biases the networks to short ones. It is interesting to investigate this feature in 'real' gene regulation networks when data become available.

The developmental instability of the critters of this evolutionary run is again through differences in the first cell division. Unequal early cleavage should, therefore be strongly selected (were it made possible in our experiments).

Selection for mutational robustness has been shown to be a hallmark of evolution dominated by neutral paths (van Nimwegen *et al.*, 1999). The two types of stepwise evolution, i.e. in maximum and in medium fitness are not yet reported (as far as I know). It is an evolutionary signature of developmental instability.

9.6 Discussion and conclusions

In biotic systems, processes do not operate in isolation. The existence of entanglement of processes at different time and space scales does not need an explanation: it is there by default; ignoring the entanglement, by artificially separating space and time scales is only a (frequently used) modelling artifact.

In this chapter, we have examined modelling approaches that do not incorporate this artifact. The aim is to uncover the consequences of the entanglement. Our approach is to study the features that emerge by combining two or more simple descriptions of fairly well-known processes, rather than to elaborate the description of these processes by including details of their implementation in biotic systems. By using the relatively simple descriptions and therewith limited number of parameters, we can focus on and study the features that arise through the interplay between these processes. We have shown that through the interplay of the processes we indeed can model not only 'generic' features, which biological systems share with many other systems, but can explain specific, experimentally observed, but previously not yet understood behaviour of particular organisms (e.g. orientation and culmination in *Dictyostelium*).

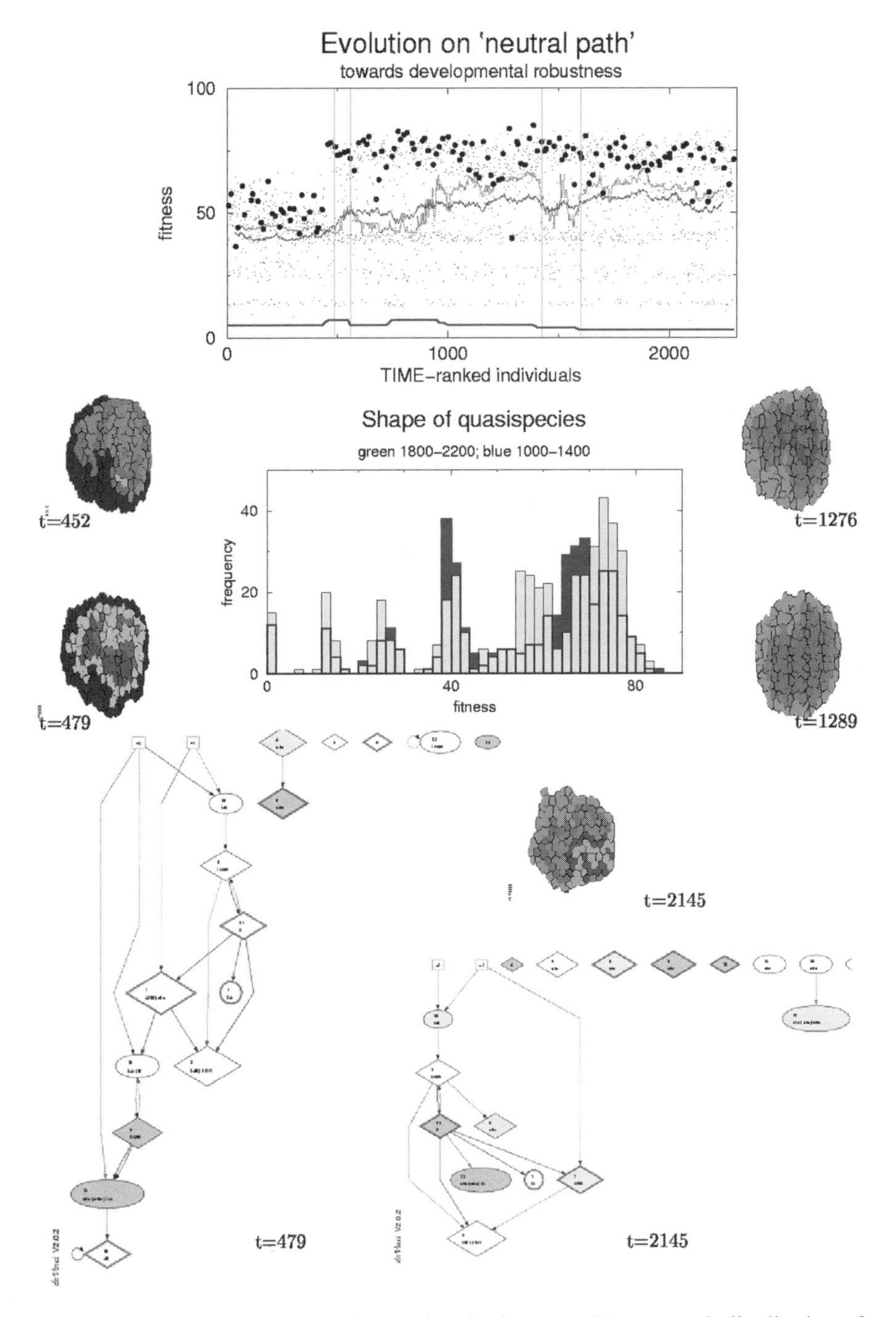

Figure 9.5 Evolution of developmental and mutational robustness. Upper panel: distribution of population fitness over time. Black circles represent the common ancestors of the last population; vertical lines: duration of fixation: Upper pale grey line: medium fitness; upper dark line: average fitness. Lower dark line: maximum path length in gene regulation network. Middle panel: Histograms of quasi-species distribution in two periods of neutral evolution; distribution change towards higher developmental and mutational robustness. Phenotypes left, transition to high fitness; right, developmental variation of identical genotypes. Lower panel: change in regulatory network during 'neutral evolution'. The shortening increases the mutational robustness.

A hallmark of biological systems is that they are physical, dynamical systems which have evolved. Nevertheless, evolution is seldom studied in interaction with 'physical' systems with dynamics of their own. In most models, genes are supposed to 'do what they do', or the physical systems are supposed to 'do what they do'. The differential adhesion based model of cell sorting with a genetic network governing adhesion parameters proved to be an interesting testbed for studying the relation between the physical dynamics of the system and the inherited information.

We have seen that, although the development of the critters is 'governed' (through the cell adhesion parameters) by the gene regulation network, inheritance of phenotype does not follow automatically. The positive feedback between gene expression and adhesion is crucial for the genes to do anything at all, but also causes sensitivity to noise. We have seen, however, that the evolutionary process will reduce this sensitivity, while exploiting the self-correcting features to which the interaction between gene expression and differential adhesion can also give rise. An efficient general way to do so seems to be the unequal cell division seen in the early cleavages of most organisms. We should note, however, that further studies are necessary to generalize the results obtained so far.

In conclusion: entanglement of processes at different time and space scales is a source of interesting, complex and life-like behaviour and should not be ignored in our attempts to model organisms. Evolution under an invariant selection regime automatically leads to increase of robustness against mutations (Huynen and Hogeweg, 1994; van Nimwegen *et al.*, 1999) and against developmental variation (this chapter). We have shown that it is feasible to formulate models that incorporate and exploit these two facts. This is done by allowing (but not rigorously predefining) entanglement of processes and by using an evolutionary processes to tune positive and negative feedback between levels automatically, as a side effect of some general fitness criterion. Thus, we come one step closer to, in some sense, 'computing an organism', although some of the desiderata of computing systems (e.g. full modularity and environmental independence) are certainly not met (or desired) in these models.

Acknowledgements

I thank Nick Savill and Stan Marée for their research on *Dictyostelium* development, and the latter also for composing Figure 9.2. I am much indebted to Roeland Merks for developing the computer program for the evolution for morphemes. I thank Ben Hesper for long-term discussions and support.

This chapter is from Hogeweg, P. (2002) Computing an organism: on the interface between informatic and dynamic processes. *Biosystems*, **64**(1–3), 97–109.

References

Chen, C.S., Mrksich, M., Huang, S. *et al.* (1997) Geometric control of cell life and death. *Science*, **276**, 1425–1428.

Clow, P.A., Chen, T., Chisholm, R.L. and McNally, J.G. (2000) Three-dimensional *in vivo* analysis of *Dictyostelium* mounds reveals directional sorting of prestalk cells and defines a role for the myosin II regulatory light chain in prestalk cell sorting and tip protrusion. *Development*, **127**(12), 2715–2728.

Conrad, M. (1995) Cross-scale interactions in biomolecular information processing. *Biosystems* **35**(2–3), 157–160.

Conrad, M. (1996) Cross-scale information processing in evolution, development and intelligence. *Biosystems* **38**(2–3), 97–109.

Dogterom, M., Maggs, A.C. and Leibler, S. (1995) Diffusion and formation of microtubule asters: physical processes versus biochemical regulation. *Proc. NatlAcad. Sci. USA*, **92** (15), 6683–6688.

Glazier, J.A. and Graner, F. (1993) Simulation of the differential driven rearrangement of biological cells. *Phys. Rev.*, **E 47**, 2128–2154.

Hogeweg, P. (1998) On searching generic properties in nongeneric phenomena: an approach to bioinformatic theory formation. In *Artificial Life VI* (Adami, C. Belew, R.K., Kitano, H. and Taylor, C.E. (eds), pp. 285–294. MIT Press.

Hogeweg, P., (2000a) Evolving mechanisms of morphogenesis: on the interplay between differential adhesion and cell differentiation. *J. Theor. Biol.*, **203**(4), 317–333.

Hogeweg, P. (2000b) Shapes in the shadow: evolutionary dynamics of morphogenesis. *Artif. Life*, **6**(1), 85–101.

Hogeweg, P. (2002) Multilevel processes in evolution and development: computational models and biological insights In Laessig M. and Valleriani, A. (eds). *Biological Evolution and Statistical Physics*, pp. 217–259. Springer.

Hogeweg, P. and Hesper, B. (1974) A model study on morphological description. *Pattern Recogn.*, **6**, 165–179.

Huynen, M.A. and Hogeweg, P. (1994) Pattern generation in molecular evolution: exploitation of the variation in RNA landscapes. *J. Mol. Evol*, **39**, 71–79.

Huynen, M.A., Stadler, P.F. and Fontana, W. (1996) Smoothness within ruggedness: the role of neutrality in adaptation. *Proc. Natl Acad. Sci. USA*, **93**, 397–401.

Kerszberg, M. and Wolpert, L. (1998) Mechanisms for positional signalling by morphogen transport: a theoretical study. *J. Theor. Biol*, **191**(1), 103–114.

Lindenmayer, A. (1968a) Mathematical models for cellular interactions in development. I. Filaments with one-sided inputs. *J. Theor. Biol.*, **18**(3), 280–299.

Lindenmayer, A. (1968b) Mathematical models for cellular interactions in development II. Simple and branching filaments with two-sided inputs. *J. Theor. Biol.*, **18**(3), 300–315.

Marée, A.F. and Hogeweg, P. (2001) How amoeboids self-organize into a fruiting body: multicellular coordination in *Dictyostelium discoideum*. *Proc. Natl Acad. Sci. USA*, **98**(7), 3879–3883.

Marée, A.F., Panfilov, A.V. and Hogeweg, P. (1999a) Migration and thermotaxis of *Dictyostelium discoideum* slugs, a model study. *J. Theor. Biol.*, **199**(3), 297–309.

Marée, A.F.M., Panfilov, A.V. and Hogeweg, P. (1999b) Phototaxis during the slug stage of *Dictyostelium discoideum*: a model study. *Proc. R. Soc. London B*, **266**, 1351–1360.

Meinhardt, H. and Gierer, A. (2000) Pattern formation by local self-activation and lateral inhibition. *Bioessays*, **22**(8), 753–760.

Nishida, H., Morokuma, J. and Nishikata, T. (1999) Maternal cytoplasmic factors for generation of unique cleavage patterns in animal embryos. *Curr. Top. Dev. Biol*, **46**, 1–37.

Prusinkiewicz, P. and Lindenmayer, A. (1990) *The Algorithmic Beauty of Plants*. Springer.

Ruoslahti, E. (1997) Stretching is good for a cell. *Science*, **276** (5317), 1345–1346.

Salazar-Ciudad, I., Sole, R.V. and Newman, S.A. (2001) Phenotypic and dynamical transitions in model genetic networks. II. Application to the evolution of segmentation mechanisms. *Evol. Dev.*, **3**(2), 95–103.

Savill, N.J. and Hogeweg, P. (1997) Modelling morphogenesis: from single cells to crawling slugs. *J. Theor. Biol*, **184**, 229–235.

Segel, L.A. (2001) Computing an organism. *Proc. Natl Acad. Sci. USA*, **98**(7), 3639–3640.

Smith, A.R. (1984) Plants, fractals and formal languages. *Computer Graphics*, **18**, 1–10.

Turing, A.M. (1952) The Chemical Basis of Morphogenesis. *Transactions of the Royal Society of London*, BR57, pp. 37–72.

van Nimwegen, E., Crutchfield, J.P. and Huynen, M.A. (1999) Neutral evolution of mutational robustness. *Proc. Natl. Acad. Sci. USA*, **96**, 9716–9720.

Section 3

The Role of Physics in Development

10 Broken symmetries and biological patterns

IAN STEWART

10.1 Introduction

The search for pattern in the natural world is one of the key motivating forces of science. Powerful techniques including deep mathematical theories, computer simulations and sophisticated experiments using complex apparatus have been developed in order to analyse patterns and how they are formed. Mostly, these methods see pattern as an accidental side effect, emerging from reductionist rules. It would represent a major step forwards if we could somehow develop a theory of pattern formation that made the role of the patterns themselves central. An example would be a theory of hurricanes in which the key conceptual element is a huge rotating spiral of gas and vapour – rather than one in which the key conceptual element is a system of equations representing the interactions of tiny packets of gas or vapour and from which the awesome reality of hurricanes emerges almost by accident.

Biology has a particular need for such a synthesis. Mainstream biology has largely concerned itself with *things* – first organisms, then cells, now the molecular structure of hormones, proteins, RNA and DNA. Because of this historical development, most biologists prefer reductionist explanations of nature, in which the forms and patterns of particular things are consequences of other things that can be found inside them. An example of such thinking is the human genome project, which was largely promoted on the assumption that DNA contains all of the 'information' needed to determine an organism, so that sequencing human DNA will tell us everything we wish to know about human biology. This view is both profound and naive: profound in that the molecular nature of genetics represents the biggest breakthrough in biological understanding since the invention of the microscope and naive in that it assumes that it is the ultimate breakthrough, and that all else will be a kind of gloss upon DNA. Recent puzzlement that the human genome contains only 35 000 or so genes, in contrast to the more than 100 000 proteins in human cells, points to a conceptual gap. This is reinforced by the recent proliferation of 'omes' – proteome, legome, transcriptome – whose avowed intention is to bridge the gaps between DNA, cells and organisms.

The idea that the explanation of biological patterns should have a mathematical element is not new. Wilhelm His (see Gould, 1987) attempted to counter Ernst Haeckel's view that 'ontogeny recapitulates phylogeny' – that embryos 'climb their own family tree' by passing through stages of development that represent evolutionary ancestors. His argued that a lot of embryo-

 ISBN 0-12-428765-4

logical development is the result of purely mechanical laws and he demonstrated it by cutting rubber tubes in various ways and folding them up to get forms very similar to those arising in embryos. His was ridiculed: Haeckel talked of a 'sartorial theory of embryology' and of the 'rag-bag' or 'rubber-tube' theory. Haeckel's own theory has turned out to be totally misconceived. His's theory may yet be rehabilitated, though in a radically different form, because many features of development include mechanical effects (Murray, 1989; Goodwin and Brière 1992; Goodwin 1994).

One of the best known early advocates of biomathematics was D'Arcy Thompson (1942), who emphasized common mathematical patterns in the inorganic and organic worlds. Conrad Waddington was less overtly mathematical, but invented concepts such as 'canalization' – a kind of inherent stability of wild-type genetics, which he explained using quasi-mathematical metaphors (Waddington, 1975). Alan Turing (1952) attempted to provide mathematical foundations for the development of form and pattern in living creatures, by specifying mathematical equations obeyed by substances called 'morphogens' which, he believed, laid down a pre-pattern for subsequent development to build on. In the 1960s the French mathematician René Thom (1975) invented 'catastrophe theory', which attempted to classify sudden changes in geometric terms – inspired to a great extent by Waddington's attempts to find a dynamic formalism for biological development. Many specific models based on his ideas were proposed by Zeeman (1977), but on the whole these did not take root among biologists. The latest episode in this long-running saga is Stuart Kauffman's work on complex adaptive systems and autonomous agents, expanding into the 'space of the adjacent possible' as rapidly as they can manage (Kauffman, 1993, 2000). There is a deep common theme: in place of molecular information and more or less arbitrary 'instructions', all of these visionaries have emphasized geometry and dynamics.

The enormous success of the approach unleashed by Crick and Watson's discovery of the DNA double-helix is undeniable and its importance should not be underestimated. However, there is a serious danger that future generations of biologists, brought up within that molecular paradigm, will ignore the associated dynamical processes of development and behaviour, believing them all to be reducible to DNA. However, DNA 'instructions' cannot be the whole developmental story, for a number of reasons, among them the following (Lewontin, 1992; Cohen and Stewart, 1994; Stewart, 1998):

- Evolution is opportunist and will exploit any structures and processes that come 'for free' as a consequence of the laws of physics; therefore these structures do not *need* complete specification within DNA and therefore they are unlikely to be so specified.
- Organisms transmit far more than just genetic material to their developing offspring. They provide a context and an environment for development – a fully functioning support system within which the genes can begin to function. Parent organisms provide materials, spatial geometry and temporal organization.
- The amount of information in DNA (i.e. the number of bases, 300 billion in humans) is insufficient to define the structure of an adult organism (e.g. the 10^{12} connections to be made in the human brain). The best it can do is define *processes* that generate the required structures. But then those processes must, to a great extent, be 'free-running'.

If these structures and processes operated transparently, then it would be reasonable to consign them to an unspecified physical 'default' background, as many biologists do. However, many crucial aspects of the biology are concealed within the subtleties of physicochemical processes and it is essential to understand how these processes work. For example, DNA specifies the amino acid sequence of a protein, but physicochemical processes prescribe what shape it folds into. Protein function is determined almost entirely by the shape, so the effect of the DNA sequence is cryptic. Most amino acid chains fold into a bewildering variety of shapes and it is virtually impossible (NP-hard, in the computing jargon) to compute the lowest-energy shape from the sequence. It appears that evolution may have favoured sequences that naturally fold in a robust manner. So here we see a deep complicity (Cohen and Stewart, 1994; Stewart and Cohen, 1997) between DNA, physics and evolution, which cannot usefully be referred back just to DNA.

Ironically, while many biologists have been rejecting the model of development as a dynamic process, seeing it as something more akin to a knitting pattern written in DNA and woven in proteins, the mathematical mainstream has been moving in precisely the opposite direction. Modern mathematical theories of dynamics and pattern-formation emphasize qualitative features and processes, rather than reductionist fine details (Langton, 1986; Stewart, 1989; Stewart and Golubitsky, 1992). Such theories may constitute a first step towards the programme of understanding pattern formation in terms of the patterns themselves – what I have elsewhere (Stewart, 1995) called 'morphomatics'. There is a growing mismatch between biologists' perceptions of mathematics (classical, rigid, deterministic, simple, quantitative, linear, rigorous) and the actuality (modern, flexible, adaptive, complex, qualitative, non-linear, speculative).

10.2 Modelling

Can this new mathematical paradigm be parlayed into a useful biomathematics? The American research funding agencies certainly hope so: they are currently calling for just that and offering a lot of money to back the call up. One reason for their enthusiasm is that the successful sequencing of most of the human genome has brought into sharp focus the uselessness of sequence information unaccompanied by any understanding of associated function. Thus *Nature*, in an article (Knight, 2002) about the need for physical insights to help biologists make sense of the vast mountains of data that they are accumulating, quotes Laura Garwin of Harvard University's Bauer Center for Genomics Research: 'What has been all too rare in biology is the symbiosis between theory and experiment that is routine in physics'.

Mathematics has been the key to physical theory and it may prove to be the key for biology too. However, mathematics designed for the problems of physics does not transfer readily to biology: DNA introduces a wild card, an element of 'control', that is not normally relevant in the physical sciences. The same *Nature* article also quotes neuroscientist John Hopfield: 'The word "function" does not exist in physics, but physicists need to learn about it, otherwise they will be playing in a sandbox all by themselves'. Ditto mathematicians. So mathematicians stand to gain a lot by thinking about biological problems and biology stands to profit from encouraging the development of appropriate mathematics.

There is, however, a perceptual gap, which is not aiding the process of developing an effective mathematical biology. Indeed, it is actively hindering that development by insisting that biology is too complicated for mathematics to be of any help. This perception is based on a misunderstanding of the role of modelling, and the techniques involved, and it is essential to clarify the issue before we proceed.

There is a tendency among biologists to insist that a mathematical model of some organic process or thing should incorporate all known features and factors and predict all behaviour under all possible circumstances. A model that emphasizes phenotype, but ignores genotype, will be dismissed because it is well known (except to ignorant mathematicians) that organisms have genes. A model that works well with large populations will be dismissed because it fails when numbers are small. A deterministic model will be criticized on the grounds that evolution is random and has no purpose; a stochastic model will be dismissed on the grounds that organisms do not throw dice.

In fact, the most important feature of mathematical models, throughout the sciences, is that they simplify a problem by idealizing it. In astronomy, planets are idealized as point particles, or perfect spheres. Real planets have craters, mountains and bulging equators. In astrophysics, the incredible complexity of a star (far more complex, biologists please note, than any organism) is simplified by making statistical assumptions about the distribution of nuclear species, assuming that the structure depends only on the distance from the centre of the star, ignoring all elements except, say, hydrogen, helium, and carbon ... whatever. It is not useful to criticize these models on the grounds of what they leave out: the test is what they predict, and how well it fits the *appropriate* aspects of reality, with what they leave in. A simple model may, for example, yield useful insights into the distribution of carbon during the early growth of a star of spectral type G3, while totally failing to capture the role of iron in its eventual destruction as a supernova. A map that is as complex as the territory is no map at all, and the same goes for models. But the price that must be paid for simplicity is a limited range of validity and a limited breadth of insight.

Admittedly – and this is important, and biologists rightly point it out – we got lucky with physics. The physical sciences reveal a lot of their secrets through idealized models; it is not clear that the same will hold for biology. The role of mathematical laws is much more overt in physics; a spherical star seems reasonable enough, whereas the proverbial spherical cow is deservedly a joke. Nonetheless, when considering the heat production of a cow and its loss through the skin, a spherical model may be entirely appropriate. The wonderful symmetric patterns of vibration that appear on a sphere, on the other hand, are unlikely to be seen on a cow. So the art of modelling is to be choosy, both about ingredients and about the type of question that the model is supposed to answer.

In this chapter, the most common idealization will be symmetry. Clearly this is absurd, taken literally. But it should *not* be taken literally, that is the whole point. A model is a metaphor. It is well known to mathematicians that a very effective way to understand an approximately symmetric system is first to idealize it to an exactly symmetric one and, only after analysing that, to take account of small variations away from exact symmetry. Often it turns out that these make no essential difference to the answer, but the idealized model puts the crucial features into sharper relief. The problem of sympatric speciation, for example, in which a population of more or less identical organisms in more or less identical environments nonetheless splits into two subpopulations with different phenotypes (Kondrashov and Kondrashov, 1999), is especially sharp if we consider a population of *exactly* identical organisms in *exactly* identical environments (Cohen and Stewart, 2000; Stewart *et al.*, 2003). That idealization does not make it *easier* for speciation to occur, but it does make the analysis of the model more tractable. Small deviations from perfect symmetry do not change the main answers, but they confuse the central issue, which turns out to be instability of the single-species state in the presence of selective forces such as assortative mating and environmental stress.

I have picked symmetry as the unifying theme here in order to restrict the discussion to one identifiable area of mathematics. It is not the only concept that might prove useful in biomathematics; it is not even the most important. The list of topics that are crying out for mathematical input is enormous: protein folding, enzyme activity, gene regulation, mechanics of the cytoskeleton, cellular dynamics, ecosystem dynamics, evolution, social behaviour ... But symmetry is a topic with which I am most familiar and it provides an accessible context for discussion of the issues and a reasonable instance of the potential for mathematical modelling in the biosciences.

10.3 Symmetry

The concept of symmetry is central to the mathematical imagery that I wish to invoke (and, on a more technical level, to the rigorous analysis that lies behind it). But what is symmetry? In 1952 the mathematician Hermann Weyl was about to retire from the Institute for Advanced Study at Princeton, and to mark the occasion he gave a series of public lectures on symmetry (Weyl, 1969). His first lecture opens with these words:

> *If I am not mistaken, the word* symmetry *is used in our everyday language with two meanings. In the one sense symmetric means something like well-proportioned, well-balanced, and symmetry denotes that sort of concordance of several parts by which they integrate into a whole. Beauty is bound up with symmetry ... The image of the balance provides a natural link to the second sense in which the word symmetry is used in modern times:* bilateral symmetry, *the symmetry of left and right, which is so conspicuous in the structure of the higher animals, especially the human body. Now this bilateral symmetry is strictly geometric and, in contrast to the vague notion of symmetry discussed before, an absolutely precise concept.*

Weyl's informal description relates to the standard mathematical characterization of bilateral symmetry: a shape is bilaterally symmetric if there exists a reflection that leaves it *invariant* – that is, unchanged in appearance. Reflection is a mathematical concept – but it is not a shape, a number, or a formula. It is a *transformation*, a rule for moving things around. So a symmetry of an object or system is a transformation that leaves it invariant.

There are many other types of symmetric shape, for example an idealized five-limbed starfish. (I make no claim that a real starfish is perfectly symmetric. Nonetheless, any discussion of its form has to come to terms with its astonishing approximate symmetry. Pointing out that the symmetry is not exact does not mean that the question of form has been answered.) The most obvious symmetry of a starfish is the transformation that makes all the limbs 'click one space on': a *rotation* through an angle of 72°. In fact there are precisely five different angles through which a starfish can be rotated without the change being detectable afterwards: 0°, 72°, 144°, 216° and 288° – the integer multiples of one fifth of a turn. Informally, we say that the starfish has *fivefold rotational symmetry*. These rotations are the most immediate symmetries of a starfish, but there are five more. Each limb is bilaterally symmetric and the limb's symmetry axis passes through the centre of rotational symmetry. Moreover, the entire starfish is bilaterally symmetric about this axis; indeed it has five different bilateral symmetries, because each of the five arms has its own symmetry axis. There are thus five distinct reflections that leave a starfish invariant and their axes are inclined at angles of 72° to each other. In total, a starfish has exactly ten symmetries (Figure 10.1).

This insight, that objects possess not just symmetry, but *symmetries*, means that symmetry can be given a quantitative aspect. For example we can prove that the symmetry of a starfish is different from that of a human being, for starfish have ten symmetries and we have only two. (Again, we are talking idealizations. Departures from symmetry are also important; for example, many medical conditions in humans involve such departures. But how do we sensibly discuss *departures* from symmetry if we do not discuss symmetry itself?)

The most important types of symmetry are *rigid motions* of space. Here the 'things' are the points of space – the line, the plane, three-dimensional space, or spaces of higher dimension. 'Rigid' means that the images of any two points are the same distance apart as the points themselves and the word 'motion' is the traditional term to describe a transformational process related to everyday experience that produces the correct end result.

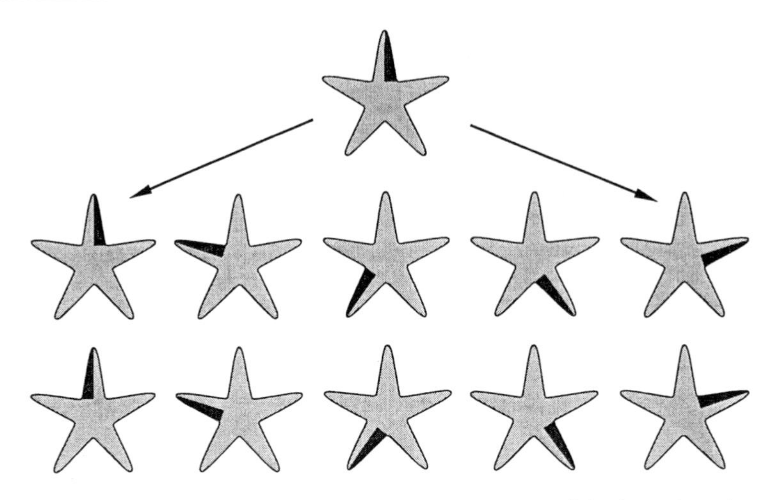

Figure 10.1 The ten transformations that leave a starfish shape invariant.

There are two types of rigid motions of a one-dimensional space – a line. A *translation* slides the line along, either to left or right. A *reflection* flips the line over, interchanging left and right. In space of two dimensions – the plane – translations slide the entire plane some chosen distance in some chosen direction. There are also *rotations* through some angle about some fixed centre. Translations and rotations keep the plane the same side up; but we can also flip it over, using reflections in some chosen axis. We can also combine a translation with a reflection. This is a new kind of motion if the translation is parallel to the mirror, when it is called a *glide reflection*. In three dimensions we again find translations, rotations, reflections and glide reflections but, in addition, if a rotation about some axis is combined with a translation parallel to that axis, we obtain a helical motion known as a *screw*.

The understanding that symmetries are best viewed as transformations dawned upon mathematicians in the early 1800s, when they realized that the set of symmetries of an object is not just an arbitrary collection of transformations, but has a beautiful internal structure. The starfish provides an example. There are ten distinct transformations that leave a starfish invariant: five rotations (one trivial) and five reflections. Suppose that two of these transformations are applied, one after the other. Each leaves the starfish apparently unchanged, so the final result also leaves it apparently unchanged. That is, *the result of performing two successive symmetry transformations must be another symmetry transformation*. For example, rotate the starfish by one fifth of a turn, and then by a further two fifths of a turn. The combination of the two rotations has exactly the same effect as rotating it three fifths of a turn. But that is just another symmetry of the starfish.

This property holds in considerable generality. For any shape whatsoever, any two of its symmetries, when combined together, must yield another symmetry. (If we leave something looking unchanged twice, then we leave it looking unchanged.) To express this fact, we say that the symmetries form a *group*. A group is a closed system of transformations: whenever two of them are combined, the result is another member of the same group. We call this particular group the *symmetry group* of the chosen object. Groups are very special sets of transformations, so the fact that the symmetries of an object form a group is significant. It leads to a natural and elegant 'algebra' of symmetry, known as Group Theory (Rotman, 1984; Sattinger and Weaver, 1986).

10.4 Symmetry breaking

According to an old principle of Pierre Curie, symmetric causes produce equally symmetric effects. The principle is defensible for exact symmetries, but in the real world, symmetries are never exact. In the presence of small asymmetric perturbations, such as thermal noise or the gravitational attraction of a small lizardoid on a planet in the Andromeda galaxy, symmetric causes can (appear to) produce less symmetric effects. This phenomenon is called *spontaneous symmetry breaking*. It is most easily grasped using a physical example and a good one is the formation of sand dunes.

Begin with a perfectly flat, uniform desert. Suppose that a uniform wind of constant speed blows across the desert in a fixed direction. This is a highly symmetric system. Modelling the desert by an infinite plane, to remove edge effects, the system is symmetric under all rigid motions of the plane that preserve the wind direction. The only state of the desert that possesses this much symmetry is the flat, uniform state.

However, deserts do not, in practice, stay flat. Typically, the sand builds up into ridges and hummocks: sand dunes. The commonest patterns of dunes are long, parallel ridges, like ocean waves; these are known as transverse dunes and they form at right angles to the wind. Other patterns include fields of crescent-shaped barchan dunes, wavy barchanoid ridges and dome dunes (Figure 10.2). All of these patterns are (in an idealization) symmetric, with differing symmetries; none of them has the full symmetry of the original flat desert. Why not?

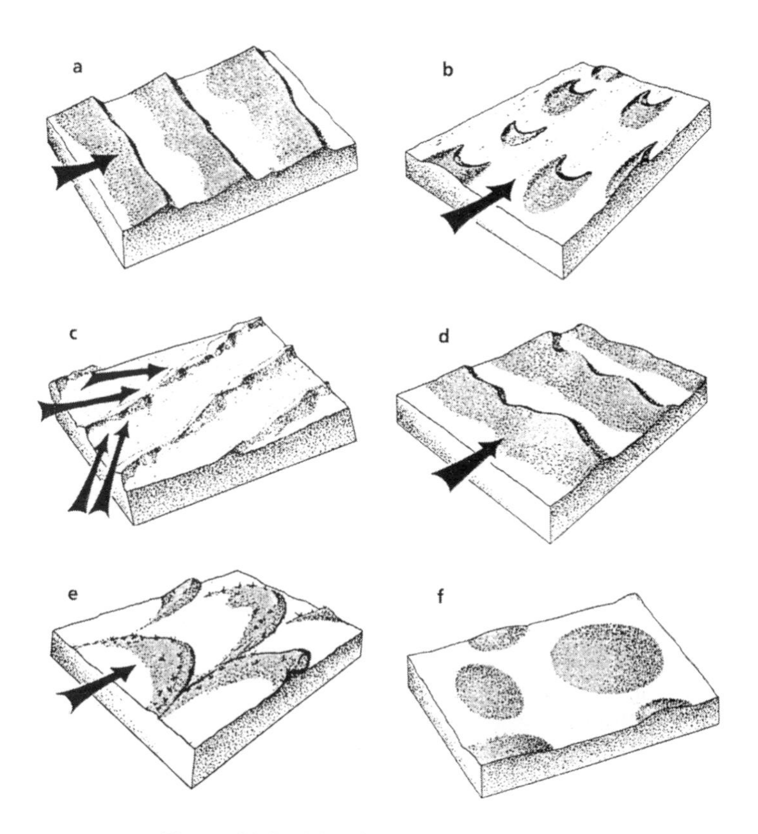

Figure 10.2 Morphology of sand dunes.

The answer is that the uniform state of the desert is unstable. The slightest deviation from flatness will grow exponentially. For instance, if one grain of sand pokes out a little, the wind will form a tiny vortex. The vortex will scoop out a hollow in the sand and pile the sand up nearby. Now we have a larger obstacle and the process continues, building up a substantial dune. Tiny non-uniformities elsewhere also build local dunes. Now the flow of wind over the system of incipient dunes and the consequent transfer of sand, couples the dunes together, so that their collective dynamic becomes more important than the small differences that triggered their initial formation. By using a combination of group theory and dynamics, it can be shown that the typical outcome is a pattern with some of the symmetry of the original system – perhaps none, but certainly not all.

The physical ingredients here are symmetric laws, plus instability of the symmetric solution. The main mathematical ingredient needed to represent the process of symmetry breaking adequately is *non-linearity*. Symmetry breaking does not occur in so-called linear dynamical systems, in which solutions of the equations are unique: in linear systems, the uniform state is a solution and, since there are no others, that is the only thing that can occur. Non-linearities create the possibility of coexisting solutions, which can have different symmetries. Now changes of symmetry become possible.

A good rule of thumb (though not, unfortunately, a generally valid theorem) states that symmetry breaks 'reluctantly'. The system retains as much symmetry as it can. As a result, we often observe a series of breakages of symmetry, in which more and more symmetries are lost as the system is more highly 'stressed'. For example, the uniform desert breaks symmetry to transverse dunes, which are symmetric under all translations perpendicular to the wind, but only under translations by multiples of a wavelength in the direction of the wind. Then the transverse dunes break up into wavy barchanoid ridges, which are no longer symmetric under all translations perpendicular to the wind, but only under translations perpendicular to the wind by multiples of a second wavelength. At each step in the cascade, the symmetry group gets smaller.

Broken symmetries are an example of the phenomenon of *bifurcation*, in which the solutions of a differential equation change in a major, qualitatively different way, when numerical parameters in the equations change continuously (Guckenheimer and Holmes, 1983). In the symmetry-breaking case, the qualitative change is a change of symmetry. Other common possibilities are the appearance or disappearance of an equilibrium, a change from equilibrium to a periodic state and various scenarios leading to chaos. A visual image here associates these different bifurcations with the boundaries between various regions in *parameter space*, a mathematical space whose coordinates are the different parameters. When the parameters vary continuously, the associated point in parameter space traces out a curve; when the curve crosses a boundary, a bifurcation occurs. So bifurcations are simultaneously rare (most points in parameter space do not lie on boundaries) and unavoidable (a typical curve is likely to cross a boundary if it is extended sufficiently far). Many conceptual difficulties associated with sudden changes of behaviour in slowly varying systems can be resolved by understanding bifurcations.

10.5 Images of morphogenesis

The dune-forming system provides a simple model not only for the formation of patterns, but for the sequential development of different patterns. The model is that of a dynamical system (with symmetries), whose state is determined by a number of 'internal' variables, and governed by a number of parameters. For simplicity of exposition we assume two timescales: a 'fast' timescale for the internal dynamic variables and a 'slow' timescale for changes to the parameters. That is, the experimentalist (or the surrounding universe) alters the parameters gradually and lets the system react to them before altering them again. In practice, this separation is probably not as clear-cut, but the end result is similar. A developmental sequence corresponds to a path through parameter space, determined by the slow changes to the parameters. In mathematics we call this a *developmental path*. Because the internal variables respond quickly, the system's state evolves through a sequence of states, each having its own characteristic types of dynamics and symmetry.

A state may be dynamically simple, the simplest of all being equilibrium. More complex states include periodic cycles, quasiperiodic dynamics in which cycles of different periods are combined and chaos. In all cases, the dynamics can be described as motion on an *attractor* (Thompson and Stewart, 1986). For equilibria, the attractor is a single point; for a periodic cycle it is a closed loop; for quasiperiodic dynamics it is a torus; for chaos it is a fractal (Mandelbrot, 1982; Peitgen *et al.*, 1992) Whatever the initial conditions, the system will rapidly approach some attractor and behave as if it is following a dynamic trajectory on that attractor. In the 'classical' cases of a point, a loop, or a torus, the resulting dynamics are very regular. In the chaotic case, where the attractor has a complex fractal structure, the dynamics are generally highly irregular and can appear random in many respects.

When the representative point on the developmental path is in the interior of one of the regions into which parameter space is divided, well away from transition boundaries, we see only a slow variation in the attractor, with gradual quantitative changes and no qualitative changes. But whenever the developmental path crosses a transition boundary, the attractor – and hence the observed dynamics and pattern – changes its qualitative nature.

As Thom (1975) noted, morphogenesis possesses exactly the features deduced from the model: generally continuous quantitative change without qualitative change, plus occasional sudden changes in the qualitative morphology. Goodwin (1994) suggests that the role of genes is to set parameter values (and perhaps to select between alternative attractors and stabilize 'useful' dynamics that might otherwise be unstable). This theory explains why the forms and patterns adopted by organisms generally resemble those that are dynamically 'natural', but why evolution is able to select virtually any pattern from the available 'zoo'. Of course genetic parameter-selection can produce very specific forms if that is evolutionarily desirable. Lewis Wolpert pointed out one example to me at a conference in Spello in 1993: the markings of birds, which are extraordinarily rich and somewhat arbitrary. Here strong evolutionary pressures (sexual selection?) may have caused the birds' genes to modify, to an unusual extent, what would otherwise be the natural dynamic of pattern formation.

Be that as it may, we can often interpret pattern formation and development as a *symmetry-breaking cascade* – a sequence of changes in the symmetry of the developing organism. We use this term somewhat loosely: symmetries may increase as well as break, and they may only be localized or approximate. Nonetheless, it is a striking image that bears a close resemblance to the development of actual organisms. Here we briefly interpret two sequences of development in this language. The first is the single-celled marine alga *Acetabularia acetabulum*; the second is a laboratory classic, the frog *Rana pipiens*.

Acetabularia begins as a spherical zygote (whose external form has spherical symmetry, although certain symmetry-breaking gradients already exist in the internal chemistry). It puts out a rootlike structure and a stalk whose cross-section is approximately circular, breaking the spherical symmetry but retaining circular symmetry. The stalk grows and generates a ring of small hairs called a *whorl*. The symmetry has broken again, replacing the continuous rotations of the circle with n-fold rotational symmetry, n being the number of hairs. The tip continues to grow from the centre of the whorl, producing a sequence of whorls that are roughly equally spaced along the stalk. Here the growth region alternately regains and loses circular symmetry, while the overall symmetry of the stalk breaks from all translations (a uniform featureless stalk) to discrete translations (repeated features, the whorls). Then it develops a substantial capped structure, which gives rise to its popular name 'mermaid's cap', which has a different discrete rotational symmetry (Figure 10.3).

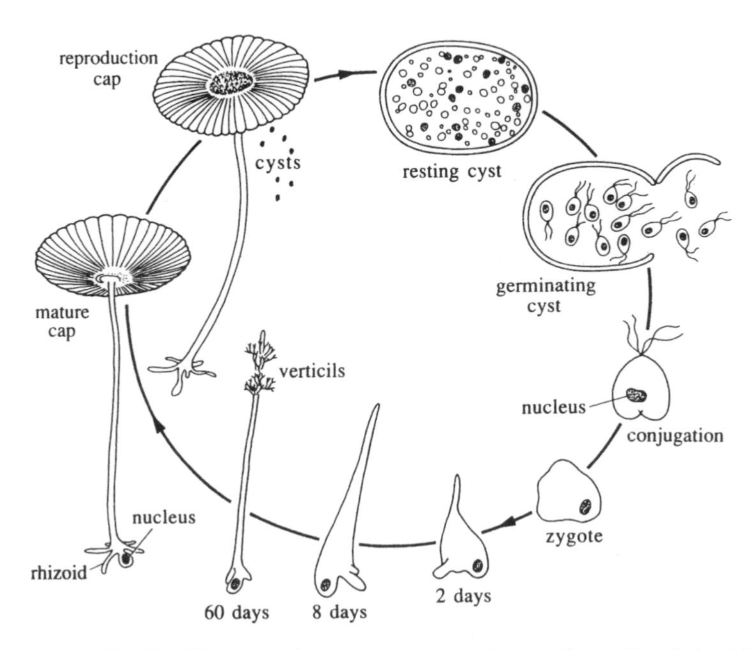

Figure 10.3 The life cycle of *Acetabularia acetabulum* (from Goodwin, 1994).

After the initial breaking of spherical symmetry to circular (or more appropriately cylindrical), most of these stages have symmetries that are much the same as those encountered in the main sequence for Couette-Taylor flow – as they should be since such patterns are universal in cylindrically symmetric systems. The undecorated stalk corresponds to Couette flow, the whorls to wavy vortices, the cap to a wavy vortex with a different number of waves. Goodwin and Trainor (1985) and Goodwin and Brière (1992) have developed equations which show that the formation of the whorl is caused by a broken symmetry in the distribution of calcium within the organism, which in turn affects its local growth rate and hence its shape. (See also Chaplain and Sleeman (1990), who apply membrane theory to tip morphogenesis.)

The development of the frog also goes through a series of well-defined changes of symmetry. The spherical symmetry of the zygote (stage 1 in Figure 10.4) breaks successively, first to circular symmetry plus an up/down reflection (stage 2), then to a discrete rotational symmetry through a right angle, plus up/down reflection (stage 3), then to right-angle rotational symmetry plus up/down reflection (stage 4), then to right-angle rotational symmetry (stage 5). After this the changes to symmetry are local (and symmetry is probably not the best terminology), until stage 9, when approximate spherical symmetry is restored. Stage 11 breaks this to circular symmetry and by stage 14 it has broken further to bilateral symmetry. Bilateral symmetry is retained right through to the adult organism, except that the internal organs sometimes break it. This time the changes are typical of those observed in systems with spherical symmetry (Thompson and Stewart, 1986).

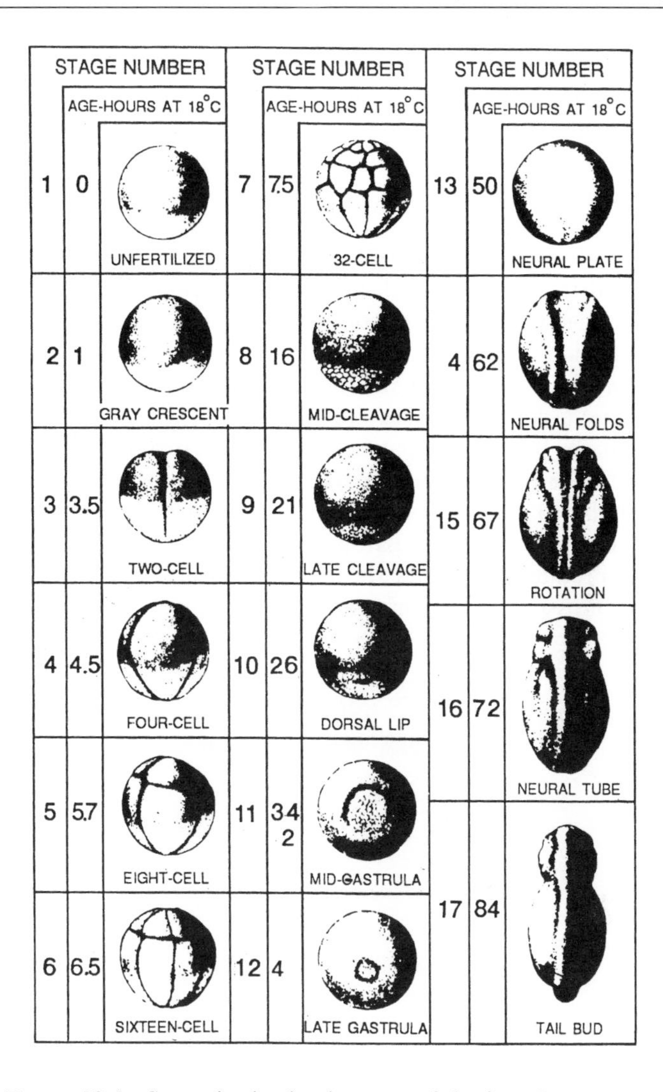

Figure 10.4 Stages in the development of the frog *Rana pipiens*.

10.6 Pattern formation

The markings of animals are also broadly organized by symmetry-breaking cascades. They often form striking geometric arrangements such as spots, stripes and dappling. Recent research on the molecular genetics of the fruit fly *Drosophila* reveals striped patterns of gene activity, broadly similar to the patterns seen in developing embryos. The genes that specify the structure of the embryo 'switch on' in a spatially patterned manner, controlled by so-called *Hox* genes.

DNA is thus in some sense the 'blueprint' for morphogenesis. However, it takes more to build a car than just an engineer's blueprint; similarly, it takes more to build a tiger than just a list of instructions for what proteins to make and where to put them. They have to be made and put. The remarkable mathematical regularities in the morphology of living creatures suggests that the laws of physics and chemistry may play a major role in morphogenesis, rather than just being passive 'carriers' for genetic instructions.

In 1917, D'Arcy Thompson (1942) attempted to explain the shape of a jellyfish by analogy with drops of gelatin falling through water. Implicitly, he was modelling jellyfish development in terms of fluid dynamics and explaining jellyfish symmetries in terms of the symmetries found in fluid patterns. Of course, it is not that simple, but the mechanical forces between jellyfish cells do play a role in the animal's shape, nevertheless. Alan Turing (1952) made the role of geometry explicit by capturing it in mathematical equations and showed that ordered patterns do not need a patterned precursor. He argued that a system of chemical substances, which he called morphogens, reacting together and diffusing through a tissue, could explain the formation of patterns. Specifically, he suggested that the distribution of chemicals within the tissue should obey so-called 'reaction-diffusion' equations. When Turing solved these equations he found that patterns form *spontaneously* (Figure 10.5). The homogeneous state (identical concentrations of chemicals everywhere) becomes unstable and any slight lack of uniformity in the distribution of chemicals grows rapidly, because the patches are coupled together by diffusion, which causes them to arrange themselves into coherent spatial patterns resembling spots, stripes and other geometric textures. It is a kind of chemical symmetry breaking. If pigments are deposited according to the peaks and troughs of parallel waves, like Taylor vortices, you get stripes. More complex waves produce spots, and so on. Experiments with a phthalocyanine reaction on a silver substrate (Ourisson, 2002) demonstrate that complex patterns, very similar to those

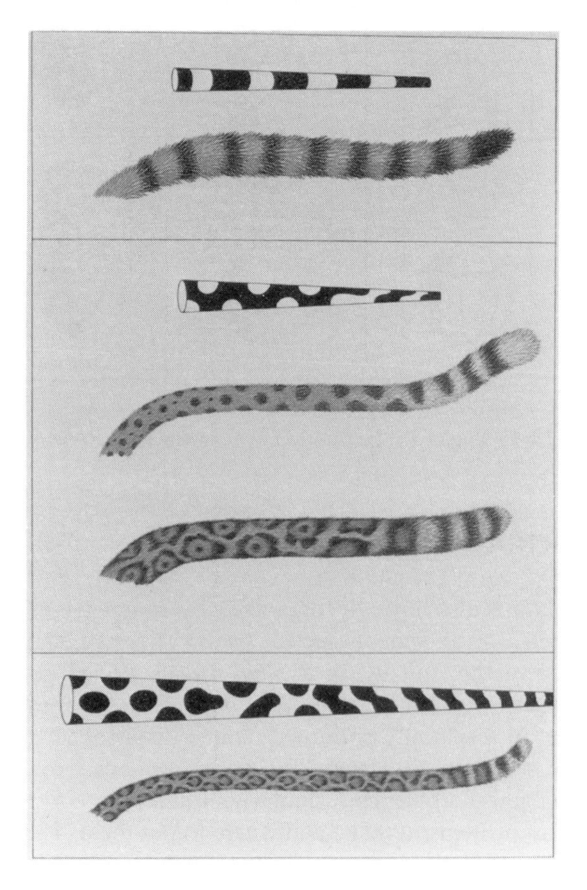

Figure 10.5 Spontaneous breaking of symmetry in models of patterns on the tails of cats (after Murray, 1989).

found in, for example, the spots on a leopard, can form via spontaneous symmetry breaking. These patterns are more subtle than the obvious, highly symmetric patterns that are most tractable mathematically: this is a form of spatial chaos in a symmetric system and it shows that the mathematics of symmetry breaking goes well beyond the relentlessly regular patterns that we usually associate with the word 'symmetry'. Form, rather than pattern, can be controlled in a similar way, by growing bumps and dents according to the chemical prepattern and again the patterns need not be perfectly regular.

Maynard Smith and Sondhi (1960) showed that hairs on the fruit fly *Drosophila* occur in a variety of Turing-like patterns whose genetic variants are *also* Turing patterns. It is hard to explain this if patterns are *arbitrary* consequences of DNA codes – why should natural selection always select Turing patterns? However, this initial success turned to failure as other systems, such as feather development, fell apart under scrutiny. For example, when feathers are grown at different temperatures, the observed quantitative changes in their patterns do not fit Turing's equations. By the 1970s most biologists had got bored with finding Turing patterns that weren't, and moved on. Instead, they concentrated on the DNA code and its implementation. Wolpert (1969) developed the theory of 'positional information', in which, effectively, each cell is equipped with a map and a book. The map tells the cell whereabouts in the organism it is and the book tells it what genes to switch on when it is in a given place. The map is provided by chemical gradients – regular changes in chemical concentrations. The book is the organism's DNA code.

However, it can be argued that the dismissal of Turing's model was based on too narrow an interpretation of its specific form. Because symmetry breaking is a universal phenomenon, *all* systems like Turing's – 'mechano-chemical' equations describing the interaction of chemical changes and tissue growth – tend to generate the same range of patterns. Indeed the very creation of the 'map' – a system of chemical gradients – is itself a form of symmetry breaking, as Scott Kelso pointed out to me in 1993. An extensive theory of this kind of emergence of pattern has been pioneered by Stuart Kauffman (1993, 2000). He criticizes the orthodox emphasis on ever more detailed aspects of DNA chemistry, and argues that:

> *Both the prepattern concept and the latter-day theory of positional information rely on a conceptual separation between positional-information assessment and the subsequent interpretation of that information by the cell, with no provision whatsoever for the obvious possibility that the very interpretation made by the cell might feed back and modify the information. Yet ... such phenomena are common, not rare.*

A variety of dynamical models of particular developmental processes are now beginning to appear. As well as Goodwin's model of *Acetabularia*, there is work by Maini *et al.* (1991, 1992) generalizing Turing's scheme to include more realistic biology. One application is the development of the skeleton in vertebrate limbs. A similar model, which conforms with certain grafting experiments, has been devised by Dillon and Othmer (1993). Three especially successful applications of Turing-like models deserve more detailed mention here. They are stripes in angelfish, spirals in slime mould and patterns on seashells. We describe them in turn.

Tropical fish are typically brightly coloured and highly patterned. Several species of angelfish display striking stripes, notably *Pomacanthus imperator*, which has black-and-white stripes running the length of its body. Kondo and Asai (1995) noticed that these patterns are seldom perfect; instead, they exhibit 'dislocations' at which a stripe may suddenly stop, or one stripe may split into two. When they investigated these dislocations from the point of view of Turing models, they were led to a surprising prediction: the stripes ought to *move*. When they photographed the same angelfish many times over a period of 3 months, they discovered that the

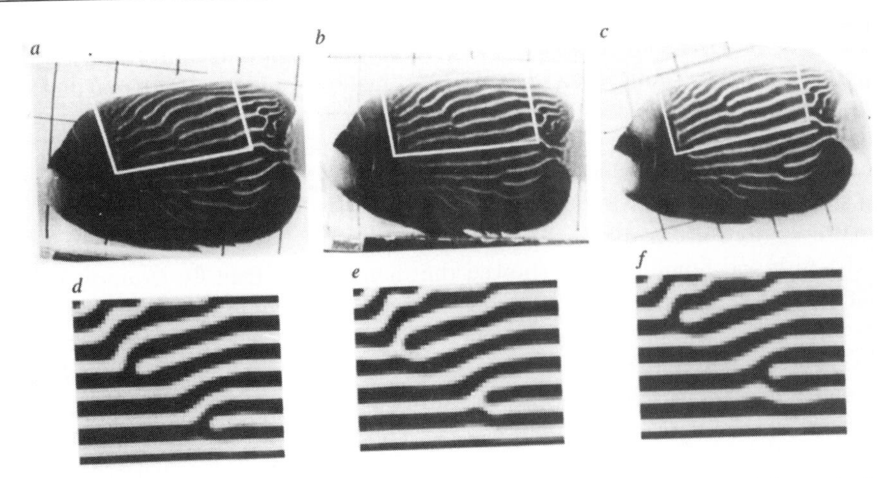

Figure 10.6 Moving stripes of an angelfish: experimental observations (above) and theoretical predictions (below) (from Kondo and Asai, 1995, reproduced by permission from Nature Publishing Group).

stripes do indeed move. Moreover, the way they move and the way the pattern changes with time is exactly as predicted by computer solutions of Turing's equations (Figure 10.6).

The slime mould is an amoeba and it is huge colonies of these amoebas that have a slimy appearance. At a certain stage in the life cycle of slime mould, groups of amoebas come together to form a mobile colony known as a 'slug'. The slug migrates to somewhere dry and puts up a round 'fruiting body' on a long stalk. Amoebas in the fruiting body dry into spores and blow away on the wind; when they land somewhere wet they turn back into ordinary amoebas and start the cycle over again. During the stage at which the amoebas are joining together to create the slug, they form characteristic spiral-like patterns (Figure 10.7). It has been known for some time that very similar patterns can form in chemical systems, especially the celebrated Belousov-

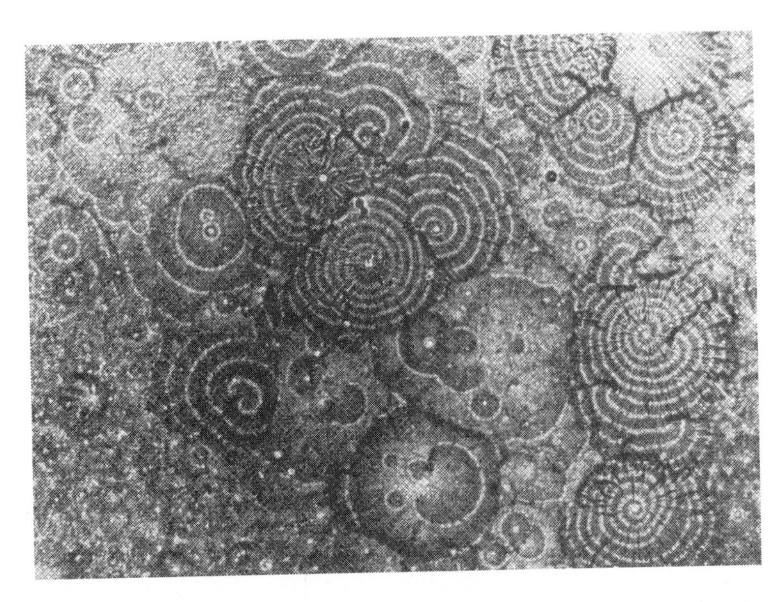

Figure 10.7 Slime mould aggregation patterns (courtesy of Peter Newell, University of Oxford).

Zhabotinskii reaction. Moreover, those systems can be well modelled by Turing equations. Höfer (1996) and Höfer and Maini (1997) discuss a system of three equations, much like Turing's, which correctly reproduces not only the spiral patterns, but the treelike 'streaming' patterns that set in later in the same process. These equations show that this part of the behaviour of the slime mould amoebas reflects natural patterns in inorganic systems, rather than being programmed in by DNA code instructions.

In his beautiful and highly illustrated book, Meinhardt (1995) has made an extensive study of the markings on seashells of many different species. He shows that virtually all of the patterns found in seashells, including various irregularities, can be accounted for using simple equations that are variations on Turing's scheme. Meinhardt (1999) discusses many aspects of development in a similar mathematical framework, along with experimental evidence. A noteworthy feature is the inclusion of gene activation by morphogens, which relates the model more closely to features of biological importance. The interplay between genes and free-running dynamics is instructive.

Kauffman (1993) has devised yet other models of morphogenesis. One of the simplest is a line of cells equipped with a rudimentary internal dynamic governing cell division, in which cells 'communicate' with each other using microhormones. Computer simulations show that regular patterns can form spontaneously in such systems. What we see here is a combination of the dynamic ideas of Turing with the controlling role of DNA, not unlike our model of genes as parameters and development as dynamics. Yet another promising approach is the idea that a developing organism is a 'cellular automaton' with real cells, which can move, divide, die, change shape ... The beginnings of a theory of these systems exists, in the form of the Cell Progamming Language of Agarwal (1995). As Agarwal says, 'These programs are an estimate of the minimal information needed to model realistically ... developmental systems'. His paper contains numerous examples of pattern formation in such systems.

10.7 Animal locomotion

A widespread case of symmetry breaking, which is clearly of dynamical origin, occurs in animal gaits (see Schöner *et al.*, 1990; Collins and Stewart, 1992, 1993a, b). Like cars, animals sometimes 'change gear'. At low speeds horses walk, at higher speeds they trot and at top speed they gallop. Some can also canter. Humans walk at moderate speeds, but run if they need to move quickly. The differences are fundamental: a trot is not just a fast walk, but a different kind of movement altogether. These distinct patterns of leg-movement are known as *gaits*.

Hildebrand (1966) observed that most gaits possess a degree of symmetry. For example when an animal *bounds*, both front legs move together and both back legs move together. Other symmetries are more subtle: for example when a camel *paces*, its left half follows the same sequence of movements as the right half, but half a period out of phase – that is, after a time delay equal to half the gait period (Figure 10.8). This is an example of symmetry breaking: the gait of a bilaterally symmetric animal need not be bilaterally symmetric.

However, such a gait has its own characteristic symmetry. If we define the *phase* of a periodic oscillation to be the fraction of a period that has passed relative to some fixed reference value, then that symmetry is 'interchange left and right sides and shift phase by half a period'. Humans use exactly this type of symmetry breaking to move around: for obvious reasons, they do not move both legs simultaneously.

Why do gaits have patterns? Recently it has become clear that there is a striking analogy between certain 'universal' symmetry-breaking patterns found in systems of coupled oscillators and animal gaits. This suggests that the patterns are natural consequences of the structure of the

Figure 10.8 The pace of the camel breaks the animal's bilateral symmetry (Gambaryan, 1974).

animal's neural circuitry and gives a few clues about how such circuits might be organized. It also opens up a simple evolutionary route for the development of locomotion.

There are two basic bipedal gaits. The two legs can be *out of phase* with each other, that is, doing the same thing but hitting the ground at different times. Walking and running are examples in humans. Alternatively, both legs can be *in phase*: they do the same thing at the same time. Two-legged hopping is an example of this.

With more legs to operate, there are more possibilities for gaits. Here are the eight commonest quadrupedal gaits,together with a potted description. The phase relationships are summarized in Figure 10.9.

- *Walk*. The legs move a quarter of a cycle out of turn in a 'rotating figure-8' wave.
- *Trot*. Diagonal pairs of legs move together and in phase. One diagonal pair is half a period out of phase with the other pair.
- *Pace*. The left legs move together and in phase. The right legs move together, half a period out of phase with the left legs.
- *Bound*. The front pair of legs move together and in phase. The back legs move together, half a period out of phase with the front pair.
- *Transverse gallop*. This gait resembles the bound, but the feet of the front and back pairs are slightly out of phase with each other.
- *Rotary gallop*. This is similar to transverse gallop, except that the left and right back legs have swapped patterns.
- *Canter*. The right front/left back legs move together and in phase. The left front/right back legs move half a period out of phase with one another, and a rather arbitrary amount out of phase with the strongly coupled diagonal pair.
- *Pronk*. All four legs move at the same time.

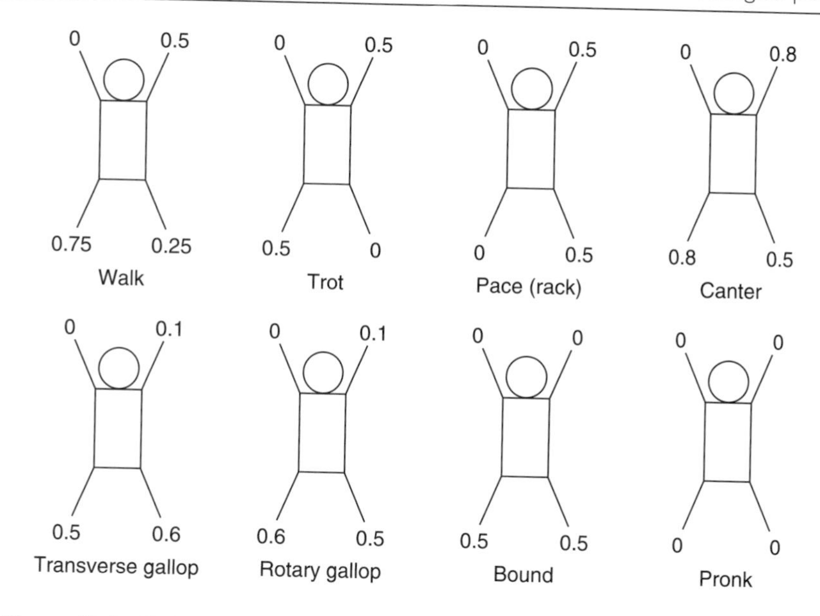

Figure 10.9 Relative phases of quadruped limbs in the eight commonest gaits.

Animals' brains control their movements by sending electrical signals along the neurons, or nerve cells. The nerves hook together a bit like the components of an electronic circuit. Theorists have suggested that locomotion could be controlled by a 'central pattern generator', or CPG. A CPG is a neural circuit – not necessarily in the brain itself – that produces rhythmic behaviour. Von Holst (1935) showed experimentally that an assembly of neurons in the spinal cord can coordinate fine movements in some fish, without receiving any messages from the brain.

What kinds of symmetry can a gait have? The prime message from the mathematical theory of symmetry breaking (Golubitsky *et al.*, 1988; Golubitsky and Stewart, 2002) is that the symmetries of oscillating systems operate both in space and in time. We thus distinguish two types of symmetry. The first, *spatial* symmetry, refers to permutations of the legs (or, more fundamentally, of the corresponding neuron systems in a CPG). The second, *temporal* symmetry, involves patterns of phase-locking. The symmetries observed in gaits specify specific phase relationships between the various legs. They are mixtures of space (swap legs) and time (shift phase).

The basic idea behind CPG models is that the rhythms and phase relations of animal gaits are determined by relatively simple neural circuits. Neural circuits can be built to any reasonable pattern, but the structure of the CPG should, to some extent, mimic the gross physiological structure of the whole animal. Networks of oscillators with various 'natural' symmetries successfully reproduce the observed gaits. The latest such network (Figure 10.10), suggested by Golubitsky *et al.* (1998, 1999), overcomes certain technical difficulties in previous such models. It has the unusual feature of using twice as many oscillators as there are legs – eight oscillators to control a quadruped, twelve to control a hexapod. This network is the only one that can overcome certain dynamical obstacles to agreement with experiment, within a particular context; however, Li (2002) has proposed a 4-unit network for quadruped gaits which overcomes these obstacles in a different way.

Figure 10.11 shows the range of gaits predicted by this network for a quadruped. Another advantage of this network is that it provides a simple explanation of the dynamical origins of less symmetric gaits such as the gallop (transverse or rotary) and the canter: they are 'mode

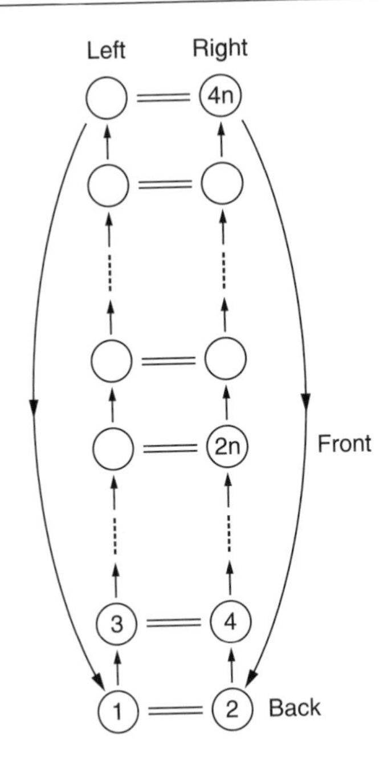

Figure 10.10 Modular network for $2n$-legged locomotion using $4n$ oscillators.

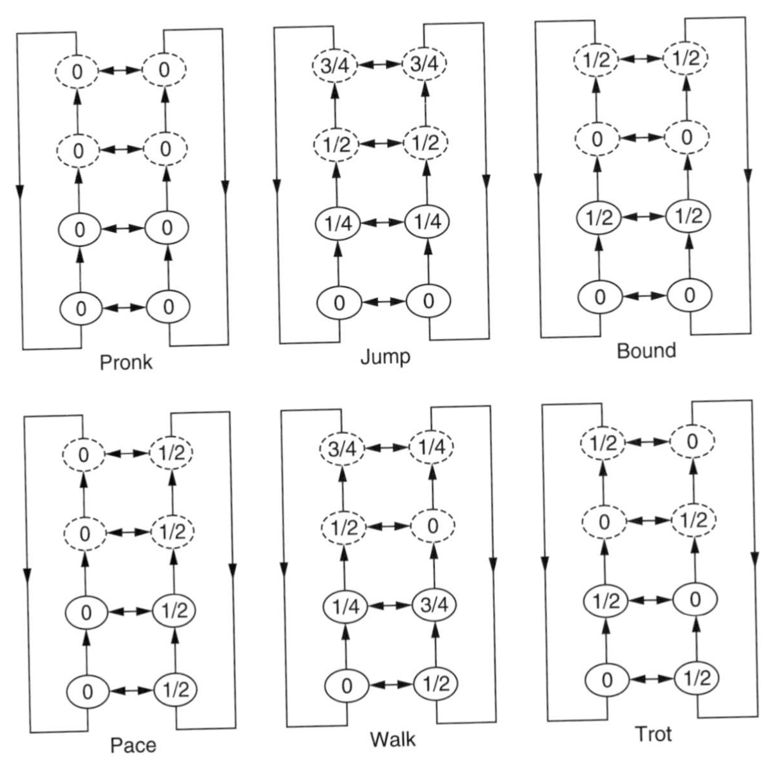

Figure 10.11 Quadruped gaits predicted by the network in Figure 10.10.

interactions', combinations of simpler and more symmetric gaits. The network applies to any number of legs, and it has a simple modular structure that may be of interest for the design of legged robots.

A further point is that symmetry-breaking transitions (occurring across boundaries in parameter space) provide a natural rationale for changes in gaits. The faster the animal moves, the less symmetry its gait has: more speed breaks more symmetry. This is consistent with the general phenomenon that the more a system is 'stressed', the less symmetry its states possess. Whatever the physiological constraints on locomotion may be, the remarkable variety of natural gaits basically boils down to the standard mathematical patterns of symmetry breaking in networks of coupled oscillators. This is a valuable unification; moreover, the idea extends to other types of movement and to other creatures. The flight of birds involves bilaterally symmetric oscillations of two wings. The theory extends readily to hexapods – six-legged creatures, such as insects. For example, the typical gait of a cockroach is the *tripod*, in which the middle leg on one side moves in phase with the front and back on the other, and then the other three legs move together, half a period out of phase with the first set. Mathematical analysis reveals this as one of the natural patterns for six oscillators arranged in a hexagon. Myriapods (centipedes and millipedes) produce rippling patterns of leg movements These can be understood as symmetric travelling waves in large networks. Some microscopic single-celled creatures propel themselves along by rotating a helical tail, or flagellum, just like the mechanical device known as an Archimedean screw, which – appropriately – has screw symmetry. So does the curious gait of the snake known as a sidewinder, which flips itself forwards and sideways like a rolling coil of a spring.

10.8 Where next?

Non-linear dynamical systems are representative of a broad variety of mathematical systems, all of which possess similar features. These include big qualitative changes brought about by small changes to parameters, symmetry breaking and the consequent spontaneous formations of pattern and structure, order and chaos occurring on an equal footing, and apparently distinct behaviour exhibited by identical individuals. All of these features arise in physical systems and there they can be directly related to conventional mathematical models based on differential equations. They are real, ubiquitous and important.

Early mathematical equations for the development of organisms were too far removed from real biology to provide accurate models. However, the current emphasis on DNA goes too far the other way: it explains the production of proteins, but it does not adequately explain how they are assembled to form an organism, or – crucially – why nature so often prefers mathematical patterns of the symmetry-breaking variety. To see the difference between the two approaches and how both fall short of reality, imagine a vehicle (corresponding to a developing organism) driving through a landscape (respresenting all the possible forms that the organism might take, with valleys corresponding to common forms and peaks to highly unlikely ones). In models like Turing's, once you have set the vehicle rolling, it has to follow the contours of the landscape. It cannot suddenly decide to change direction and head uphill if the 'natural' dynamic is to continue straight ahead into the nearest valley. In contrast, the current view of the role of DNA sees development as an arbitrary series of instructions: 'turn left, then straight ahead a hundred metres, then turn right; stop for ten seconds; reverse five metres; turn left ...' *Any* destination is possible given the right instructions, and no particular destination is preferred.

The true picture, however, must combine genetic 'switching' instructions and free-running mechano-chemical dynamics. If a car driving through a fixed landscape follows an arbitrary series of instructions, it is very likely to drive into a lake or off the edge of a cliff – and it has little chance of reaching the top of a mountain. On the other hand, a car with a driver has more freedom in selecting destinations than a free-running vehicle without any controls. In the same manner, an organism cannot take up any form at all: its morphology is constrained by its dynamics – the laws of physics and chemistry – as well as by its DNA instructions. But the DNA instructions can make arbitrary choices between several different lines of development that are consistent with the dynamical laws.

The new mathematical models are finally beginning to put these two aspects of development together. It is not DNA alone, or dynamics alone, that controls development. It is both, interacting with each other, like a landscape that changes shape according to the traffic that passes through it. And one of the most important universal features of the landscape is symmetry breaking. Dynamic mathematical models, giving rise to symmetry-breaking cascades, can reproduce many of the observed patterns, forms, movements and behaviours observed in nature. Some of these models are biologically plausible, others are distinctly speculative. All attempt to provide a counterbalance to conventional reductionist theories by treating the large-scale patterns and forms as entities in their own right. Many of them suggest feasible and interesting experiments that could test their validity. Amid the rush to acquire ever more biological 'information', it may be worth asking what we intend to do with it when we've got it and what else we need to know to achieve those goals.

Acknowledgement

This chapter is adapted from Symmetry-breaking cascades and the dynamics of morphogenesis and behaviour, *Science Progress*, **82** (1999) 9–48. It is reproduced with permission of the publishers.

References

Agarwal, P. (1995) The cell programming language. *Artificial Life*, **2**, 37–77.
Chaplain, M.A.J. and Sleeman, B.D. (1990) An application of membrane theory to tip morphogenesis in Acetabularia. *J. Theor. Biol.*, **146**, 177–200.
Cohen, J. and Stewart, I. (1994) *The Collapse of Chaos*. Viking.
Cohen, J. and Stewart, I. (2000) Polymorphism viewed as phenotypic symmetry-breaking. In *Nonlinear Phenomena in Biological and Physical Sciences* Malik, S.K., Chandrasekharan, M.K. and Pradhan, N. eds. Indian National Science Academy. pp 1–63.
Collins, J.J. and Stewart, I.N. (1992) Symmetry-breaking bifurcation: a possible mechanism for 2:1 frequency-locking in animal locomotion. *J. Math. Biol*, **30**, 827–838.
Collins, J.J. and Stewart, I.N. (1993a) Coupled nonlinear oscillators and the symmetries of animal gaits. *J. Nonlin. Sci.*, **3**, 349–392.
Collins, J.J. and Stewart, I.N. (1993b) Hexapodal gaits and coupled nonlinear oscillator models. *Biol. Cybern.*, **68**, 287–298.
Dillon, R. and Othmer, H.G. (1993) Control of gap junction permeability can control pattern formation in limb development. In *Experimental and Theoretical Approaches to Biological Pattern Formation* Othmer, H.G., Maini, P.K. and Murray, J.D. eds. Plenum Press.
Gambaryan, P. (1974) *How Mammals Run: Anatomical Adaptations*. Wiley.

Golubitsky, M. and Stewart, I. (2002) *The Symmetry Perspective*. Birkhäuser.
Golubitsky, M., Stewart, I. and Schaeffer, D.G. (1988). *Singularities and Groups in Bifurcation Theory*, vol. 2, Applied Mathematical Sciences, **69**, Springer-Verlag.
Golubitsky, M., Stewart, I., Buono, L. and Collins, J.J. (1998) A modular network for legged locomotion, *Physica D*, **115**, 56–72.
Golubitsky, M., Stewart, I., Buono, L. and Collins, J.J. (1999) Symmetry in locomotor central pattern generators and animal gaits. *Nature*, **401**, 693–695.
Goodwin, B.C. (1994) *How the Leopard Changed Its Spots*. Weidenfeld & Nicolson.
Goodwin, B.C. and Brière, C. (1992) A mathematical model of cytoskeletal dynamics and morphogenesis in *Acetabularia*. In *The Cytoskeleton of the Algae* Menzel, D. ed. pp. 219–238. CRC Press.
Goodwin, B.C. and Trainor, L.E.H. (1985) Tip and whorl morphogenesis in *Acetabularia* by calcium-regulated strain fields. *J. Theor. Biol.*, **117**, 79–106.
Gould, S.J. (1987) *Ontogeny and Phylogeny*. Harvard University Press.
Guckenheimer, J. and Holmes, P. (1983) *Nonlinear Oscillations, Dynamical Systems, and Bifurcations of Vector Fields*. Springer-Verlag.
Hildebrand, H. (1966) Analysis of the symmetrical gaits of tetrapods. *Folio Biotheoretica*, **4**, 9–22.
Höfer, T. (1996) *Modeling* Dictyostelium *Aggregation*. PhD thesis, Balliol College Oxford.
Höfer, T. and Maini, P.K. (1997) Streaming instability of slime mold amoebae: an analytical model. *Phys. Rev. E*, **56**, 2074–2080.
Kauffmann, S.A. (1993) *The Origins of Order: Self-Organization and Selection in Evolution*. Oxford University Press.
Kauffmann, S.A. (2000) *Investigations*. Oxford University Press.
Knight, J. (2002) Bridging the culture gap. *Nature*, **419**, 244–246.
Kondo, S. and Asai, R. (1995) A reaction-diffusion wave on the skin of the marine angelfish *Pomacanthus*. *Nature*, **376**, 765–768.
Kondrashov, A.S. and Kondrashov, F.A. (1999) Interactions among quantitative traits in the course of sympatric speciation. *Nature*, **400**, 351–354.
Langton, C.G. (1986) Studying artificial life with cellular automata. *Physica D*, **22**, 120–149.
Lewontin, R.C. (1992) The dream of the human genome. *New York Review of Books*, 28 May, 31–40.
Li, Y.-X. (2002) A minimal network model for quadrupedal locomotion based on symmetry and stability, preprint, Depts. of Mathematics and Zoology, University of British Columbia.
Maini, P.K. and Solursh, M. (1991) Cellular mechanisms of pattern formation in the developing limb. *Internat. Rev. Cytol.*, **129**, 91–133.
Maini, P.K., Benson, D.L. and Sherratt, J.A. (1992) Pattern formation in reaction-diffusion models with spatially inhomogeneous diffusion coefficients. *IMA J. Math. Appl. Med. Biol.*, **9**, 197–213.
Mandelbrot, B. (1982) *The Fractal Geometry of Nature*, 2nd edn. Freeman.
Maynard Smith, J. and Sondhi, K.C. (1960) The genetics of a pattern. *Genetics*, 1039–1050.
Meinhardt, H. (1995) *The Algorithmic Beauty of Sea Shells*. Springer-Verlag.
Meinhardt, H. (1999) On pattern and growth. In *On Growth and Form*. Chaplain, M.A.J., Singh, G.D. and McLachlan, J.C. eds. pp. 129–148. Wiley.
Murray, J.D. (1989) *Mathematical Biology*. Springer-Verlag.
Ourisson, G. (2002) Vers une chimie maîtrisée. *Pour La Science*, **F**, 126–129.
Peitgen, H.-O., Jürgens, H. and Saupe, D. (1992) *Chaos and Fractals*. Springer-Verlag.
Rotman, J.J. (1984) *An Introduction to the Theory of Groups*. Allyn and Bacon.
Sattinger, D.H. and Weaver, O.L. (1986) *Lie Groups and Algebras with Applications to Physics, Geometry, and Mechanics*. Springer-Verlag.
Schöner, G., Yiang, W.Y. and Kelso, J.A.S. (1990) A synergetic theory of quadrupedal gaits and gait transitions. *J. Theoret. Biol*, **142**, 359–391.
Stewart, I. (1990) *Does God Play Dice? – the Mathematics of Chaos*. Blackwell, Penguin.
Stewart, I. (1995) *Nature's Numbers*. Weidenfeld & Nicolson.
Stewart, I. (1998) *Life's Other Secret*. Wiley.
Stewart, I. and Cohen, J. (1997) *Figments of Reality*. Cambridge University Press.
Stewart, I. and Golubitsky, M. (1992) *Fearful Symmetry – is God a Geometer?* Blackwell.

Stewart, I., Elmhirst, T. and Cohen, J. (2003) Symmetry-breaking as an origin of species. *Conference on Bifurcations, Symmetry, Patterns, Porto 2000*. (In press).
Thom, R. (1975) *Structural Stability and Morphogenesis*. Benjamin.
Thompson, D'A.W. (1942) *On Growth and Form* 2nd edn. Cambridge University Press.
Thompson, J.M.T. and Stewart, H.B. (1986) *Nonlinear Dynamics and Chaos*. Wiley.
Turing, A.M. (1952). The chemical basis of morphogenesis, *Phil. Trans. R. Soc. London B.*, **237**, 37–72; reprinted in 1992. *Collected Works of Alan Turing: Morphogenesis*. Saunders, P.T., ed. North-Holland.
von Holst, E. (1935) Erregungsbildung und Erregungsleitung im Fischrückenmark. *Pflügers Arch.*, **235**, 345–359.
Waddington, C.H. (1975) *The Evolution of an Evolutionist*. Edinburgh University Press.
Weijer, C.J. (1999) The role of chemotactic cell movement in *Dictyostelium morphogenesis*. In *On Growth and Form*. Chaplain, M.A.J., Singh, G.D. and McLachlan, J.C., eds. pp. 173–199. Wiley.
Weyl, H. (1969) *Symmetry*. Princeton University Press.
Wolpert, L. (1969) Positional information and the spatial pattern of cellular differentiation. *J. Theoret. Biol.* **25**, 1–47.
Zeeman, E.C. (1977) *Catastrophe Theory: Selected Papers 1972–77*. Addison-Wesley.

Using mechanics to map genotype to phenotype

11

MARK A MIODOWNIK

I want to make the case that the engineering perspective on biology is not merely occasionally useful, not merely a valuable option, but the obligatory organizer of all Darwinian thinking, and the primary source of its power.

Daniel C. Dennett (1995)

11.1 Introduction

The genomes of many organisms have now been sequenced. This linear code is known to determine the behaviour and physical characteristics of an organism. However, it is not a blueprint; it is not a plan of an organism. Rather it is akin to a set of building instructions. We know that by altering the code we change the final shape and behaviour of the organism. For instance changing a particular gene might change the construction process in such a way as to alter eye colour.

Unfortunately, this sort of direct correlation between genotype and phenotype is the exception. In general, it is very difficult to predict the consequences of changing a single gene. This is because most genes have more than one role in building an organism and most characteristics are coded for by more than one gene. Without a step-by-step knowledge of nature's construction mechanism we cannot hope to interpret the significance of a particular gene. In other words, it is not enough to know the genotype, we must know how an organism processes this information into a three-dimensional physical form, how it maps genotype to phenotype.

11.2 Construction methods

Let us consider the way we build things. Most man-made objects are built using assembly construction. This is an approach in which an object is assembled from component parts. Houses, cars, computers and furniture are all built in a similar fashion; even silicon chips and tiny micro-machines (Figure 11.1) are built this way, assembled layer by layer. Ultimately this method of construction could be reduced to a simple list of tasks, an instruction code. Perform the list of tasks, in the correct order, and you can assemble even the most complex of objects, such as a mobile phone, without having actually to understand anything about the design of the

 ISBN 0-12-428765-4

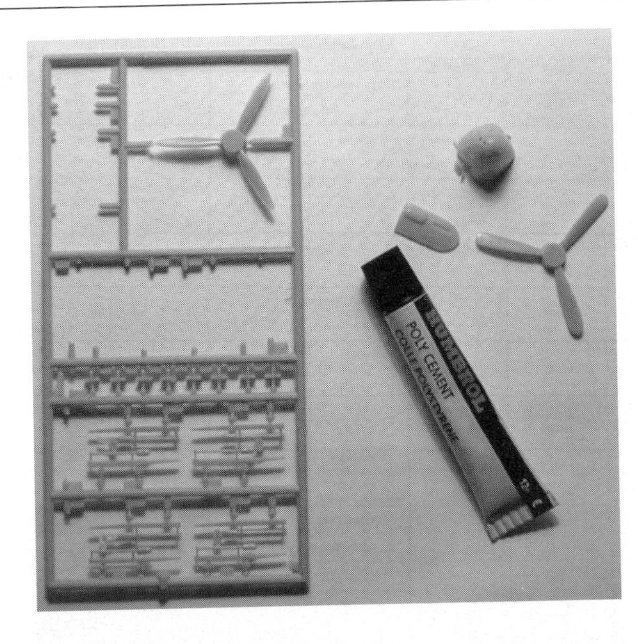

Figure 11.1 Most man-made objects are assembled from components. Part of an Airfix kit for a 'De Havilland Mosquito' (top). A man-made microchain and gears, fabricated by Sandia National Laboratories (bottom).

phenotype. Clearly different objects require very different instruction lists, materials, components, skills and equipment. Building a mobile phone is a much more complex task than building a bookcase, but the principle for both is the same.

In nature, the key difference is that the instruction list only specifies tasks that a cell can carry out. The cell is nature's only assembly unit, only component and only equipment. The materials required are only materials that can be self-generated by cells, or absorbed and transformed by cells. All that is needed to grow an organism is a cell, the instruction set and the right conditions. Natural forms are limited to what nature can persuade cells to do through shape change, division and growth. Nature starts from a single cell, whether the result is a whale or a petunia.

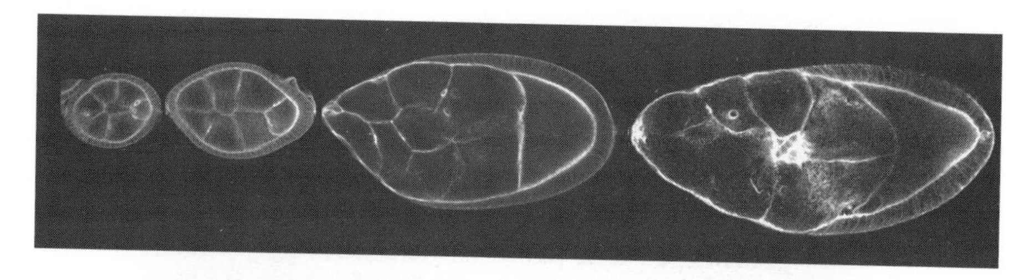

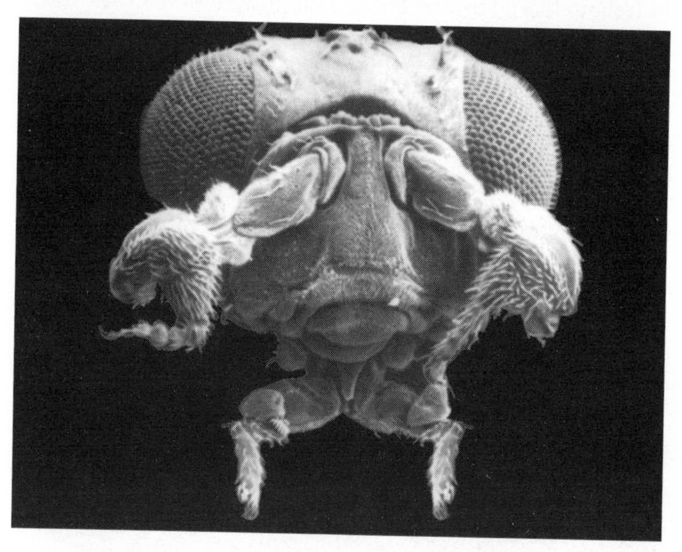

Figure 11.2 An adult fly starts life as a single cell and develops complexity through division and growth. The early stages of egg development, a few cells divide and grow (top). Courtesy of B. Baum. An adult fly, a complex multicellular organism (bottom). Courtesy of Ernest Orlando, Lawrence Berkeley National Laboratory.

Nature builds everything from a thinking, speaking New Yorker to a blade of grass, all from a single cell. The natural world in all its rich beauty is dominated by multicellular organisms that have been grown from scratch (e.g. see Figure 11.2).

But we still do not really know how biology pulls off this amazing trick. It is easy to say that you can build anything from cells – but biology is not Lego. To build a complex 3D organism just from cells that divide, combine and cooperate is a profoundly mysterious process. Turning information into a 3D object with no outside help, without aid of clamps, welding machines or cranes, seems very difficult. Nature has to be its own engineer, not only knowing what goes where, but internally generating the forces to assemble and transform the structure.

The central question explored here is this: is biology's approach an evolutionary relic, clever, but ultimately to be surpassed by component construction? Or are there hidden advantages to nature's construction method, tricks that make it superior? Will we end up growing bridges rather than building them?

11.2.1 Natural form

Some forms are easier to construct than others, for example, naturally occurring forms for which there is no design or plan. All they require are the right physical conditions. Galaxies emerge from clouds of interstellar gases and their various forms can be understood in terms of a competition between opposing forces. Stars are sucked together to form galaxies by gravitational forces which are opposed by kinetic forces that would have the stars spinning out into space. The combination of forces is responsible for the shape of the galaxy, but from within, its form does not appear particularly ordered or symmetrical. This is why if you look up into the night sky, the pattern of stars seems random. You have to have a pretty active imagination to join the dots to get the signs of the Zodiac. In fact pattern and order is rare in the universe. The Giant's Causeway in Ireland has a striking hexagonal pattern in the rock. But most other lumps of rock are utterly unremarkable.

It is not until you get down to the microscopic scale that naturally occurring order and form becomes more common. Even the lumpiest rock is made of crystals arranged in intricate patterns. These structures self-assemble, so that given the correct conditions they will always form. Let a hot, sweet cup of tea cool down before drinking it and you will find little crystals of sugar at the bottom of the cup (Figure 11.3). The sugar, which is less soluble in cold tea than hot tea, has crystallized out. Sugar molecules that were randomly diffusing in the hot liquid have

Figure 11.3 Scanning electronic microscope image of sugar crystals left at the bottom of a cup of tea (top). Courtesy of Internet Microscope. Nanotubes, these structures self-assemble when grown by chemical vapour deposition (bottom). Courtesy of NASA.

suddenly decided to club together in an exact, ordered crystal structure at the bottom of a teacup.

The form of natural structures is very dependent on physical conditions. Ice crystals form in clouds but can grow into a hailstone or an exquisite snowflake. Yet, even if conditions were perfect, you could not make a ten-foot snowflake, because as the crystals grow in size different forces come into play that upset the balance that is responsible for their form. It is a complex process and most examples of perfection in naturally occurring physical forms are rare precisely because getting the conditions right is so difficult. Some physical processes that produce structure are more controllable and repeatable than others. The success of nanotechnology hinges on this ability and the self-assembly of bucky balls and nanotubes are striking success stories (Figure 11.3) (Nalwa, 2001).

It has long been suggested that biology may harness these physical pattern-forming mechanisms to build organisms. The mathematician and zoologist D'Arcy Thompson was convinced of this and catalogued the tell-tale signs that he observed in nature, to highlight the evidence of the mathematical and physical origins of biological form. The evidence and his analysis were first published in his book *On Growth and Form* (Thompson, 1961). The book is encyclopaedic in its scope, including chapters on *The forms of cells, The shapes of horns, and of teeth or tusks; with a note on torsion, On leaf-arrangement, or phyllotaxis* and *On form and mechanical efficiency*. He gives striking examples of the simple mathematics behind the spiral form of shells, the physics of frog embryos and jelly fish and the numerology of plant growth.

However D'Arcy Thompson's arguments run counter to the Darwinian explanation for biological form, which is that natural selection determines the form best suited for an organism's survival, irrespective of its mathematical beauty or simplicity. For this reason, *On Growth and Form* is ignored by most biologists. Yet, as emphasized earlier, we still know relatively little about the mechanisms by which nature turns information into organisms. So far no one has pinpointed how nature's myriad of beautiful patterns and symmetries are directly coded for in the genome. In the absence of this evidence many researchers have found it difficult to resist continuing to explore the simple mathematical and physical explanations for the shape and patterns of some organisms (Murray, 1990; Prusinkiewicz and Lindenmayer, 1990; Meinhardt, 1995; Stewart, 1998; Ball, 1999).

Furthermore, the subject has gained an important analytical tool since *On Growth and Form* was published: computer simulation. This has enabled much more rigorous testing and exploration of D'Arcy Thompson's hypothesis. It has now been established that some complex shapes such as spirals, fractals, spots and stripes can be produced very easily by many different types of simple rules. It seems very likely that nature has discovered this trick to build structures such as skin patterns (Kondo and Asai, 1995), snail shells (Koch and Meinhardt, 1994; Meinhardt, 1995) and vascular systems (Ball, 1999). Importantly, an organism would not need to manipulate equations to build these complex patterns. For example: walk one step ahead, turn to the left and walk two steps in that direction, turn to the left and walk three steps in that direction, and so on. Keep doing this and you will walk in a spiral without needing to know anything about the equation that is governing your motion.

The exquisite shapes of shells can be built with similarly basic rule sets. The spiral form, complex to our eyes, is in fact simple in terms of the information that encodes it. However, the simplicity relies on the method of construction. Building a spiral in this way depends on starting small and building up. A spiral-shaped house, built using conventional construction materials and methods would be altogether more complicated, because you would need to know the mathematical equation that defines that shape. So structural complexity is relative. We find rectangular structures easy to build using our component assembly building methods and nature finds spirals easy to build using a growth method.

11.2.2 The scale of the cell

Nature builds organisms using a self-assembly unit, the cell. A cell is a very complex and controlled environment, with an internal skeleton, organs and a nucleus all protected by a semipermeable membrane. It is able to send and receive signals and to process information. Cells can change shape and move around, divide and grow.

When nature wants to make bigger organisms it does not make bigger cells. Instead it builds multicellular organisms. Cells in these organisms are all roughly the same size. Elephants do not have bigger cells than mice. The size scale of cells is in the range one thousandth to one tenth of a millimetre and this is therefore an extremely significant scale for all living organisms. This is the size at which cells organize, communicate, divide and grow and at which development and cell differentiation happen.

Building organisms clearly involves turning information into form, which requires communication between cells. There are many intercellular signalling methods, but one of the most important is diffusion of signalling molecules. Essentially diffusion is an atomic scale process where individual molecules move by random walk. These molecules are thousands of times smaller than a cell, so getting across a cell by random movement is a slow process. Drop ink into a glass of water and you can see how slowly it spreads. Diffusion only really works as an effective signalling mechanism at the scale of cells precisely because they are so small. It would be useless if cells were much bigger. Growth and development of an organism are related to how quickly signals can be passed. This brings together the spatial and temporal issues and provides a hint as to why nature does not choose to build bigger organisms from bigger cells.

The argument for why cells are not much smaller has been linked to the size and nature of atoms. Atoms behave in a fundamentally statistical manner and it is only at the level of collections of atoms that ordered behaviour can be obtained. Clearly this crucial size is relative to the size of atoms. In other words, to paraphrase Erwin Schrödinger, 'Why are atoms so small? Because we need to be so much bigger' (Schrödinger, 1969).

Size is relative and even though cells are big by atomic standards, they are microscopic by our standards. Cells inhabit a world with which we are not familiar. At the microscale gravity is almost an irrelevance. Here, the dominant forces are surface tension, viscosity, electrostatic forces, elastic forces and pressure. Building organisms at this scale and in the aqueous conditions of cellular environments is a very different process to assembling macroscopic objects. But this is the natural environment in which all living organisms are patterned and grown. Understanding the physics at this scale is therefore key to understanding nature's construction mechanism.

Consider the effects of surface tension. A cell experiences aqueous environments in much the same way we would experience a pool of honey. Does this make life difficult for cell organization? Interestingly, the exact reverse is true. Surface tension is now known to be involved in many self-patterning and self-organizing phenomena (Hildebrandt and Tromba, 1996; Ball, 1999). Is it possible that nature is harnessing this ability? This will be the main focus of the rest of this chapter.

11.3 Surface tension

The term surface tension invokes images of small insects tip-toeing across the surface of a calm lake. You know they can do it because of the surface tension of the water. But what is surface tension and why does it allow insects to waltz around where we can only swim?

Surface tension arises because the molecules in liquids like to band together. They are much happier surrounded by each other because this lowers the energy of the system; low energy is a state to which all physical systems aspire. Water molecules on the surface of the lake interface with molecules in the air as well as other water molecules. The surface is characterized by the water–air interaction, and it has a characteristic energy associated with it. This is called surface energy and it quantifies how unhappy the water molecules are being next to molecules in the air as opposed to water molecules. Surface tension is a specific measure of this surface energy and relates to the energy of a unit area of interface.

Why is it called surface tension? How do you get a force from a surface? A force is a way of defining how a system will behave. In the case of water droplets for instance, the surface tension greatly affects their behaviour. A falling droplet will try to assume a spherical shape to minimize its area, thus minimizing the water–air interface. What drives this behaviour is the fact that the water–air interface has an energy associated with it and minimizing area means minimizing energy. Similarly, soap bubbles are spherical because this minimizes their energy. Try blowing one through a square or triangular hole; you will still get a spherical bubble. Even if you distort this sphere it will wobble around until it regains its shape. In this way, a force that is actually the minimization of surface energy manifests itself like a tension at the level of the interface and from thence comes the name – surface tension.

When a falling water droplet hits something, it often sticks to it. A rain drop hits the windscreen and does not bounce off it but streams down the window. Why does it do that? You are having a cup of tea and as you lift the cup up the saucer comes with it. There is no glue, just a film of tea between the cup and the saucer. What is happening?

Surface energy is behind this behaviour. When a rain drop hits a glass surface, it goes from a state where it has only one interfacial energy, the water–air interface energy, to suddenly having two interface energies, a water–glass interface energy and a water–air interface energy. Now the shape of the drop must change in order to minimize the total interface energy. This shape will depend on the relative magnitudes of the two interfacial energies. If the water–glass surface energy is small then the water droplet will like being in contact with it and spread out (hydrophilic glass). But if the water–glass surface is high then it will try to minimize its contact with it and bunch up into a ball (hydrophobic glass) (Figure 11.4). It all depends on the surface energies involved. Either way, the energy is lower when the droplet is in contact with the glass and this lowering of the energy appears as an attractive force. This is the key to understanding why rain sticks to a windscreen. When two droplets come into contact they suddenly rush together. Again it is a lowering of the surface energy, but it looks like a force of attraction between them. The same phenomenon is responsible for the embarrassing teacup episode. The surface energy of the tea sticks the surfaces of the cup and saucer together.

The point of these examples is to show that surface energy alone can cause shape changes in droplets and cause attractive forces. At the scale of droplets gravity also plays a role but not a great one. As you go smaller, towards the size of a cell, gravity has less and less influence and surface energy becomes a bigger deal. This is because smaller things have more surface and less mass. Gravity acts on mass, surface tension acts on surfaces. Furthermore, it is not just surfaces that are important, but all interfaces.

11.3.1 Foams and cellular networks

Soap froths are easy to make because surface tension does all the work for you. Take a soap bubble, add another soap bubble and they are immediately attracted to each other because by being together they minimize their surface energy (Figure 11.5). Add a few more bubbles and

Figure 11.4 An illustration of the power of surface tension to dictate the shape of water droplets. Water droplet on hydrophilic glass (top), water droplet on hydrophobic glass (bottom).

you notice something else. It is not just that the bubbles stick together; they actually rearrange their interfaces to create a minimum internal surface area. How do we know? The characteristic 'Y' junctions where three bubbles meet have perfect 120 degree angles. You never get four bubbles meeting with 90 degree angles. It does not matter how big or small the bubbles are, you always get these 'Y' junctions. It can be proved that all these observations are a direct result of minimization of interface energy. However, the important thing is that these soap froths are not just a jumble of bubbles. They have form and the form is dictated by surface energy considerations (Hildebrandt and Tromba, 1996).

Further analysis of the effect of surface tension on soap froths is complicated by the fact that air diffuses through the soap membranes. Luckily, there is another natural system that acts in a similar way, minus these complications and can take us much further in exploring the relationship between surface energy and form. Many things like metals, rocks and ceramics are made of microscopic crystals, but since they are opaque materials this is not obvious. The internal structure of many metals looks like a soap froth, but instead of soap boundaries there are crystal boundaries (Figure 11.6). These crystal boundaries have surface tension and so the crystals rearrange themselves to minimize the interface energy. This is at first surprising. Metals are solid and we do not think of them as being able to rearrange their internal structure.

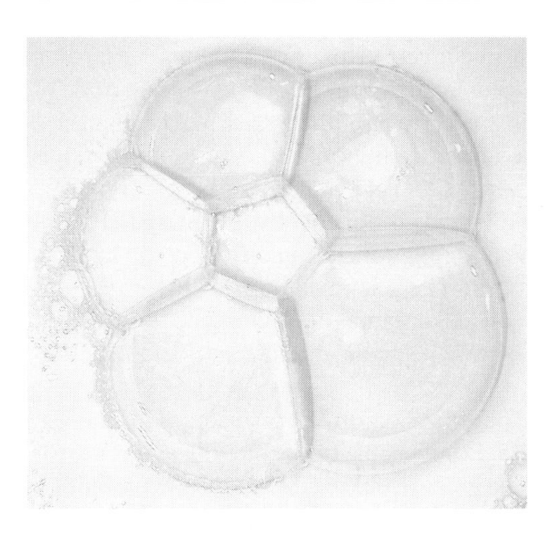

Figure 11.5 Soap bubble foams. A group of five soap bubbles showing the Y internal junctions that naturally form (left). The structure of a general foam is fixed by interfacial energy considerations (right).

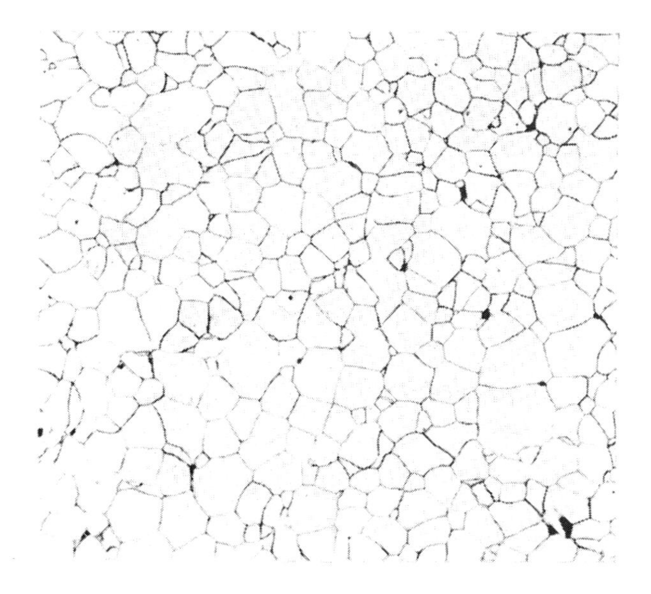

Figure 11.6 The internal microstructure of an aluminium alloy, notice the cellular structure of the aluminium crystals and the Y internal junctions that naturally form as a result of interfacial energy minimization.

In fact, it happens very slowly at normal temperatures because the atomic bonds are so strong. Give the atoms some energy by increasing their temperature and it is a different story – they can really move. This is not melting, but creep, a process by which solids can and do change their shapes in response to forces. Not every solid can do this. Some burst into flames or disintegrate before they get hot enough, but many can. For lead, glass and ice even normal temperatures are enough to allow them to rearrange their atoms, albeit over many years. This is why old panes of glass are thicker at the bottom, lead piping sags and glaciers flow down valleys.

Most metals have to be heated up before they can behave in this way and surface tension plays a big role in changing their microstructures. During heating, the metal recovers and one of the processes is that the crystal boundaries move in order to minimize their surface energy. Figure 11.7 shows how a crystal network can evolve under such circumstances. Some crystals grow while others shrink, but the pattern remains very elaborate. It looks like an organized rearrangement of the crystal network, but it can be explained completely by assuming that each crystal boundary is just trying to minimize its area. No communication between the crystals takes place. No external regulation controls the pattern. The self-regulation of these types of pattern is driven solely by surface tension. This type of self-organization phenomenon is extremely common in metals and is manipulated to control the mechanical and electrical properties of many alloys (Miodownik, 2001).

Let us consider one final example of the effect of surface tension on microstructure formation. Rocks and ceramics have such high melting points that it is unfeasible to use melting as a route for manufacture. Instead small crystals are bound together with a sticky substance called a binder. Clay is a naturally occurring combination of organic binding agents and ceramic crystals. When

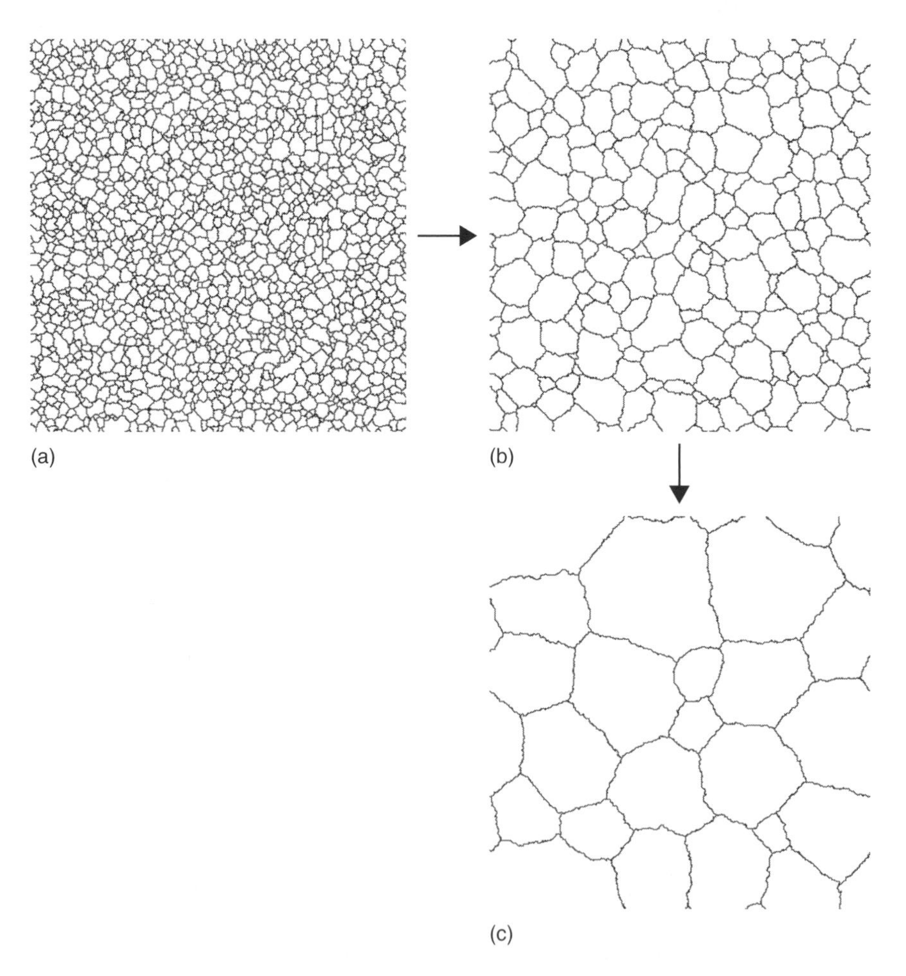

Figure 11.7 An illustration of the process of crystal growth in metal alloys during annealing. (a) 0 hours. (b) 1 hour. (c) 10 hours. The structure of the crystal network remains self-similar and appears to organize itself during growth. The process is a result of interfacial energy minimization.

heated up, the binding substance burns off, leaving an agglomerate of small ceramic crystals all touching each other. If the temperature is high enough then the surface tension of these particles acts to try to minimize the surface area of the agglomerate. The crystals act like droplets of water, coalescing together and shrinking as the space between then becomes smaller. All this is done while it is solid, so you can mould the clay into whatever shape you want, a teapot for instance and the shape is retained even though the material properties are transformed. Figure 11.8 shows a sequence in which the crystals diffuse together, reducing their surface area and finally becoming fully dense. This process, called sintering, is used to manufacture all ceramics from exquisite bone china to bricks and at its centre is a surface tension assembly mechanism.

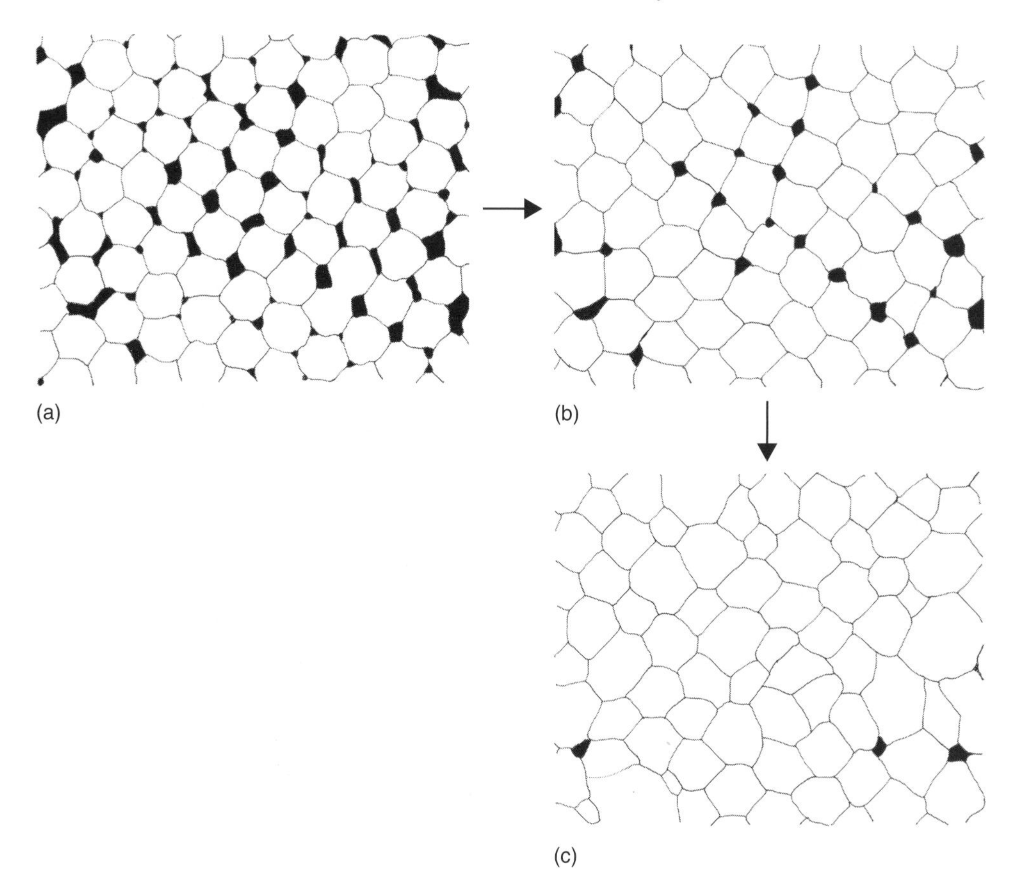

Figure 11.8 A schematic showing the sintering process. Ceramic crystals are clustered together, the gaps between the crystals are shown as dark regions. As sintering progresses the initially spherically-shaped crystals coalesce to form a bulk material. This process is driven by surface energy minimization (a) 0 hours. (b) 0.5 hours. (c) 1 hour.

11.3.2 Surface tension and organism construction

The form of a single cell can be greatly influenced by the surface energy of its interfaces. If a single cell was only surrounded by a thin membrane then, in isolation, we would expect it to be spherical. More complicated cell-like objects, called vesicles, with bilayer membranes can be created with special surfactants. The shapes of these vesicles can be changed by modifying the surface energy and it has been shown that the shapes obtained mirror many types of biological cell shape, such as the shape of the red blood cell (Lipowsky, 1991). However, biology does not generally leave cell shape in the hands of surface tension. In the case of plants there is a cell wall that governs cell shape and in animals an internal skeleton determines shape. That is not to say that surface tension is not important for individual cells. Cell division, for instance, relies on surface tension to reconfigure the cell membrane from an hourglass shape at the point of division, into the membranes of two complete cells after division. This is a complicated morphological transformation that surface tension completes with amazing ease, as can see if you pinch one bubble into two (Hildebrandt and Tromba, 1996).

Cell membranes have a surface energy which is more often called cell adhesion, a term which emphasizes cells' ability to stick to each other. This stickiness is exactly the surface tension force that we saw in action earlier, binding together ceramic crystals to make teapots. In the case of biology it is immensely more complicated because there are many different types of glycoproteins that sit on the cell surface and thus change the surface energy. Not only that, but cells can manipulate their surface energy by changing the cell adhesion molecules on their surfaces (Bard, 1990; Slack, 2001; Wolpert, 2002). This means that each individual cell can manipulate its surface energy in order to influence the surface tension forces acting on it. The corollary of this is that by fine-tuning their surface energy, cells can self-sort.

This sorting ability has been demonstrated in experiments in which cells from different tissues were mixed together. For instance if you take inner cells from an early amphibian blastula endoderm and bind them to cells from the ectoderm, a smooth sphere will initially form. However, after a while the two types of cell will start to repel each other, resulting in two separate agglomerates with a narrow strip joining them together (Figure 11.9). This is entirely a surface tension effect, driven by the fact that the ectoderm–endoderm cell interfaces are high energy. In trying to minimize the internal energy the cells rearrange to minimize the ectoderm–endoderm interfacial area (Gumbinger, 1996).

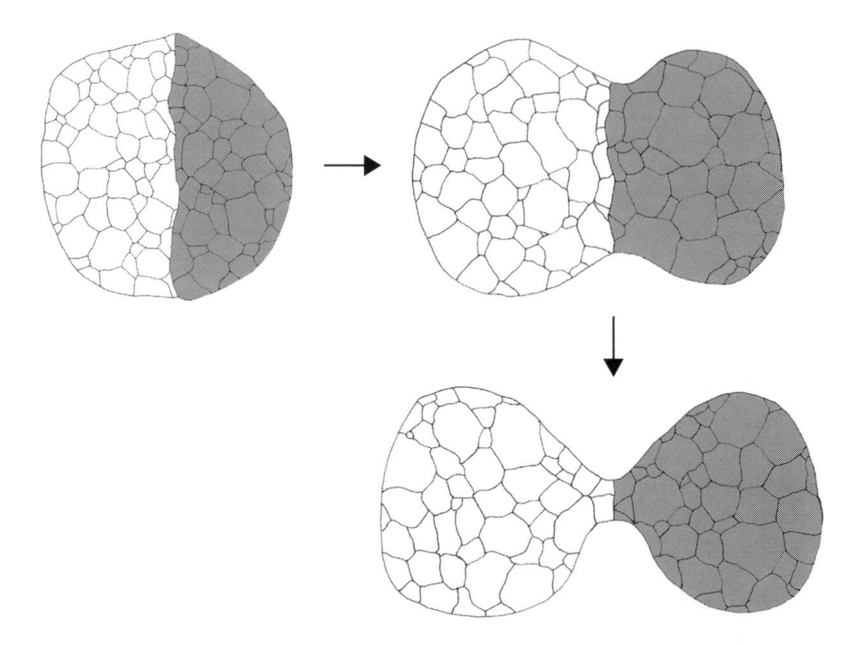

Figure 11.9 Showing the separation of embryonic tissues with different adhesive properties (shown as grey and white cells). The two different cell types rearrange to minimize the interfacial area.

Another example of surface tension driving cell sorting can be illustrated by an experiment carried out on tissues taken from later amphibian embryos. If the cells from the presumptive epidermal cells are mixed up with neural plate cells, then the initial disordered soup of cells can undergo spontaneous ordering (Wolpert, 2002). A sphere of cells is formed with the epidermal cells found on the outside surrounding the neural cells (Figure 11.10). The effect is again driven by surface tension. The sorting out of the cells occurs because it minimizes the interfacial energy of the system. How do the cells know that this is the lowest energy? They do not need to know. The cells jiggle about randomly and are constantly changing their neighbour cells. When the interface between them and their neighbours is a low energy interface, it is much harder for them to move away from that neighbour. Conversely, if they have a neighbour with which they share a high energy interface then movement to a new neighbour with a low energy interface is preferable. Clearly swapping neighbours when there is no energy change is unbiased. That such a simple mechanism can lead to cell sorting may seem at first unlikely, because it involves a large degree of cell rearrangement; cells have to be constantly changing their neighbours, always seeking a lower energy state. Although this can be made to happen in an experiment where cells can be separated and manipulated, how likely is it to happen during development of an organism?

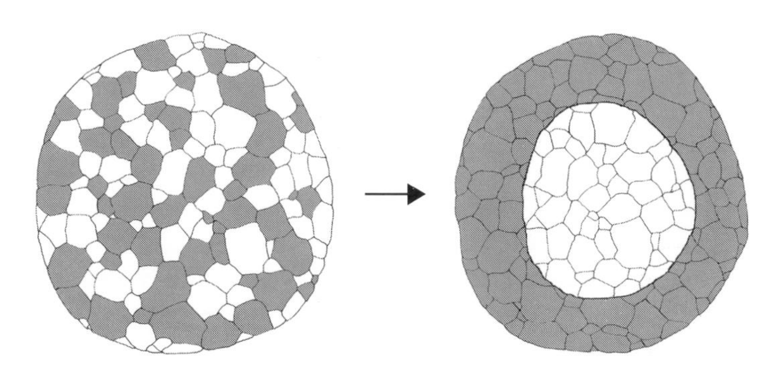

Figure 11.10 Showing cell sorting when different cell types (shown as grey and white cells) are mixed. The initial disordered soup of cells can undergo spontaneous ordering driven by the minimization of the interfacial energy of the system.

Shuffling of cells in a random manner may not happen much in living systems but there are other ways in which different cells sample new neighbours and reorganize themselves. For example, during growth, cellular systems naturally sample different neighbours as a result of having to reorganize their spatial organization. Growth of such systems with conserved numbers of cells can show very rapid cell sorting behaviours without the need to invoke random shuffling. Figure 11.11 shows snapshots from simulations in which cell agglomerations with two types of cells, with different growth rates and different surface energies, spontaneously undergo cell sorting behaviour. The smaller cells end up on the outside, with the bigger cells on the inside in a form that minimizes surface energy but results in a bimodal structure. There was no enforcing of structure formation or of neighbour shuffling in the simulations. All that was specified was the surface energies and the growth rates. Interestingly, without growth, ordering did not occur because cell shuffling alone did not allow the cells to sample enough neighbours in order to find the minimum energy configuration.

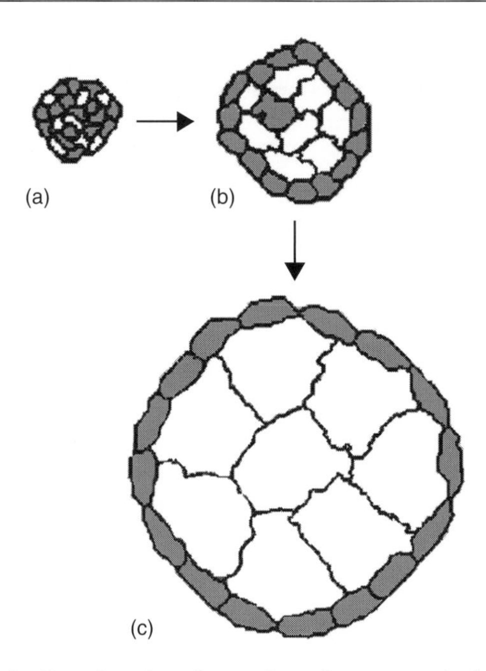

Figure 11.11 Simulations of cell sorting via anisotropic surface energy during growth of a simple multicellular ensemble. (a) 0.5 hours. (b) 1 hour. (c) 5 hours. There was no enforcing of structure formation or neighbour shuffling in the simulations. All that was specified was the surface energies and the growth rates of the grey and white cells.

One final example of the surface tension effects in morphogenesis is the observation of convergent extension. This is where an active group of cells, initially equiaxed, elongate, align and intercalate with each other (Figure 11.12). The elongation increases the overall length of the group of cells and the cells become aligned. At the same time the cells intercalate with each other in the direction of the alignment. The net result is that the cells stack on top of each other like bricks in a wall. This basic mechanism is extremely common in the morphological development of many animals. Although there are various theories about the mechanism, it seems clear that surface energy effects are important and, in fact, much of the sorting behaviour can be explained in terms of changes in the surface energy of individual cells.

The key idea is that different parts of the cell surface have different surface energies. In the case of convergent extension, all that is required once the cells have flattened, is that the surface energy of flattened parts of the cells is lower than the short ends. In this way the surface tension effects would immediately push the ensemble in a direction to minimize the end–end junctions and maximize flat–flat junctions. The net result is a transformation of the structure into a pile of pancakes, consistent with the convergent extension (Zajac *et al.*, 2000).

11.3.3 Robust construction

Clearly surface tension is a major force at the cellular level and nature has to deal with it if it wants to build organisms. We have seen from the examples in the previous section that as well as having to contend with surface tension, organisms can actively manipulate surface tension effects to construct and rearrange cellular material. Is this evidence that organisms use surface tension as a natural self-assembly mechanism?

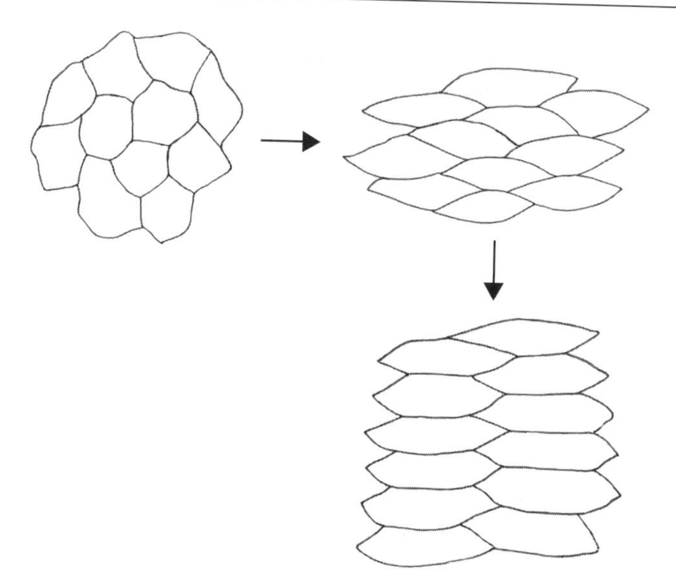

Figure 11.12 Cell movements during the process of convergent extension. An active group of cells, initially equiaxed, elongate, align and intercalate with each other. The net result is that the cells stack on top of each other like bricks in a wall. This basic mechanism is extremely common in the morphological development of many animals. Although there are various theories about the mechanism, it seems clear that surface energy effects are important.

Unfortunately, this is not a straightforward question to answer. Physical forces tend to push systems into a state of equilibrium, which is generally in one direction only; this is the essence of the second law of thermodynamics. But the development of organisms is a dynamic process in which cell rearrangement is ongoing. The cell sorting examples outlined above are demonstrations of the fact that cells obey physical laws. Yet only in the case of the convergent extension have we postulated that these forces are actually harnessed to change the morphology of an organism. For nature to harness surface tension it needs not only to change the surface tension of different parts of the cells but also to time and control the mechanism. This is a much more sophisticated task than just using naturally occurring surface tension forces to pull cells together.

A sailing analogy will serve to further illustrate the point. Sailing with the wind behind you is easy, any kind of sail will get you moving forward. However, you can also use the wind to take you back from the direction you came. To sail into the wind, you need to tack. Tacking is not easy. It involves turning at the right moments and setting the sails in exactly the right way. It is a complicated technique but, nevertheless, uses the same driving force, the wind, to move a boat in the opposite direction. Tacking requires the boat crew to time their actions with respect to their motion, the boat's direction and the wind's direction and strength. The decisions about when to tack require more than just timing; they require feedback about the current conditions and knowledge about the boat's behaviour. In other words they require information processing.

The same is probably true in biological systems. The patterning mechanisms issuing from surface tension forces are probably used in many different ways to help build organisms. Furthermore, the way in which the organism controls the force will almost certainly require feedback loops and information processing. These control mechanisms will also have an impact on how repeatable the transformations can be. It is all very well showing that a change in the surface energy of these cells at this time means they will self-assemble into a layer. But how consistent is it? A layer may always form, but how quickly does it form and are there always the same number of cells in that layer? In other words, biological organisms need robust mechanisms. These are not necessarily as simple as they may at first appear (Barkai and Leiber, 1997; Eldar *et al.*, 2002).

11.4 Concluding remarks

Nature builds organisms from cells and the scale of the cell is an important factor in determining nature's construction method. Surface tension, for instance, is dominant at that scale and may be closely involved in the development of an organism's cellular structure. If we took surface tension away, would it be easier or harder for nature to construct an organism? This is the key question implicit in our discussion of surface tension self-assembly mechanisms. If it is easier, then nature clearly does not make much use of surface tension's organizational abilities. If it is harder, then we know that surface tension is closely integrated into nature's construction mechanism.

Unfortunately, we cannot turn off physics. How else can we explore the complex interplay of physical forces and information processing that characterizes nature's construction mechanism? We could try to mutate a simple multicelluar organism so that the cell size becomes macroscopic not microscopic. Cells the size of tennis balls would be ideal. At this different scale surface tension would be much less dominant and gravity would start to play a big role. Comparing the growth of such an organism with its counterpart built from microscopic cells would tell us much about the role of surface tension in building organisms. The drawback of this approach is that by changing the size of the cells we change many other physical properties of the cells. It would be difficult to extract which effects were solely due to surface tension.

Alternatively, we can build a universe in which surface tension can be turned on and off at will. This may sound far-fetched, but these universes exist. They are called computer simulations. At the moment they are very simple and they do not simulate much of the sophistication of the physical world. However, there is no reason why in the future they will not be able to simulate the physics that exists at the scale of the cell. This is an exciting prospect. It will allow us to quantify how much physics we need to include in the simulation in order to model nature's construction method. Then we will be able to turn off the physics to test our hypothesis. This virtual experimental method is in its infancy at present. However, it is growing in importance as a tool to explore the complex interplay between physics and information processing in multicellular organisms (Davidson *et al.*, 1995).

Finally, since nature builds all organisms using cells, the mapping of genotype to phenotype must take into account the mechanics of cell organization. Whether this is implicitly or explicitly included in the genotype depends on how much of nature's construction mechanism relies on natural self-assembly methods. At the moment this is very much an open question.

Acknowledgements

I would very much like to thank Buzz Baum, Kathy Barrett, Lewis Wolpert, Helen McNeill and Manda Levin for useful discussions and feedback.

References

Ball, P. (1999) *The Self-Made Tapestry*. Oxford University Press.

Bard, J.B.L. (1990) *Morphogenesis. The Cellular and Molecular Processes of Developmental Anatomy*. Cambridge University Press.

Barkai, N. and Leiber, S. (1997) Robustness in simple biochemical networks. *Nature*, **387**, 913.

Davidson, L.A., Koehl, M.A.R., Keller, R. and Oster, G. F. (1995) How do sea urchins invaginate? Using biomechanics to distinguish between mechanisms of primary invagination. *Development*, **121**, 2005.

Dennett, D.C. (1995) *Darwin's Dangerous Idea, Evolutions and the Meanings of Life*. Simon and Schuster.

Eldar, A., Dorfamn, R., Weiss, S. *et al.* (2002) Robustness of the BMP morphogen gradient in *Drosophila* embryonic patterning. *Nature*, **419**, 304.

Gumbinger, B.M. (1996) Cell adhesion: the molecular basis of tissue architecture and morphogenesis. *Cell*, **84**, 345.

Hildebrandt, S. and Tromba, A. (1996) *The Parsimonious Universe*. Springer-Verlag.

Koch, A.J. and Meinhardt, H. (1994) Biological pattern formation: from basic mechanisms to complex structures. *Rev. Mod. Phys.*, **66**, 1481–1507.

Kondo, S. and Asai, R. (1995) A reaction-diffusion wave on the skin of the marine angelfish *Pomacanthus*. *Nature*, **376**, 765.

Lipowsky, R. (1991) The conformation of membranes. *Nature*, **349**, 475.

Meinhardt, H. (1995) *The Algorithmic Beauty of Sea Shells*. Springer.

Miodownik, M.A. (2001) Grain growth, Uniform. In *Encyclopedia of Materials: Science & Technology*, Buschow, K.H.J., Cahn, R., Flemings, M.C., Ilschner, B., Kramer, E.J. and Mahajan, S. vol. 4, pp. 3636–3541. Elsevier.

Murray, J.D. (1990) *Mathematical Biology*. Springer.

Nalwa, H.S. (2001) *Nanostructures Materials and Nanotechnology*. Harcourt Publishers Ltd Professional Publications.

Prusinkiewicz, P. and Lindenmayer, A. (1990) *The Algorithmic Beauty of Plants*. Springer-Verlag.

Schrödinger, E. (1969) *What is Life? & Mind and Matter*. Cambridge University Press.

Slack, J. (2001) *Essential Development Biology*. Blackwell Science Ltd.

Stewart, I. (1998) *Life's Other Secret. The New Mathematics of the Living World*. Wiley.

Thompson, D'Arcy (1961) *On Growth and Form*. Cambridge University Press.

Wolpert, L. (2002) *Developmental Biology*. Oxford University Press.

Zajac, M., Jones, G.L. and Glazier, J. (2000) Model of convergent extension in animal morphogenesis. *Phys. Rev. Let.* **85**, 2022–2025.

How synthetic biology provides insights into contact-mediated lateral inhibition and other mechanisms

12

KURT W FLEISCHER

12.1 Introduction

This chapter presents two examples of using simulation as a tool for exploring complicated biological systems – the synthetic biology approach. The first example shows how the approach helps reach insights into the behaviour of a contact-mediated lateral inhibition system and compares the results to a subsequent mathematical analysis of the same system by Collier *et al.* (1996). The second example shows how a short series of simulated experiments in a rich simulation framework provides insight about how motility, cell shape and rigidity affect differential adhesion.

Updates from the original 1995 work

This is a revised version of Chapter 6 of my thesis (Fleischer, 1995), where I advocated one of the central tenets of the then nascent field of synthetic biology (a.k.a. artificial life): we can gain valuable insights into biological systems by examining them in comprehensive simulations. Upon graduation, I decided to pursue other interests (computer animation). Looking back now with a fresh perspective, I still find significant promise in this approach and believe it will one day be commonplace for biologists to work with these tools alongside their experimental and analytical tools. This updated chapter contains additional material so it can stand alone, as well as some other examples to highlight the range of capabilities of the multiple-mechanism simulator.

How do we evaluate work in synthetic biology?

The preface to Murray's book *Mathematical Biology* gives a nice description of the criteria for evaluating models:

> *Mathematical biology research, to be useful and interesting, must be relevant biologically. The best models show how a process works and then predict what may follow. If these are not already obvious to the biologists and the predictions turn out to be right, then you will have the biologist's attention.*
>
> (Murray, 1993, pp. v–vii)

 ISBN 0-12-428765-4

These same criteria apply to synthetic biology and any theoretical analysis for biology: the results must be relevant and non-obvious. If you are lucky, they will be right sometimes, though of course there is no guarantee of that for any method of generating hypotheses about the natural world.

The examples of biological modelling in Murray's book are indicative of the style of mathematical biology in general use today. A mathematician provides a theoretical explanation of a biological phenomenon, usually reducing the system to a small number of parameters and identifying regimes of behaviour via mathematical analysis.

A different approach is examined here. Instead of trying to simplify the system sufficiently to enable thorough mathematical analysis, we try to include as many of the relevant phenomena as is computationally feasible. The resulting simulation system represents more features of the real system being modelled. It will be harder to analyse conclusively. This is a trade-off.

We then use the simulation to gain *insights* into and *intuitions* about the biological system. This is the proposed *synthetic approach to biology*. This approach is an outgrowth of work in mathematical biology and shares many techniques in common. We place a greater emphasis on the use of the simulator as an exploratory tool to help generate intuitions about biological systems. If these are promising, they can later be examined more closely and subjected to more rigorous analysis.

In this approach, the user creates a developmental system from scratch by specifying the properties of a cell and its interactions with other cells and the environment. The difference between learning by *building* a system (engineering) versus learning by observing and dissecting a system (biology) is substantial. Each approach brings a different point of view and is thus likely to emphasize and discover different things. By enabling biologists to take this engineering approach of constructing simulated organisms, we add a new tool to their repertoire. This does not replace experiments, nor does it replace mathematical analyses; it is complementary to both.

A multiple-mechanism developmental simulator

This simulator was created at Caltech as part of my thesis work during 1991–1995 (Fleischer, 1995). It is a multiple-mechanism simulator, providing:

- discrete motile cells
- mechanical interactions (collision and adhesion)
- intracellular processes: gene transcription and regulation, kinetics, transport and metabolism
- model of cell surface proteins (homophilic and heterophilic)
- cell lineage and cell cleavage plane control mechanisms
- neurite growth and electrical activity.

The original goal of the simulator was to create a testbed for evolving artificial neural networks. As such, the genetic representation is crafted to be a fertile representation for evolutionary operations such as crossover and mutation. The representation uses fragments of differential equations that are summed together (notation: +=). The biological motivation for this is that a gene expresses the rate of production of a protein (easily and appropriately represented by a differential equation). Mechanisms of inter-gene regulation affect this rate. All of these can be conveniently represented by terms in differential equations; we developed a small and powerful language containing sigmoidal functions (essentially a smoothed step function, denoted ~>) for activating and deactivating terms in the differential equations. The total rate of production of a given product is a sum of contributions that are attenuated by sigmoidal functions. This formulation is closely related to those used by Mjolsness *et al.* (1991) and Kauffman (1993) to represent gene regulation.

The resulting system is general and well-suited for a variety of biological simulations, such as those discussed in this chapter. (For more information on the evolutionary aspects of this model, see Fleischer, 1995, Ch. 7.) A similar and powerful system, the cell programming language, is described by Agarwal (Agarwal, 1994). It uses a different cell-shape model and is based on a discrete simulator instead of continuous differential equations.

The inclusion of multiple mechanisms enables a rich set of biological models to be explored. In the first example below, the ability to represent chemical as well as contact and growth models is important. In real systems, all of these mechanisms (and more) are present and the interplay between them can be crucial. Morphogenesis comes about via the expression of genes within an environment and the characteristics of the environment are a critical component. Evolved developmental systems depend on several mechanisms operating at different times and locations and at different time and distance scales, to achieve morphogenesis. A multiple-mechanism simulation aids us to study them.

12.2 Simulation experiments

The first suite of simulations examine contact-mediated lateral inhibition. Lateral inhibition is a type of cell–cell interaction whereby a cell that adopts a particular fate inhibits its immediate neighbours from doing likewise (Collier *et al.*, 1996). This competition creates a pattern of cells where one cell is activated (differentiated) and causes its neighbours to be deactivated. 'Contact-mediated' indicates that the inhibition proceeds only between cells in contact; there is no diffusing inhibitory chemical. The inhibitory activity occurs due to interactions of surface chemicals between adjacent cells. In this set of simulation experiments, we found that it is possible to get long-range patterning via this mechanism if there is some temporal agent to regulate the patterning process.

The second suite explores cell sorting via differential adhesion and finds that further conditions apply for cells with limited shape deformation capabilities. A few more examples are shown later to illustrate the breadth of possibilities with the multiple-mechanism simulator.

In each case, the work of setting up and running the initial experiment was a few hours and the entire suite of simulations was done part-time over the course of a few days. Individual simulation runs took a few minutes for the first example and a couple of hours for the second (on 1994-era workstations).

12.2.1 Long-range inhibition via contact

During conversations with biologist Ajay Chitnis in 1994, then at the Salk Institute, we considered the possibility of large scale patterning due to contact-mediated lateral inhibition. This investigation was motivated by experiments involving the Delta and Notch genes in *Drosophila* (a fruitfly). The inhibition is due to a surface chemical (the ligand) on one cell that binds specifically to another surface chemical (the receptor) on an adjacent cell. The ligand and receptor are a pair of complementary molecules that bind together if they are expressed on the membranes of adjacent cells. Thus a signal triggered by this binding can only be transmitted between cells that are in direct contact. Although cell contact is inherently local, can the inhibition be passed on from cell to cell to create a long-range inhibition? What properties would such a system have? These were some of the questions posed by Chitnis that we set out to examine.

There are two important mechanisms operating in this system:

1. a cellular switch that turns on if the inhibition is low, indicating the cell has differentiated into a new cell type, and
2. an inhibition that is passed on from cell to cell via interaction of surface chemicals (not diffusion). The inhibition acts to prevent the switch from activating. An inhibited cell will have a different fate (type).

The first simulation, shown in Figure 12.1, verifies that inhibition via a contact mechanism can indeed propagate across multiple cells. For this simulation, one seed cell in the centre is initialized as 'on' and exhibits high levels of the inhibitor; no other cells are allowed to differentiate in this simulation. Figure 12.1 illustrates the falloff of contact inhibition: the amount of inhibition sensed by each cell is proportional to the size of the grey circle drawn within it. Cells several neighbours away from the centre still sense the inhibition, even though they are not in direct contact with the single differentiated cell.

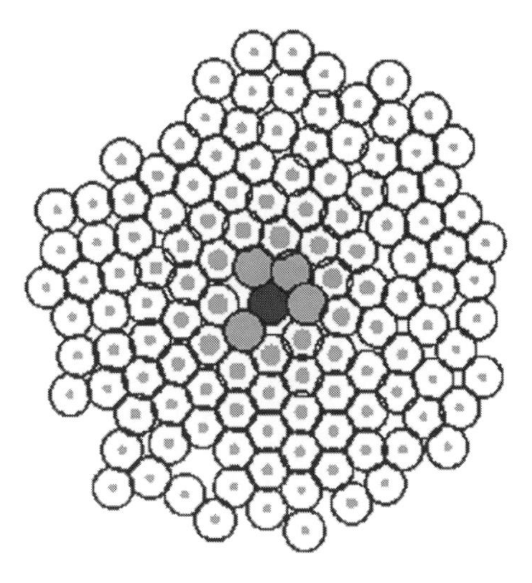

Figure 12.1 Propagation of inhibitory signal via surface chemicals. Differentiated cell shown in black. The amount inhibition felt by a cell is indicated by the size of the grey circle within it.

When we add to the simulation the ability for all cells to differentiate, we quickly found that different results arise depending on temporal ordering, as shown in Figures 12.2 and 12.3. In Figure 12.2, the cells first divided to form a sheet, and then began to differentiate roughly in unison. In Figure 12.3, the cells differentiated as they were dividing and the sheet was growing.

Cell state equations for this experiment

Let z_d be a state variable representing the concentration of some chemical that we will use to indicate that a cell has differentiated when it reaches a threshold value. This is shown in dark grey in the images. The concentrations of the two surface chemicals will be denoted z_r (receptor) and z_l (ligand).

The cell senses e_r, the amount of receptor bound *per unit area of contact*. This value is independent of the number of cells in contact and reports an approximation to the average concentration of ligand expressed by its neighbours (see Fleischer (1995) for details on this). The

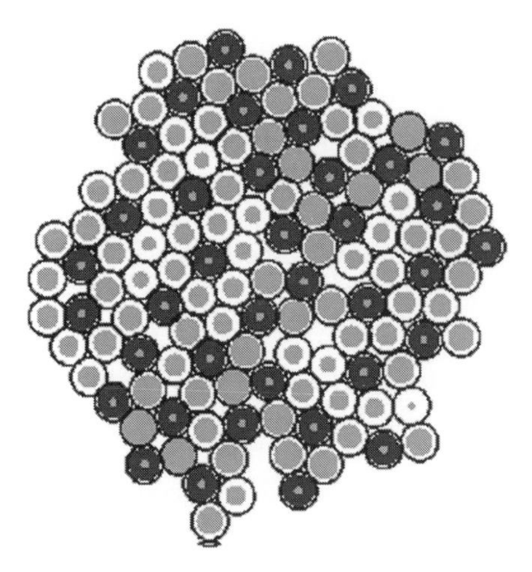

Figure 12.2 Short-range inhibition via surface chemicals.

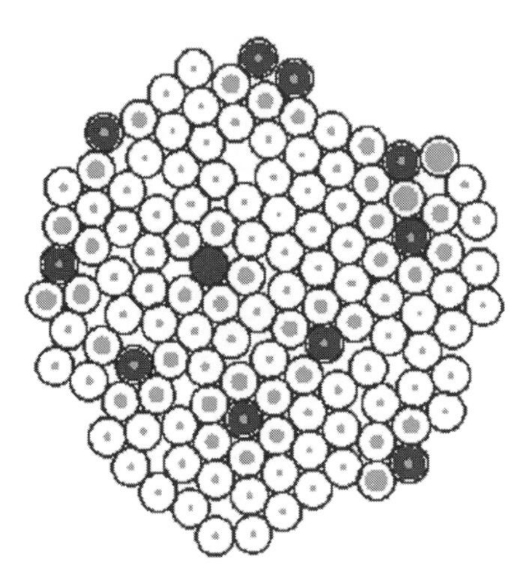

Figure 12.3 Long-range inhibition via surface chemicals.

equations for the genetic switch and the inhibitory surface chemicals are given here (and explained afterwards).

$$\begin{aligned}
\frac{dz_d}{dt} &= -20e_r + 10\frac{z_d^2}{1+z_d^2} - 0.5z_d + 13 \\
\frac{dz_l}{dt} &+= 0.95e_r \\
\frac{dz_l}{dt} &+= (z_d > 3)z_d \\
\frac{dz_l}{dt} &+= -z_l \\
\frac{dz_r}{dt} &+= 5 - z_r
\end{aligned} \tag{1}$$

(**1**) gives the cell state equations for experiments with inhibition via surface chemicals. The $+=$ notation indicates that the terms are summed to get the full contribution to z_l. The $>$ is a [0,1] sigmoid. r is a receptor for the ligand l. The surface chemical r is expressed equally on all cells; they all have the same amount of receptors, enough to bind all the ligands they may contact.

Our model system contains additional equations determining the rate of cell division, small motions of the cells, etc., but they are not germane to the present discussion. The complete experiment file can be found in Fleischer (1995, Appendix A.2); we cover the salient points in the following paragraphs.

Let us examine the ligand equations for z_l. Consider first what happens for a cell which has z_d low. This cell is undifferentiated ('off') and $(z_d > 3)$ evaluates to zero, removing one term from the equation. The remaining terms give:

$$\frac{dz_l}{dt} = 0.95e_r - z_l$$

As we compute this differential equation forward in time, we soon reach a state where l is expressed at 95% of what a cell senses via the receptors. This means that it attempts to inhibit its neighbours at 95% of the inhibition it gets from them. Thus the inhibitory influence is passed along between non-activated cells. This can be seen in Figure 12.1, where all of the cells apart from the one seed cell are undifferentiated (z_d is low). The inhibition diminishes for cells further from the differentiated (dark grey) cell.

For a cell which has z_d high (i.e. it is 'on' = it has differentiated), the conditional term $(z_d > 3) = 1$, causing an extra term to be added into the equation for z_l:

$$\frac{dz_l}{dt} = 0.95e_r - z_l + z_d$$

The output inhibition then approaches 95% of what the activated cell *sees*, *plus* more, proportional to how 'on' the cell is. Thus a differentiated cell inhibits its neighbours even more strongly and they pass this inhibition on to their neighbours.

The remaining equation is for the genetic switch, which implements the differentiation process by making z_d large if the inhibition e_r is small.

$$\frac{dz_d}{dt} = -20e_r + 10(z_d^2)/(1 + z_d^2) - 0.5z_d + 13$$

This equation is similar to those discussed in Murray (1993, Chapter 5). It has both elements of the standard reaction-diffusion equations: autocatalysis (a reaction which tends to grow) and lateral inhibition (inhibit the neighbours), but differs in that they are computed in discrete cells and without the use of continuous diffusion.

Many stable states are possible; temporal sequence is important

There are many stable states for this system and different regimes of stable states as well. This became obvious quite quickly when beginning to work with this simulation experiment.

The different regimes shown in Figures 12.2 and 12.3 are due to the discrete nature of the lateral inhibition in this experiment. *It can only be passed on from cell to cell if the cells are in direct contact.* A non-activated cell cannot express enough inhibitor to deactivate a neighbour which has already differentiated.

If it could do so, then it could cause the differentiated cell to reverse state and the system would not be stable. Thus, if cells ever achieve the state in Figure 12.4(ii), where every other cell is 'on', they will remain in that state. This generates patterns with short-range inhibition.

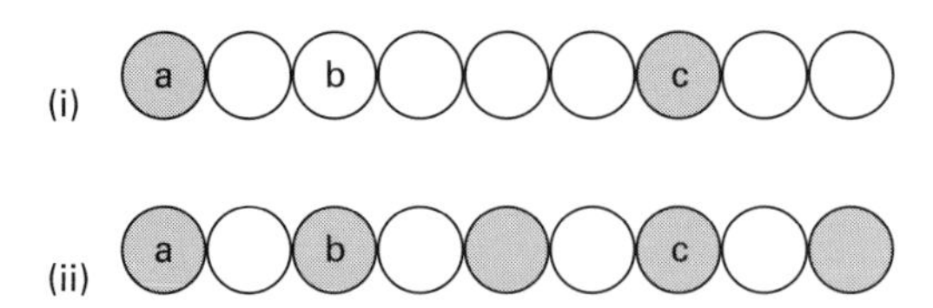

Figure 12.4 Cells shown in dark grey have differentiated and inhibit their neighbours from differentiating. There are many different stable states for this system, two of which are shown in (i) and (ii). Which state is reached depends on initial conditions; (i) corresponds to the long-range inhibition shown in Figure 12.3 and (ii) corresponds to the short-range inhibition shown in Figure 12.2.

But what about long-range inhibition? Can it also be accomplished with this mechanism? Yes, as shown in Figure 12.3, but it depends on the temporal order in which cells differentiate. Cells which differentiate early can inhibit their neighbours for a large distance, given enough time. That is the central issue — *is there time for the inhibitor to propagate to the other cells before they differentiate?* In simulation terms, this is similar to discussing initial conditions. In a developing organism, one must be careful when discussing 'initial conditions'. It is probably not the case that a sheet of cells is held at some set of initial values and then suddenly, synchronously, they all begin to differentiate. It is more plausible to consider that the cells are gradually reaching the stage when they are ready to differentiate and then the competition to differentiate is triggered by some internal or external signal: based on an internal cellular clock, a diffusing agent, or a propagating signal such as the morphogenetic furrow sweeping across the *Drosophila* retina.

One way to ensure state Figure 12.4(i) (long-range inhibition) is to have an additional chemical signal that is required for a cell to activate. Let this signal propagate in time, first enabling cells on the left and later the cells to the right. If the rate at which the enabler is progressing is slower than the rate of the inhibition, then the state shown in Figure 12.4(i) can be reached reliably. This type of temporal effect is intimately related to the ability of surface chemicals to exhibit long-range inhibition.

Consider a sheet of cells as it grows – there is a natural temporal ordering. The cells which exist first have first crack at differentiating. As new cells are formed at the edge of the sheet, they are inhibited from differentiating by the previous cells. New cells will only be able to differentiate when the edge of the sheet has progressed far enough from the previously differentiated cells. Thus, a developing sheet can exhibit long-range inhibition using the surface inhibition mechanism in conjunction with growth at the edge of a sheet. This was the method used to generate the simulation shown in Figure 12.3.

Did it make relevant, non-obvious predictions?

We learned that temporal ordering is important for long-range inhibition using surface chemicals. It recently came to my attention (thanks to Dr Chitnis) that about a year after I had done this work, Collier *et al.* tackled this same problem using the more rigorous tools of mathematical biology (Collier *et al.*, 1996). They reached much the same conclusion. Regarding the activator-inhibitor model, they reported that although it is stable once it was in the long-range pattern, that model alone is unable to generate long-range patterns. Something else is needed:

> *Where the scale of the pattern is on the order of many cell diameters, our arguments imply that other mechanisms instead of, or in addition to, the Delta-Notch signalling considered here must be operating.*
>
> (Collier *et al.*, 1996)

Within my simulation environment, I was able to follow that insight a bit further and develop a simulation that did indeed generate the pattern, by using cell division to stage the rate of differentiation. Note that the cell division used to generate Figure 12.3 was occurring during the differentiation; as the sheet of cells grew outwards, cells were differentiating and continuing to inhibit each other[1]. Collier *et al.* cite an earlier result that indicates the cell-division method is not the mechanism used in the Delta-Notch situation, since the timescale for the differentiation is much shorter than the cell cycle (Hartenstein and Posakony, 1990). So this part of our hypothesis turns out to be incorrect. Perhaps the diffusable modulator we postulated above is a better prediction.

Nonetheless, simply the realization that there must be an additional element at play to set up the long-range patterning is significant. As Collier *et al.* (1996) say in their abstract:

> *It is not clear under precisely what conditions the Delta-Notch mechanism of lateral inhibition can generate the observed types of pattern, or indeed whether this mechanism is capable of generating such patterns by itself.*
>
> (Collier *et al.*, 1996)

Our investigation with the synthetic biology approach helped us quickly understand that the Delta-Notch contact mechanism alone is insufficient to produce the pattern. Further, we postulated that a temporal regulator is likely to be present. We arrived at this insight without having to perform a thorough mathematical analysis. The subsequent investigation by Collier *et al.* (1996) indicates that they also found this result to be relevant and non-obvious.

This rough simulation experiment is a means of gaining experience with a model system. It is not as 'real' as an experiment, but not as abstract as the back-of-the-envelope calculations or a simplified mathematical model. I found it instructive for this problem and others.

12.2.2 Relationship between cell shape and the differential adhesion hypothesis

In this next suite of simulated experiments, we quickly discovered some implications of cell shape for Steinberg's *differential adhesion hypothesis* (Steinberg, 1964, 1970; Armstrong, 1989). The differential adhesion hypothesis is a model to describe cell sorting (the separation of different types of cells when mixed either *in vitro* (Armstrong, 1989; Gilbert, 1991; Steinberg and Takeichi, 1994) or *in vivo* (Crawford and Stocum, 1988; Gilbert, 1991). It states that

> *in any population of motile, cohesive cells, weaker cell attachments will tend to be displaced by stronger ones, and this adhesion-maximization process ... (produces a) ... configuration in which the total intensity of cell bonding is maximised.*
>
> (Steinberg and Takeichi, 1994)

[1] This is distinct from another way of generating the same pattern which would be to first create a short-range inhibition pattern, then allow a few generations of cell division by one of the cell types, thus 'filling in' between the activated cells.

It is also assumed that the motility is 'non-directed.' The outcome of this process is that a mixture of two types of cells can rearrange itself into a particular configuration due solely to the adhesive properties of the two cell types. Typically, cells with similar adhesion will end up clumped together.

The initial motivation for this exploration was to show differential adhesion as another capability of the simulator. After encountering some unexpected behaviours (e.g. cells wandering away, see Figure 12.9), we were led to consider the discrepancies between our model, Steinberg's model and the actual biological phenomena.

In the current implementation of our simulator, cells are spheres and are not capable of large deformations. Other models, such as the lattice model proposed by Glazier and Graner are able to represent this transition more accurately and hence they achieve much nicer results when simulating differential adhesion (Glazier and Graner, 1993)[2].

Our simulated experiments with this model indicate that the cell sorting behaviours depend on additional conditions on cell shape and/or the magnitudes of the motile forces relative to the adhesive forces.

The essence of our result is shown in Figure 12.5. If the cells are able to deform sufficiently to remain in contact throughout the transformation, then they can achieve the state shown in Figure 12.5(c). This is compatible with the models for cell movement mentioned by Armstrong:

> *Two mechanisms of cell movement have been considered for sorting: active pseudopod-generated locomotion and associative movement. The latter process envisions cell movement to be the consequence of a 'zippering up' of homotypic contacts and a resultant reduction in the area of contact between cells. The process does not require the activities of pseudopods as being responsible for cell locomotion.*
>
> (Armstrong, 1989)

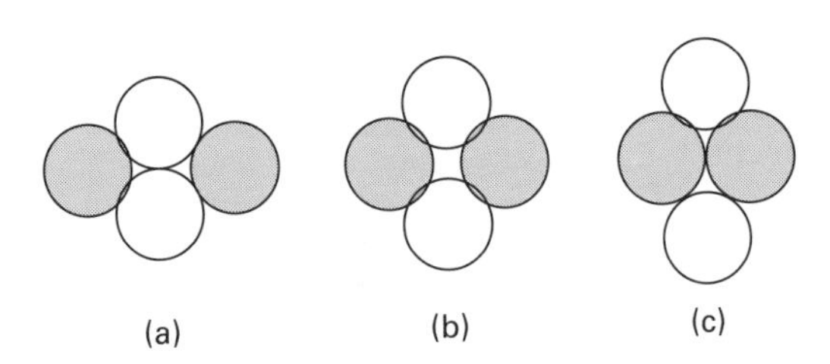

Figure 12.5 A configuration of four cells which cannot reach the minimum surface-energy configuration without going over an energy 'hill'. This is sometimes called a T1 transition (Glazier and Graner, 1993, Figure 1). Assume that the dark–dark cell adhesion is stronger than the light–light cell adhesion or the dark–light adhesion. The four cells in (a) cannot reach the minimum surface energy configuration in (c) without either deforming substantially, or passing through state (b), which requires work. Note: an analogous situation can be constructed in 3D.

However, if the cells cannot deform in this manner, then the light cells in Figure 12.5(a) must lose contact due to their 'non-directed' motility. This is necessary to escape a local peak in the energy landscape. But if cells have this ability to break an adhesive bond with a neighbour, then some cells will (with some probability) eventually break all bonds and wander off alone. This can be observed in several of the simulation experiments below.

[2] We consider the possibilities of extending our model to use lattice methods for shape representation in Fleischer (1995, 9.3.1).

In summary, we found one of the following additional conditions must hold for the differential adhesion hypothesis to hold:

1. cells must be capable of extensive shape deformation, or
2. a cell's motility must be strong enough to break its bonds with neighbouring cells.

Did it make relevant, non-obvious predictions? Differential adhesion simulation experiments

Figures 12.6 to 12.11 illustrate the experiments that led to the preceding analysis. In the first simulation (Figure 12.7) light-coloured cells adhere to each other and dark-coloured cells adhere to each other, but dark and light do not adhere. Even with a mixed initial condition, the clumps of light and dark cells form and do not adhere to each other, as one would expect.

As we introduce varying amounts of adhesion between the two cell types, we get patterns like those in Figures 12.9 to 12.11. The light-coloured cells have weaker adhesion between each other than the darker cells, which causes them to end up on the outside. After trying several parameter settings, it became apparent that the ratio of the cell motility to the adhesion was critical in determining the patterns. This realization prompted the insights discussed above.

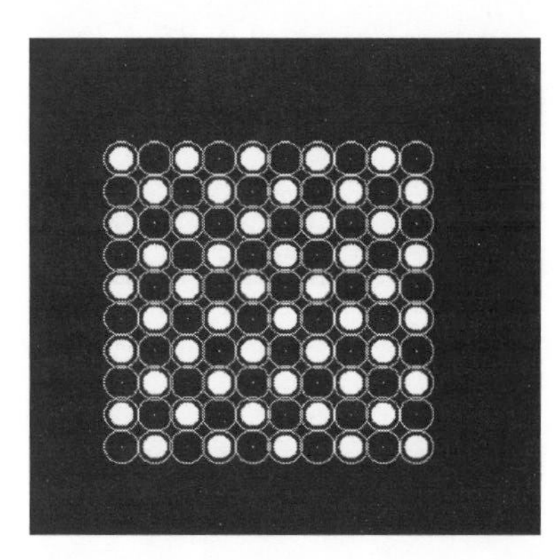

Figure 12.6 Differential adhesion starting state.

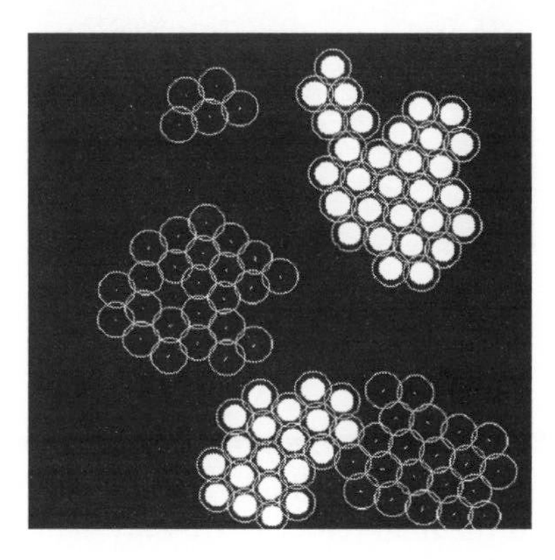

Figure 12.7 Differential adhesion later state. Cells of each type do not adhere to the other at all in this simulation.

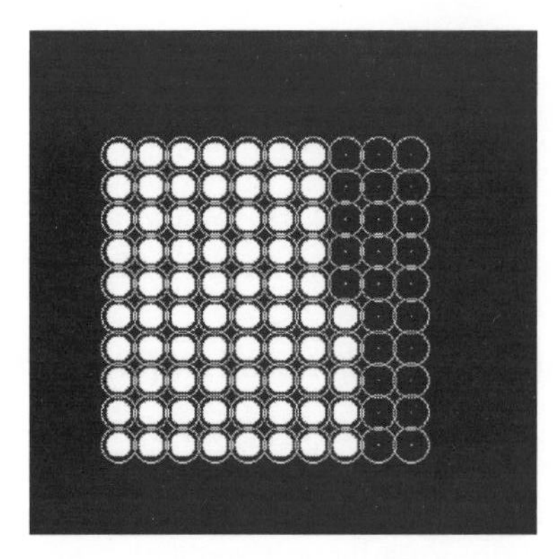

Figure 12.8 Differential adhesion starting state for remaining experiments.

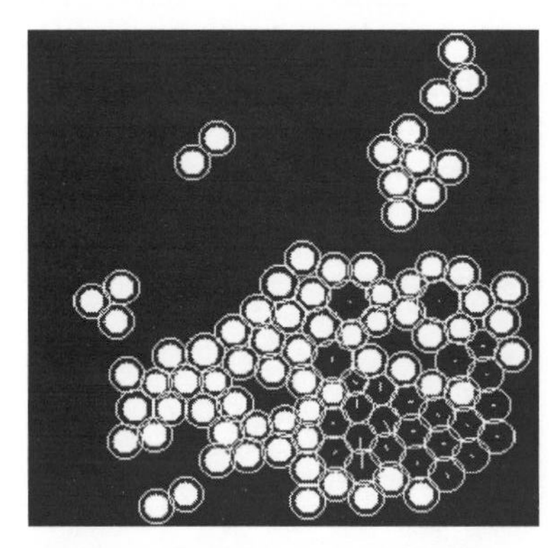

Figure 12.9 Slight adhesion for a–a, more for a–b, lots for b–b.

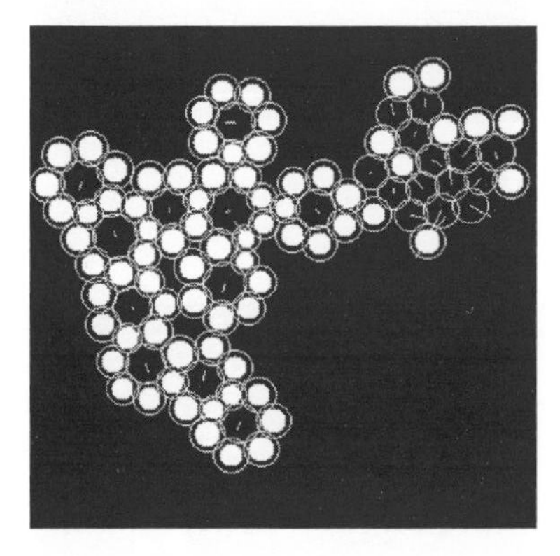

Figure 12.10 Slight adhesion for a–a, more for a–b, lots for b–b, run for longer time.

Each simulation run for the differential adhesion took on the order of 2 hours (on 1994-era HP workstations). This long run-time is partially due to stiffness arising from the collision equations. Another factor is that the final patterns are only achieved after a substantial amount of random motion of the cells. The interaction between the collision forces and the random motion causes small solver steps.

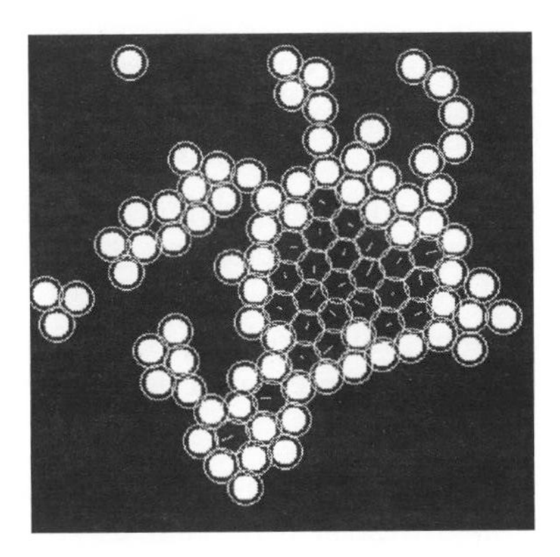

Figure 12.11 Same as Figure 12.10, with slightly different compression/expansion forces. (See Fleischer, (1995, Ch. 9.3.1) for details.)

Did it make relevant, non-obvious predictions?

In this set of experiments, the result was non-obvious, but only marginally relevant to the situation we are interested in. Cell shape and deformation are important components of differential adhesion. Thus the synthetic biology approach was again useful for generating further understanding of the phenomenon.

While this is true, it is not particularly interesting for morphogenesis. There do exist real cells that are nearly spherical with limited deformation, like those of our model (e.g. various bacteria). However, the differential adhesion hypothesis is of particular interest as a possible means of creating structure in animal tissues, where cells are typically capable of substantial deformation.

This is a case where we applied a model that was not quite appropriate. It is worth noting that doing the simulation experiments gave us more insight into the regimes where our model is applicable. This is a valuable and essential part of biological investigation – as we develop theories and models, we need to verify that they are sufficient to represent the phenomena in question.

Recalling that the initial motivation for this exploration was to show capabilities of the simulator, it is not so surprising that the insights we have gained here are less relevant than for the first suite, which were motivated directly by a biological question.

12.3 Discussion

The two suites of experiments on lateral inhibition and differential adhesion show us that a short time spent with a simulator can give us insights into biological phenomena and into our models of them.

In recent years, several factors have made it possible to consider a more comprehensive simulation approach: increasing computational power, improved graphical display, better understanding of user-interfaces and greater accessibility of computers. To be most useful, a system like this could be used directly by biologists, aiding them to explore their own hypotheses and models. This goal remains challenging, but not out of reach.

As we mentioned at the outset, the key to the success of this approach will be if it can satisfy the criteria indicated by Murray: biological relevance and making non-obvious predictions. In this chapter we have given examples of this on a small scale. Novel features of the approach include:

- provides an alternative 'constructivist' way of thinking about biological models
- enables the study of combinations of different mechanisms and their interactions in space and time
- encourages users to consider how 'initial conditions' might arise, or how a model of one interaction might fit into the larger context of a developing organism.

As with any simulation or mathematical analysis, this approach enables experiments which would be difficult or impossible *in vivo* or *in vitro*, enables very controlled experiments to be done and provides the ability to study parameter variations.

12.3.1 Challenges for synthetic biology

A large hurdle to overcome is creating a system that enables a biologist to explore a model without needing detailed mathematical and numerical expertise. I believe this can be done using appropriate User Interface techniques. If the benefit is significant enough, even before this stage we will increasingly see computational specialists working in biology labs (as we do in the DNA analysis area).

Another major challenge is that the base simulation system needs to contain a wide range of pre-existing models to build upon. The system used to create the examples in this chapter had simple models for contact, adhesion, chemical kinetics with the cell, gene regulation and more. This made it easy to do a range of simulated experiments. However, we still run into limitations because the components are simple and there is typically only one component for any given phenomenon. As a concrete example, in the differential adhesion investigation discussed above we ran into issues with the spherical cell-shape model. It is desirable to have a choice of cell-shape models that could be plugged in when desired, such as that used in Agarwal's CPL system. For some situations, spherical cells are adequate and compute faster. For other situations, a more detailed model of cell shape and deformation is needed.

A particular model is valid only in a particular domain. The computer does not understand intent, so it cannot determine if the model is being applied appropriately. Thus it is critical that the user understand the model and its range of applicability. Working with the simulator can help us to understand the behaviours, and limitations, of our models.

Biological investigation can be viewed as a process of creating and refining mental models of biological phenomena. Synthetic biology gives us a tool to formalize our models and to verify through simulation that they are sufficient to produce the phenomena. Of course, this does not imply that a particular model accurately represents a phenomenon – that can only be verified by experiment.

12.3.2 Synthetic biology applied to archival model representation

Another potentially powerful role for simulation is to serve as a tool for communicating and storing biological knowledge. As our understanding grows, we create more and more complex models of biological phenomena. For example, the current state of knowledge of the developmental process in *Drosophila* retina is kept in the minds of researchers, and, indirectly, in many journal papers, books, etc. With systems like the retina, many genes are involved and many of their roles and interactions are known or postulated. It is difficult to keep track of all of the hypotheses and to detect inconsistencies. A suitably flexible simulation framework could serve as a book-keeping mechanism for the research of such phenomena. This would also be a valuable educational tool. Along these lines, Jim Bower's group at Caltech has started a database of compartmental models for pyramidal cells (Beeman *et al.*, 1997) and Hiroaki Kitano's group has embarked on an ambitious and interesting project to simulate *C. elegans*. (Kitano *et al.*, 1998).

12.4 Examples indicating the range of simulator

We draw the chapter to a close with several examples showing the range of behaviours one can create with this multiple-mechanism simulator. This is further evidence that we are nearing the point where we can simulate detailed enough models to capture the essence of a complicated developmental system.

Colour plates 12.1 to 12.7 show a variety of developmental motifs that were created with the multiple-mechanism simulator. These experiments were done to test the simulator as well as to gain familiarity with programming in a genetic developmental language. No evolution is shown here; these are all created by hand.

First and foremost, one must deal with size regulation. In colour plate 12.1, we start with a single cell that divides under genetic control, halting when a diffusing chemical reaches a certain threshold. Cell adhesion causes the cells to form chains. Colour plate 12.2 regulates its size via a different mechanism – there is a limited number of cell divisions. In this simulation, we also introduce differentiation. An initial pattern is formed using a combination of a diffusing inhibitor and cell migration. Then these yellow precursor cells divide several times to create the purple cells that fill in between them, creating a compartmental pattern. A traditional reaction-diffusion pattern is used to prepattern cell growth and migration in colour plate 12.3.

Colour plate 12.4 is an example of a hierarchical structure that can regenerate itself. Killing off of a cell or two in one of these will trigger cells to divide again, re-creating the original shape. The mechanisms used are primarily surface chemicals – each cell senses its neighbours and will divide to create more neighbours of the proper type if it senses they are missing.

Colour plate 12.5 introduces more mechanical aspects. Cells on one side of the chain shrink, causing the entire chain to curl. The chain is held together by adhesive surface chemicals. This chain begins as a single cell and the shape is determined via cleavage plane and cell division. These are controlled via the internal cell chemicals represented as differential equations, just as in all the simulation experiments described here.

How does nature wire up a one-to-one network? You might suppose that the target cells simply emit some diffusing attractor and the neurites follow that. A little more thought, or a short simulation experiment, shows that this does not work so well. Because the neurites are simply attracted to the concentrations of diffusing chemicals, they are likely to head for the first neurons they can find, rather than innervating the entire area equally. This problem can be addressed in various ways, one of which is shown in colour plate 12.6.

The small neural net shown in colour plate 12.6 is in the process of sending out neurites from one population of cells on the right to those on the left. In order to avoid all neurites innervating the first cells, the target cells change state when they are innervated. This state change causes changes in surface chemical expression, which in turn makes the neurites no longer stick to these cells. So the first one sticks and later neurites continue on, finding cells deeper in the left-hand colony. This set of interacting mechanisms enables the innervation to spread throughout the cells on the left. After all cells are innervated, neurites that did not find a connection die off, leaving us with a one-to-one network. Due to the time it takes for the intracellular processes, sometimes there will be a slight error – two neurites to one cell, or a cell without a neurite. These are the sorts of errors that occur in real systems as well and further mechanisms can be postulated to address them. Applying similar strategies, a hierarchical network can be created as well (colour plate 12.7).

Finally, colour plates 12.8 and 12.9 show applications of developmental models to computer graphics, creating organic detail automatically (Fleischer *et al.*, 1995).

12.5 Conclusions

A comprehensive simulation environment can be a useful tool for a biologist exploring models of development. It encourages us to consider how initial conditions arise and how multiple mechanisms work together to form developmental patterns. The constructive nature of a simulation exploration promotes an alternative viewpoint that may encourage insights that are less obvious via other approaches.

Acknowledgements

Thanks to Dr Alan Barr and to Caltech, the CNS department, and the members of the graphics lab for collaborating with me and for supporting this research. This work was supported in part by grants from Apple, DEC, Hewlett Packard, and IBM. Support was provided by NSF (ASC-89-20219) as part of the NSF/ARPA STC for Computer Graphics and Scientific Visualization, by DOE (DE-FG03-92ER25134), the Beckman Foundation, the Parsons Foundation, the National Institute of Health as part of their Training Grants programme, and by the National Institute on Drug Abuse and the National Institute of Mental Health as part of the Human Brain Project.

References

Agarwal, P. (1994) The cell programming language. *Artificial Life*, **2**(1), 37–77. Thesis (1993): http://www.cs.nyu.edu/csweb/Research/theses.html.

Armstrong, P.B. (1989). Cell sorting out: the self-assembly of tissues *in vitro*. *CRC Critical Reviews in Biochemistry and Molecular Biology*, **24**, 119–149.

Beeman, D., Bower, J.M., Schutter, E.D. *et al.* (1997) The genesis simulator-based neuronal database. In *Neuroinformatics: An overview of the Human Brain Project*, Chapter 4, Koslow, S.H. and Huerta, M.F. eds. In particular, see: http://www.genesis-sim.org/hbp/GOOD/node5.html, 57–80. Lawrence Erlbaum Associates.

Collier, J.R., Monk, N.A.M., Maini, P.K. and Lewis, J.H. (1996) Pattern formation by lateral inhibition with feedback: a mathematical model of delta-notch intercellular signalling. *Journal of Theoretical Biology*, **183**(4), 429–446.

Crawford, K. and Stocum, D.L. (1988) Retinoic acid coordinately proximalizes regenerate pattern and blastema differential affinity in axolotl limbs. *Development*, **102**, 687–698.

Fleischer, K.W. (1995) *A Multiple-Mechanism Developmental Model for Self- Organizing Structures*. PhD. dissertation, Caltech, Department of Computation and Neural Systems, Pasadena, CA, 91125.

Available online at http://caltechcstr.library.caltech.edu/perl/search (search for Fleischer). Color images online at http://www.gg.caltech.edu/ kurt/avail.html.

Fleischer, K.W., Laidlaw, D.H., Currin, B.L. and Barr, A.H. (1995) Cellular texture generation. *Proceedings of ACM SIGGRAPH '95*, Los Angeles, CA.

Gilbert, S. (1991) *Developmental Biology*. 3rd edn. Sinnauer Associates.

Glazier, J.A. and Graner, F. (1993) Simulation of the differential adhesion driven rearrangement of biological cells. *Physical Review E*, **47**(3), 2128–2154.

Hartenstein, V. and Posakony, J.W. (1990) A dual function of the notch gene in *Drosophila sensillium* development. *Developmental Biology*, **142**, 13–30.

Kauffman, S.A. (1993) *Origins of Order: Self-Organization and Selection in Evolution*. Oxford University Press.

Kitano, H., Hamahashi, S. and Luke, S. (1998) The perfect *C. elegans* project: an initial report. *Artificial Life*, **4**(2), 141–156.

Mjolsness, E., Sharp, D. and Reinitz, J. (1991) A connectionist model of development. *J. Theo. Bio.*, **152**(4), 429–454.

Murray, J.D. (1993) *Mathematical Biology*. 2nd edn. Springer-Verlag.

Steinberg, M.S. (1964) The problem of adhesive selectivity in cellular interactions. In Locke, M., ed. *Cellular Membranes in Development*. Academic Press.

Steinberg, M.S. (1970). Does differential adhesion govern self-assembly processes in histogenesis? Equilibrium configurations and the emergence of a hierarchy among populations of embryonic cells. *J. Exp. Zool.*, **173**, pp. 395–434.

Steinberg, M.S. and Takeichi, M. (1994) Experimental specification of cell sorting, tissue spreading and specific spatila patterning by quantitative differences in cadherin expression. Proceedings of the National Academy of Science (USA), 91, pp. 206–209. *Developmental Biology*.

Section 4

Developmental Biology Inspired Computation

The evolution of evolvability

13

RICHARD DAWKINS

13.1 Introduction

A title like The Evolution of Evolvability ought to be anathema to a dyed-in-the-wool, radical neo-Darwinian like me. Part of the reason it is not is that I really have been led to think differently as a result of creating, and using, computer models of artificial life which, on the face of it, owe more to the imagination than to real biology. The use of artificial life, not as a formal model of real life but as a generator of insight in our understanding of real life, is one that I want to illustrate in this chapter. With a program called *Blind Watchmaker*, I created a world of two-dimensional artificial organisms on the computer screen (Dawkins, 1986). Borrowing the word used by Desmond Morris for the animal-like shapes in his surrealistic paintings (Morris, 1987), I called them biomorphs. My main objective in designing *Blind Watchmaker* was to reduce to the barest minimum the extent to which I designed biomorphs. I wanted as much as possible of the biology of biomorphs to *emerge*. All that I would design was the conditions – ideally very simple conditions – under which they might emerge. The process of emergence was to be evolution by the Darwinian process of random mutation followed by non-random survival. Once a Darwinian process gets going in a world, it has an open-ended power to generate surprising consequences: us, for example. But, before any Darwinian process can get going, there has to be a bare minimum of conditions set up. These were the conditions that I had to engineer in my computer world.

The first condition, one that I have emphasized sufficiently before (Dawkins, 1976, 1982), is that there must be *replicators* – entities capable, like DNA molecules, of self-replication. The second condition is our main concern in this chapter. It is that there must be an embryology; the genes must influence the development of a phenotype; the replicators must be able to wield some phenotypic power over their world, such that some of them are more successful at replicating themselves than others. The type of embryology that we choose for our artificial life is crucial. This is another way of stating the key message of this chapter.

13.2 Genetics versus embryology

The fundamental principle of embryology in real life (and one that I decided was worth imitating in artificial life) was formulated by Weismann (1893). It is illustrated in Figure 13.1, which

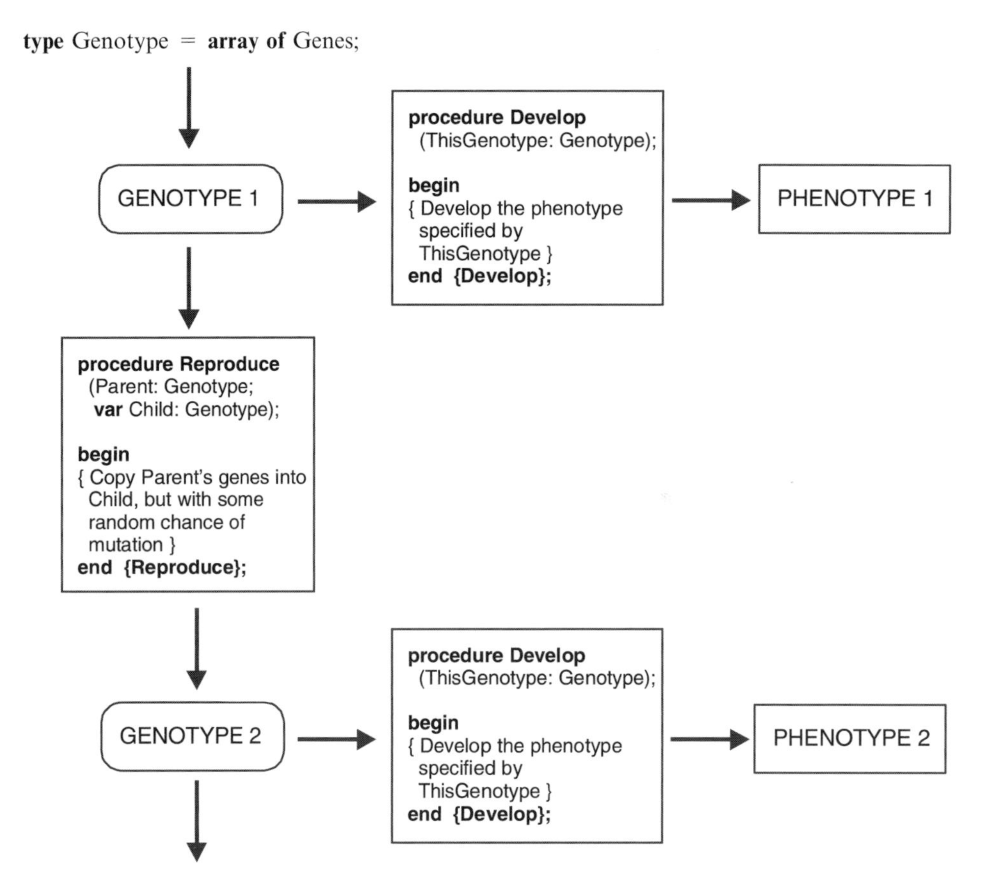

Figure 13.1 Weismann's continuity of the germ plasm (expressed in Pascal).

covers a period of two generations preceded and followed by an indefinite number of generations. I have expressed part of it in Pascal, in anticipation of my explanation of how the Blind Watchmaker program closely follows Weismann's doctrine.

The key point is that, in every generation, it is only genes that are handed on to the next generation (I shall leave it to pedants to fuss about the sense in which this is not strictly true). In every generation, the genes of that generation influence the phenotype of that generation. The success of that phenotype determines whether or not the genes that it bears, a set that largely overlaps with the genes that influenced its development, shall go forward to the next generation. (In *The Extended Phenotype* (1982) I explore the far-reaching consequences of the fact that these two sets do not necessarily *have* to overlap.) Any individual born, therefore, inherits genes that have succeeded in building a long series of successful phenotypes, for the simple reason that failed phenotypes do not pass on their genes. This is natural selection. It is why organisms are well adapted and it is why we are all here.

It is important to understand that genes do two quite distinct things. They participate in embryology, influencing the development of the phenotype in a given generation; and they participate in genetics, getting themselves copied down the generations. It is too often not realized – even by some of those that wear the labels geneticist or embryologist – that there is a radical separation between the disciplines of genetics and embryology. Genetics is the study of the vertical arrows in Figure 13.1, the study of the relationships between genotypes in successive generations. Embryology is the study of the horizontal arrows, the study of the relationship between genotype and phenotype in any one generation. If you doubt that the separation between the two disciplines is fundamental, consider the matter methodologically. You could do perfectly respectable embryology on a single individual. Genetics on a single individual would be meaningless. Conversely, you could do perfectly respectable genetics, but not embryology, on a population of individuals, each one sampled at only one point in its life cycle.

13.3 Implementing embryology

I resolved to maintain the separation between genetics and embryology in my artificial life. To this end, I wrote the program around two strictly separate procedures called **Reproduce** and **Develop** (Figure 13.1). Another part of the program presented an array of phenotypes for selection, each one drawn by the procedure **Develop** under the influence of genes that would be held responsible for its success or failure. In any generation the phenotype chosen (by some criterion) as successful would be the one whose genotype went forward via **Reproduce** (with some possibility of random mutation) to the next generation. The selection criterion itself I was content to leave, for the moment, to the aesthetic taste of a human chooser. The model would therefore be, at least in the first implementation, a model of artificial selection (like breeding cattle for milk-yield) not natural selection. As a didactic technique this has an honourable history. Charles Darwin made persuasive use of artificial selection as a metaphor for natural selection.

Now to flesh out the bare bones of Figure 13.1. We must write some code in the two procedures, to specify the details of genetics and embryology respectively. Genetics is straightforward. However we choose to represent genes, it is obviously easy to copy them from parent to child and it is obviously easy to introduce some minor random perturbation in the copying to represent mutation. It is embryology that we have to think about more carefully. What shall we write between the { } brackets in the procedure **Develop**, to specify the relationship between genotype and phenotype?

13.3.1 Pixel peppering

The first naive idea that might occur to us is to go for maximum generality. We know that the phenotypes in our artificial world are all going to be two-dimensional pictures on a Macintosh screen. The Macintosh screen is an array of $340 \times 250 = 85\,000$ pixels. If we give our biomorphs genotypes of 85 000 genes, each having a value of 1 or 0, we know that any conceivable phenotype in our artificial world can be represented by a specific genotype. Moreover, any pixel can mutate to its opposite state and the resulting picture might be selected, or not, in preference to its parent. It follows, therefore, that we could in theory ‘breed’ any picture from a random starting pattern (Figure 13.2a) or, indeed from any other picture, getting from, say, Winston Churchill to a Brontosaurus, by scanning every generation hopefully for slight resemblances to the target picture.

But only in theory. In practice we would be waiting till kingdom-come. This really would be a very naive way of writing **Development** and it would produce a very boring kind of artificial life. It is a kind of zero-order embryology, the kind of embryology we must improve upon. Our improvements will take the form of constraints. Constrained embryologies are improvements over naive pixel-peppering, not because they have greater generality but because they have less. Naive pixel-peppering can produce all possible pictures, including the set that anyone might

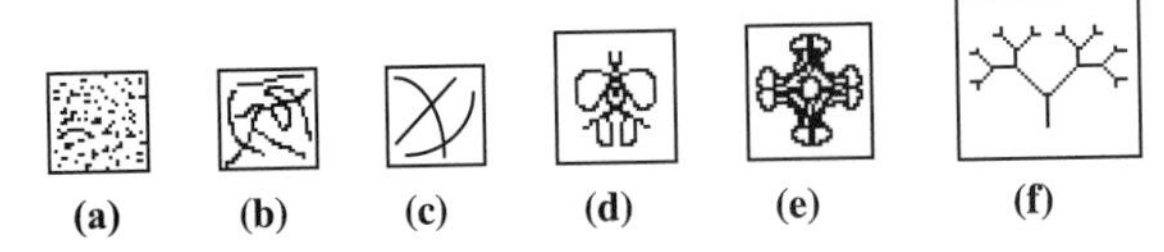

Figure 13.2 Breeding from a random starting pattern (a); random lines (b); lines of mathematical families (c); mirror algorithms (d); letting genes determine the presence or absence of mirrors in various planes of symmetry (e); and 'archetypal' body form generated by *Blind Watchmaker's* artificial embryology (f).

regard as biological. The trouble lies in the astronomical number of nonsense pictures that it can also produce. Constrained embryologies have a restricted set of phenotypes that they can generate, and they will be specified by a smaller set of genes, each gene controlling a more powerful drawing operation than colouring a single pixel. The task is to find an embryological procedure whose phenotypes are restricted in biologically interesting directions.

13.3.2 Lines and mirrors

So, what kinds of constrained embryologies might we think of as improvements over naive pixel-peppering? A slight improvement would be gained if, instead of drawing random pixels, we draw random lines (Figure 13.2b). Pixels, in other words, tend to pop up next to one another rather than just anywhere. We might further specify that the lines should belong to recognized mathematical families – straight lines whose length and angle is specified by genes; curves whose shape is specified by a polynomial formula whose coefficients are specified by genes (Figure 13.2c). Yet another constraint might be one of symmetry. Most animals are, as a matter of fact, bilaterally symmetrical, though some show various kinds of radial symmetry, and many depart from their basic symmetry in minor respects. We could use mirror algorithms in writing **Development** (Figure 13.2d), letting genes determine the presence or absence of mirrors in various planes of symmetry (Figure 13.2e).

13.3.3 Recursive trees

But though all these embryologies are obvious improvements over naive pixel-peppering, I did not tarry long over them. Right from the start of this enterprise, I had a strong intuitive conviction about the kind of embryology I wanted. It should be *recursive*. My intuition was based partly upon the generative power of recursive algorithms well known to computer scientists; and partly upon the fact that the details of embryology in real life can, to a large extent, be thought of as recursive. I can best illustrate the idea in terms of the procedure that I ended up actually using.

```
procedure Tree(x, y, length, dir: integer; dx, dy:, array [0..7] of integer);
{Tree is called with the arrays dx & dy specifying the form of the tree, and
length specifying the starting value of length. Thereafter, tree calls itself recursively
with a progressively decreasing value of length until length reaches 0};
var xnew,ynew: integer;
begin if dir < 0 then dir:=dir + 8; if dir >= 8 then dir:=dir -8;
xnew:=x + length * dx[dir]; ynew:=y + length * dy[dir];
MoveTo(x, y); LineTo(xnew, ynew);
```

```
if length > 0 then {now follow the two recursive calls, drawing to left and right respectively}
  begin
  tree(xnew, ynew, length − 1, dir − 1) {this initiates a series of inner calls}
  tree(xnew, ynew, length − 1, dir + 1)
  end
end {tree};
```

What this procedure actually draws depends upon the starting value of the parameter 'length', and the values of *dx[0]* to *dx[7]* and *dy[0]* to *dy[7]* that are plugged in. A particular setting of these values, for instance, draws a tree like Figure 13.2(f), which I think of, somewhat arbitrarily, as the basic, 'archetypal' body form generated by my artificial embryology. The sequence of pictures in Figure 13.3 shows the sequence of lines by which the tree is drawn by the recursive algorithm.

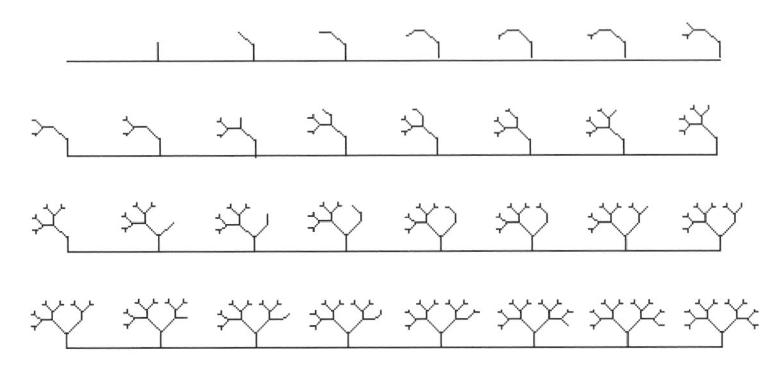

Figure 13.3 Recursive tree-drawing sequence.

Real-life embryology is quite like this, and very unlike the pixel-peppering embryology that we thought about before. Genes do not control small fragments of the body, the equivalent of pixels. Genes control growing-rules, developmental processes, embryological algorithms. Powerful though they are, an important feature of these growing-rules is that they are local. There is no grand blueprint for the whole body. Instead, each cell obeys local instructions for dividing and differentiating, and when all the local instructions are obeyed together a body eventually results. Each little local region of the tree growing in Figure 13.3 is like other local regions and also (though this does not necessarily have to be true of all trees in my artificial world) like a scaled-down version of the whole tree. If, instead of branching a mere four times (this number is controlled by the starting value of *length*), we let it branch, say, 10 times, an apparently complicated structure results (Figure 13.4). Look carefully at this tree, however, and you will see that it is built up from fundamentally the same local drawing rules. The individual twigs do not 'realize' that they are part of an elaborate pattern. This is all-of-a-piece with the extreme simplicity of the procedure **Tree**. Probably there is an important similar sense in which real-life embryology, too, is simple.

Tree, then, was to be the basis of my embryology. Since the arrays *dx* and *dy* and the parameter *length*, determine the shape of a tree, these were clearly the numbers that should be controlled by genes. On the face of it this suggests that there should have been 15 genes, 7 for *dx*, 7 for *dy*, and 1 for *length*. However, I wanted, for biological reasons, to add one more constraint. Most animals, as remarked above, are bilaterally symmetrical. Such a requirement could be built into the biomorphs by constraining certain members of the *dx* and *dy* parameters

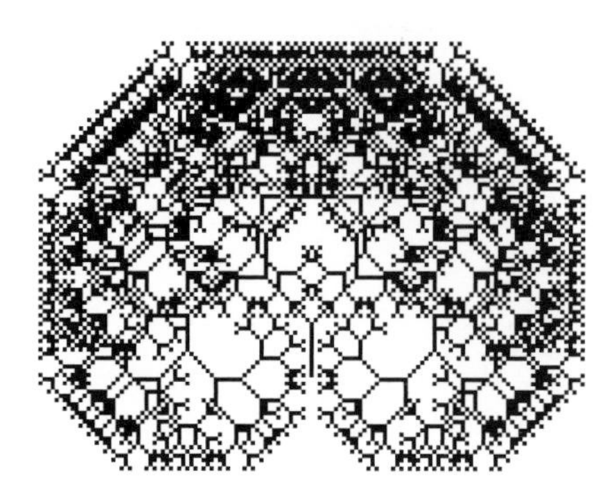

Figure 13.4 Basic tree with high-order branching.

to be equal to one another, sometimes with opposite sign. Instead of 15 genes, therefore, I ended up with 9. Genes 1 to 3 control clusters of the *dx* array. Genes 4 to 8 control clusters of the *dy* array. And Gene 9 controls the starting value of *length*, the 'order' of the recursive tree or in other words the number of branchings. The details are as follows:

```
procedure PlugIn(ThisGenotype: Genotype);
{Plugin translates genes into variables needed by Tree}
begin
  order := gene[9];
  dx[3] := gene[1]; dx[4] := gene[2]; dx[5] := gene[3];
  dx[1] := −dx[3]; dx[0] := −dx[4]; dx[2] := 0; dx[6] := 0; dx[7] := -dx[5];
  dy[2] := gene[4]; dy[3] := gene[5]; dy[4] := gene[6]; dy[5] := gene[7]; dy[6] := gene[8];
  dy[0] := dy[4]; dy[1] := dy[3]; dy[7] := dy[5];
  end {PlugIn};
```

The call of Tree then follows:

```
Tree(Startx, Starty, order, Startdir, dx,dy);
```

and the appropriate biomorph is drawn.

Plugin and **Tree,** then, are called in sequence within **Develop,** to draw any particular biomorph. **Reproduce** is then called a dozen or so times, breeding a litter of mutant progeny whose phenotypes are drawn, by **Develop**, on the screen (Figure 13.5). A human chooser then chooses one of the litter for breeding, its genes are fed into **Reproduce**, the screen is cleaned and a new litter of progeny drawn and the cycle continues indefinitely. As the generations go by, the forms evolve gradually in front of the chooser, who witnesses true evolution by Darwinian (artificial) selection.

13.4 Evolving biomorphologies

As I said before, I had the intuitive feeling that some kind of recursive procedure would prove to be both morphologically prolific and biologically interesting. But I deliberately did not give much thought to the precise details of the recursive algorithm, because I wanted as much as

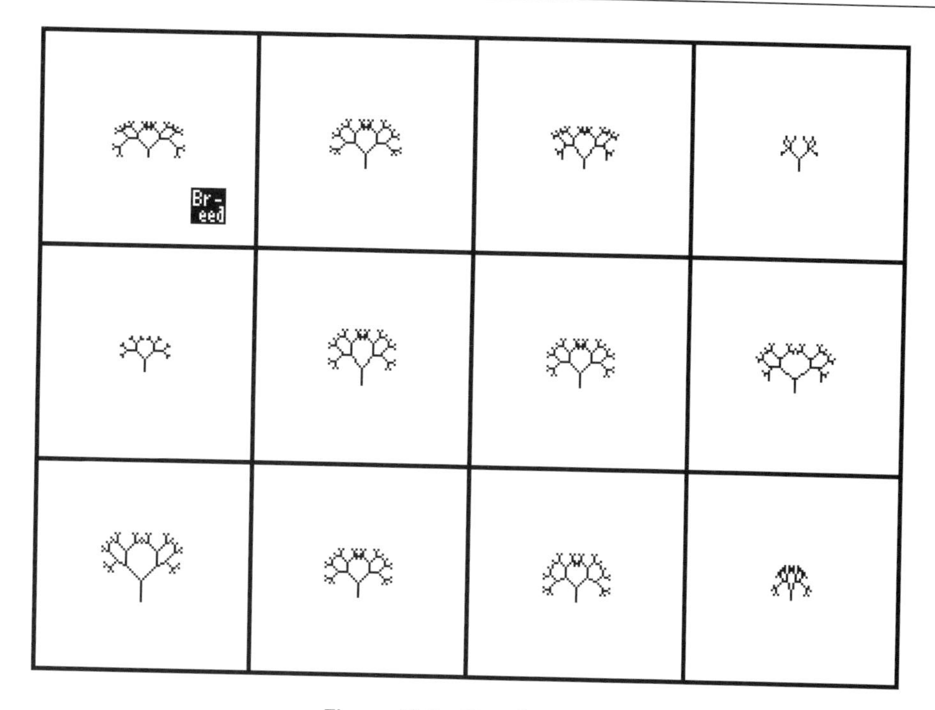

Figure 13.5 Breeding screen.

possible to emerge rather than be designed. I had the feeling that it should not really matter how the genes affected development, provided the development procedure was some kind of recursive drawing-rule. In the event, my intuition proved to have been a considerable underestimate. I was genuinely astonished and delighted at the richness of morphological types that emerged before my eyes as I bred. Figure 13.6 shows a small sample portfolio. Notice how un-treelike many of the biomorphs are. They have evolved under the selective influence of a zoologist's eye, so it is not surprising that many of them resemble animals. Not *particular* animals that actually exist, necessarily, but several of the specimens in Figure 13.6 would not look out of place in a zoological textbook. Some were bred to resemble other things, for instance the spitfire at middle left and the silver coffeepot at top right. It was biomorphs of the type shown in Figure 13.6 that I used (Dawkins, 1986) to illustrate the power of gradual cumulative selection to build up morphologies. The rest of this chapter is about other families of biomorphs with more elaborate embryological principles.

In introducing the superiority of constrained embryologies over 'pixel-peppering', I implied that constrainedness was a virtue. So it is, but you can have too much of a good thing. Having arrived at our basic recursive embryology, and found it good, are there any judicious relaxations or additions we can make to it, which will improve its biological richness? Remember that in doing this we must resist the temptation to take the easy route and build in known biological details. Our watchword is that as much as possible must emerge rather than being designed. But having seen the range of phenotypes that emerge from the basic program, can we think of any modifications to the basic program that seem likely to unleash opulent flowerings of new emergent properties? In seeking such powerful enrichments we need not fear to make use of general biological principles. All that we must avoid is building in detailed biological knowledge.

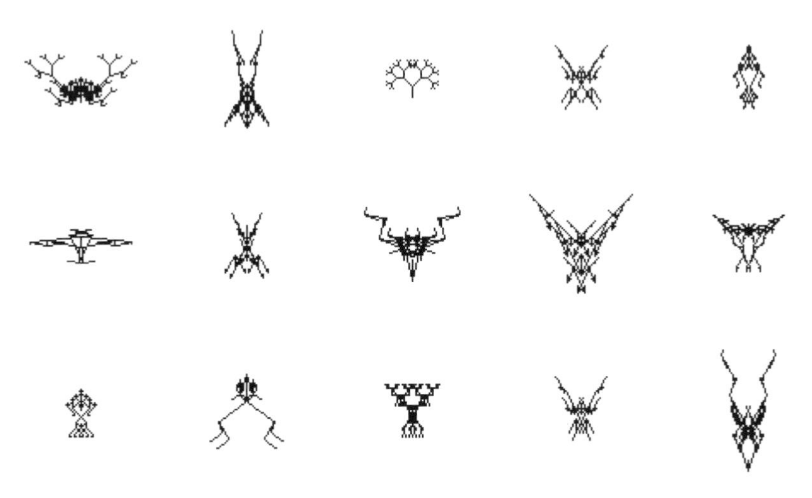

Figure 13.6 Portfolio of biomorphs varying according to nine genes.

13.4.1 Symmetry

I have already mentioned symmetry as an important constraint in real biology. The basic program produces biomorphs that all have to be bilaterally symmetrical. What if we relax this constraint, and allow asymmetrical biomorphs? It now becomes possible to breed forms like Figure 13.7a. Not very interesting in itself, but let us allow our biomorphs this kind of asymmetry nevertheless, because it may interact interestingly with other relaxations of the basic embryology that we shall allow. Let us, then, invent a new gene, one with only two values, on and off, which switches bilateral symmetry on or off.

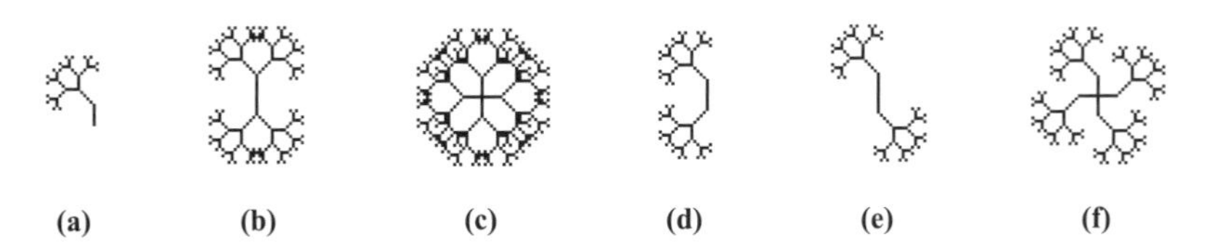

Figure 13.7 Asymmetrical biomorph (a); up-down symmetry (b); four-way radial symmetry (c); up-down symmetry as a reflection in a horizontal mirror (d); up-down symmetry by rotation (e); and left-right asymmetry with four-way radial symmetry (f).

Left/right is not the only plane of symmetry that real animals play about with in their evolution. Suppose we take our basic *Blind Watchmaker* program but allow a new gene to switch on and off symmetry in the up/down direction. We can then breed shapes like Figure 13.7b. We can give this new gene an additional possible value, to enable it to switch on full, 4-way radial symmetry (Figure 13.7c). Now let us come back to our other gene, for left/right asymmetry, and combine it with up/down symmetry. Here we have a decision to make. Do we consider that up down symmetry is achieved by reflection in a horizontal mirror, in which case we can obtain a picture like Figure 13.7d? Or do we consider that it is achieved by rotation, in which case we shall obtain a picture like Figure 13.7e? I arbitrarily decided on the latter. When we combine left-right asymmetry with 4-way radial symmetry we get a picture like Figure 13.7f.

These symmetry genes, it seems to me, have the potential that we seek[1]. They seem to have the power to add rich emergent properties without the programmer having built in a lot of contrived design. I shall return to give further examples of the additional flowerings of biomorph structure that these two symmetry genes permit.

13.4.2 Segmentation

Symmetry is not the only such principle that is suggested by a general knowledge of real zoology. Segmentation is another widespread biological phenomenon that lends itself to biomorphic treatment. Animals belonging to three of the most successful phyla – vertebrates, arthropods and annelids – organize their bodies along segmented lines. Segments are repeated modules, more or less the same as each other, running from head to tail like the trucks in a train. In an annelid worm such as an earthworm, or some arthropods such as millipedes, the segmentation is extremely obvious. A millipede really is rather like a train. It is easy to imagine that the developmental program of a millipede has the instructions for building a single segment, then sticks those instructions in a **repeat** loop. Segmentation is equally obvious in some vertebrates, such as snakes and fish when viewed internally. In mammals, such as ourselves, it is less prominent, but it is clearly seen in the backbone. Not only are the vertebral bones themselves 'repeated' down the backbone; so are a whole series of associated blood vessels, muscles and nerves. Even the skull was originally segmented, but the traces of segmentation are now so well hidden that uncovering it was one of the triumphs of comparative anatomy. It is probably fair to say that the invention of segmentation was one of the major innovations in the history of life, an invention that was made at least twice, once by an ancestor of the vertebrates and once by an ancestor of the annelids and arthropods (who are probably descended from a segmented common ancestor).

How might we change the basic *Blind Watchmaker* program to allow biomorphs to be segmented like millipedes? The obvious way is to use the basic program to generate a single segment. Then, just as I speculated for millipedes above, enclose it in a repeat loop. So I added this feature to the program, with a new gene controlling the number of segments, and another new gene controlling the distance between segments. Figure 13.8a shows a series of biomorphs, identical except with respect to the value of the first of these genes. And in Figure 13.8b is a series, identical except with respect to the value of the second gene, the one controlling the distance between segments.

13.4.3 Scaling gradients

Finally, I introduced genes controlling segmental *gradients*. The segments of a millipede may all look pretty much the same, but many segmented animals taper, being broadest at the front and narrowest at the back. Others, for instance woodlice and many fish, are narrow at each end and broad in the middle. The segments all down the body follow the same basic plan but get progressively larger, or smaller, over stretches of the animal's length from front to rear. It is as though the developmental **repeat** loop includes scaling factors that are incremented or decremented each time the program passes through the loop.

[1] Note added in 1997. The whole matter of animal symmetry in various planes is explored further, using the analogy of a kaleidoscope, in the chapter Kaleidoscopic Embryos of *Climbing Mount Improbable* (1996) Penguin, London, and W.W. Norton, New York.

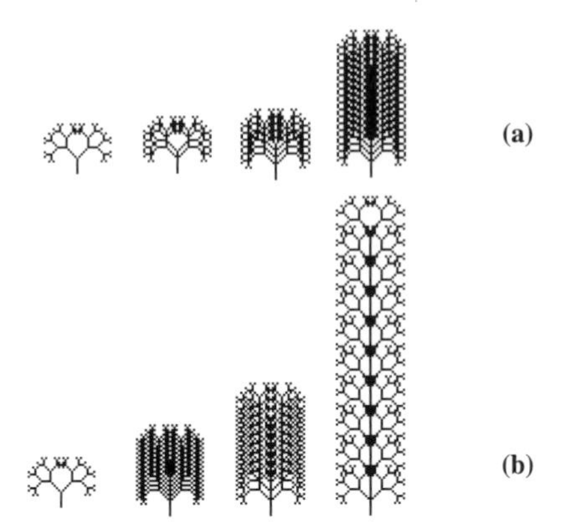

Figure 13.8 (a) Shows a series of biomorphs which are identical except with respect to the value of the segment-number gene and (b) is a series which is identical except with respect to the value of the segment-distance gene.

In introducing segmental gradients into the biomorph program, I aimed for greater generality than could be achieved with a single scaling factor. I allowed for gradients affecting the expression of each of the nine basic genes and the intersegment distance gene, separately. Figure 13.9 shows what happens if you put a gradient on Gene 1 and on Gene 4. The left hand biomorph has no gradients and all the segments are the same. The middle biomorph is the same except that, as you move from front to rear, Gene 1's expressivity increases by one unit per segment. The right hand biomorph shows the same for Gene 4.

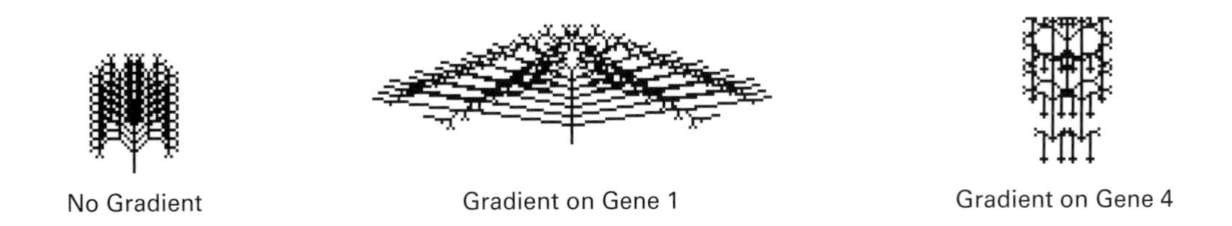

Figure 13.9 Effect of gradient on two genes.

Figure 13.10 is a portfolio of segmented biomorphs, most of them with gradients. These are biomorphs that have been bred by selection, in the same way as those in Figure 13.6, but with the possibility of mutation in the two segment-controlling genes. You can think of the Figure 13.6 biomorphs as having had those two genes set to zero. If you compare Figure 13.10 with Figure 13.6, I think you will agree that segmentation has added to the zoological interest of the specimens. An embryological innovation, in this case segmentation, has allowed the evolution of a whole new range of types. We may conjecture that much the same thing happened in the ancestry of the vertebrates and in the separate ancestry of the annelids and arthropods.

13.4.4 Asymmetry and segmentation

But the animals in Figure 13.10 are all bilaterally symmetrical. What happens if we combine segmentation with asymmetry? At this point in the programming, I introduced another arbitrary

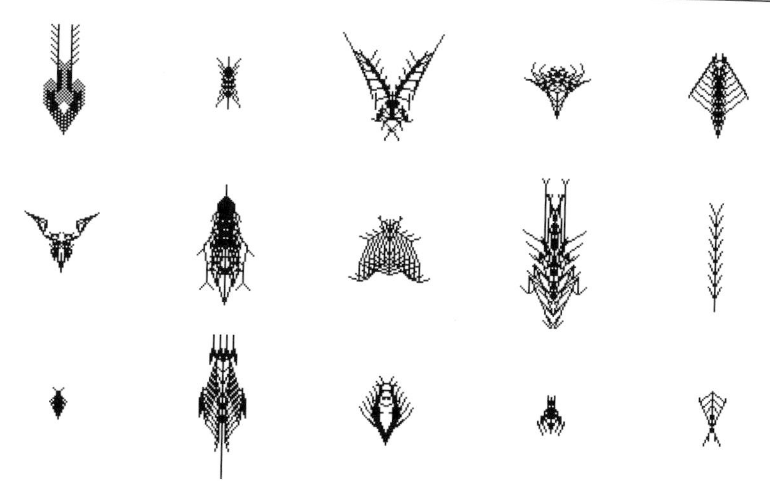

Figure 13.10 Portfolio of segmented biomorphs.

constraint. I simply decreed that when a one-sided animal like Figure 13.11a became segmented, instead of each segment simply repeating the asymmetry as in Figure 13.11b, the successive segments should always be asymmetrical in alternate directions, as in Figure 13.11c. There was no particular reason why I should have done this. I think I did it partly because I wanted to use segmentation to help recreate, at the level of the whole body, the symmetry that had been lost at the level of the individual segment. And I also think I did it partly to make the biomorphs more botanically interesting. Many plants send off alternating buds. Indeed, my portfolio (Figure 13.12) of asymmetrical segmented biomorphs may interest botanists more than zoologists. When I was breeding them by artificial selection on the screen, I frequently had plants in my mind. The biomorph at the top left could be any of a wide variety of plant species. The one next to it could be an inflorescence, or perhaps a colonial animal such as a siphonophore. While looking at the top row, notice the barley and the cedar of Lebanon. The next row begins with DNA and contains a dollar sign which also looks rather like a different view of DNA.

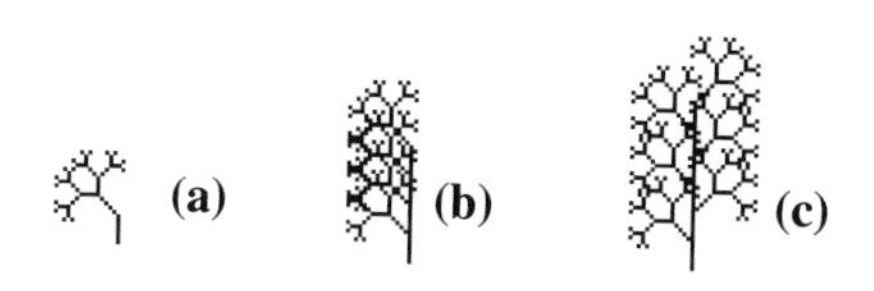

Figure 13.11 One-sided animal (a); segmented with repeating symmetry (b); and with successive segments asymmetrical in alternate directions (c).

13.4.5 Pimpernels, 'echinoderms' and alphabets

Finally, let us switch on our other newly invented gene, the one that controls up/down symmetry and radial symmetry. Figure 13.13 is a portfolio of biomorphs bred with this gene, the segmentation genes and the lateral symmetry gene all permitted to mutate. Many of these are more like human artifacts, ornaments or regalia than like living organisms. But there is a nice scarlet pimpernel at the right of the middle row and the 8-pointed star at the bottom left suggested to me that I should try to breed a portfolio of echinoderms (starfish, sea urchins, brittle-stars etc.).

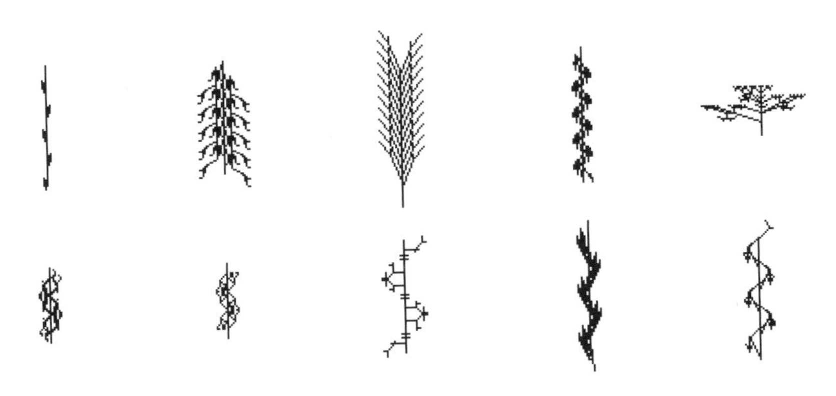

Figure 13.12 Portfolio of asymmetrical segmented biomorphs.

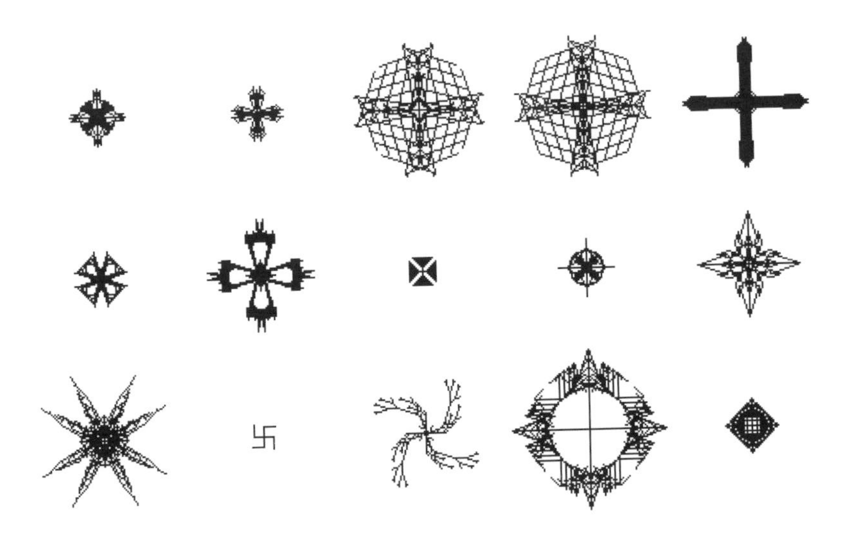

Figure 13.13 Portfolio of radially symmetrical biomorphs.

I present my 'echinoderm' portfolio (Figure 13.14) as another illustration of all my 'new' mutation types – all the various symmetry mutations and the segmentation and gradient mutations. Any zoologist will instantly spot the trouble with these 'echinoderms'. They have 4-way radial symmetry rather than 5-way symmetry. The present program is not capable of producing 5-way radial symmetry. Once again, we are brought back to our major theme. Huge vistas of evolutionary possibility, in real life as well as in artificial life, may be kept waiting a very long time, if not indefinitely, for a major reforming change in embryology. Which brings me to the problems of the biomorph alphabet.

If we can breed animals and plants, I thought, why not any arbitrarily designated shapes? You cannot get much more arbitrary than the alphabet, so how about an alphabet of biomorphs? While breeding, wandering around in 'Biomorph Land' (Dawkins, 1986), I would sometimes encounter biomorphs that had a slight look of one of the letters of the alphabet. I would then try to perfect the resemblance by selective breeding, preserving intermediate results and returning to try again whenever I had an idle moment. In this way I gradually built up an album of alphabetic characters. My aim was eventually to sign my name legibly in biomorphic

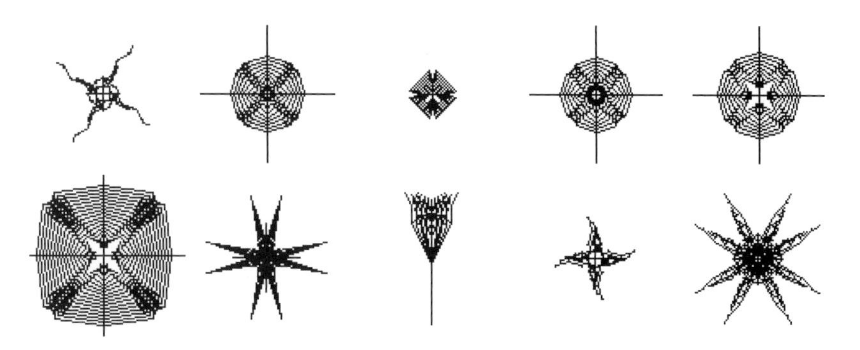

Figure 13.14 Portfolio of 'echinoderms'.

script. Figure 13.15a shows that I have not quite achieved this yet, although I had more luck (Figure 13.15b) in using biomorph script to pay tribute to the uniquely brilliant microcomputer which made all this work so easy. The problems are not trivial. Some letters are perfect, like I and N. Others are a little odd, but still not unpleasing, for instance S and A. D is pretty horrible, and I cannot seem to get rid of the irritating little upward-pointing tail. As for K, I despair of ever breeding a proper K. I simply had to fake it in my name by running part of a K into the preceding W, an obvious cheat.

Figure 13.15 Biomorphs that resemble letters of the alphabet.

And that is the point. There are some shapes that certain kinds of embryology seem incapable of growing. My present *Blind Watchmaker* embryology, that is the basic 9 genes plus segmentation with gradients and symmetry mutations, is, I conjecture, forever barred from breeding a respectable K, or a capital B. Or if I am proved wrong in this particular conjecture, I am fairly sure there are *some* kinds of shape that the present *Blind Watchmaker* program can never breed. Just as we extended the basic program by adding segmentation and symmetry genes, there are presumably other extensions that could be made, which would make difficult letters of the alphabet become easy. For instance, it is not far-fetched to guess that, if we relaxed the 'alternation' constraint on segmented asymmetrical biomorphs, K would become easy. The way to do this, in keeping with our earlier extensions to the embryology, would be to add an 'alternation gene' which could mutate itself on or off. Then all present biomorphs would be a subset of the larger set that would become possible – the subset with the alternation gene permanently turned on. K would also be easy to breed if we had taken the decision to achieve up/down symmetry by mirror reflection rather than by rotation (see above). Leaving the biomorph alphabet on one side, many other new kinds of mutations could be implemented, including genes controlling colour. The colour version described in *Climbing Mount Improbable* produces results spectacular beyond my previous imaginings, though so far the aesthetic appeal of these coloured biomorphs has overshadowed their biological interest.

13.5 Evolution of evolvability

Finally, let us return to the evolution of evolvability. The point I have been trying to make so far in this chapter is that certain kinds of embryology find it difficult to generate certain kinds of biomorphs; other kinds of embryology find it easy to do so. It is clear that we have here a powerful analogy for something important about real biology, a major principle of real life that is illustrated by artificial life. It is less clear which of several possible principles it is! There are two main candidates, which I must take some time to expound in order to explain why I favour one rather than the other.

In order to explain the first one, we need to make a preliminary distinction between two kinds of mutation: ordinary changes within an existing genetic system and changes to the genetic system itself. Ordinary changes within an existing genetic system are the standard mutations that may or may not be selected in normal evolution within a species. One allele is replaced by an alternative allele at the same locus, as in the famous case of industrial melanism where a gene for blackness spread through moth populations in industrial areas (see any biology textbook). This is how all normal evolutionary change happens. But it is an inescapable fact that different species, to a greater or lesser extent, have different genetic systems from one another, even if this only means that they have different numbers of chromosomes. 'The same locus', when we are talking about an elephant and a human, may not even be a meaningful thing to say. Humans and elephants employ the same *kind* of genetic system, but they do not have the same genetic system. They have different numbers of chromosomes and you cannot make a locus-for-locus mapping between them like you can between two individual humans. Yet humans and elephants undoubtedly have a common ancestor. Therefore, during their evolutionary divergence, there must have been changes to the genetic systems, as well as changes within the genetic systems. These changes to genetic systems must have been, at least in one sense, major changes, changes of a different order from the normal allele-substitutions that go on within a genetic system.

Now, the changes to the biomorph program that I have been talking about in this chapter – the addition of symmetry mutations and segmentation mutations – are, as it happens, changes of just this character, changes to the genetic system itself. They constituted major rewritings of the program to increase the 'chromosome' size from 9 to 16 genes[2]. But I want to argue that this is incidental. This is not the analogy that I want to draw between artificial life and real life. Changes in genetic systems must, indeed, be fairly commonplace in the history of life: changes in chromosome number are nearly as common as initiations of new species and the number of species initiations that have occurred in the history of life on earth is probably to be counted in the hundreds of millions. So, although changes in genetic systems are much rarer than allelic substitutions within genetic systems, they are not very rare events on the geological timescale.

[2] Not all the 16 have been discussed here. For details of what the remaining genes do, see the Instruction Manual supplied with the disc of the Macintosh program, which is marketed in America by W W Norton & Co, 500 Fifth Avenue, New York 10110. The Instruction Manual is also printed as an Appendix to the Norton chapterback edition of *The Blind Watchmaker*, which also gives details on how to obtain the program at a reduced price. A version of the program for IBM-compatible computers, with only the original nine genes, is also available from the same source. Outside America, both versions of the program are marketed by Software Production Associates, PO Box 59, Tewkesbury, Glos, England.

What I want to argue is that there is another class of evolutionary innovations which *are* very rare on the geological timescale and which I shall call evolutionary watersheds. An evolutionary watershed is something like the invention of segmentation which, as we have seen, may have occurred only twice in history, once in the lineage leading to annelids and arthropods and once in the lineage leading to vertebrates. A watershed event like this may or may not have coincided with a change in the genetic system such as a change in chromosome number. In any case that is not what is interesting about watershed events. What is interesting about them is that they open floodgates to future evolution.

I suspect that the first segmented animal was not a dramatically successful individual. It was a freak, with a double (or multiple) body where its parents had a single body. Its parents' single body plan was at least fairly well-adapted to the species' way of life, otherwise they would not have been parents. It is not, on the face of it, likely that a double body would have been better adapted. Quite the contrary. Nevertheless, it survived (we know this because its segmented descendants are still around), if only (this, of course, is conjecture) by the skin of its teeth. Even though I may exaggerate when I say 'by the skin of its teeth', the point I really want to make is that the individual success, or otherwise, of the first segmented animal during its own lifetime is relatively unimportant. No doubt many other new mutants have been more successful as individuals. What is important about the first segmented animal is that its descendant lineages were champion *evolvors*. They radiated, speciated, gave rise to whole new phyla. Whether or not segmentation was a beneficial adaptation during the individual lifetime of the first segmented animal, segmentation represented a change in embryology that was pregnant with evolutionary potential.

Not all evolutionary watersheds are as dramatic in their magnitude or in their evolutionary consequences as the invention of segmentation. There may be many changes in embryology which, though not dramatic enough in themselves even to deserve the title watershed, nevertheless are, to a lesser extent than the invention of segmentation, evolutionarily pregnant. Suppose we rank embryologies in order of evolutionary potential. Then as evolution proceeds and adaptive radiations give way to adaptive radiations, there is presumably a kind of ratchet such that changes in embryology that happen to be relatively fertile, evolutionarily speaking, tend to be still with us. New embryologies that are evolutionarily fertile tend to be the embryologies that characterize the forms of life that we actually see. As the ages go by, changes in embryology that increase evolutionary richness tend to be self-perpetuating. Notice that this is not the same thing as saying that embryologies that give rise to good, healthy individual organisms tend to be the embryologies that are still with us, although that, too, is no doubt true. I am talking about a kind of higher-level selection, a selection not for survivability but for evolvability.

It is all too easy for this kind of argument to be used loosely and unrespectably. Sydney Brenner justly ridiculed the idea of foresight in evolution, specifically the notion that a molecule, useless to a lineage of organisms in its own geological era, might nevertheless be retained in the gene-pool because of its possible usefulness in some future era: 'It might come in handy in the Cretaceous!' I hope I shall not be taken as saying anything like that. We certainly should have no truck with suggestions that individual animals might forgo their selfish advantage because of possible long-term evolutionary benefits to their species. Evolution has no foresight. But with hindsight, those evolutionary changes in embryology that *look* as though they were planned with foresight are the ones that dominate successful forms of life.

Perhaps there is a sense in which a form of natural selection favours, not just adaptively successful phenotypes, but a tendency to evolve in certain directions, or even just a tendency to evolve at all. If the embryologies of the great phyla, classes and orders of animals display an 'eagerness' to evolve in certain directions, and a reluctance to evolve in other directions, could these 'eagernesses' and 'reluctances' have themselves been favoured by a kind of natural selection? Is the world filled with animal groups which not only are successful, as individuals, at the business of living, but which are also successful in throwing up new lines for future evolution?

If that were all there was to it, it would be simply another case, like sieving sand, of what I have called 'single-step selection' (Dawkins, 1986) and therefore not very interesting, evolutionarily speaking. It is only *cumulative* selection that is evolutionarily interesting, for only cumulative selection has the power to build new progress on the shoulders of earlier generations of progress, and hence the power to build up the formidable complexity that is diagnostic of life. I have been in the habit of disparaging the idea of 'species selection' (e.g. Gould, 1982; Eldredge, 1985) because, as it is normally presented, it is a form of single-step selection, not cumulative selection, and therefore not important in the evolution of complex adaptations (Dawkins, 1982). But selection among embryologies for the property of evolvability may, it seems to me, have the necessary qualifications to become cumulative in evolutionarily interesting ways. After a given innovation in embryology has become selected for its evolutionary pregnancy, it provides a climate for new innovations in embryology. Obviously the idea of each new adaptation serving as the background for the evolution of subsequent adaptations is commonplace and is the essence of the idea of cumulative selection. What I am now suggesting is that the same principle may apply to the evolution of evolvability which may, therefore, also be cumulative.

Others have pointed out that we should speak of 'species selection' only in those rare cases where a true species-level quality is being evolved (Maynard Smith, 1983). Species selection should not, for instance, be invoked to explain an evolutionary lengthening of the leg, since species do not have legs, individuals do. It might, on the other hand, be invoked to explain the evolution of a tendency to speciate, since speciating is a thing species, but not individuals, do. It now seems to me that an embryology that is pregnant with evolutionary potential is a good candidate for a higher-level property of just the kind that we must have before we allow ourselves to speak of species or higher-level selection.

The world is dominated by phyla, classes and orders whose embryology equipped them to diverge and inherit the earth. Although I am sure I have always been dimly aware of this, I think it is true to say that it is the biomorph program – writing it, playing with it and, above all, modifying it to increase its evolutionary potential – that has really drummed it into my innermost consciousness. So what started out as an educational exercise – I was trying to develop a tool to teach other people – ended up as an educational exercise in another sense. I ended up teaching something to myself about real life. There could be less worthy uses of artificial life.

Acknowledgements

I thank Alan Grafen, Ted Kaehler and Helena Cronin for help and discussions of various kinds; Chris Langton for inviting me to Los Alamos to one of the most stimulating conferences I have been to; and Apple Computer Inc. whose generosity made it possible for me to go to Los Alamos.

This chapter is from Dawkins, R. (1989) in *Artificial Life* edited by Langton. Perseus Books.

During the 15 years since I first used the phrase Evolution of Evolvability, other authors have adopted it, not always in quite the same sense. Their papers may be found by searching for the phrase on the world wide web.

References

Dawkins, R. (1976) *The Selfish Gene*. Oxford University Press.
Dawkins, R. (1982) *The Extended Phenotype*. W.H. Freeman.
Dawkins, R. (1986) *The Blind Watchmaker*. Longman.
Eldredge, N. (1985) *Unfinished Synthesis*. Oxford University Press.
Gould, S.J. (1982) The meaning of punctuated equilibrium and its role in validating a hierarchical approach to macroevolution. *Perspectives on Evolution*, Milkman, R., ed. Sinauer.
Maynard Smith, J. (1983) Current controversies in evolutionary biology. *Dimensions of Darwinism*, Grene, M. ed. Cambridge University Press.
Morris, D. (1987) *The Secret Surrealist*. Phaidon.
Weismann, A. (1893) *The Germ Plasm: a theory of heredity*, translated by Parker, W.N. and Rönnfeldt, H. W. Scott.

Artificial genomes as models of gene regulation

14

TORSTEN REIL

14.1 Introduction

The pace of genome sequencing has vastly accelerated over the past decade. As a consequence, the full genetic code of many organisms, from the worm *C. elegans* (Wilson, 1999) to humans (Bentley, 2000) is now known.

Not surprisingly, while the immense amount of available genetic information has allowed us to (at least start to) answer many questions, it has also raised many new ones. For example, the degree of genetic similarity between quite obviously different organisms such as yeast (Mewes *et al.*, 1997) and humans is astonishing, as is the fact that the latter can be built with seemingly much less genetic information than previously assumed: the current mean estimate for the number of genes contained in the human genome is around 61 000 (Stewart, GeneSweep, 2002), with some estimates being as low as 35 000 (Ewing and Green, 2000).

It is becoming increasingly clear that nature has a relatively small repertoire of genetic tools. The crucial difference between organisms is not which tools (i.e. genes) are present, but when and how they are used. As a consequence, there has been a growing interest in the mechanisms and dynamics of how genes are switched on and off: the regulation of gene expression.

There is now a considerable number of mathematical and simulation models of gene regulation (Reil, 2000). These differ in focus and range of applications – a brief overview is given below.

One recently introduced model is artificial genomes (AGs) (Reil, 1999a). Unlike most other approaches, such as Kauffman's random Boolean networks (Kauffman, 1993), AGs model genomes at the sequence level and employ template matching as the basic regulatory mechanism. It is argued that these properties make them an attractive system for investigating questions regarding genome structure and evolution.

14.2 Molecular basis of gene regulation

Gene expression is the process of reading and interpreting a given stretch of DNA to make functioning protein. While the control of gene expression and activity can take place at several

 ISBN 0-12-428765-4

different stages, most models of gene regulation focus on the transcriptional control component and, more specifically, transcriptional control in eukaryotes.

The structure of a typical eukaryotic gene is shown in Figure 14.1. The actual gene product, the protein, is coded for by the *coding region*. For the gene to be expressed, the coding region is transcribed by the enzyme *RNA polymerase* into messenger RNA (mRNA), which in turn is translated into a string of amino acids, eventually yielding a three-dimensionally folded protein.

The initiation of transcription is dependent on a number of factors, most importantly the presence of regulatory proteins called *transcription factors* (*TFs*). As an essential component, *general transcription factors* must bind to a stretch always immediately preceding the coding region, called the *promoter* (more specifically, they bind to a short sequence within the promoter called the TATA box, which is rich in the nucleotides thymidine and adenine). Once the general transcription factors have assembled, RNA polymerase can dock to transcribe the coding region.

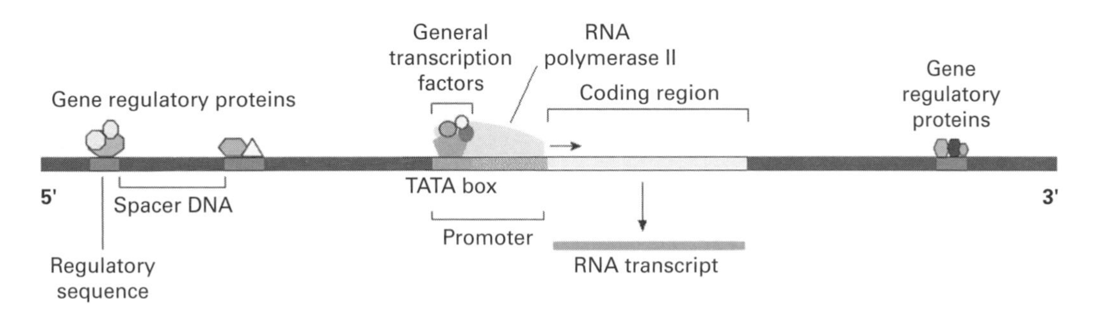

Figure 14.1 Simplified structure of a typical eukaryotic gene.

In addition to the promoter, the control region of a gene typically contains a number of additional regulatory sequences (also called *cis-elements*). These can be located before (*upstream*) or, less frequently, after (*downstream*) the actual coding region. Regulatory sequences need not be close to the gene, but can in fact be located several thousand bases away.

Similar to the TATA box, regulatory sequences act as binding sites for transcription factors (also called *trans-elements*). The probability of a binding event between a given sequence/TF pair is determined by the 3-dimensional fit between the DNA and the protein structure. Hence, TFs tend to be sequence specific.

Once bound, transcription factors can influence the expression of specific genes (typically by physically interacting with the promoter complex of the gene or with other transcription factors). DNA folding enables interactions even over several thousand bases. Broadly, two types of regulators are distinguished: *enhancers* increase the probability that a given gene is expressed, *inhibitors* decrease it.

Transcription factors regulate the presence of structural proteins needed to build and maintain an organism. However, transcription factors themselves, as proteins, are of course subject to the same gene regulatory processes as all other proteins. In other words, transcription factors regulate the expression of transcription factors. It follows that gene regulatory systems typically take the shape of complex dynamic networks of interacting transcription factors, whose output is the switching on and off of structural genes.

14.3 Models of gene regulation

The complex dynamics of gene regulatory networks have led a number of researchers to address the problem through mathematical and computer simulations. The types of models differ considerably, for example in terms of whether time is discrete or continuous, whether updating is synchronous or asynchronous, whether gene states are continuous, quantized or binary and whether stochastic elements are included. Moreover, models can be broadly categorized on the basis of whether they are seeking to simulate a particular model system (such as the *Drosophila* embryo), or whether they are investigating more general properties of genetic regulatory networks (such as robustness or evolvability).

Stuart Kauffman (Kauffman, 1969, 1993, 1995) was one of the first to realize that gene regulatory networks are amenable to considerable abstraction. In Kauffman's random Boolean networks (RBNs), genes are represented as nodes with two possible states: *on* (1) and *off* (0). Each node is connected to (i.e. its state is regulated by) a pre-determined number (K) of other nodes.

The way a node responds to the nodes that regulate it is determined by a random Boolean function. Essentially, these express logic of the following type: 'IF node 3 is ON, node 8 is OFF, and node 10 is OFF, THEN the next state of node 5 is OFF'. Each node requires as many Boolean functions as there are possible state combinations of the nodes connected to it (2^K). For example, a node regulated by three other nodes ($K = 3$) has 8 Boolean functions (Figure 14.2).

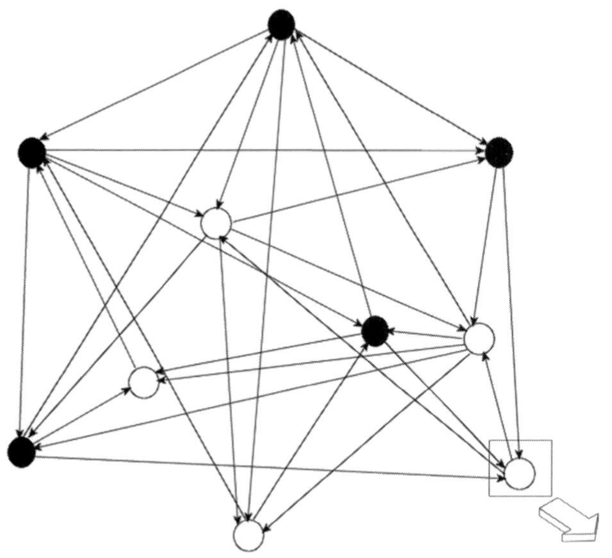

Node 1	Node 2	Node 3	This Node
0	0	0	**0**
0	0	1	**1**
0	1	0	**1**
0	1	1	**0**
1	0	0	**1**
1	0	1	**1**
1	1	1	**0**
1	1	0	**0**

Figure 14.2 Random Boolean network with $N = 10$, $K = 3$ (black: on; white: off). The Boolean function for one node is shown.

To investigate the dynamics of a given RBN, the nodes are updated synchronously in discrete time steps. For this, the network is initiated with either a random or predetermined set of states. Basic RBNs depend on only two parameters: the number of nodes N and the degree of connectivity K.

Kauffman (Kauffman, 1993) showed that networks with K = 2 typically exhibit cyclic behaviour if allowed to unfold over time: a net quickly settles into one of only a small number of recurring sequences of states. Each of these state cycles is relatively short: for example networks with 100 000 nodes settle down into state cycles with, on average, just 317 states (out of $2^{100\,000}$ possible states). State cycles were also found to show a certain degree of homeostasis, in that they are robust to moderate transient disturbances. However, with large or sustained external inputs, a transition from one state cycle to another is observed. (Following the terminology of complexity theory, the state cycles of RBNs can be viewed as *limit cycle attractors*; Reil, 1999b.)

Kauffman argued that the above properties mirrored those found in biological development: biological organisms exhibit a relatively small number of cell types despite a vastly larger number of possible gene expression states. Ontogeny is robust against (moderate) disturbances. Thirdly, cells have the ability to differentiate into various cell types during development.

Moreover, the experiments on random Boolean nets appeared to indicate that the above properties (or 'order') of gene regulatory systems may emerge spontaneously in certain conditions, without the apparent need of natural selection (Kauffman, 1993).

Despite the relative simplicity of the model, the dynamics of random Boolean networks are not straightforward to analyse mathematically, but can be visualized for small N: for example, Wuensche (Wuensche, 1999) has developed the Discrete Dynamics Lab software (DDLab, Figure 14.3) which identifies the attractors for a given network, as well as the possible transi-

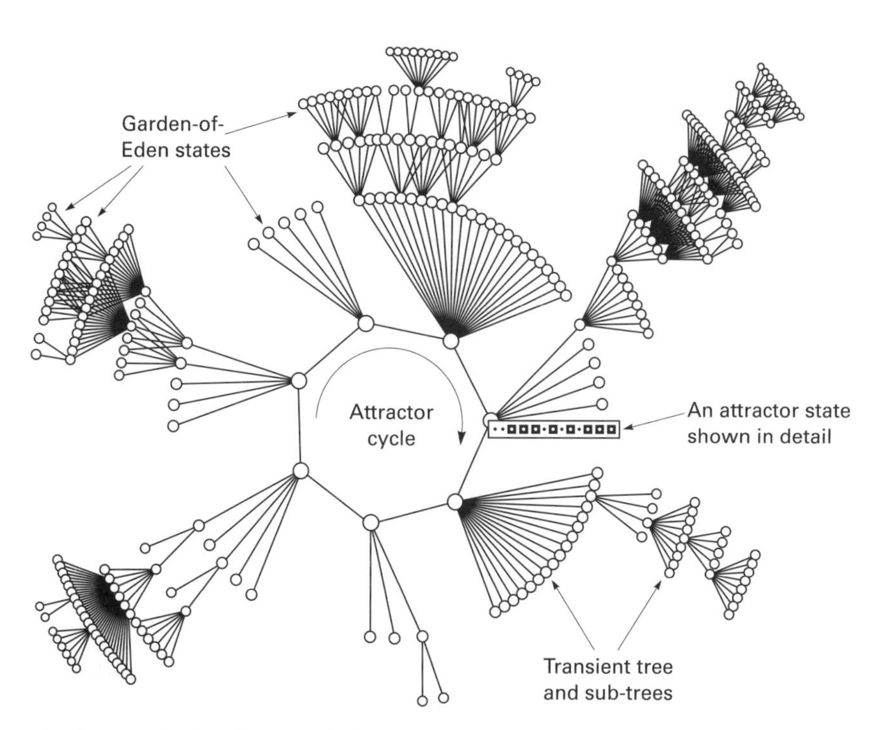

Figure 14.3 A DDLab visualization of a basin of attraction of an RBN with N = 3 and K = 3 (Wuensche, 1999).

tions between them. Following Kauffman, these elements are considered to be the equivalent of cell types and cell differentiation paths, respectively.

Like other models of gene regulation, random Boolean networks have been subject to criticism on several grounds. For instance, Harvey and Bossomaier (Harvey and Bossomaier, 1997) argued that the quasi-quantization brought about by synchronous updating is not observed in biological gene regulation and therefore introduces artifacts to the dynamics of RBNs. Indeed, if updated asynchronously and stochastically (which Harvey and Bossomaier argue is biologically more plausible), RBNs lose the properties listed earlier (Harvey and Bossomaier, 1997), the most prominent casualty being state-cycle attractors (Figure 14.4).

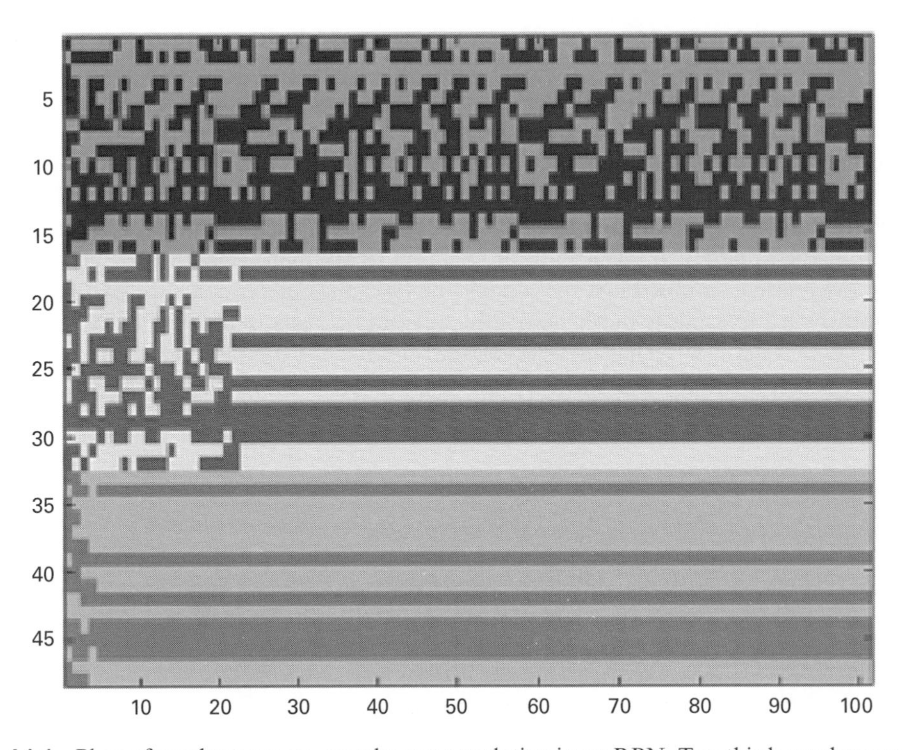

Figure 14.4 Plots of synchronous vs. asynchronous updating in an RBN. Top third: synchronous updating. Middle and bottom third: asynchronous updating. Only the synchronously updated network exhibits cyclic activity (Harvey and Bossomaier, 1997).

The importance of the cyclic nature of RBN attractors is itself contentious – to date, there is little evidence (for example, from microarray experiments) that cell-type specific gene expression has a strong cyclic component (there is, of course, cyclic activity in the genes involved in mitosis). Finally, random Boolean networks (in their original form) employ other simplifications that potentially compromise their usefulness as biological models: for example, gene activity is binary (on/off), each gene is regulated by the same number of other genes and genes are not represented structurally: primarily, nodes in RBNs represent gene interaction rather than the genes themselves. This, as is argued below, limits the application of RBNs in research areas such as genome evolution.

Following Kaufmann's work, a considerable body of research on modelling gene regulatory networks has emerged. Eggenberger (Eggenberger, 1997) was the first to combine a model of gene regulation with a simulated developmental system. It uses a genome implementation which, unlike RBNs, represents genes structurally and, moreover, it includes various classes of gene products (such as transcription factors, cell adhesion molecules and receptors). In combination with a genetic algorithm, Eggenberger was able to evolve multicellular tissues of various shapes (Figure 14.5). Following Kauffman, cell types are defined by having a distinct pattern of gene expression.

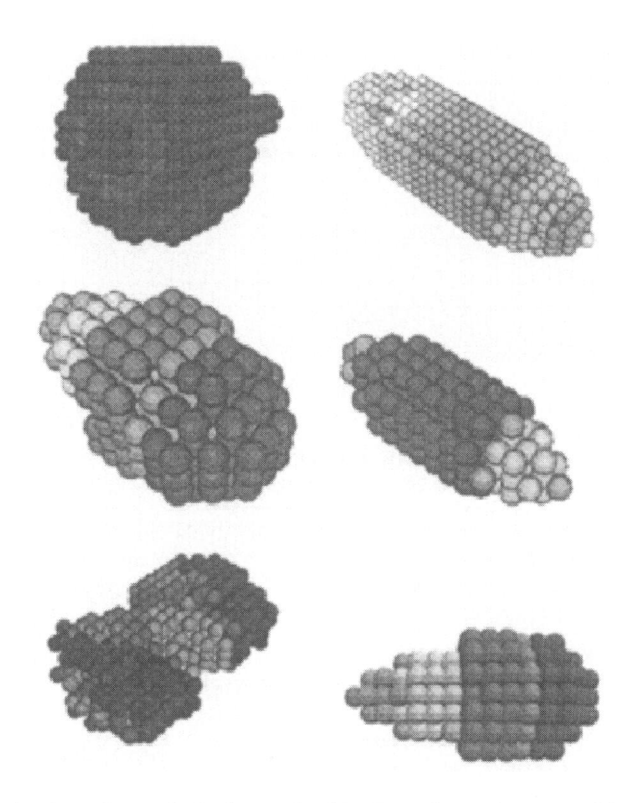

Figure 14.5 A selection of morphologies evolved in Eggenberger's system (Eggenberger, 1997).

Bongard (Bongard, 2002) took the approach of combining gene regulation and ontogeny further to develop an artificial ontogeny system capable of producing multicellular agents in a physically simulated environment. Similar to Eggenberger's approach, the system contains a number of hard-coded gene products (both regulatory and structural); genes are represented structurally and contain an array of real numbers, coding for properties such as type of gene product or the type of transcription factor to respond to.

Bongard showed that the genetic regulatory system, when evolved to create multicellular agents with locomotory capabilities (Figure 14.6), exhibited a strong tendency towards modularity. That is, the genomes developed strong integration (co-regulation) of gene groups involved in the same functional units, but only weak co-regulation between functional gene groups. The importance of these functional units for genetic regulatory networks has previously been pointed out elsewhere (e.g. Wagner and Altenberg, 1996; Thieffrey and Romero, 1999). Bongard concludes that the artificial ontogeny system may be used to test hypotheses about biological genetic networks, in particular the *Hox* family of regulatory genes (Bongard, 2002).

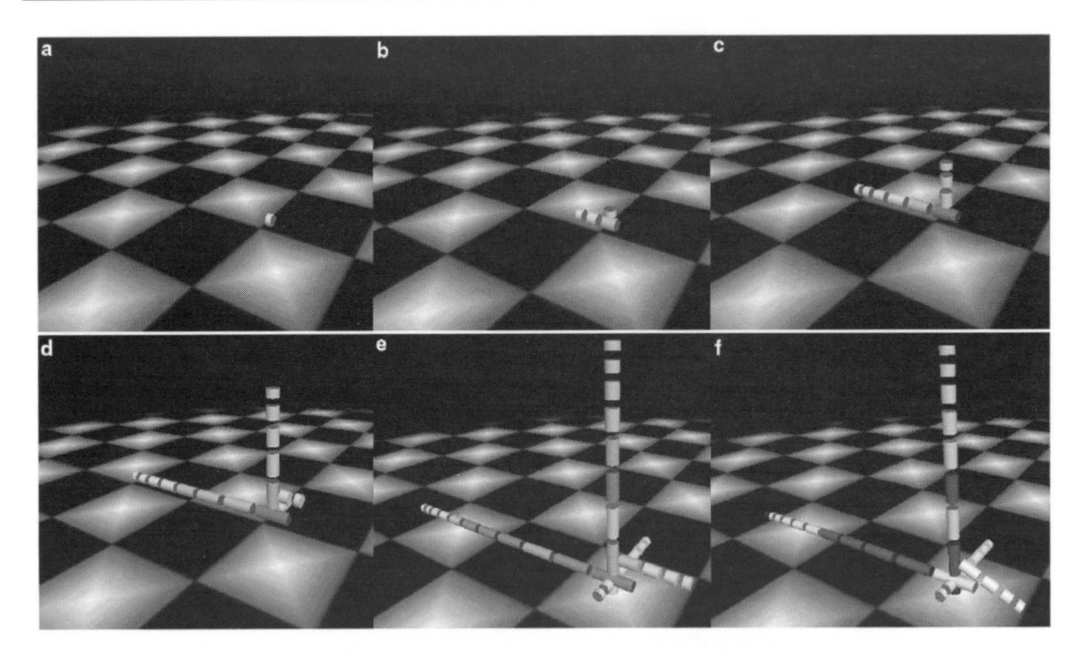

Figure 14.6 The growth phase of an evolved agent using Bongard's gene regulatory system (Bongard, 2002).

Despite their obvious differences, Kauffman's, Eggenberger's and Bongard's models have in common that they try to capture the fundamental concepts of gene regulation, rather than trying to simulate a particular biological system. Several researchers, however, have followed the latter route of using models of gene regulation to analyse a given biological gene network. For example, Mendoza and colleagues (Mendoza *et al.*, 1999) used published genetic and molecular data to build a model of the gene network underlying the control of flower morphogenesis in the weed *Arabidopsis thaliana*. Unlike many other regulatory networks, that of flower morphogenesis is relatively well understood, and the key genes and interactions are known (Figure 14.7; Coen and Meyerowitz, 1991).

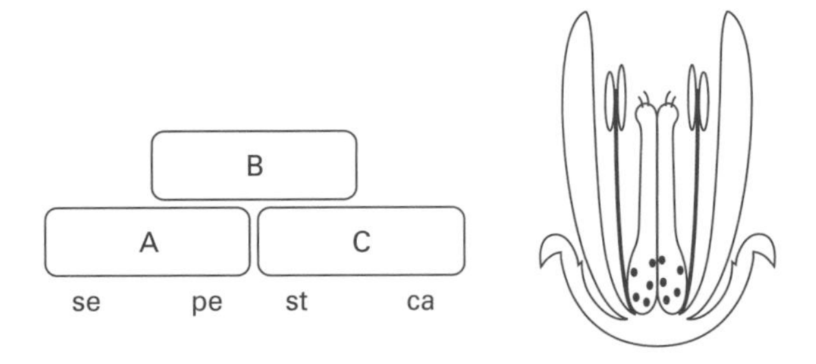

Figure 14.7 The ABC model of *Arabidopsis* flower morphogenesis. The three gene expression domains (A, B, C) determine four whorl identities (sepal, petal, stamen, carpel). For example, simultaneous expression of genes of groups A and B result in the development of petals (from Mendoza *et al.*, 1999).

Mendoza and colleagues formalized the latter through gene interaction matrices (Figure 14.8). Using the logical formalism introduced by Thomas (1978), they were able to reproduce the gene expression patterns known to be present in the wild-type, simulate mutant ones and predict the existence of one new pattern of gene expression, which had not yet been found experimentally.

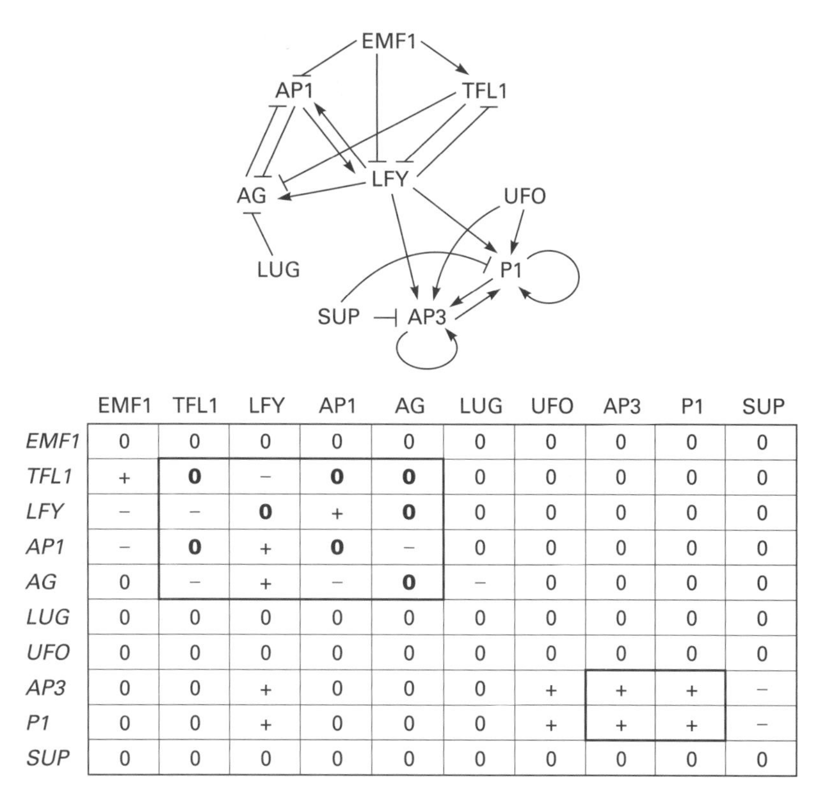

	EMF1	TFL1	LFY	AP1	AG	LUG	UFO	AP3	P1	SUP
EMF1	0	0	0	0	0	0	0	0	0	0
TFL1	+	**0**	–	**0**	**0**	0	0	0	0	0
LFY	–	–	**0**	+	**0**	0	0	0	0	0
AP1	–	**0**	+	**0**	–	0	0	0	0	0
AG	0	–	+	–	**0**	–	0	0	0	0
LUG	0	0	0	0	0	0	0	0	0	0
UFO	0	0	0	0	0	0	0	0	0	0
AP3	0	0	+	0	0	0	+	+	+	–
P1	0	0	+	0	0	0	+	+	+	–
SUP	0	0	0	0	0	0	0	0	0	0

Figure 14.8 Gene interaction matrix of *Arabidopsis* flower morphogenesis. '+': positive regulation; '–': negative regulation; '0': no regulatory influence (Mendoza *et al.*, 1999).

Gene regulatory networks of another popular model organism, the fruit fly *Drosophila melanogaster*, have also been modelled. Kyoda and Kitano (Kyoda and Kitano, 1999) used a computer simulation to model the interaction of eight major genes involved in leg disc formation. Similar to Mendoza *et al.*'s approach, gene interaction matrices were employed (Figure 14.9), using known data where available. In the model, gene activation was regulated by predefined transcription factors, the binding of which was modelled stochastically and dependent on concentration. In addition, the system supported pre-defined diffusing morphogens. This setup successfully reproduced the gene expression patterns found in the *Drosophila* leg disc (Figure 14.10), shed light on the formation of the coaxial pattern in the leg disc and predicted the nature of interaction between the genes CI and dpp (Kyoda and Kitano, 1999).

	en	*ci*	*hh*	*dpp*	*wg*	*al*	*dll*	*dac*
Engrailed (En)		–	+	–	–			
Cubitus interruptus (Ci)			–	(–)				
Hedgehog (Hh)				+	+			
Decapentaplegic (Dpp)					–	+	+	+
Wingless (Wg)				–		+	+	+

Figure 14.9 Gene interaction matrix for five genes involved in *Drosophila* leg disc development. Brackets represent a hypothetical interaction (Kyoda and Kitano, 1999).

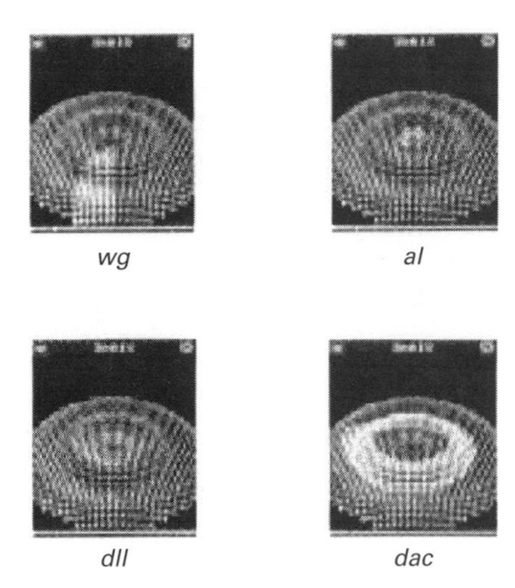

Figure 14.10 Simulated expression patterns of four genes in the *Drosophila* leg disc (Kyodo and Kitano, 1999).

Finally, even where only few published data on a biological gene regulatory system are available, models have been drawn upon to reproduce the overall gene expression dynamics. Marnellos and Mjolsness (Marnellos and Mjolsness, 1998) used an optimization algorithm (simulated annealing) to produce gene interaction matrices in early *Drosophila* neurogenesis. Specifically, the algorithm was used to optimize the gene interaction matrix in such a way that the resultant gene expression patterns matched those found in the differentiation process from proneural cells to neuroblast or sensory organ precursor cells (Figure 14.11).

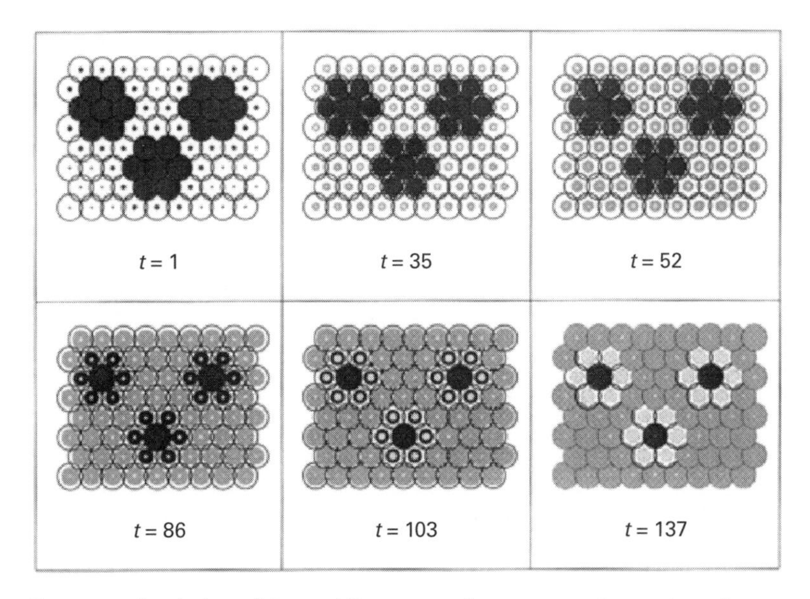

Figure 14.11 Computer simulation of *Drosophila* proneural gene expression and resultant neuroplast and SOP differentiation using a two-gene model (Marnellos and Mjolsness, 1999).

While the validity of the resultant gene interaction matrix could not be determined, several general conclusions were drawn from the model: for the typical cluster differentiation to take place, lateral cell interactions with only the immediate neighbourhood were sufficient. Secondly, the model indicated that cells may be strongly committed to a particular fate (i.e. the future cell type), even when their gene expression patterns provide yet little indication of the commitment (Marnellos and Mjolsness, 1998).

In conclusion, models of gene regulation are being used to investigate the general dynamics of gene regulatory systems and the associated ontogenic process, and simulate specific biological gene regulatory networks, with a view better to understand the dynamics of the system and fill remaining knowledge gaps.

Artificial genomes, which are discussed in more detail in the next section, fall into the first of the two categories. They aim to model gene regulation from first principles and, unlike most other models, use template matching as their basic mode of action.

14.4 Artificial genomes

Artificial genomes seek to model biological gene regulation by capturing its fundamental underlying principle, template matching. Template matching occurs in DNA-based genomes in the form of transcription factors binding to specific stretches of DNA which (as discussed earlier) act as regulatory sequences. Unlike most other models of gene regulation, artificial genomes do not contain pre-specified genes, regulatory sequences or regulatory interactions (such as Boolean functions or interaction matrices).

14.4.1 The model

The core of an artificial genome is a string of digits of length L (Figure 14.12). This is randomly created and contains all the information present in the model. At any one position, digits can range from 0 to B-1, where B is called the *base* parameter. Typically, 4 is chosen as the base, giving a total of four different possible digits (0,1,2,3).

Genes in an artificial genome are identified by the sequence '0101', which can be viewed as a loose equivalent to the TATA box of a promoter. Note that the sequence '0101' will occur by chance (given a sufficiently large genome) – it is not specifically inserted (Figure 14.12). For example, an artificial genome of 20 000 digits and a base of 4 will contain an average of 78 genes.

The N digits immediately following the '0101' promoter are defined as the coding region of the gene in question. This short sequence codes for a gene product; to obtain the latter, each digit of the coding sequence is incremented by 1, so that the gene sequence '021333' becomes '132000' (if a resultant digit exceeds the base, i.e. the allowed digit range, it is set to 0). A gene product is only produced if the gene is active (see below).

Gene products in artificial genomes are the equivalent of transcription factors and, when expressed, bind to matching sequences in the genome. For example, a '320122' gene product binds to all stretches of '320122' in the artificial genome string (Figure 14.13).

When bound to a regulatory sequence, a gene product activates (enhances) or inhibits the next gene downstream in the genome string. Which type of action a gene product has depends on its last digit. With the inhibition parameter (I) set to '2', for example, all gene products ending in '2' or a lower-value digit act as inhibitors. In the simplest artificial genome imple-

```
31331033231210010210303213121001232023011203010200212200313010202033010311200112
33110002313322311003001310000220200121331021131310021310310110312123221113222211
10233102001333200102121023222010133201322001132001233120323301200020012213322203
00223220003012233023013221133331203101303112023231230312100300120333232333001303
32002001232211210330332223122020001131332232230231001210310102032132311202102203
01320333202030200121322323023121221003310311022113031030232332220221103110233 23
23110321032221013120001113302123002331033011022301112200303210212133201011210302
12121121313230010330100323222221003321311230301231210112100332212121222302300310
01020003233232231110113110313000133011031000020002111312332313120122122032313201
23211300121012023123122203200030131023003103011310322022021200332032331020301301
23020323231232303203012310200302200020023000201113220103113232300002200003332202
00100130031323022031302331323021120311113230003123130302022233022003220231013331
33001222332321112110013122123222231032103131021102233311001233310132230023222302
22121301020301201022112000223313310220021103300202201211333230221001320101201023
20022303100221112302000330133221132233010012221321121323231300101031311031203000
20002232031033111313210210210311001032122130123201332011033102302000222132120220
10022232000130002000000122233132202201002021310202313201033220121002300211003320
33022023220331133321012223102213220223220203322301333010111112003102101211033122
12233110000121332000330330102010233320131132132003111323030031132011202002022313
20102221232303021100022320013223121302320020202320010132131112021102303213221023
20100032013110012223213110030320213320120013331002030310011122332121003210111321
23203110303103300323223332333302100333220211303022103302003003213030001021003223
00202300312321003200220003312003213003011021320120302203322222102000310002313110
31103211223010113221132311203030003330300121200012313300213300001333103322002032
11322311113210101120010220103213221031121222313010131003012321122110033021332110
13203233200332202212310102211131312031200021031301223233320331112303302330332032
31033102212313003012303330201010323011301133001233310201013103123020011321030110
12032222122223010232232102300233013211130232003230122321000211102223000332013111
31133021030020333100020000013111223212313121102303320120021013031013002301020322
00302023121213200323213112120212221100131232311121223200022131102223201001131003
31202022013213111201330200120020123003330220010002302220020201020201330012210232
00212002213121123211022221223312010320131221000130021000123332111220000230023313
03232031211321310322230010001200302013100023301220213103002030311310230010301212
32130201001122032022113332332203123013130222302030322013123232230211011212321010
33212021030133100020102112102012212203301223300100020011010113103210331202133301
11301111200020102323331101122232211312210022131222100300103313133213203313122123
21313021033321302010013213333320000102202022130331320233031103010130230203030230
```

Figure 14.12 Excerpt from an artificial genome with N = 5 and B = 4. Promoters are shaded light grey, coding sequences are dark grey.

mentation, inhibitors take precedence over enhancers. That is, if at least one inhibitor is bound upstream of a gene, any enhancers also present are ignored.

Through template matching each gene has the potential to regulate other genes through its gene products. The number of genes that each genes regulates varies (as it is dependent on the number of times the gene product has a matching sequence in the entire string). The average connectivity of a genome is described by the parameter K, and can be calculated once all genes have been identified and matching sites for the gene products have been localized.

In short, an artificial genome is initialized as follows: a string of L digits is created and all genes are identified. The gene products are calculated, all corresponding regulatory sequences are identified and the resultant gene interactions are determined. The result is a gene regulatory network as depicted in Figure 14.14.

The goal of the artificial genome model is to keep the amount of predetermined information low. Instead, genes and their regulatory interactions are determined stochastically and are broadly influenced by parameter settings in the following way:

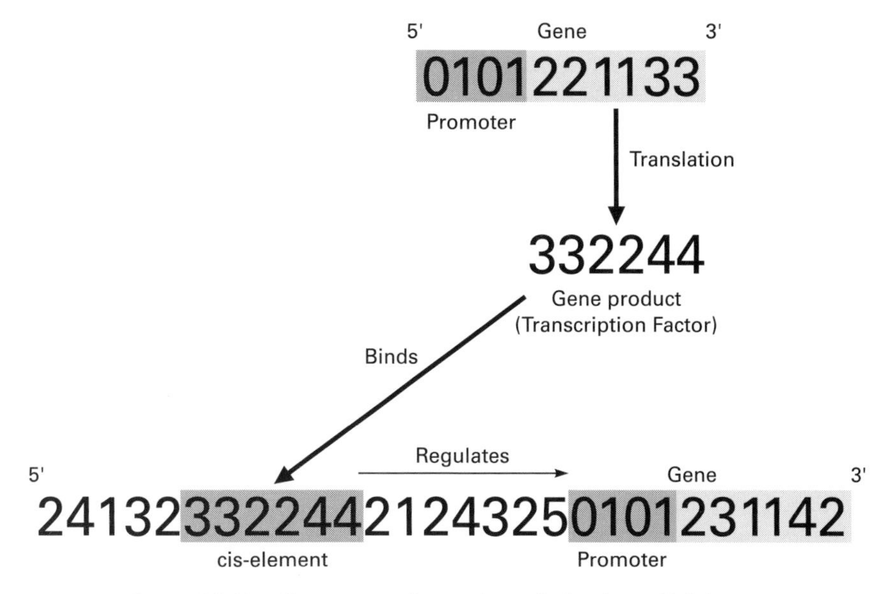

Figure 14.13 Gene expression and regulation in artificial genomes.

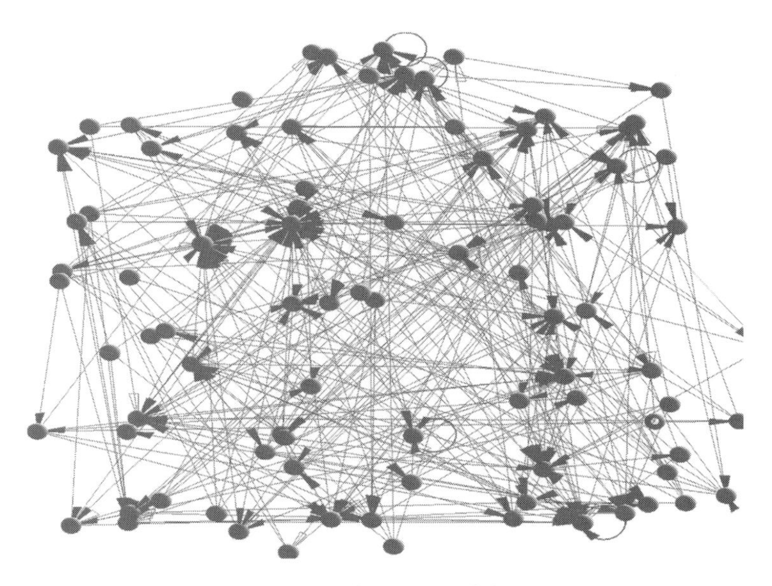

Figure 14.14 A gene regulatory network based on an artificial genome (L = 20 000; N = 6; B = 4, I = 1) (visualized using Watson's artificial genome software, Watson, 2002).

- The length L of the genome is positively correlated with the number of genes and with the average connectivity of the network K.
- The length N of a gene is negatively correlated with K.
- The base B of the genome string is negatively correlated with the number of genes as well as with K.
- The inhibition parameter I is negatively correlated with gene activity.

14.4.2 Dynamics

The artificial genome implementation discussed here uses binary gene states (on/off) and discrete time steps. Similar to random Boolean networks, all genes are updated at each time step. Those genes whose net regulatory input is *enhancement* will be *on* at the next time step (i.e. their gene product is expressed), those whose input is *inhibition* will be *off*.

Consequently, the unfolding dynamics of artificial genomes take the form of a gene expression pattern over time. Because genes cannot be active by default, the system needs to be seeded with at least one initially active gene. The expression pattern of artificial genomes are conveniently displayed in expression graphs. These depict discrete time steps on the x axis, and genes on the y axis. Figure 14.15 shows a gene expression graph with 86 genes over 200 time steps. The behaviour of the system is highly dependent on the parameter settings. Three regimes can be distinguished, each exhibiting distinctly different dynamics: ordered, complex and chaotic.

As discussed elsewhere (Reil, 1999a), gene expression is classified as *ordered* when genes are continuously active or inactive throughout a run (allowing for an initial settling phase, Figure 14.16). Ordered gene expression is correlated with low average connectivity in the network ($0 < K < 2$).

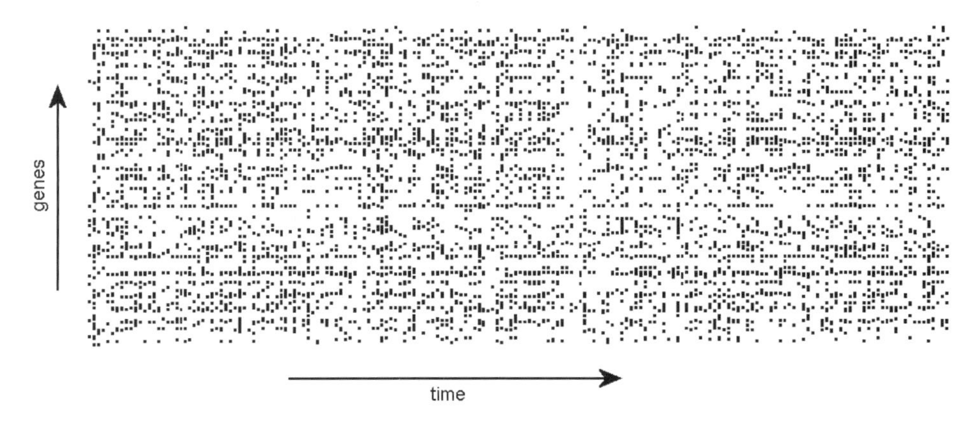

Figure 14.15 Expression graph depicting a gene expression pattern of an artificial genome. y-axis: genes; x-axis: time steps.

Figure 14.16 Ordered gene expression.

On the other side of the spectrum lie chaotic gene expression patterns, as depicted in Figure 14.17. Here, gene expression appears to be completely random. Chaotic behaviour is observed if the number of genes and regulatory sequences is large and the degree of connectivity is high (typically $K > 5$).

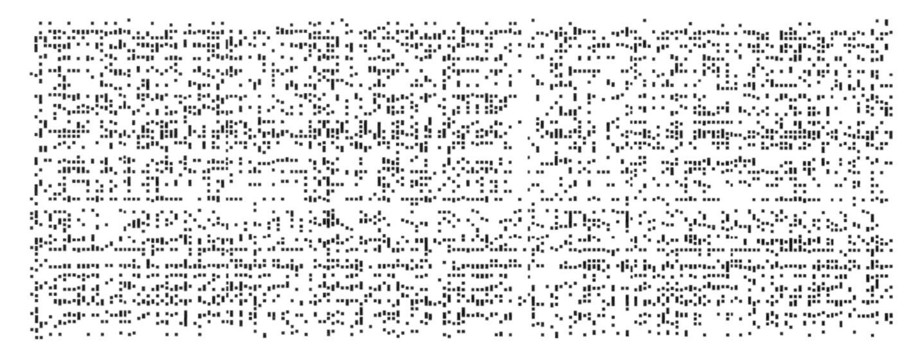

Figure 14.17 Chaotic gene expression.

The intermediate, complex, regime exhibits the richest dynamics of the three expression dynamics categories. In the artificial genome implementation discussed here, complex dynamics are observed if the number of genes is moderate (typically between 50 and 100) and the average connectivity K is above 3.

The hallmarks of artificial genome dynamics in this regime are:

- cyclic gene expression
- multiple distinct expression patterns per genome
- homeostasis
- differentiation.

14.4.3 Cyclic gene expression

Gene expression is classified as cyclic if a gene expression state at a given time step is revisited within a moderately small period. Figure 14.18 (page 270) depicts a gene expression pattern with a period of 44 (parameters: $L = 20\,000$, $N = 5$, $B = 4$, $I = 2$, number of genes: 84, $K = 7.3$), Figure 14.19 (page 270) shows the distribution of cycle lengths for an artificial genome with $L = 100\,000$.

14.4.4 Multiple expression patterns

Artificial genomes in the complex regime typically exhibit a number of distinct expression patterns. Which pattern is displayed depends on the initial conditions (i.e. the active genes at the start of the run). Figure 14.20 (page 271) shows four different expression patterns contained in a genome with 79 genes and $K = 8.3$.

Importantly, while multiple patterns are contained within a single genome, the number of these distinct expression patterns per genome is relatively low. For example, the average number of patterns per genome in the runs depicted in Figure 14.16 is 5.425.

Figure 14.18 Cyclic gene expression of an artificial genome in the complex regime.

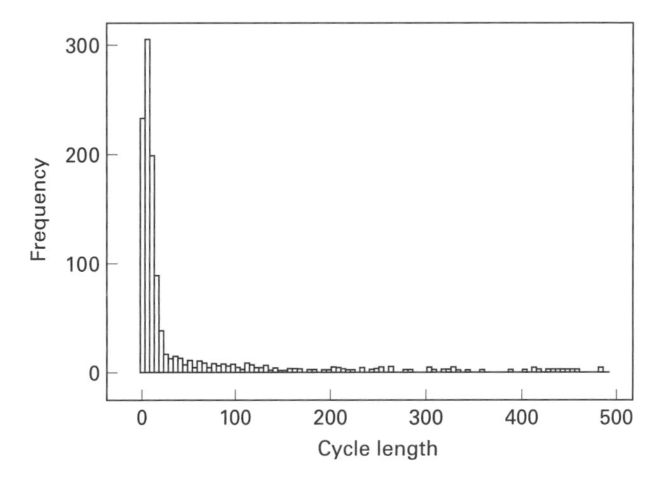

Figure 14.19 Distribution of limit cycle lengths of 200 artificial genomes. L = 100 000; B = 5; N = 6; I = 2). The average number of attractors per genome is 5.425.

A closer look at the underlying dynamics reveals that the same cyclic expression pattern is converged upon from different initial conditions (Figure 14.21, page 272). In analogy to RBNs, the properties outlined above lead to the classification of cyclic expression patterns exhibited by artificial genomes as *limit cycle attractors*.

14.4.5 Homeostasis and controlled differentiation

Homeostasis and controlled differentiation are related to the limit cycle attractor dynamics discussed in the previous section. Specifically, homeostasis refers to the robustness against external disturbances, which is exhibited by artificial genomes in the complex regime. Figure 14.22 (page 273) shows a gene expression pattern which is transiently disturbed (the state of gene 8 is briefly switched) and which recovers to the previous cyclic pattern within 23 time steps.

Similarly, Figure 14.23 (page 273) depicts a pattern, whose overall dynamics (in terms of the period and active sets of genes) are not affected by the permanent knocking out (deactivation) of one gene.

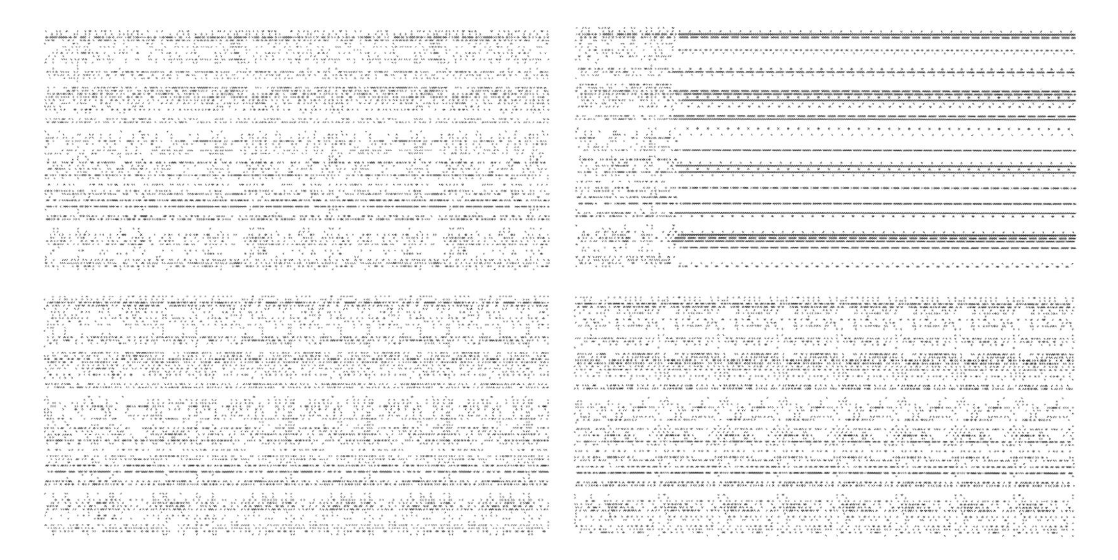

Figure 14.20 Multiple limit cycle attractors in an artificial genome. L = 20 000; N = 5, B = 4, I = 2.

Differentiation on the other hand is the phenomenon that can be observed when transient disturbances do not lead to the eventual recovery of the original expression pattern, but lead instead to the transition into one of the alternative attractors of the artificial genome (Figure 14.24, page 274).

Note that robustness to transient disturbances is closely related to the dynamics shown in Figure 14.21 (convergence upon the same attractor from different initial conditions), whereas differentiation is analogous to the dynamics of Figure 14.20 (different attractors from different initial conditions).

In complexity terms, the former behaviour is produced if the disturbed state remains in the basin of attraction of the original limit cycle attractor (Figure 14.25a, page 274); conversely, differentiation occurs if the disturbed state is outside the original basin of attraction, and inside an alternative one (Figure 14.25b, page 274).

14.4.6 Artificial genomes as models of gene regulation

The goal of the artificial genome framework is to model closely the structure and mode of action of biological genomes. Table 14.1 (page 274) lists the essential similarities between biological and artificial genomes.

Of particular interest is the fact that such a constructed system produces dynamics that closely resemble those of another major gene regulatory model, random Boolean networks (Reil, 1999a). Specifically, these are: cyclic gene expression, multiple distinct expression patterns per genome, homeostasis and differentiation. In contrast to random Boolean networks, however, artificial genomes achieve these dynamics through structurally representing the genome, genes, gene products and regulatory sequences.

This makes artificial genomes a suitable model for investigating questions concerning genome structure. For example, because coding genes and regulatory sequences are clearly distinguished, it is possible to examine the relative importance of each in the process of evolution. It was

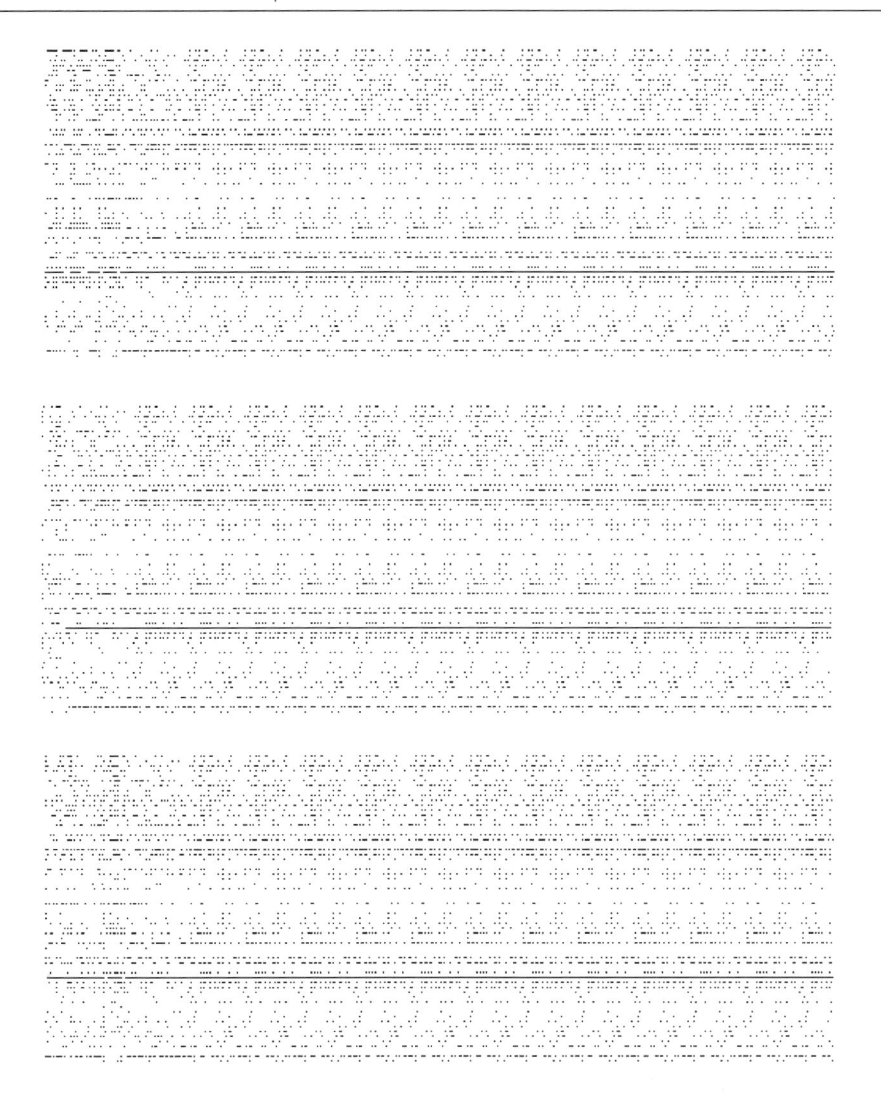

Figure 14.21 Convergence onto same limit cycle attractor after initial settling phase L = 20 000; N = 5, B = 4, I = 2.

argued at the beginning of this chapter that genes (and their products) are conserved throughout the evolutionary process because they are used as a set of basic tools, with the main differentiator between species being when and how these tools are used. The argument behind this is that tools themselves cannot be modified without knock-on effects on functions in which they are involved; the application of tools, on the other hand, can be modified or augmented without affecting already established structures.

This is one of the hypotheses that can be tested in the artificial genome framework. As a first step, initial experiments have been conducted that use genetic algorithms to evolve artificial genomes with increasing limit cycle attractor lengths. Figure 14.26 (page 275) shows an evolutionary run in which the cycle length was successfully increased from a period of 10 (at generation 0) to a period of 60 in generation 140 (population size: 50). (Mutations are implemented as a changed digit at a random position in the artificial genome.) Mirroring biological DNA

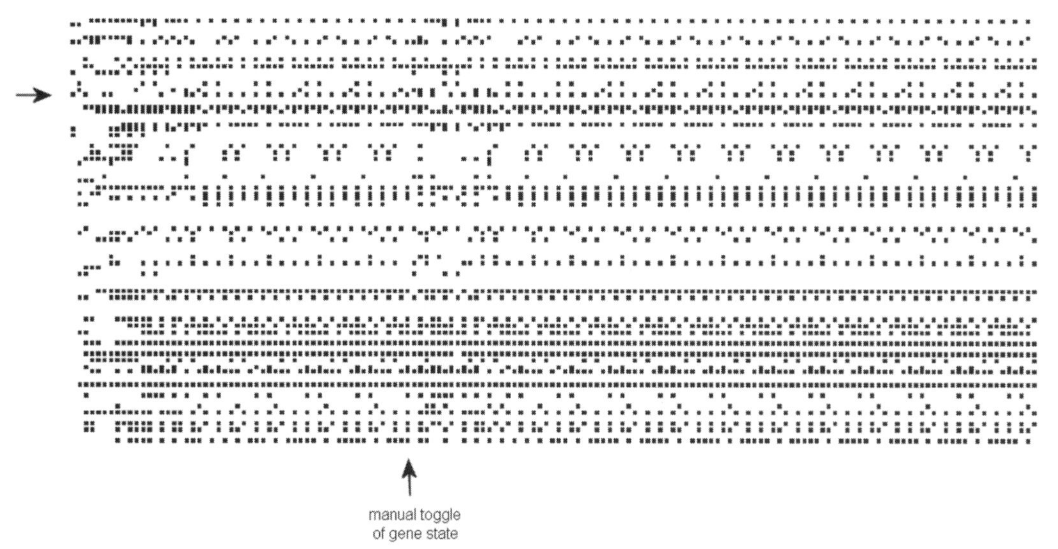

Figure 14.22 Robustness against a transient disturbance. L = 20 000; N = 6, B = 4, I = 1.

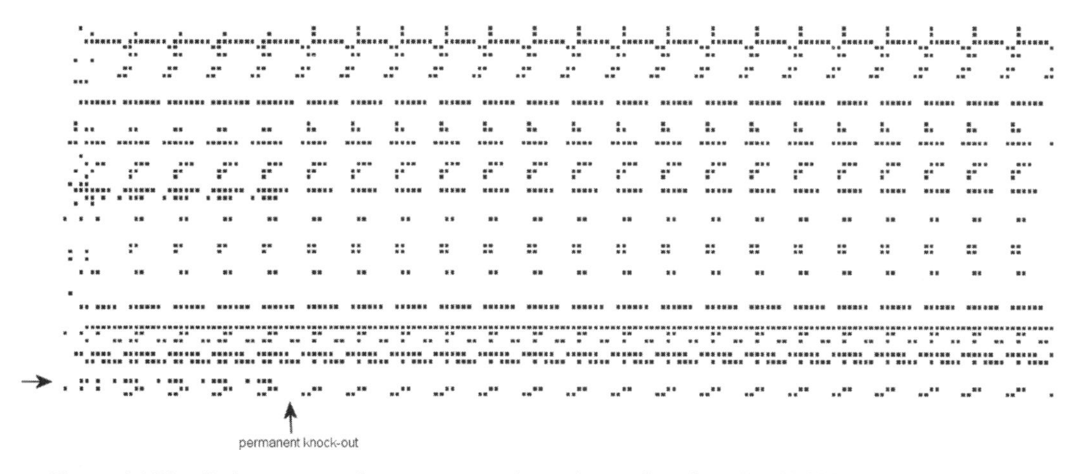

Figure 14.23 Robustness against permanent loss of gene function. L = 20 000; N = 6, B = 4, I = 1.

evolution, all runs had in common that substitutions (i.e. fixed mutations) were accumulated much faster in non-coding regions of the artificial genome than in coding and regulatory ones (a phenomenon that is mirrored in biological DNA evolution).

While successful, evolutionary experiments of this type will not directly reveal the functional difference between regulatory and coding regions in the course of evolution, as they put no conservative constraint on the evolutionary process. A more suitable fitness function is therefore the evolution towards an increasing number of attractors for a given genome, while simultaneously not compromising the expression pattern of those already present. If the hypothesis outlined above is correct, most substitutions (fixed mutations) should take place in the regulatory sequences, whereas coding sequences should be preserved. Corresponding experiments

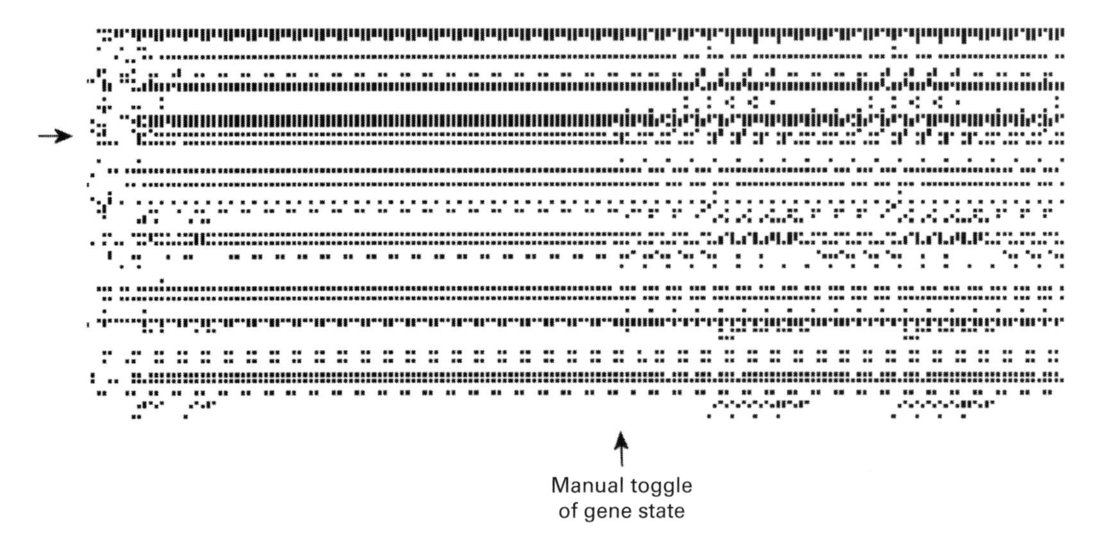

Figure 14.24 Transition between two attractors in response to a transient disturbance of gene expression. L = 20 000; N = 6, B = 4, I = 1.

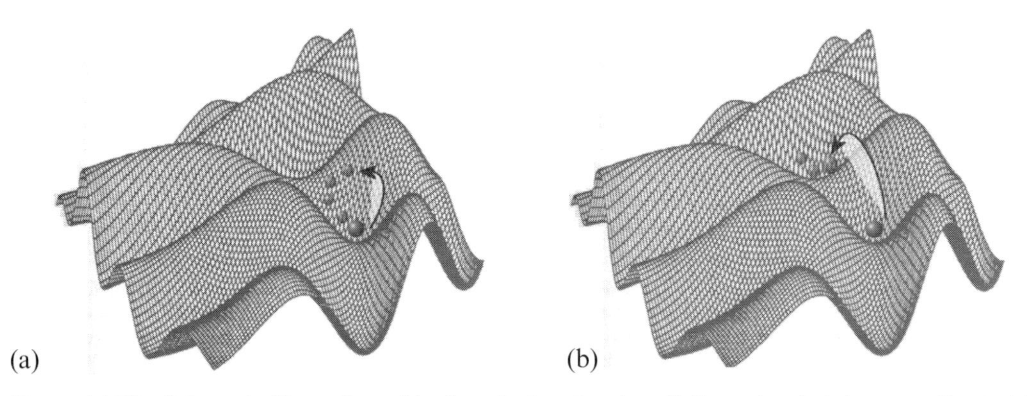

Figure 14.25 Schematic illustration of basins of attraction in a 2-dimensional state space. See text for explanation.

Table 14.1 Similarities between biological and artificial genomes

	Biological genome	*Artificial genome*
Information	Nucleotide sequence	Digit sequence
Gene structure	Promoter-coding sequence (exons + introns)	Promoter-coding sequence
Global structure	Large regions of non-coding DNA interspersed with coding sequence	Large regions of non-coding DNA interspersed with coding sequence
Translation	DNA-mRNA-protein	Coding sequence – gene product
Regulation	Template matching	Template matching

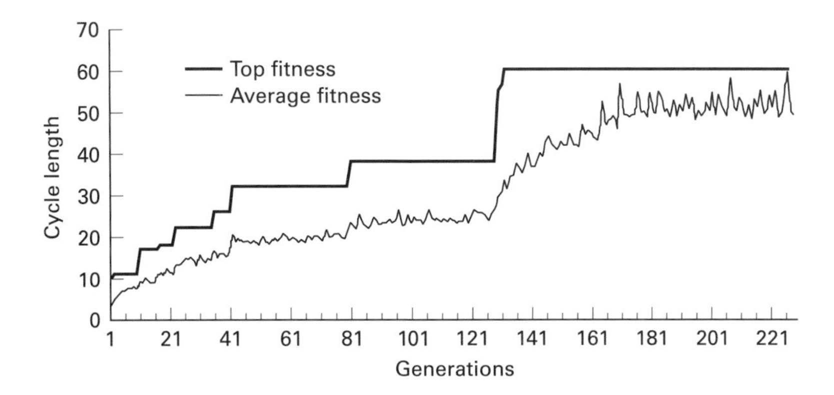

Figure 14.26 Fitness graph of evolved artificial genome. Fitness function: limit cycle attractor length. *x*-axis: generations; *y*-axis: fitness (cycle length). Population size: 50.

using the current implementation have not yet identified this correlation. In fact, in the form discussed here, artificial genomes have proven to show low evolvability with respect to gradual, conserved attractor number evolution.

One possible reason for this is the assumptions regarding gene states and updating. Like random Boolean networks, artificial genomes use binary gene states, concentration independent transcription factor binding, synchronous updating, and non-stochasticity. All four assumptions are simplifications of their biological equivalents and are potential sources of dynamics artifacts.

A new implementation of artificial genomes is therefore being developed. Specifically, this features non-binary gene activation and transcription factor expression. Instead, transcription factor concentrations increase with continued gene expression and decrease without it. Moreover, transcription factors now bind stochastically to regulatory sequences; the probability of a binding event is determined by the concentration of the transcription factor. Preliminary results suggest that the introduction of concentration dependence and variable binding durations solve the problems introduced by stochastic and asynchronous updating, i.e. limit cycle attractor dynamics are observed. Figure 14.27 shows the dynamics of an artificial genome with stochastic updating.

It is anticipated that such enhancements of the implementation presented here will further improve the suitability of artificial genomes as models of gene regulation and shed light on the

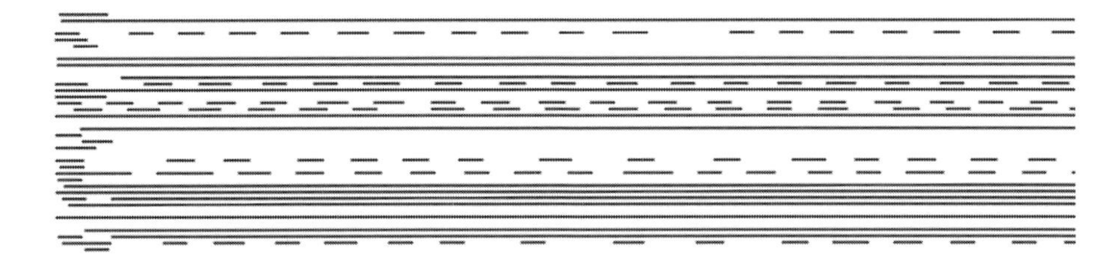

Figure 14.27 An artificial genome with stochastic TF binding, concentration dependence and variable (but increased) TF binding durations. At each time-step, there is on average only a 40% chance of a transcription factor binding to a matching regulatory sequence. Note that despite this noise periodicity is maintained.

dynamics (evolutionary and ontogenic) of one of the most fascinating complex systems in biology.

14.5 Conclusions

Artificial genomes are models of gene regulation that seek to capture the fundamental mode of operation of biological gene regulatory networks: template matching. To this end, the entire information of the system is contained in a string of digits, in analogy to the nucleotide sequence of DNA. Genes, regulatory sequences and gene interactions are determined on the basis of template matching and are not predefined.

Artificial genomes structurally resemble biological DNA in that they contain long stretches of non-coding spacer sequences, in which the frequency of substitutions is much higher during evolution than in coding regions. Moreover, the dynamics of simple artificial genomes mirror those observed in random Boolean networks. Specifically, artificial genomes in the complex regime exhibit cyclic gene expression, multiple distinct expression patterns per genome, homeostasis and differentiation. Like RBNs, artificial genomes in their simplest implementation have binary gene states and are updated synchronously; they are therefore subject to similar interpretations and criticisms with regards to the biological plausibility of the resultant dynamics.

Initial experiments with enhanced artificial genome models, which feature concentration dependence, variable TF binding times and stochastic updating, indicate that these limitations can be overcome. This, as is argued here, makes artificial genomes a suitable model for the study of the dynamics and structural evolution of genomes and genetic regulatory networks.

References

Bentley, D.R. (2000) The human genome project – an overview. *Medicinal Research Reviews*, **20**(3), 189–196.

Bongard, J. (2002) Evolving modular genetic regulatory networks. *Artificial Life*, **8**.

Coen, E.S. and Meyerowitz, E.M. (1991) The war of the whorls: genetic interaction controlling flower morphogenesis. *Nature*, **353**, 31–37.

Eggenberger, P. (1997) Evolving morphologies of simulated 3D organisms based on differential gene expression. *Proceedings of the 4th European Conference on Artificial Life*, Husbands, P. and Harvey, I. eds. MIT Press.

Ewing, B. and Green, P. (2000) Analysis of expressed sequence tags indicates 35 000 human genes. *Nature Genetics*, **25**, 232–234.

Harvey, I. and Bossomaier, T. (1997). Time out of joint: attractors in asynchronous random Boolean networks. *Proceedings of the 4th European Conference on Artificial Life*, Husbands, P. and Harvey, I. eds, pp. 67–75. MIT Press.

Kauffman, S.A. (1969). Metabolic stability and epigenesist in randomly constructed genetic nets. *Journal of Theoretical Biology*, **22**, 437–467.

Kauffman, S.A. (1993). *The Origins of Order*. Oxford University Press.

Kauffman, S.A. (1995). *At Home in the Universe*. Oxford University Press.

Kyoda, K. and Kitano, H. (1999) Simulation of genetic interaction for *Drosophila* leg formation. *Pacific Symposium on Biocomputing*, **4**, 77–89.

Marnellos, G. and Mjolsness, E. (1998) A gene network approach to modelling early neurogenesis in *Drosophila*. *Pacific Symposium on Biocomputing*, **3**, 30–41.

Mendoza, L., Thieffry, D. and Alvarez-Buylla, E.R. (1999) Genetic control of flower morphogenesis in *Arabidopsis thaliana*: a logical analysis. *Bioinformatics*, **15**(7/8), 593–606.

Mewes, H.W., Albermann, K., Bahr, M. *et al.* (1997) Overview of the yeast genome. *Nature*, **387**, 7–8.

Reil, T. (1999a) Dynamics of gene expression in an artificial genome – implications for biological and artificial ontogeny. *Proceedings of the 5th European Conference on Artificial Life* Floreano, D., Nocoud, J.-D. and Mondada, F. eds, pp. 457–466. Springer Verlag.

Reil, T. (1999b) An Introduction to Complex Systems, http://users.ox.ac.uk/~quee0818/complexity/complexity.html

Reil, T. (2000) Models of gene regulation – a review. In Maley, C.C. and Boudreau, E. eds, *Artificial Life 7 Workshop Proceedings*, pp. 107–113, MIT Press.

Stewart, D., GeneSweep (2002), Cold Spring Harbour Laboratory, www.ensembl.org/Genesweep.

Thieffry, D. and Romero, D. (1999) The modularity of biological regulatory networks. *BioSystems*, **50**, 49–59.

Thomas, R. (1978) Logical analysis of systems comprising feedback loops. *Journal of Theoretical Biology*, **73**, 631–656.

Wagner, G. and Altenberg, L. (1996) Perspective: complex adaptations and the evolution of evolvability. *Evolution*, **50**(3), 967–976.

Watson, J. (2002) http://www.itee.uq.edu.au/~jwatson/software.html

Wilson, R.K. (1999) How the worm was won – the *C. elegans* genome sequencing project. *Trends in Genetics*, **15**, 51–58.

Wuensche, A. (1999) Classifying Cellular Automata Automatically; finding Gliders, Filtering and Relating Space-Time Patterns, Attractor Basins and the Z-Parameter. *Complexity* vol 4, no 3, 47–66.

15 Evolving the program for a cell: from French flags to Boolean circuits

JULIAN F MILLER and **WOLFGANG BANZHAF**

15.1 Introduction

The development of an entire organism from a single cell is one of the most profound and awe inspiring phenomena in the whole of the natural world. The complexity of living systems itself dwarfs anything that man has produced. This is all the more the case for the processes that lead to these intricate systems. In each phase of the development of a multicellular being, this living system has to survive, whether stand-alone or supported by various structures and processes provided by other living systems. Organisms construct themselves out of humble single-celled beginnings, riding waves of interaction between the information residing in their genomes – inherited from the evolutionary past of their species via their progenitors – and the resources of their environment.

Permanent renewal and self-repair are natural extrapolations of developmental recipes, as is adaptation to different environmental conditions. Multicellular organisms consist of a huge amount of cells, the atoms of life, modular structures used to perform all the functions of a body. It is estimated that there are of the order of 10^{13} cells in the human body. Some of them are dying and being grown again constantly, and it seems miraculous that an organism manages to remain stable. The developmental process supports an amazing variety of individual organisms of a species, yet these vast colonies of cells have the same basic body plan.

The contrast cannot be stronger to human-made systems and machines. The manner in which living systems are built is the antithesis of the way that we construct things. Functionally invalid until the very last part has been inserted, machines await construction from outside agents who follow plans the machines themselves have no clue about. Thrown into the world of action, machines are usually functioning in a very narrow band of environmental conditions, quickly wearing down if those conditions are not favourable. Being unable to repair themselves, they stop functioning once the weakest part breaks. This is true also for electronic systems and software. Transistors, the non-linear elements of our computer world, have a defined working point. Environmental conditions leading to a deviation from this working point will quickly destroy electronic systems. Similarly, input to software systems that has not been thought about beforehand and prepared for, will usually throw a software system out of order or cause an exception. Thus, many of the simple things that living organisms do naturally, are fiendishly difficult for a computer program to accomplish.

 ISBN 0-12-428765-4

Nowadays it is commonplace for computer chips to be built on a sub-micron ($< 10^{-6}$ metres in size) scale. As we try to miniaturize electronic circuits further we start to meet problems that arise purely for reasons of size. First, the wires get in the way and their density becomes unmanageable, secondly, we find ourselves facing immense problems in verifying that the circuits work according to their specification. Imagine building circuits on a molecular scale, where we might have to specify the locations of the order of 10^{20} molecules. How will we get them to the correct position? The real difficulty associated with extreme miniaturization is the problem of getting information from the world we live into the tiny world of molecules.

Living systems have used an entirely different strategy: the information is in the smallest unit to begin with! Cells can be thought of as incredibly sophisticated robots that eat food, interact with the intercellular environment and build copies of themselves. The problem that researchers in developmental biology face is to try to understand how cells can accomplish these feats. If we want to learn from the natural example, an important question arises: How can we eliminate the many things that are incidental to the development process in biology from those that are fundamental? This is where computer scientists may be able to help. Already, a number of attempts have been made by computer scientists to create models inspired by biological development. There are two main directions to the effort in this field. Some researchers are trying to model (as accurately as possible) biological processes to help developmental biologists understand how development works. Others are exploring the degree to which developmental approaches may help us solve certain problems in computer science. There is no doubt that both fields of work will ultimately be of great benefit to humanity.

This chapter is about our attempts to explore the capabilities of a highly idealized developmental computer algorithm. Although we hope that the ideas discussed here may be of benefit to developmental biologists (at least on a conceptual level) we expect our work to be of more interest to computer scientists. We hope to ask some of the questions that concern developmental biologists, but ask them of developmental computer algorithms. Two of the questions we address particularly are the following: (i) How can we define programs that run inside cells that construct complex structures, when each cell runs an identical program? and (2) How can we obtain structures that are self-regulating (e.g. once mature, remain at a fixed size)?

In this book we have seen various attempts at modelling developmental phenomena on computers and computer algorithms that are inspired by, or attempt to utilize, developmental ideas. In Section 15.2 we describe a form of genetic programming that has been recently created and applied to developmental algorithms. We make no claims as to its particular suitability to this problem, although it has been found to be effective on a number of problems. In Section 15.3 we apply it to the problem of growing computer programs in the form of directed graphs. In Section 15.4 we apply it in a very different way to the problem of growing two-dimensional arrays of 'cells'. Much of the work reported here is new and therefore by definition immature, however we feel that it has produced some surprising and even beautiful results.

15.2 Cartesian genetic programming

In this chapter we have used a form of genetic programming known as Cartesian genetic programming (CGP) (Miller and Thomson, 2000). CGP is a particular form of genetic programming that allows the automatic evolution of computer programs (Koza, 1992, 1994; Banzhaf *et al.*, 1998; Koza *et al.*, 1999).

CGP encodes graph-like computer programs. The particular graphs used here are directed and acyclic graphs. Essentially the program is a mapping function, mapping input data to output data. In Section 15.3 the programs map integers to integers, while in Section 15.4, the programs map binary data to binary data. To clarify how this works we give a simple example of an encoded program that represents the mathematical function:

$$g = x^4 + 2x^3 + x^2 + x$$

where the variable x stands for a numerical value. This can be represented using a set of primitive operations listed below:

$$f_0(a, b) = a + b$$
$$f_1(a, b) = ab$$

Now suppose that the value of x is made available to us. We will think of this value being supplied from a node labelled 0 (i.e. node 0 contains x). Function g can now be represented as an acyclic directed graph using the input values x and the functions f_0 and f_1 (Figure 15.1). Figure 15.2 shows this in graph form.

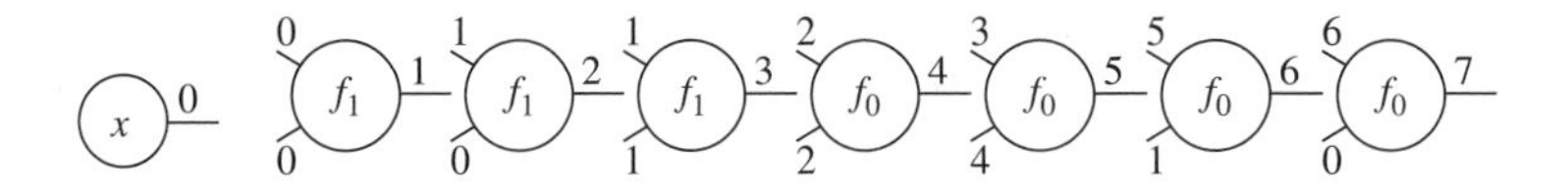

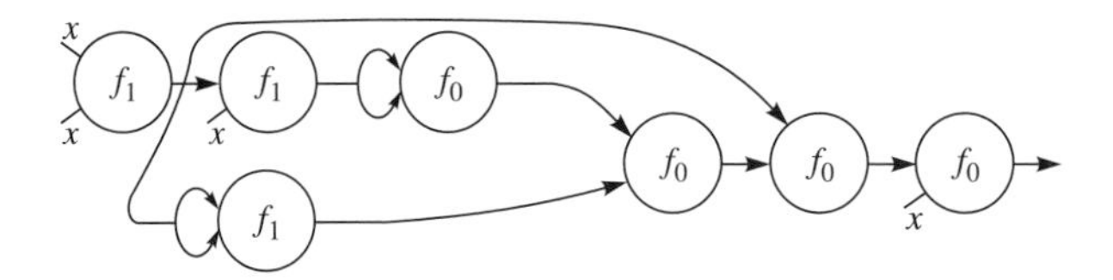

Figure 15.2 The same function drawn as a graph.

Now let us look at the first function node. It is function 1 so it multiplies whatever is presented at its two inputs (labelled 0), these are both connected to the node that produces x, so this means that the value leaving the node is x^2. This value is labelled 1 (meaning the output of node 1). Node 2 is again a multiplication node and multiplies one input (x^2) by the other (x), thus producing x^3. The process continues in this way until we get to node 7 which outputs the function g. Note that the outputs of all nodes always connect to the input of a node on the right, this is called feed-forward (hence directed, acyclic). In CGP, graphs of this sort are represented by a list of integers. In this case each node requires three integers that describe what its inputs are connected to and the function of the node. The graph above would be represented by the following list of integers, which we call a chromosome, for reasons that will become apparent shortly (the node functions are represented by the function index, in boldface).

0 0 **1** 1 0 **1** 1 1 **1** 2 2 **0** 3 4 **0** 5 1 **0** 6 0 **0**

Now suppose that we altered one of these integers in the group 3 4 **0** to 3 2 **0**, something interesting happens. Node 4 is now not part of the graph and the function the graph represents has changed to g′:

$$g' = x^4 + x^3 + x^2 + x$$

The $2x^3$ node (node 4) was disconnected and replaced the x^3 node (node 2). Now that node 4 is disconnected it could be changed without affecting the graph that was encoded by the chromosome. We will refer to node 4 as inactive. Changing a single integer in the chromosome is called a *mutation*. It is not difficult to guess now how we might go about *evolving* chromosomes of this sort. Next, suppose that somebody gave us a table of values for x and $h(x)$, where we are told that $h(x)$ is another polynomial, but we are not told its form. If we generate a chromosome at random (respecting, of course, that every third integer from the left must be a valid function label, and the two first integers in each triplet must be an integer less than the node label) we will always obtain some polynomial which we can evaluate to give a table of values for x. We could then compute the difference between the value for h predicted from our chromosome and the value for h given in the table. If we computed these differences for all the values in the table we could form a measure of how close our polynomial is to the unknown one. A measure of this sort is called a *fitness function*. We could proceed to solve this problem by evolving chromosomes of the form described. In all the experiments described in this paper we use the following algorithm:

1. Generate 5 chromosomes randomly to form the population
2. Evaluate the fitness of all the chromosomes in the population
3. Determine the best chromosome (called it *current_best*)
4. Generate 4 more chromosomes (offspring) by mutating the *current_best*
5. The *current_best* and the four offspring become the new population
6. Unless stopping criterion reached return to 2.

Step 3 is a crucial step in this algorithm: if more than one chromosome is equally good then always choose the chromosome that is not the *current_best* (i.e. equally fit but genetically different). In a number of studies this step has been proved to allow a genetic drift process which turns out be of great benefit (Vassilev and Miller, 2000; Yu and Miller, 2001, 2002). In later sections we may refer to the mutation rate, this is the percentage of each chromosome that is mutated in step 4. It should be noted that on some occasions (especially in Section 15.4) the program that we are trying to evolve has many outputs. These outputs are taken consecutively from the rightmost node in the chromosome towards the left. The chromosome length is always made to be larger than the number of program outputs.

15.3 Developmental Cartesian genetic programming

In this section we describe a new form of CGP which is developmental in character. The philosophical standpoint taken is that *the cell is the basic unit of biology*. We see evolution as a process of evolving a cell. The cell is a very clever piece of machinery that can, in cooperation with an environment, self-replicate and differentiate to form a whole organism. Thus in developmental Cartesian GP (DCGP) we attempt to evolve a cell that can construct a larger program by iteration of the cell's program in its environment. Since we are trying to construct graphs made of nodes it is clear that we need to identify a cell with a node and different nodes could be cells that have differentiated from each other. The environment of a node could be its position and its connections. Imagine that we start with a 'seed' node that can only be connected to the program inputs. The seed node will have a function also and a position. To clarify the idea let us suppose that the program we wish to build has two inputs, x and y (labelled 0 and 1). Let us also suppose that we have two node functions (or cell types) f_0 and f_1 each with two inputs, as in Section 15.2, let f_0 add numerically the data presented to its inputs and f_1 multiply its inputs. Figure 15.3 shows one possible seed node:

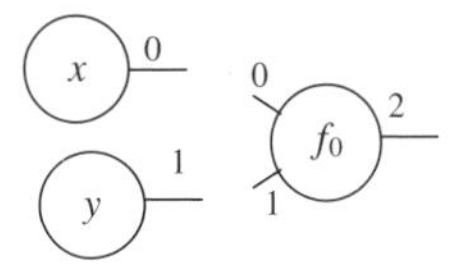

Figure 15.3 A possible seed node for a program.

Now we want to run a program inside the node that uses information about the node to construct a new one, indeed, we also want it to be able to replicate itself, so that we can grow a larger program. The node needs four pieces of information, its two connections, its function and its position and the program inside the node needs to take this information and create a new node with connections and a function and whether the node should be duplicated. So the situation is now as shown in Figure 15.4.

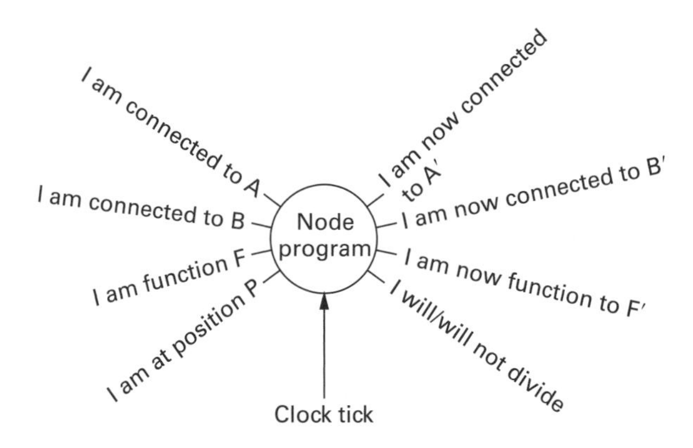

Figure 15.4 A node of developmental Cartesian GP.

What might the program inside the node look like? Well, we know that it must take in four integers and output four integers (the divide output could be just 0, for no replication, and 1, for replication). Thus we need to think up a way where a program (which we want to evolve) can do this. Well, if the program used the operations of addition and multiplication that would do fine. We have already discussed how to evolve programs of that form! However, we need to take care of one more point. We are using this program to construct a graph of nodes, that is feed-forward so the integers that come out of the node program must take the right values (i.e. the connections must be to nodes on the left of our current position, the functions must be ones on the list of valid functions, and finally the divide must be 0 or 1). Unfortunately, it is very difficult to construct programs that will automatically do that irrespective of the node position, so we need to resort to something that is a little inelegant. After we have run the node program we will need to carry out some sort of operation to bring the numbers into the correct ranges. A simple way to do this is to apply a modulo operation. Thus the new picture is shown in Figure 15.5.

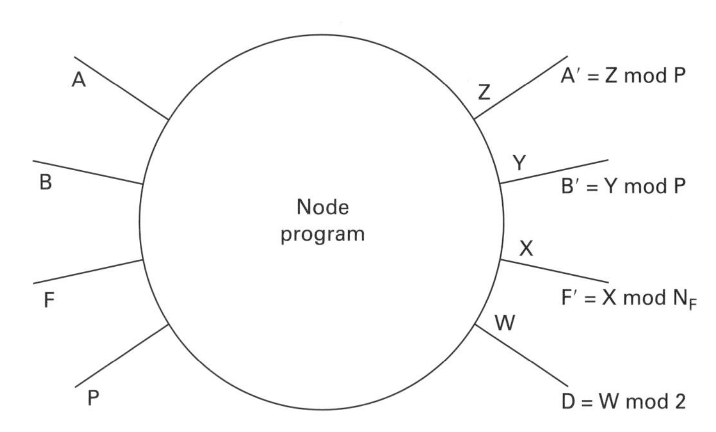

Figure 15.5 A node of developmental Cartesian GP with its modulo operation.

Let the outputs of the node program be W, X, Y, Z then if we divide Z and Y by the position P and take the remainder we will get two integers A' and B' which will be valid connections to nodes on the left of the current position P. If we have N_f possible node functions then the modulo operation will once again map the output integer X to a valid function type F'. Finally, we will need to take the integer, W, that decides whether the node is to be replicated, and find out whether it is even or odd using the modulo 2 operation.

So far we have not defined a node program. Here we will construct a simple one and apply it just to show that the whole idea actually works. Suppose the node program is defined by the following rules:

$$\begin{array}{ll} Z = (2A + B), & A' = Z \bmod P \\ Y = (Z + F), & B' = Y \bmod P \\ X = Y + P, & F' = X \bmod N_f \\ W = ZX, & D = W \bmod 2 \end{array}$$

Now we use the seed node defined earlier ($A = 0, B = 1, F = 0, P = 2$). Applying the above rules we get

$$\begin{array}{ll} Z = (0 + 1) = 1, & A' = 1 \bmod 2 = 1 \\ Y = (1 + 0) = 1, & B' = 1 \bmod 2 = 1 \\ X = 1 + 2 = 3, & F' = 3 \bmod 2 = 1 \\ W = 3, & D = 3 \bmod 2 = 1 \end{array}$$

The divide signal is 1 so the new node is replicated and we obtain the graph shown in Figure 15.6.

The graph in Figure 15.6 represents the function $g = y^2$. The n program outputs are assumed to be taken from the n nodes with the largest output labels (i.e. here we have one output taken from node 3).

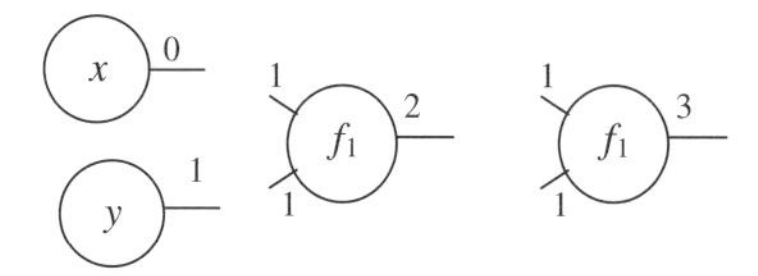

Figure 15.6 Graph obtained from the node program.

Now applying the node program in each node we obtain

First node (P = 2)		Second node (P = 3)	
$Z = (2 + 1) = 2,$	$A' = 2 \bmod 2 = 0$	$Z = (2 + 1) = 2,$	$A' = 2 \bmod 3 = 2$
$Y = (2 + 1) = 3,$	$B' = 3 \bmod 2 = 1$	$Y = (2 + 1) = 3,$	$B' = 3 \bmod 3 = 0$
$X = 3 + 2 = 5,$	$F' = 5 \bmod 2 = 1$	$X = 3 + 3 = 6,$	$F' = 6 \bmod 2 = 0$
$W = 2 \times 5 = 10,$	$D = 10 \bmod 2 = 0$	$W = 2 \times 6 = 12,$	$D = 12 \bmod 2 = 0$

So we now obtain the graph shown in Figure 15.7. The nodes have connected themselves together and the graph represents the function $g = x^2 y$.

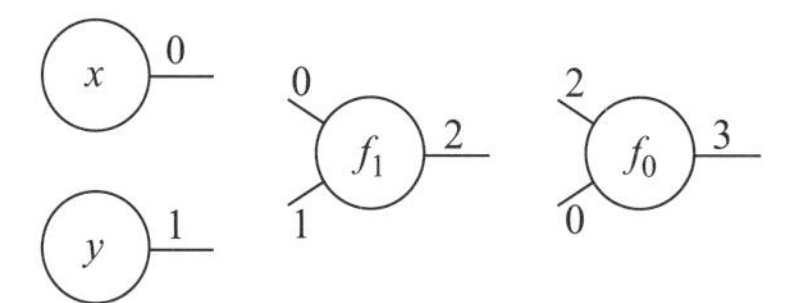

Figure 15.7 Graph obtained after applying the node program in each node.

How can we obtain the rules that define how nodes get transformed? Remember that the whole idea here was to evolve the rules inside a node, then apply those rules inside every node of the graph so that a new graph is built. But we already know how to represent rules that will map integers to new integers! We use a Cartesian program inside the nodes. So the picture now is as shown in Figure 15.8.

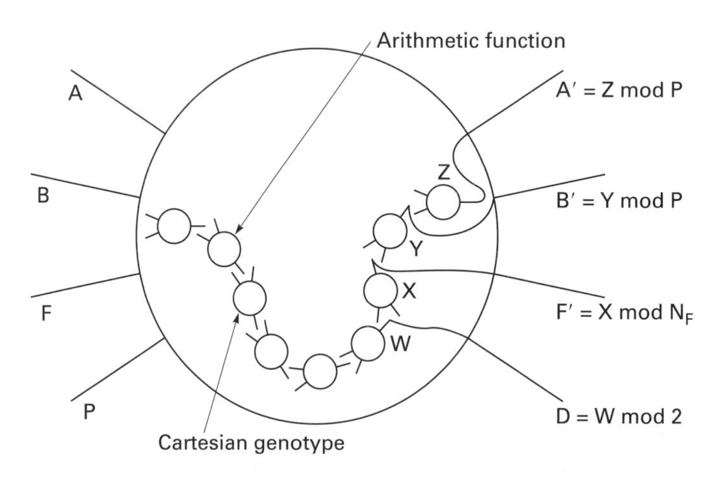

Figure 15.8 A node of developmental Cartesian GP with its cartesian genotype.

Thus all the nodes of the developing graph have identical chromosomes inside them that represent the node program rules. For instance, the rules we used in our example could be represented by a chromosome of the form:

0 0 **0** 4 1 **0** 5 2 **0** 6 3 **0** 7 5 **1**

Chromosomes of this form could themselves be subject to evolution using the algorithm defined in Section 15.2.

Let us summarize what we have achieved here. We described a way of building programs by running identical programs in nodes. We can iterate these programs and build enormous programs that would have required very large amounts of information to specify. We can evolve the node programs to try to obtain solutions to many problems. The way the method works is similar in some respects to real biology where cells replicate and differentiate to create enormous collections of cells that we call organisms. The hope is that it might prove easier to evolve relatively small programs and then iterate those programs to build very large programs rather than evolve the very large programs directly. Some early experiments have been performed using these ideas and it appears that for simple problems it is easier to evolve a chromosome that represents a program using a non-developmental approach (see Section 15.2). Although, certainly, it would be inconvenient to have to evolve enormously long chromosomes it is not obvious that it would necessarily take more generations to evolve a good solution than evolving smaller developmental chromosomes. This is because as the size of a program grows there are more ways that it can solve a problem (Langdon, 1999). There are obvious reasons why nature chooses to build things in a developmental way. It is much easier for large organisms to carry little packets of information around that under the right circumstances can grow into a mature organism. Just think of a tree, its seeds can be blown or carried for miles by the wind or inside the gut of birds. Thus there are strong energetic reasons in favour of development.

The techniques described in this section have benefitted from a number of aspects of developmental biology, however, it is not clear how the field of developmental biology can benefit. In the next section we will discuss another form of computer problem solving using developmental ideas that we think will hold some promise of contributing to developmental biology, inasmuch as it may allow the exploration of ideas that are relevant to that domain.

15.4 Evolving growing two-dimensional maps

In Section 15.3 we described a method for evolving the program for a cell, where the cell was a node in a graph. As we continued to run the identical programs in the cells, graphs of increasing size could be constructed. After a pre-determined number of iterations of this process the graph (or organism) constructed was tested against a user specification of the problem that was to be solved. We found that it was necessary to impose modulo conditions on the outputs of the cells so that valid programs or graphs could result. Although the ideas behind the method were inspired heavily from biological development we wished to extend the ideas to an invented world that was in some sense closer to the real biological world. In this way we hoped that our work would potentially shed some light at a conceptual level on real problems of biological development and, at the same time, allow us to address fundamental issues in the evolution of computer programs.

15.4.1 How cells and chemicals are represented

In this section we describe our attempts to create a two-dimensional world of cells of different types, each carrying identical genotypes that may grow, die, change colour and receive and emit chemicals. Here we evolve the genotype of the cell so that after a period of time we obtain a two-dimensional map that meets the user's requirements. We emphasize that we have attempted to construct the simplest possible way of doing this. Our aim is to understand the importance of various factors in an embryological growth process and we feel that an overly complicated model would make it difficult to assess the relative importance of these factors. We wish to shed light on questions such as: How can collections of independent cells interact and coordinate their activity so that they build an organism? How can the cells build increasingly larger versions of that organism that are faithful to the same body plan? What is the relative importance of direct cell interaction and global chemical signalling? How can growth be regulated?

The world we envisage is binary, so that amounts of chemical, cell types and actions can be denoted as binary codes. The genotype is a representation that encodes a Boolean circuit (that implements the cell program) that maps input conditions to output conditions. A cell sees its own state and the states of its eight immediate neighbours. It also sees the amount of chemical (at present a single chemical) at the location of each of its eight neighbours. From this information the cell program decides on the amount of chemical that it will produce, whether it will live or die, whether it will change to a different cell type at the next time step, and how it will grow. Unlike real biology, when a cell replicates a single copy of itself, it is allowed to grow in any or all of the eight neighbouring cells simultaneously (this is done to speed up growth, mainly for reasons of efficiency). The situation is described in Figure 15.9.

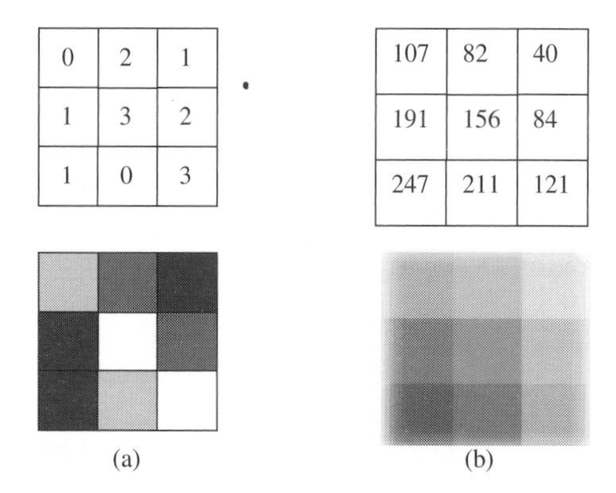

Figure 15.9 (a) Shows a white cell surrounded by 6 neighbours (0 denotes an empty or dead cell, 1 a blue cell, 2 a red cell and 3 a white cell), (b) shows the chemical environment at the location of the cells.

Let us suppose that a chemical is represented by an eight bit binary code and that there are four cell types (dead, blue, red, white). The cell program reads simultaneously the eight bit codes for all the neighbouring locations, the two-bit code signifying the cell's own type, followed by the two-bit codes representing the types of all the neighbours in a clockwise fashion (as shown in Figure 15.10). The cell's program defined by its genotype (see below) then outputs the following bits: an eight bit chemical code indicating the new strength of the chemical to be produced by the cell or daughter cells at the next time step, two new cell type bits, followed by an eight bit

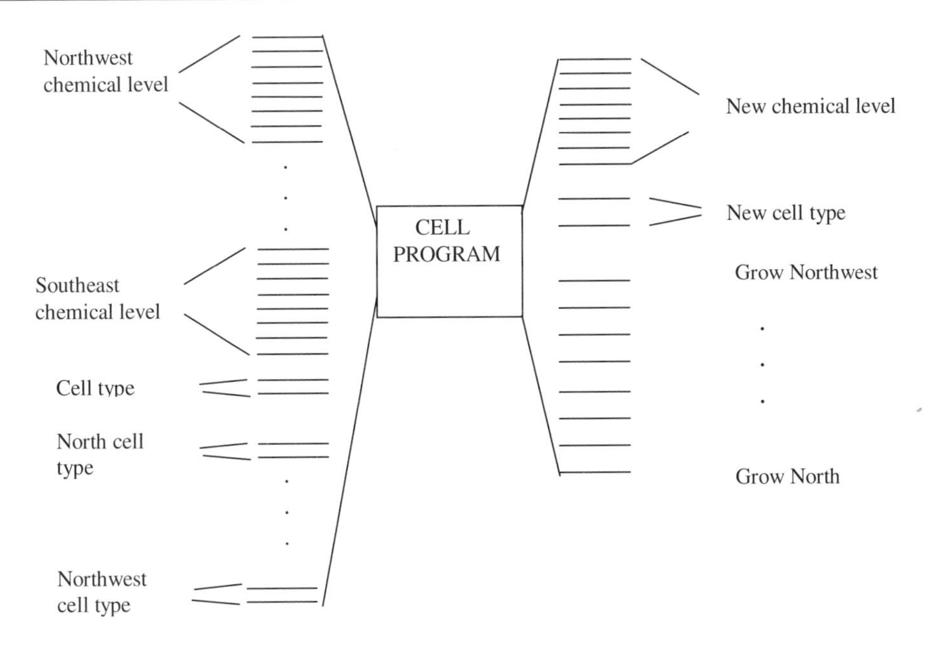

Figure 15.10 How data from the environment are read by the cell and result in change at the next time step.

growth code that indicates where the cell will replicate itself, starting at NW and then denoting places to grow in an anticlockwise manner.

If the cell program dictates that the cell should grow but that its new type is 0 then the cell will still grow into its chosen locations, but it and its replicated offspring will then die. Also the chemical level dictated by the cell program only applies to locations where the cell grew. If the cell dies, the chemical level on that square will be the previous value.

Chemicals can diffuse according to a number of user selected fixed diffusion rules. For all of the experiments reported in this chapter the diffusion rule is given below: Let N denote the neighbourhood with neighbouring position k, l, the chemical at position i, j at the new time step is defined by the update rule:

$$(c_{ij})_{new} = \frac{1}{2}(c_{ij})_{old} + \frac{1}{16}\sum_{k,l \in N}(c_{kl})_{old}$$

In the absence of cells, this rule ensures conservation of the chemical. However cells have no 'energy' constraints placed upon them and can pump out the maximum amount of chemical allowed (i.e. 255 if there are eight chemical bits). This diffusion rule is updated at the same time step (for the whole map) as the cells. Other diffusion rules have been implemented that allow chemicals to diffuse over a wider area but, as yet, we have no definitive results as to their efficacy. The way the chemical and cell maps are updated is as follows: the original maps are scanned cell by cell from the top left corner of the map. The updated cell growth, differentiation or death is recorded on the new map. Cells that are later in the scan sequence may overwrite an earlier cell. This introduces an unavoidable bias towards cells at the bottom right (since these are last to be scanned). The drawbacks of cell overwriting will be discussed later.

15.4.2 How a cell program is encoded into a genotype

In Section 15.2, we discussed how one could represent a program or circuit in the form of a string of integers (the Cartesian genotype). In all the experiments reported here we used a genotype of 400 integers. Each group of four integers defines a type of 2 to 1 multiplexer and its three connections. A 2 to 1 multiplexer is a three input Boolean logic gate that expresses the following function:

$$f(A, B, C) = A.\bar{C} + B.C$$

where A, B, C are binary inputs and the dot represents the AND logic function, the + the inclusive OR function and the bar the NOT operation (inversion). We employed four types of multiplexers, corresponding to whether the inputs A and B were inverted. It is well known in digital electronics that any Boolean function can be constructed from multiplexers defined by f above (allowing the four types simultaneously increases the complexity of Boolean functions that can be built from a given number of gates).

15.4.3 Experimental parameters

The algorithm used to evolve the cell genotypes has been described in Section 15.2. The particular experimental parameters used in the experiments reported in this chapter are shown in Table 15.1 (unless otherwise stated).

Table 15.1 Experimental parameters of the system

Experimental parameter	*Value*
Population size	5
Number of generations	30 000
Number of runs	10
Mutation rate	1%
Maximum number of multiplexers	200
Number of chemical bits	8
Number of cell types	4
Iterations at which fitness tested	Variable

15.4.4 Tasks for the cellular maps

The French flag model of Lewis Wolpert (Wolpert, 1998) was the inspiration for the task the maps of cells were to achieve. In particular we looked at two problems. We wished to evolve a cell program that, over a given period of time, would construct a French flag. Moreover, we wanted the flag to grow with time but always be recognizable as a French flag. For the second problem we wanted to grow a French flag of a given size and have the flag remain that size subsequently. One can think of the French flag as an organism, a rectangular creature with three vertical stripes, blue, white and red. We were interested in whether it would prove possible to evolve the cells' program to construct such a creature. If such a creature were constructed, how would it behave when it was damaged, or particular cells were transplanted into other regions? In this toy world we reasoned that we could conduct 'classic experiments' that have been demonstrated in the real world of biological development. In addition, by seeking to evolve programs that could be indefinitely iterated over time we wished to investigate methods of evolving programs that could grow structures without limit.

Flags or maps were defined as a simple grid of cells that we wished the initial seed cell to grow into. Figure 15.11 shows how two French flags were defined.

Initially a single cell was placed at a position in the map and the task was defined so that after a given (user defined) number of iterations of the cell program in every position where the cell was alive we required the collection of cells to be as close as possible to the configuration defined in Figure 15.11. Initially we could also define an initial chemical map, typically we would set the chemical to the maximum allowed value at the same location as the initial cell. To try to direct the evolution of cell programs to the desired end we could state a list of maps together with the iterations at which they were expected. At each test point the organism was compared with the desired map. A simple numerical sum was calculated of all locations in the map where the grown map agreed with that required (including empty or dead cells in the map). The sum was calculated for all the desired maps. Later we discuss various possible shortcomings of this definition of a fitness of a particular cell program.

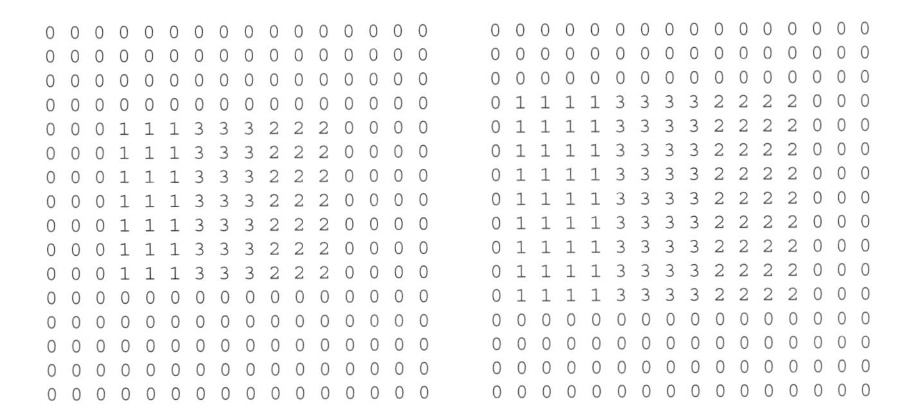

Figure 15.11 The specification for some French flags (medium, large).

15.5 Experiments and results

In the first experiment we required that the cells develop into the medium sized flag at iteration 7 (iteration 0 means the starting map) and the large flag at iteration 9. The maximum score was 512. The best result we obtained in the ten runs had a fitness of 491 and the average fitness of the 10 runs was 456. In Figure 15.12 we show the history of development of the best solution up until iteration 9.

It is curious how, even though the initial white cell is correctly placed with regard to the French flag, it nevertheless grows into two red cells. At iteration 5 white cells appear and blue cells make a first appearance at iteration 6, only one iteration away from the first test point! The corresponding developing chemical map is shown in Figure 15.13. At early iterations there appears to be virtually no chemical present, however levels are so low that they are barely perceptible in the grey background.

It should be stressed that during the evolution of the cell genotype the assessment of the cell's phenotype (the cellular map) was only ever made at iterations 7 and 9, so that different evolved genotypes are likely to display very different growth behaviours. In our next investigation we

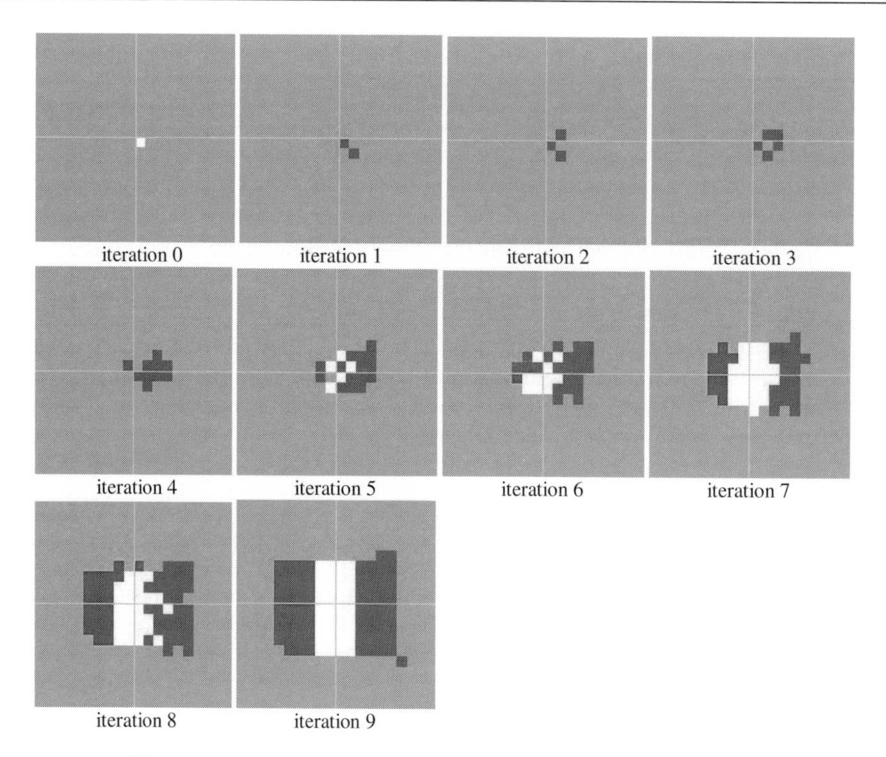

Figure 15.12 The developmental history of a cell program.

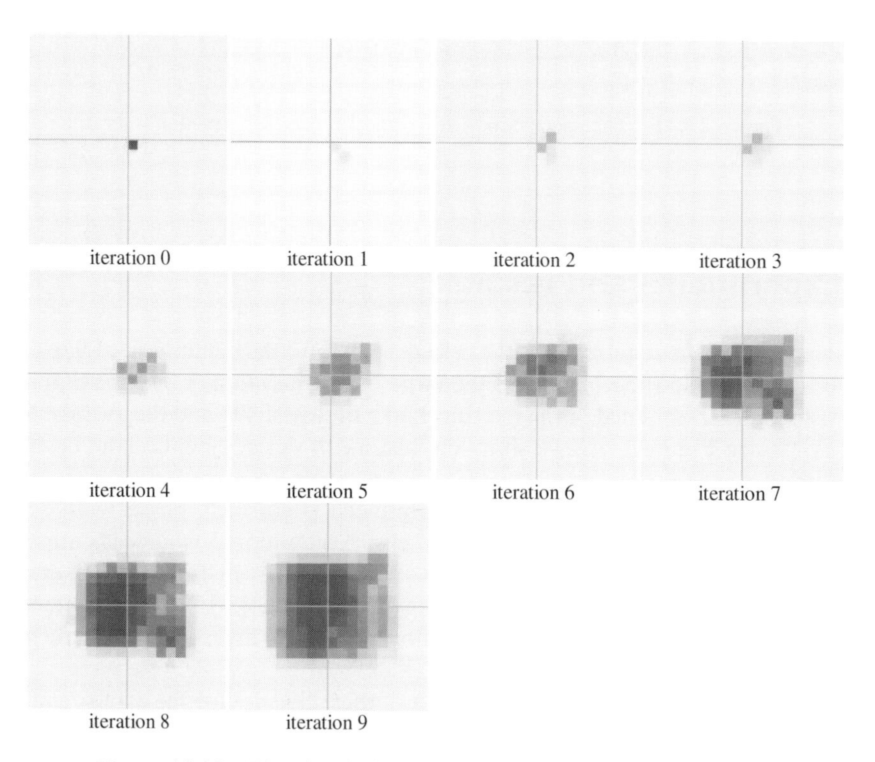

Figure 15.13 The chemical history of the growing cellular map.

examined the longer-term behaviour of the growing cellular map. What would the embryo look like beyond nine iterations? Would the map become so distorted that it was no longer recognizable as a French flag? In Figure 15.14 we show how the cellular map develops over a further 11 iterations.

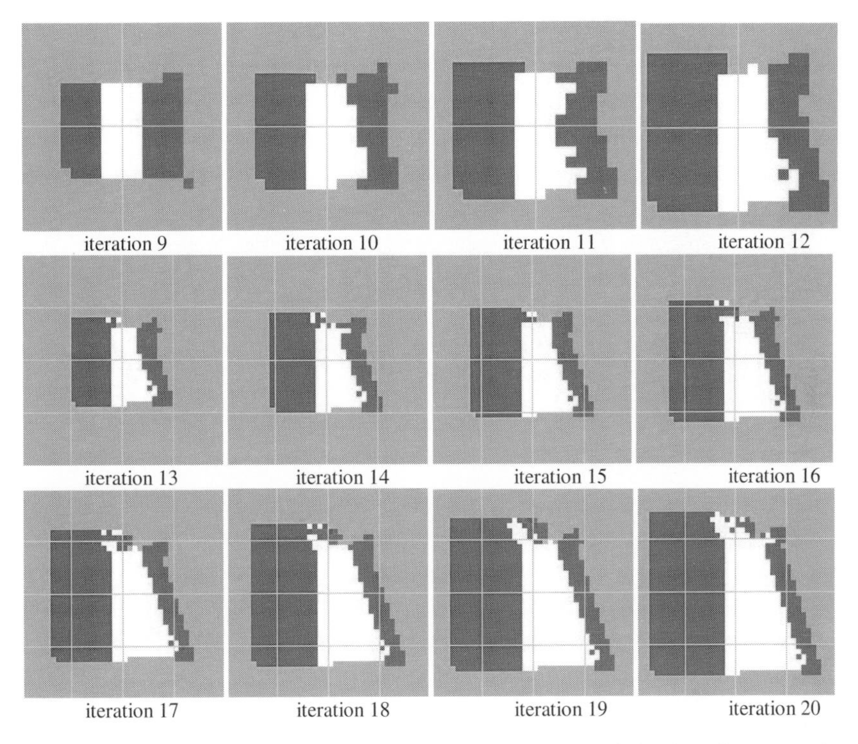

Figure 15.14 The developmental history of the cellular map beyond the last fitness testing point at iteration 9.

One of the striking observations about the maps after iteration 9 is the neat regularity of the blue section of the flag. Notice the single missing blue cell in the bottom left corner is present in all the maps, also that three sides of the blue region are straight as desired. At iteration 13 two white cells appear at the top right of the blue section. What has caused this behaviour? The cellular neighbourhoods in this region are fairly constant so it may be some underlying change in the chemical signature of the region. The diagonal red region may be caused by the asymmetry of the cell update rule, this bears further investigation. The chemical milieu is an important factor in the behaviour of the cells. This is easy to show by comparing the cellular map at iteration 10 when all the chemicals are removed from the chemical map at iteration 9 (Figure 15.15)

The next series of experiments revealed some fascinating parallels with real developmental biology. If we remove some parts of the embryonic French flag organism, what will happen? We removed the white and red cells from the cell map at iteration 9 and then proceeded to continue to grow the map (leaving the chemical map at iteration 9 unchanged). Since the blue region of cells are growing in such a regular way it seemed natural to expect that we would obtain a growing blue region of cells. However, the results, shown in Figure 15.16, quickly revealed a much more interesting behaviour.

The blue region tries to grow into the French flag again! Though it cannot quite manage it. The chemical map used for the blue region was the same as the chemical map that had been

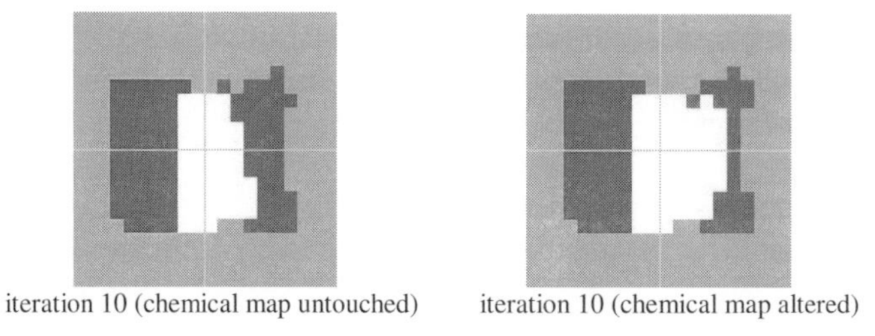

Figure 15.15 Differences caused in growing cell map due to imposed chemical changes.

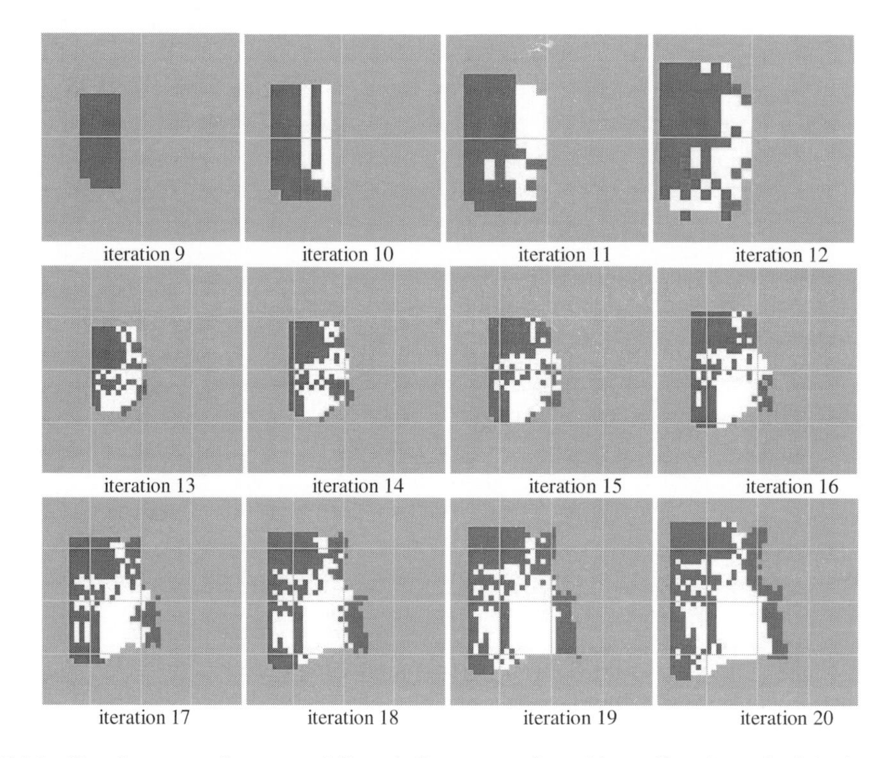

Figure 15.16 Development of corrupted French flag grown from blue cell region of original.

established for the whole complete map. It is possible that different behaviour may have occurred had we used an initial chemical map that was only local to the blue region. It is clear that the computer experiments have some parallels to the experimental problems that occur when developmental processes are altered in that alterations to the chemical conditions in altered embryos may lead to developmental effects (rather than solely being due to the alteration of the embryo alone).

In the first of two further experiments we cut a large hole in the centre region of the flag (at iteration 9) and iterated to see what would happen (Figure 15.17) and also, we cut the flag diagonally (Figure 15.18).

Once again we see that the cells gradually rebuild the map. Interestingly, the red region of the map at iteration 15 is identical to the red region on the original uncorrupted map at iteration 15. It seems that the section of red cells removed was of little importance.

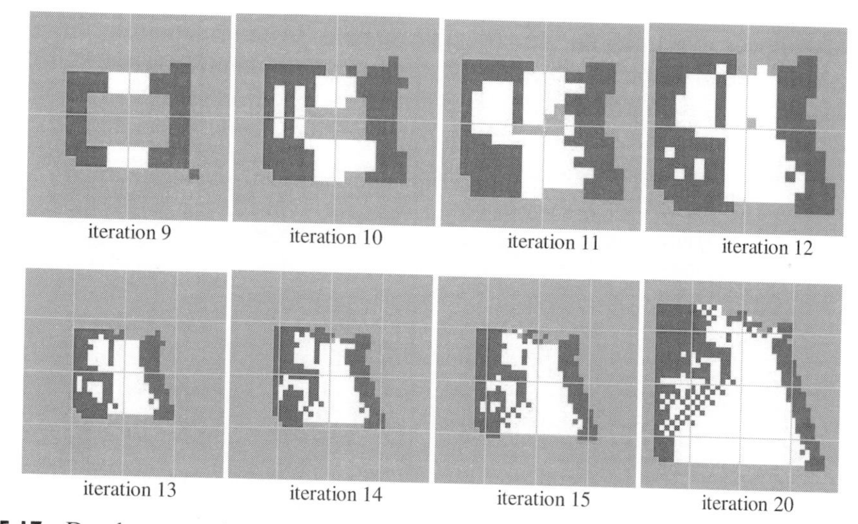

Figure 15.17 Development of cells when a rectangular section is removed from the embryonic map

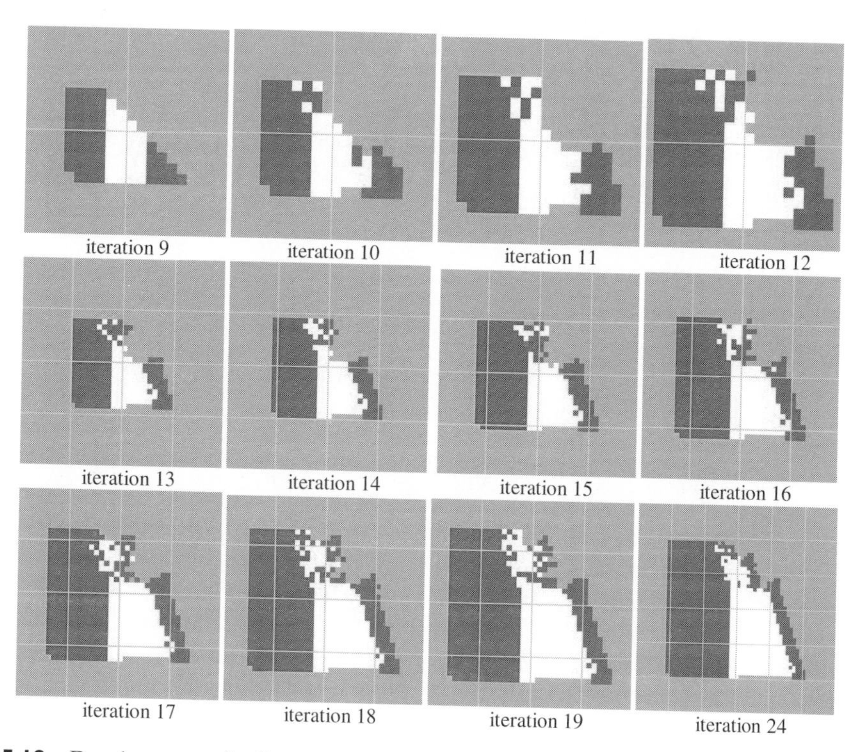

Figure 15.18 Development of cells when a diagonal section is removed from the embryonic map.

After embryonic cells are removed (iteration 9) we see that the blue region remains untouched, once again we see the familiar growing rectangular blue shape (with the bottom left corner missing). Red cells begin to appear at the top right section of the map, while the remaining red region at the bottom right begins to grow upwards until by iteration 24, the regions are almost joined. The white cells initially invade the growing blue region but eventually (iteration 24) appear to retreat. As time progresses the embryonic array starts to recover completely from the initial damage and takes on an appearance similar to the cell maps of the undamaged embryo.

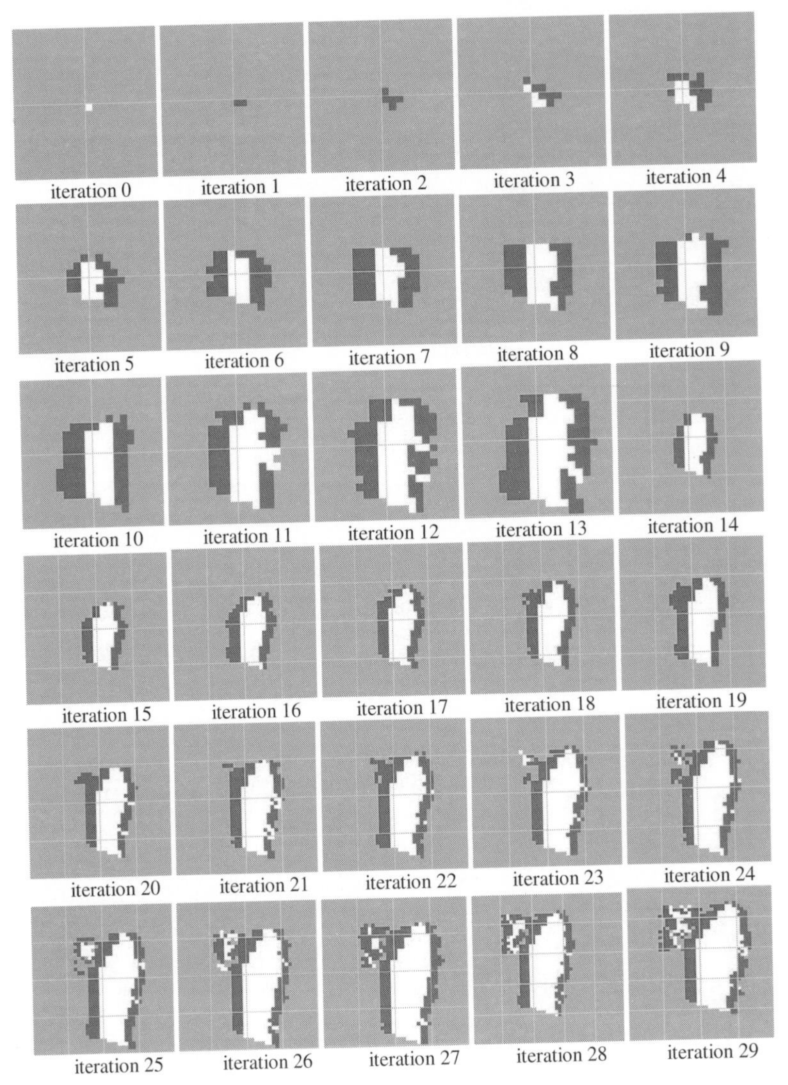

Figure 15.19 Development of cells required to grow to a fixed size map (at iterations 5,6,7,8) and then iterated over a much longer period (note the change in scale at iteration 14).

In all the experiments described so far we have seen continued growth of the embryonic cellular map. Of course, in biological development growth ultimately ceases. The next experiment attempts to create an embryonic cell that matures to a French flag appearance of a given size and then subsequently remains at that size irrespective of the number of further iterations of the cells' program. It is very hard to see how such a task could be accomplished without chemical signalling, as there are two phases: growth, followed by stasis. Something would have to change in the environment to make this possible. It was hoped that the cells might be able to use some characteristics of the chemical map to make this possible.

The results shown in Figure 15.19 are very encouraging. First, although the growth of the embryonic array of cells does not halt, it is considerably slowed (this is evident when one compares iteration 20 with the same iteration in Figure 15.14). Additionally, the appearance of the cellular map is much closer to the French flag, in that the coloured bands are more regular and are approximately vertical. The longer-term behaviour of the cellular map is more complex. Figure 15.20 shows the situation at iteration 46.

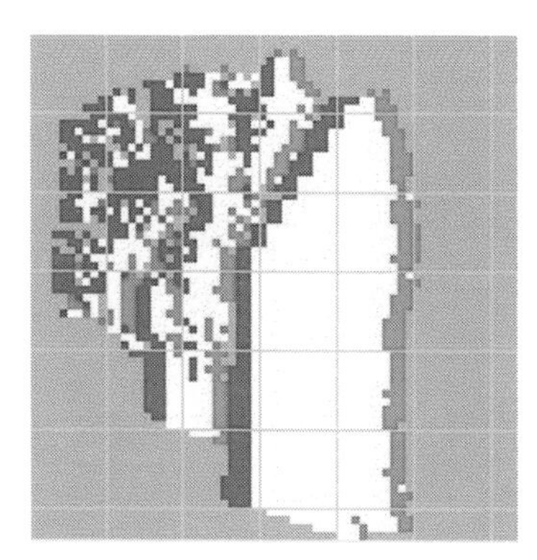

Figure 15.20 Cellular map at iteration 46 of developing cell program shown in Figure 15.19.

In the next experiment the *same* cell program was run but with all chemical levels set to zero. In this way we could investigate the sensitivity of the cell program to the chemical environment during growth. We could also investigate the importance of the chemical to slowing down growth after maturity. The results are shown in Figure 15.21.

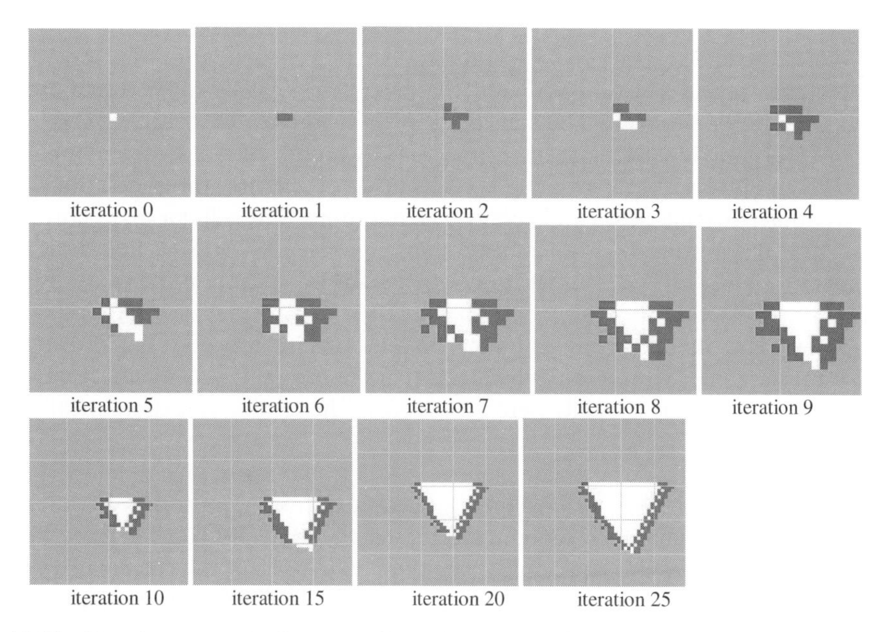

Figure 15.21 Development of cells grown from the program in Figure 15.19 but without any chemical environment.

At first the embryonic cell map develops in the same way (until iteration 2), but then gradually departs from the same behaviour as its chemical-rich counterpart, eventually assuming a triangular shape. Note that the cell mass appears to be growing at approximately the same rate. It is not possible to conclude from these data that chemical sensitivity of cells is essential to prevent growth. However, further experiments revealed that the presence of chemicals is an important factor in obtaining a cell program with a high fitness. Two sets of ten runs of the evolutionary algorithm were performed under identical conditions except that in one there were no chemicals. The runs with chemicals had an average fitness of 438 (standard deviation = 26.04), while the runs without chemicals had an average fitness of 400 (standard deviation = 14.06).

In the final experiment we wanted to see whether we could obtain a regular series of spots. We chose to require that the embryonic array of cells assume the pattern shown in Figure 15.22 (with perfect fitness 196) after the fifth iteration (the initial being 0).

```
0 0 0 0 0 0 0 0 0 0 0 0 0 0
0 0 0 0 0 0 0 0 0 0 0 0 0 0
0 0 3 3 3 3 3 3 3 3 3 3 0 0
0 0 3 3 3 3 3 3 3 3 3 3 0 0
0 0 3 3 1 1 3 3 1 1 3 3 0 0
0 0 3 3 1 1 3 3 1 1 3 3 0 0
0 0 3 3 3 3 3 3 3 3 3 3 0 0
0 0 3 3 3 3 3 3 3 3 3 3 0 0
0 0 3 3 1 1 3 3 1 1 3 3 0 0
0 0 3 3 1 1 3 3 1 1 3 3 0 0
0 0 3 3 3 3 3 3 3 3 3 3 0 0
0 0 3 3 3 3 3 3 3 3 3 3 0 0
0 0 0 0 0 0 0 0 0 0 0 0 0 0
0 0 0 0 0 0 0 0 0 0 0 0 0 0
```

Figure 15.22 Required 'spots' pattern using white cells and blue spot cells.

We still allowed red cells. The experimental parameters were as before (Section 15.5.3). We obtained three runs out of ten that achieved the perfect score. The three runs produce very different solutions to the problem! The runs are shown in Figures 15.23 and 15.24.

In Figures 15.23 and 15.24 we see an extraordinary variety of ways of accomplishing the given task. The embryological development is identical for the first three iterations. They all show an alternating pattern type on odd versus even iterations. The cellular maps corresponding to run 3 are quite extraordinary in their beauty and consistency. At even iterations there are repeating units of the pattern seen in iteration 4. A magnified section in the centre is shown in Figure 15.25.

How important are the chemical emissions to this precise process? We investigated the cellular program for run 3 without any chemicals (i.e. the program was identical, however, all chemical outputs produced by the cells were set to zero). The results were dramatically different (Figure 15.26).

This is an amazing result. The cellular map for run 3 is extremely precise and regular yet it depends on a chemical map with its inevitable fluctuations from iteration to iteration. Figure 15.27 shows the chemical maps corresponding to the cellular maps for run 3 (Figure 15.23). It is important to note there are small, apparently random fluctuations of the chemical level that are not visible in the figure. It appears that the extremely precise cell behaviour is insensitive to this minor variation. A magnified picture of the cellular and corresponding chemical map at iteration 17 are shown in Figure 15.28.

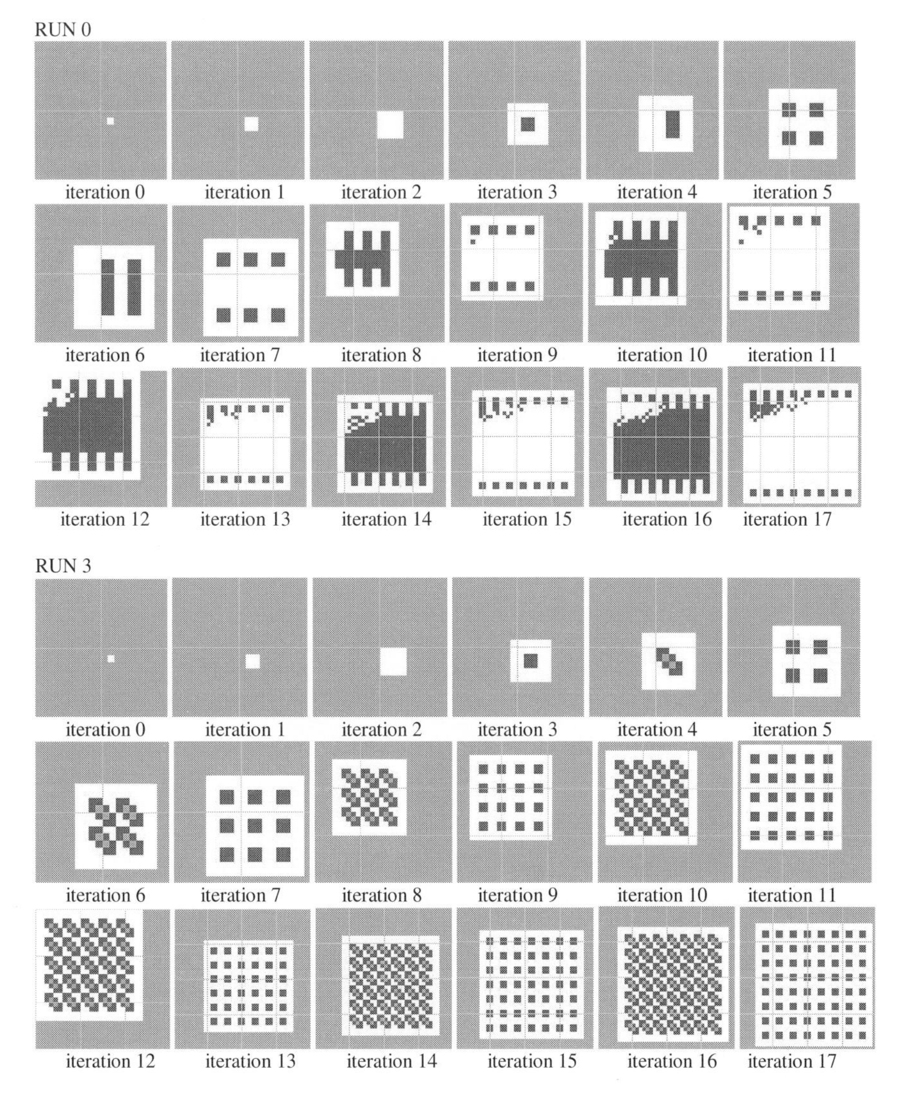

Figure 15.23 Development of cells that solve perfectly the required task of 4 blue spots at iteration 5.

When the cell is evolved with no chemicals at the outset it achieves an average fitness of 181.7 and the best solution reached has a fitness of 192 (out of 196), while under the original scenario the average fitness is 190.6 (with the three perfect solutions described).

15.5.1 Analysis of the genotype

One potential advantage of the techniques presented in this chapter is that the genotype is explicitly inspectable (i.e. a system of rules). The genotypes have had 200 nodes (strings of 400 integers). Typically only half of the nodes are active. At present the programs that produce the interesting cell growth and behaviour discussed have not been analysed. We hope that we might learn a lot from such an analysis. How can a program produce the elegant behaviour seen in Figure 15.23? We hope to report our findings in due course.

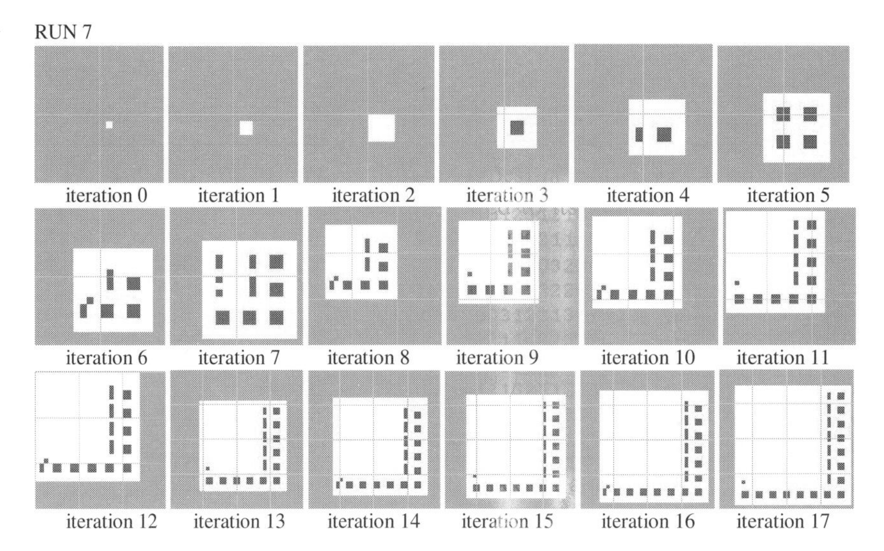

Figure 15.24 Third run that solves perfectly the required task of 4 blue spots at iteration 5.

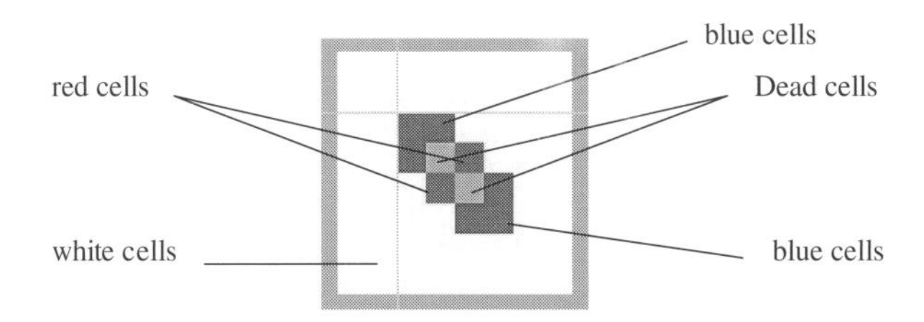

Figure 15.25 Magnified section of cellular map for run 3 iteration 4.

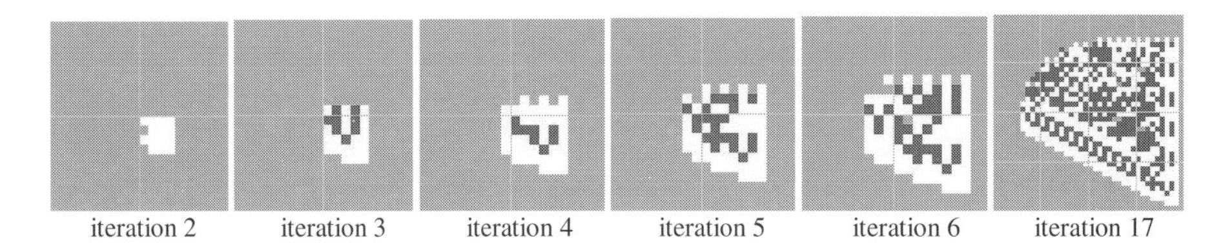

Figure 15.26 Cell program of run 3 with no chemical environment.

15.6 Discussion

There are many aspects of this work that warrant further investigation. First, the basic technique described could possibly be improved. It is likely that the cell update procedure being used may be making it more difficult to evolve a cell program that exhibits the desired behaviour. At present the cell map is scanned in a fixed order row by row starting at the top left position and finishing in the bottom right position. Cells may overwrite each other. This is illustrated in Figure 15.29.

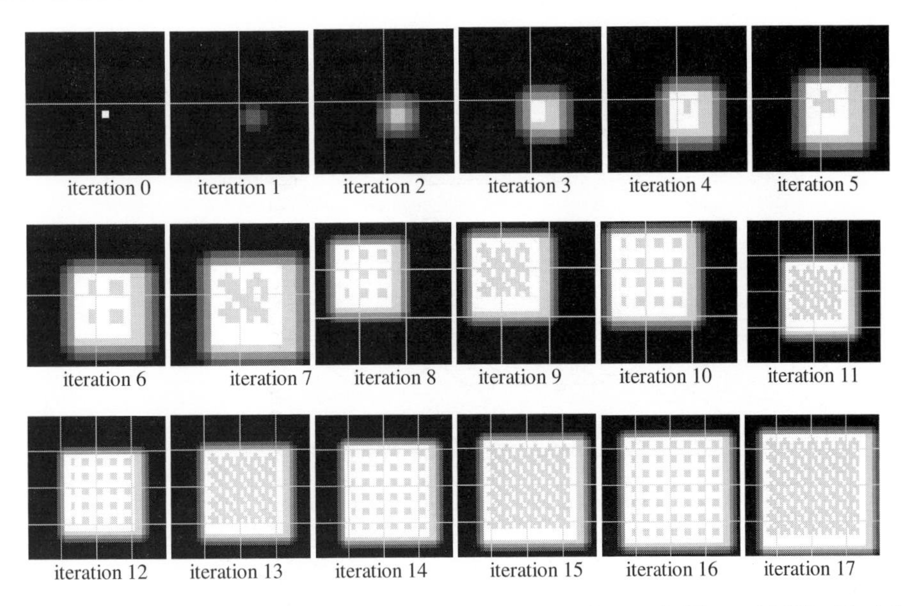

Figure 15.27 Chemical maps corresponding to the cellular maps for run 3 (Figure 15.23) – black indicates an absence of chemical. The almost white regions correspond to a chemical level of 228 and the light grey regions to a chemical level of 221.

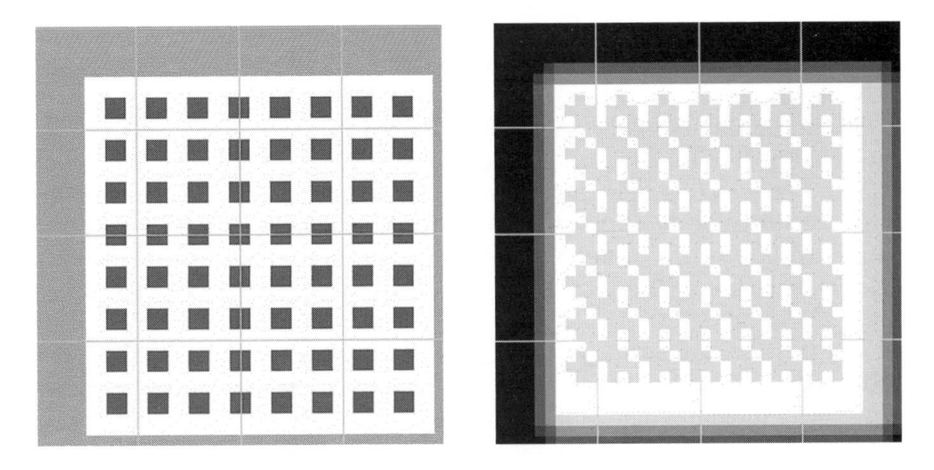

Figure 15.28 Cellular and corresponding chemical map at iteration 17.

In the real world, of course, cells occupy space and cannot be located at the same place, or overwrite each other. Therefore, one idea would be to include, within the cell's own genotype, an adjudication function that decided how the cell would grow into the new location depending on the demands of the other cells that could possibly grow into the same location. Ideally, one would implement some sort of cell shuffling procedure so that overwriting was not possible, however, this would be difficult and time consuming in practice.

In real biological development there are stem cells that gradually differentiate into more specialized types of cells. In the simulations reported here all cells are stem cells. One could, of course, create restrictions so that cells could only replicate cells of the same type and intro-

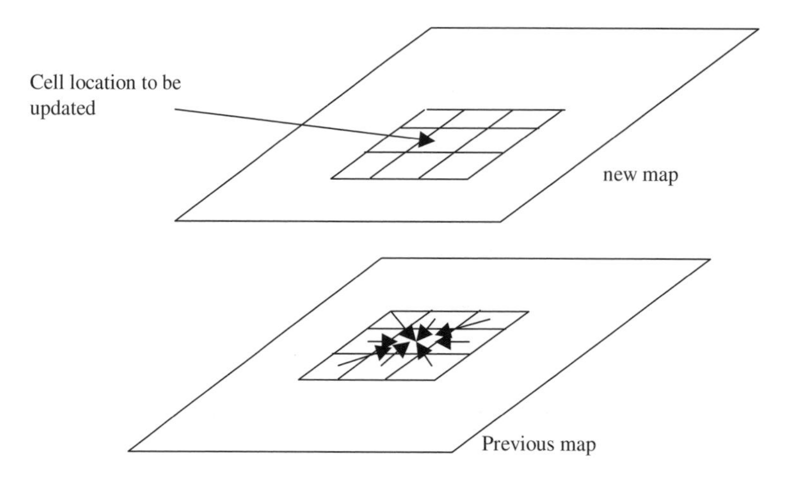

Figure 15.29 The cell update problem.

duce a special stem cell type that could replicate and then specialize. The idea of allowing simulated cells to differentiate gradually would be difficult without dictating exactly how they should do so. One of the motivations for the work reported here is to give evolution as much freedom as possible.

Another line for future work will be to allow a number of different chemicals, perhaps emitted by particular cells (i.e. blue, red and white chemicals). One could perhaps observe the emergence of morphogens.

One of the real difficulties of the approach described here is the imposition of the fitness function. It would be nice to see if a fitness function could be defined that tested the growing cellular map in an orientation independent way. This would be likely to speed up the rate at which complex maps could be evolved. Further work is already underway on adapting the methods described here to the problem of growing electronic circuits (in simulation). In this scenario one would be testing the *function* of a particular map, this would be less dependent on orientation.

We saw how these embryonic arrays could repair themselves after damage. This observation suggests a further investigation of such behaviour. One could impose damage sporadically and try to evolve cellular programs that naturally repaired themselves. Currently, in a field known as Evolvable Hardware there is strong interest in self-repairing circuits.

In the work described in Section 15.4 we have used digital multiplexers as our basic computational units. More complex functions could be employed (such as digital adders, comparators, multipliers), this would probably increase the speed of evolution and the complexity of resulting cellular maps.

In preliminary work we excluded cell movement as part of the cell program output, but this deserves further investigation. Chemotaxis could be evolved and investigated by placing a chemical source at a point in the map and defining the cells' task as one of moving (and/or growing) toward the source.

There is a bias toward growth in the representation currently used, as there are eight growth bits, all of which must be zero for zero growth to occur.

In the experiments reported here we have begun with a single cell, however, it would also be interesting to investigate multiple seed cells.

15.7 Concluding remarks

Most researchers in computer science who are using developmental processes are trying to define genotypes which they can evolve that lead to solving enormously complex problems. At present there is still only tentative evidence that direct encodings are less effective than developmental ones. Part of the problem is that researchers have still not reached a consensus about what are complex problems. Almost by definition one is imagining problems whose smallest reasonable solution is of enormous size, these problems are inevitably time consuming and so are relatively unexplored. There is a great need within the community to define clearer milestones for achievement. The new researcher is faced with a difficult question at the outset: Should I try to create as realistic simulation of real biology as possible or should I try to build the *simplest* system that can generate complexity? How can I define complexity? It is our conviction that if we are to make any contribution to thinking about biological development we need to look for the simplest and most general principles. We should be trying to define much more clearly how a form of simulated development that is evolvable can contribute to computer science. It is clear that real biological development is making use of enormous physical complexity. Some researchers are trying to enrich their developmental systems so that novel exploitable features may emerge (Bentley, 2002), indeed, some are thinking even further and trying to use a computer to control and evolve developmental processes that are embedded in the physical world.

References

Banzhaf, W., Nordin, P., Keller, R.E. and Francone, F.D. (1998) *Genetic Programming. An Introduction.* Morgan Kaufmann.

Bentley P. (2002) Evolving Fractal Proteins, Late breaking papers. *Genetic and Evolutionary Computation Conference*, New York.

Koza, J.R. (1992) *Genetic Programming: on the Programming of Computers by Means of Natural Selection.* MIT Press.

Koza, J.R. (1994) *Genetic Programming II: Automatic Discovery of Reusable Subprograms.* MIT Press.

Koza, J., Bennett III, F.H., Andre, D. and Keane, M.A. (1999) *Genetic Programming III. Darwinian Invention and Problem Solving.* Morgan Kaufmann.

Langdon, W.B. (1999) Scaling of program fitness spaces. *Evolutionary Computation*, **7**, 399–428.

Miller, J.F. and Thomson, P. (2000) Cartesian genetic programming. In *Proceedings of the Third European Conference on Genetic Programming. LNCS*, **1802**, 121–132.

Vassilev, V. and Miller, J.F. (2000) The advantages of landscape neutrality in digital circuit evolution. *Third International Conference on Evolvable Systems: From Biology to Hardware, Lecture Notes in Computer Science*, **1801**, 252–263.

Wolpert, L. (1998) *Principles of Development.* Oxford University Press.

Yu, T. and Miller, J. (2001) Neutrality and the evolvability of Boolean function landscape. In *Proceedings of the Fourth European Conference on Genetic Programming*, pp. 204–217, Springer-Verlag.

Yu, T. and Miller, J. (2002) Finding needles in haystacks is not hard with neutrality. In *Proceedings of the Fifth European Conference on Genetic Programming*, pp. 13–25, Springer-Verlag.

Combining developmental processes and their physics in an artificial evolutionary system to evolve shapes

16

PETER EGGENBERGER HOTZ

16.1 Introduction

There is an increased interest in combining evolutionary algorithms with biological concepts in the field of evolutionary computation (EC). As simple direct encoding schemes, where each primitive of the phenotype is represented by a single gene, no longer work for complex evolutionary tasks, new concepts have to be found to tackle such problems. It is not yet clear that fusing concepts of artificial evolution with biology will ultimately be the right way to go. Yet, as nature has already managed to solve most of the problems encountered in EC, I feel it worthwhile to investigate those principles by simulation and in this chapter I report on investigations done for morphogenesis and their abstracted implementation, which led to a simulator able to evolve shapes. In order to simulate and evolve morphogenetic processes producing, in turn, shaped cell clusters, a parsimonious choice of concepts had to be made. This choice was guided by screening through the biological literature to spot important concepts, which had afterwards to prove their importance in simulation.

Morphogenesis refers to those processes which are able to perform the amazing feat of developing form through an intricate web of interactions between genes, proteins, signals, cells and physical forces. These processes consist of diverse chemical and mechanical events such as cell division, differential cell proliferation, cell migration, change of cell size and shape, cell death, loss and gain of cell adhesion resulting in the dynamic shaping of an animal during embryogenesis (Edelman, 1988; Gerhart and Kirschner, 1997; Wolpert, 1998). All these events are controlled by the dynamics of the genetic regulatory networks, their products and the cells' behaviours. Morphogenetic processes involve the movement of parts of the developing system and therefore involve the action of physical forces. These forces are introduced in the proposed model of morphogenesis by adhesion molecules. At the core of the simulator is a differential genetic regulatory network (Eggenberger 1996, 1997a, b, 2000) controlling different developmental processes, but is coupled to a physical simulator allowing the use of physical mechanisms (tension, shears etc.) during development and therefore reducing the genetic load, because evolution can rely on these mechanisms. For instance, the cell's position has not to be specified by genetic parameters, but signals from other cells can inform a whole group of cells on what to do and where to go. Such mechanisms are able to reduce the number of parameters of primor-

 ISBN 0-12-428765-4

dial importance for evolution, because the number of genetic parameters determines mainly how big the search space is and how long it will take until a solution is found.

In the first step of the implementation of such a simulator a judicious level of abstraction had to be chosen; which turned out to be the cellular level. This means that the cells had to be endowed with basic mechanisms enabling them to grow shapes and which, in addition, were under genetic control permitting evolution. As development results in general from the coordinated behaviour of cells, developmental biologists often discern basic cellular mechanisms such as: cell division, cell motion, cell death, cell adhesion, change of cell shape, cell differentiation and induction. The interesting cellular processes are those that are used by the cell for developmental tasks and not those that provide 'housekeeping' functions to keep the cell alive. It would have been possible to choose a lower level of abstraction than the cellular one; however, as cellular biologists cannot yet separate the cellular genes into those two groups (Wolpert, 1998), a biochemical level of abstraction in our opinion is not yet realistic due to the high load of computation. Although one might argue that biochemistry is 'simpler' than the cellular level and the cellular behaviour would be an emergent property of the defined chemicals and their dynamics, the sheer number of possible molecular interactions makes such an approach currently unfeasible.

In contrast, a higher level of abstraction than the cellular level also has its drawbacks, because tissues or organs contain already specialized cells with much less flexibility than on the cellular level. If one wants to simulate developmental processes, it seems to us that starting with already differentiated modules will soon get cumbersome if many different mechanisms have to be implemented. On the cellular level this higher level function is an emergent property of the cells and far fewer specifications are needed. Also the implementation of tissular properties into a module is less straightforward than in the cellular approach, because their functions are much more complex.

In the proposed implementation each cell contains a continuous genetic regulatory network that controls the cell's behaviour. Each gene linked to a function or a product (analogue of a structural gene) is regulated by one or several regulatory units. Depending on the subset of active genes, a cell will perform its functions. In the current study the following mechanisms were implemented at the cellular level: cell division, cell death, cell migration, cell adhesion and cell differentiation. On the internal milieu or environmental level: the simulation of cellular interactions as physical forces (implemented as viscoelastic elements in and/or between the cells).

The importance of the genetic regulatory networks relies on the fact that they allow for intercellular communication on which cell differentiation and positional information are based. In this sense the mechanism of positional information is an emergent property of gene regulation and cell communication, and is on a different level of abstraction than the cellular processes of cell division or migration, for example. Of special noteworthy importance is cell adhesion, because it permits a link between genes and the physics, allowing the cells to exert forces on each other (Edelman, 1988; Wolpert, 1998).

The chapter is organized as follows: first an overview of related work is given, then the proposed model is explained and a description of the evolutionary model is given. Simulation results then illustrate the use of developmental processes and some examples of different shapes. Finally, the results and future research directions are discussed.

16.2 Background

Although other authors propose to combine developmental processes with evolutionary algorithms (de Garis, 1991; Fleischer and Barr, 1992; Gruau and Whitley, 1993; Belew, 1993;

Cangelosi *et al.*, 1994; Vaario and Shimohara, 1997; Cho and Shimohara, 1998; Kodjabachian and Meyer, 1998; Bentley and Kumar, 1999), here I discuss only those models that tried to evolve genetic regulatory networks to control developmental processes. Kitano (1994) extended Kaneko's model (1992) by additional interactions, coupled the system with a genetic algorithm and introduced an artificial metabolism of cells, cell division and neurogenesis. The regulation of the genes consisted of one single if-rule for one gene. Although the model points in the right direction, the use of neurogenesis failed to obtain any functional neural networks.

Dellaert (1995) proposed a model based on Boolean networks to evolve autonomous agents. The dynamics of interactions between genes are modelled with NK-Boolean networks. Briefly, NK-Boolean networks are networks of N nodes that have K different binary inputs. A function is defined for every node N describing what will happen with the states of the node depending on the K binary inputs (Kauffman, 1993). He uses a genetic algorithm to specify Boolean functions of the NK-Boolean network. The Boolean functions are dependent on different cell products, which are able to activate a gene. If a gene is activated, one or two different substances are produced. Two different models of neurogenesis were developed: one simple and one complex. With the complex model, the authors succeeded in designing an autonomous agent by hand-coding a genome.

Eggenberger (1996, 1997a, b, 2000) proposed the use of genetic regulatory networks coupled with developmental processes to use in the field of artificial evolution and was able to evolve simple shapes and simple neural networks. Bongard and Pfeifer (2001) and Bongard (2002) combined a physical simulator with a genetic regulatory network able to simulate simple organisms such as a box pusher. Their level of abstraction was higher than a cellular one, endowing their primitives with neural networks, sensors and motors, but they could evolve nice behaviours.

In contrast to all of the above approaches, the proposed artificial evolutionary system (AES) includes a continuous implementation of an extendible gene-regulatory network, which is based on specific interactions between different simulated molecules. The concept of site-specific biochemical addressing (McAdams and Shapiro 1995) is an important ingredient for the control of gene activity, the simulation of receptor–ligand interactions on the cell surface well as cell adhesion, creating local force fields of attraction or repulsion.

16.3 The model

The biological concepts that have been implemented are discussed and illustrated by simulation results in the following sections. By combining cell division, cell lineage, cell adhesion and cell differentiation in the cells and calculating the positions of the cells and the forces between them, it was possible to simulate simple morphogenetic processes. Some shapes created by varying the genetic parameters were easily obtained and these simulated results are illustrated in Figure 16.6. An overview of the AES algorithm is provided below.

Set population size
Set initial gene number
Create n * n * n grid
Repeat
Initialize first cells
Diffuse organizing gradients in the grid
Repeat
  Do for every grid point
    Update cell

```
            Update gene activities
            Update gene products
            Call developmental processes
            Update physical environment
            Update diffusing signals
            Update physical interactions
        End Do
    Until designer set end is reached
        Calculate fitness
        Reproduce and select
        Put fitness in sorting table
        Sort table
        Reproduce the fittest individuals
    Until stop criterion
```

16.3.1 Gene regulation

Seven parameters encode the properties of each structural gene. The first parameter determines what happens if a gene is activated. In the current implementation an active gene either produces a substance such as a signalling molecule or it can activate a predefined function such as cell division. The designer can choose how many of these functions he wants to investigate. The substances and functions that were used for the experiments are described below. The second parameter is used to calculate the probability of an interaction with a partner molecule. This affinity parameter, *aff*, determines which molecules (signalling molecule or axonal receptor) interact with which partner (regulatory unit or receptor). A real valued number and a function is assigned to each molecule in the simulation, to calculate a binding affinity between the molecules. A simple example of such a function is given later. The third parameter determines the sign of the effect eij, i.e. inhibitory or excitatory. The fourth parameter specifies the threshold μ. This parameter determines how high the sum of the products of all the affinities aij between the signalling molecules and the regulators times their concentrations has to be in order to turn on or off the associated structural gene. The fifth parameter designates the decay di rate for the product. The sixth parameter is used to store the affinity parameter *aff*. The seventh parameter is used as the diffusion parameter $D1$.

One or several regulatory units control a structural gene. Regulatory units are switches that control the activity of the structural gene. Active regulatory units influence the activity of the structural gene, but only an activated structural gene is able to perform a specific function such as cell migration or the production of a receptor. Two parameters are assigned to every regulatory unit: an affinity $affRU$ and a threshold. The affinity $affRU$ has the same use as the affinity parameter in a structural gene. A signalling molecule is defined by the parameters encoded in the structural gene. Both affinity parameters are used to calculate the probability for an interaction between a regulator and a signalling molecule. Both factors are variables of the affinity function $affTot = faff\ (affT;\ affS)$ and its value will influence the probability of a gene's activation. The threshold defines the limit of the minimal impact able to activate a gene: the product of the affinity $affTot$ and the concentration of the signalling molecule has to exceed the threshold's value.

Whether a given gene at position (i, j, k) in a cell on the grid will be activated depends on the affinity and concentration of all the signalling molecules at that position. All these influences are summed up and if this sum exceeds the gene's threshold the gene will be activated or inhibited according to the sign of the effect. All these parameters are varied by the evolutionary strategy and used to explore the interaction space for useful developmental processes able to solve the designer defined tasks. The gene activity of the i-th gene depends on parameters of the structural gene and its regulatory units. The activation of a gene leads to two types of responses: either a simulated molecule is produced or a function (implemented as a procedure) is executed. The link between the activation of a gene and its response depends on the first of the seven parameters of a structural gene. The following responses were implemented for the experiments in this chapter:

1. Production of chemical substances
   (a) a signalling molecule is produced to communicate between the cells
   (b) a cell adhesion molecule (CAM) is produced to connect the current cell to another one
   (c) receptors are produced for signalling molecules, axons and synapses
2. The activation of a gene calls a predefined function of the following types:
   (a) cell division
   (b) cell death.

The number of different substance classes as well as the number of functions is designer specified and depends on the problem the designer wants to solve. The possible interactions of a chemical substance with a partner molecule depend on the affinity between them and it is therefore possible that a signalling molecule can interact with different partners. Figure 16.1

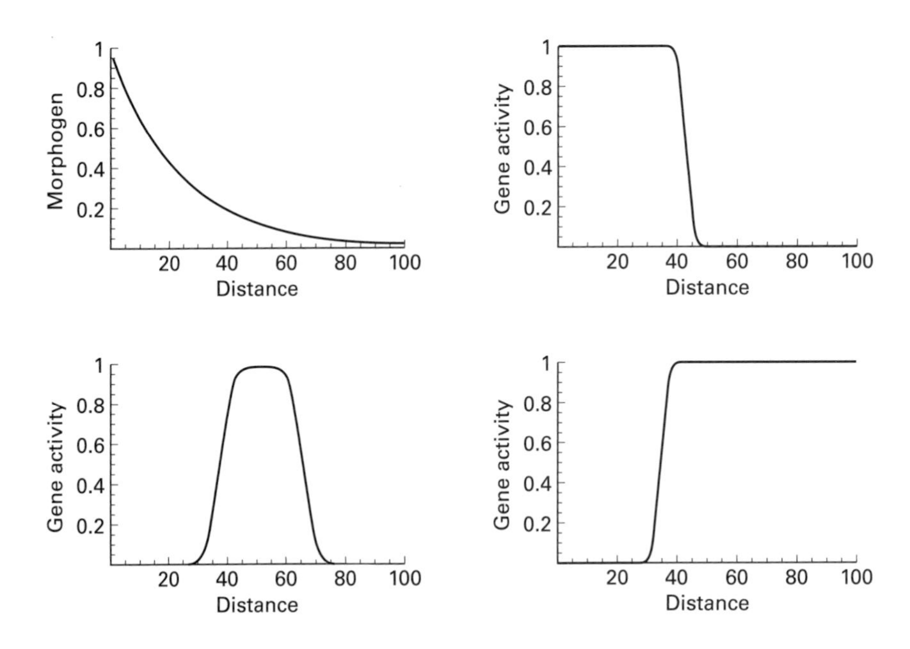

Figure 16.1 Basic principle of cell differentiation due to a gradient of a morphogen. The upper, left picture shows a gradient (MORPHOGEN-axis: concentration of the morphogen, DISTANCE-axis: is the distance from the source positioned at point 0). In the next three pictures 100 cells were lined up along the axis DISTANCE and the activity of a gene in each cell is plotted on the axis GENE ACTIVITY. Three typical example activity patterns of the gene in differently positioned cells can be seen. The differences in the activities are due to differences in the genetic regulation of the genes. This mechanism constitutes a general purpose patterning mechanism that is used in nature to position legs in precisely defined locations along the body axis (in insects, centipedes, etc.)

shows a gradient of a signalling molecule. The cells depending on the reading mechanisms of the cells for the gradient will respond differently to the concentration and three examples are shown of possible gene activity patterns induced by the gradient. To illustrate these principles the following experiments in *Mathematica*®[1] were performed to get an intuition of how a typical structural gene with one and two regulators in a static gradient behave. The *Mathematica*® program is shown below.

```
Clear[a,b,c,alpha,freq,upperlim,lowerlim];
signum[x_]:=1.0/(1.0+1*Exp[-100.0*x]);
c={};
cellnum=100;
genomelen=100;
reglen=2;
genome=Table[Random[Real,{0,1}],{i,genomelen},{j,reglen}];
geneact=Table[0,{i,cellnum},{j,genomelen}];
eff=Table[2*Random[Integer,{0,1}]-1,{j,genomelen},{i,reglen}];
grad=Table[2.0*Exp[-0.05*j],{j,cellnum}];
grad=Table[(cellnum-j)/cellnum,{j,cellnum}];
thresholds=Table[Random[Real,{0.,1.0}],{i,genomelen},{j,reglen}];
thresh=Table[Random[Real,{0.,0.5}],{i,genomelen}];
For[k=1,k<=cellnum,k++,
    For[j=1,j<=genomelen,j++,{
        geneact[[k,j]]=signum[
            Sum[eff[[j,i]]*
            signum[grad[[k]]*genome[[j,i]]-thresholds[[j,i]]],
            {i,reglen}]/reglen-thresh[[j]]];
        }
        ]
]
b11=ListDensityPlot[geneact,FrameLabel->{"X","Y"}]
    Display["C:/mathfiles/geneactivities.eps",b11,"EPS"]
```

Although the implementation of genetic regulatory networks looks quite similar to that of artificial neural networks (cis-regulators can be thought of as synaptic weights and an active gene is analogue to an active neuron), there is a fundamental difference: an active gene in contrast to an active neuron is linked to different functions and the outcome of an activation of a cell division gene or an apoptosis gene will have a totally different effect. As described above a set of different structural genes were defined, which control the developmental processes by either directly calling a predefined function such as cell division or producing a chemical such as a receptor. The interactions between different molecules are not predefined, but under evolutionary control by varying the affinity parameters which determine which substances can interact with one another. An example would be a receptor having an affinity parameter α and a signalling molecule having an affinity parameter β. These two parameters are compared and a function will return an interaction coefficient. If this coefficient is low there will be no interactions, if the coefficient is high the signalling molecule can interact with the receptor and so influence the cell's behaviour. Figure 16.2 shows an example of the gene activity of 100 structural genes.

[1] *Mathematica*® is a product of Wolfram Research.

Affinity between molecules

In more detail, the measure of affinity between two molecules, which are abstracted as parameters *aff*1, *aff*2 was implemented as follows:

$$a_{12} = f_{aff}(aff_1, aff_2) = Exp^{-\alpha(aff - aff_2)^2} \tag{1}$$

where aff_1, aff_2 are the real valued numbers representing the geometric properties of the substances and are encoded in the parameter set of the evolutionary strategy. α is the affinity parameter with positive values. If α is high, the two substances have to be very similar (i.e. $aff_1 \gg aff_2$) to get a high functional f_{aff} value, if α is low, the substances can be more different to get still high f_{aff} values. Molecules compete for a docking site (cis-regulator, receptor) and their success of binding depends on the affinity between the molecule and the docking site and on the concentrations of the competitors.

Gene regulation mechanism

The genetic regulatory mechanism is implemented according to the following equations:

$$\begin{aligned} G_i(c_{smo}, \ldots, c_{smm}) &= \frac{1}{2}(1.0 + \tanh(x)) \\ x &= \sum_{j=0}(\Theta(a_{ij} * c_{sm_j} - \vartheta_j)) \\ \Theta(x) &= \begin{cases} 1.0 : if x > 0 \\ 0.0 : otherwise \end{cases} \end{aligned} \tag{2}$$

where G_i is the activity of the i-th gene, $tanh(x)$ is the hyperbolic tangent, a_{ij} affinity to encode the effect between the regulatory unit i and the signalling molecule j (also referred to as transcription factors if they regulate the gene activity), c_{smj} concentration of the signalling molecules j, ϑ_j is a threshold value.

The regulatory units function as reading heads for the concentrations of signalling molecules (inside the cell they are also called transcription factors), which will then determine the state of a gene. The differential equations describing the whole system are:

$$\frac{dg_i(x, y, z, t)}{dt} = G_i(c_{smo}, \ldots, c_{sm_m}) + D_1 \nabla^2 g_i(x, y, z, t) \tag{3}$$

where: $gi(x, y, z, t)$: concentration of substance i at grid point (x, y, z) at time t, m: number of signalling molecules, $D1$: diffusion constant, $G_i(c_{sm0}, \ldots, c_{smm})$: see formula (**2**), $D1\nabla^2 gi(x, y, z, t)$: diffusion term.

16.3.2 Cell division

From a practical point of view cell division, especially the implementation of cleavage divisions, allows the construction of an artificial evolutionary system able to control cellular behaviour in detail (Figure 16.3). It corresponds in many aspects to a direct encoding scheme allowing the specification, in detail, of cells and it is possible to evolve recursive developmental schemes. Some authors in evolutionary computation contrast recursive and developmental approaches, but in fact, developmental processes can also be recursive, if needed, as shown with this example.

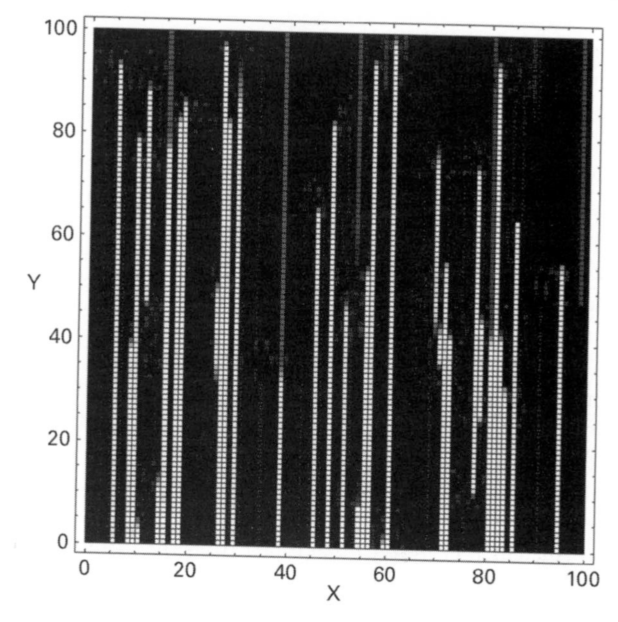

Figure 16.2 Gene activity of one hundred structural genes. Each of these genes is controlled by two regulatory units under the influence of a gradient of a signalling molecule (the same gradient as the one shown in Figure 16.1). The x-axis shows the index of all one hundred genes (that are contained in each cell). The figure demonstrates that, for example, the positioning of organs or limbs along the body axis is easy. The parameters of one hundred genes were randomly generated and their activities plotted (y-axis). Level of gene activity: white = high activity, grey = intermediate, black = inactive.

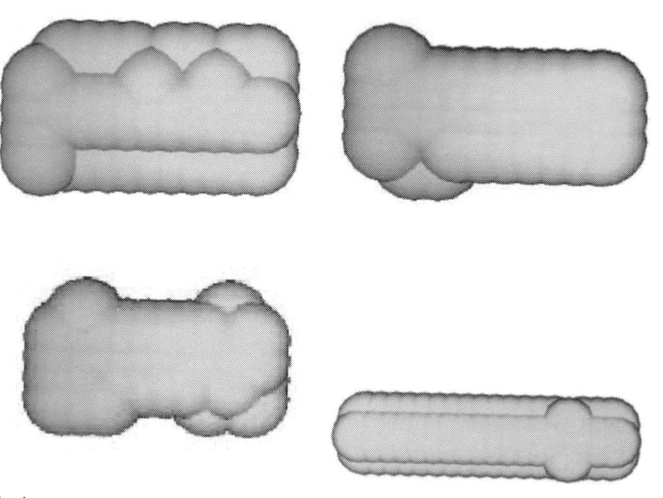

Figure 16.3 Simulation results of cell division. Artificial cells divide and produce patterns. These cell associations were created by randomly exploring the genetic parameters of the genome controlling cell growth. Although the space of genetic parameters is explored randomly, the mechanisms of cell division produce highly structured non-random assemblies. Because these cell assemblies provide initial structures (e.g. a body axis which is evident in the lower, left panel), they can speed up evolution.

A cell contains asymmetrically distributed cytoplasmic factors, which by the following cell division may be distributed in different amounts to the two daughter cells and therefore influence their fates in two different directions. The importance of cell lineage is its potential to get started during evolution, because information can be used specifically for one single cell. Again there is no contrast between direct encoding schemes and developmental methods, because the latter is a superset of it and cell lineage is one of the methods to specify the fate of a single cell, which can be crucially important for certain problems, but not for others. Cell division and the position of cytoplasmic factors are coordinated in such a way that these factors are asymmetrically distributed among the cells. In contrast to cell signalling this mechanism is important to get the cell differentiation process started.

How is it possible that cell lineage can create recursive algorithms? The genome of each cell contains genes controlling the plane of cell division a cell can perform. The choice of the first axis is free and can be done randomly; in nature gravity or the penetration point of the sperm in the egg is used as a cue to define such a plane. Therefore, the first axis was just chosen to be the x-axis. If the first gene, the X gene, is active the plane of cell division will be perpendicular to the x-axis and the daughter cell is positioned in the direction of the positive axis. A cytoplasmic factor is in the first cell and will be diluted during cell division. This factor has two effects: it activates the first gene and inhibits a second gene, the Y gene, which controls cell division perpendicular to the plane and perpendicular to gravity. After several divisions the cytoplasmic factor is so low, that the X gene is inactive and the Y gene is no longer inhibited and the Y gene starts a second sequence of cell divisions. Cell divisions in the direction of the y-axis dilute a second factor necessary for the activity of the Y gene. This second factor inhibits an activator of the X gene, which is again produced when the concentration of the second factor is too low. If this happens the X gene is active again, both cytoplasmic factors are produced again and the cycle starts again. The result is illustrated in Figure 16.4.

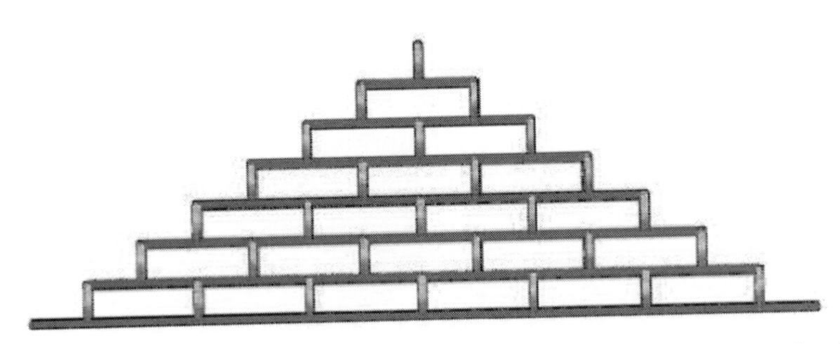

Figure 16.4 Recursive cell structures. This figure demonstrates that the general mechanism of cell division is also able to build recursive cell structures. In this case there is genetic control of the geometric orientation of the cell planes, which divide a cell during cell division to produce two daughter cells.

16.3.3 Cell adhesion

In order to show the possible advantages of this concept a simple simulator was written allowing the simulation of a large number of cells interacting by viscoelastic elements consisting of passive springs and active dash-pots. Each cell was connected to its six nearest neighbours. Cell movement and the production of pressure, etc. displace the cells in the direction of the applied force, but also create contracting forces which restrict the movements. The link to the genome was established by assuming that the amount of produced cell adhesion molecules was proportional, for instance, to the spring constant allowing the AES to explore different mechanical interaction patterns between the cells.

To each cell a set of differential equations is assigned, which are solved by an ordinary differential equation solver using a fourth-order Runge-Kutta method. Small changes in the genome result in small changes to the shape due to concurrent effects of the cell's behaviour and its physical properties. The reduction of parameters for the formation of certain shapes (obviously, for random shapes no reduction is possible) is due to the physics taking care of the positioning of cells, which react on tensions and pressures. Again, a physical process takes care of the shaping, but the genes can influence and change the outcome by changing the interaction forces between the cells.

16.3.4 Physical simulation

In order to test a large number of interacting cells or modules, a simulator was developed able to calculate tensions and pressures between the cells. As surface molecules and intracellular structures (microtubuli) are able to exert forces between the cells, a simulator should be able to calculate these forces. As currently available commercial physical simulators such as *Vortex*®, *Mathengine*® or ODE are unable to simulate large numbers of interacting cells, a simple simulator based on interacting mass elements was developed. Each cell contains genetic parameters encoding the properties of the viscoelastic elements between its neighbours. These parameters are then used to calculate the velocities of the cells as well as the forces exerted on them. The simulator is implemented as a mass-spring system, which is also used in the field of computer graphics, because it provides a simple means of generating physically realistic motion. Computer graphics people often claim that even complex behaviours of bodies can be frequently approximated by mass-spring systems and these are therefore a simple, but versatile simulation tool. Particles are objects that have position, velocity and mass and respond to forces. A wide variety of non-rigid structures (necessary to simulate developmental processes) can be approximated by connecting the particles with simple damped springs.

The dynamics of a particle is described by the equation of Newton $x:: = F/m$. This second order differential equation (ODE) is first transformed in a first order pair of first order differential equations and then solved by an ODE solver. As the differential equation is not stiff in the illustrated simulations, it was sufficient to use a fourth order Runge-Kutta method with fixed step size.

In more detail, dynamics simulation is used to calculate the movement of the cells resulting from adhesion forces or pressure. The cellular interactions are implemented as viscoelastic elements.

$$m\frac{d^2L}{dt^2} = F_{elastic} + F_{viscous} + F_{external} = -k(L - L_0) - \mu\frac{dL}{dt} + F_{external} \tag{4}$$

Equation 4 is Newton's equation of dynamics in which elastic, viscous and external forces determine the acceleration. The implemented viscoelastic element consists of a spring term with spring constant k and rest length L_0; a viscous parameter is also implemented. These three parameters can be changed in sensible limits also by the evolutionary algorithm to change the interactions between the cells. This equation is integrated for each mass point by a fourth order Runge-Kutta method.

In the next simulator other methods with adaptive step-size are implemented and compared. Experiments with the commercially available physical simulators showed that cell numbers higher than 100 connected cells with joints are not allowed. This was the main reason to write a simulator with minimal assumptions. Figure 16.5 shows four snapshots of a simulation of a mat of 3*69*69 cells connected to each other with springs. Each cell is connected to its nearest neighbours. A perturbation in the middle of the mat will produce a wave and deform it. Note the deformations of the borders. The shades of grey are proportional to the magnitude of the forces on a cell. The main reason to simulate development in this kind of physical simulator is the reduction of genetic parameters and the knowledge that in biology this is of course also true and we hope to get, in this way, more interesting shapes and structures.

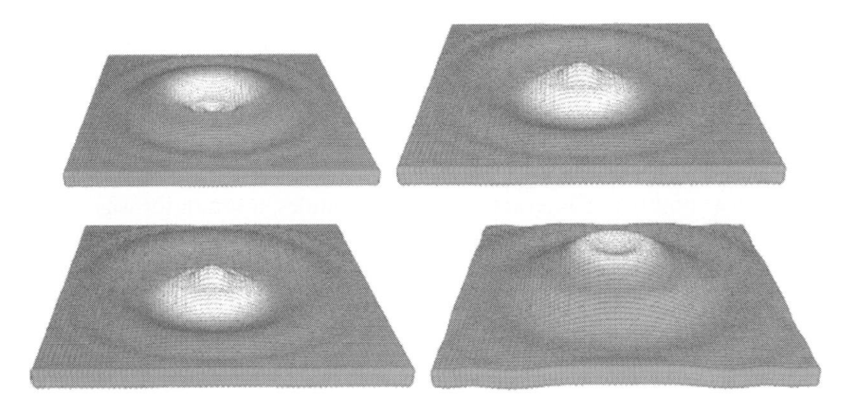

Figure 16.5 Dynamics of an oscillating mat of cells. Each cell is connected to its nearest neighbours by viscoelastic elements. A disturbance in the centre will produce a wave that travels outwards and eventually deforms the border as well (lower, right panel). The dynamic equations assigned to each cell were solved using a fourth order Runge-Kutta method. This simulator was also used to illustrate some morphogenetic processes (Figure 16.6).

16.3.5 Evolutionary strategy

To vary and select the genomes an evolutionary strategy (ES), developed by Rechenberg (1994), was used. The justification for having chosen an ES as opposed to other evolutionary algorithms was simply that the proposed AES uses floating-point numbers to encode the continuous variables used in the differential equations, whereas genetic algorithms typically encode binary variables. An evolutionary programming approach could also have been taken as it also relies on floating-point numbers. The experiments used a (μ, λ)-evolution strategy with one global step size for each individual. The notation indicates that the procedure generates λ offspring from μ parents and that it selects the parents only from the best offspring for the next generation. A mutation operator adds random numbers to the parameters, which are normally distributed with a mean of zero and standard deviation . The step size is inherited from the parents to the offspring and is modified by multiplying log normally distributed random numbers.

16.4 Simulation results

In this section different examples are shown in which the artificial genome, the simple and simulated chemistry and the physical simulator are linked together. This allows the AES to mimic simple morphogenetic processes such as deformation or invagination of cell sheets (see Figures 16.6 and 16.9). The idea is to let the AES exploit the physics during the developmental

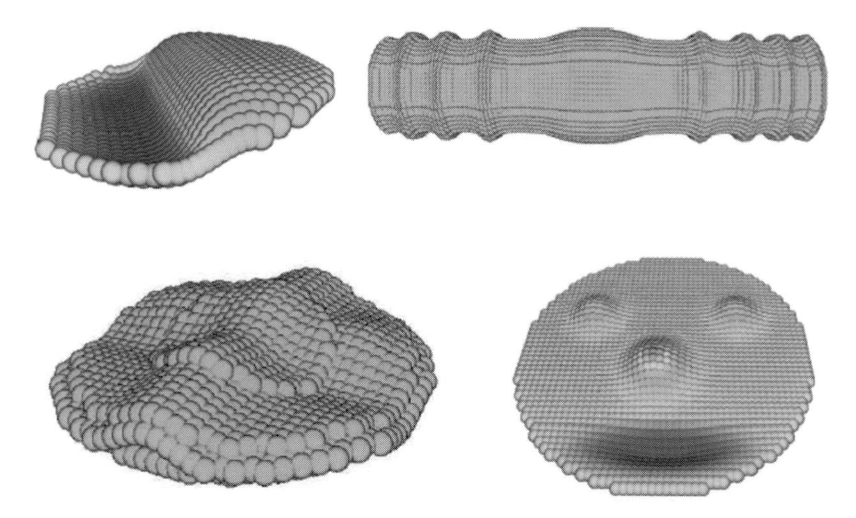

Figure 16.6 Creation of shapes by morphogenetic mechanisms. The cells are linked together as described in the section on cell adhesion. By varying the genetic parameters representing the intercellular interactions, different shapes are easily produced by changing the forces between the cells. As artificial evolution can now rely on physical processes, the parameter space can be reduced drastically compared to a direct encoding scheme because intercellular signals can change cellular properties for whole groups. Nevertheless, if more precise information is needed at a given location, the communication can be changed in such a way that a new signalling source will influence the shaping of form.

process. As will be shown with the following examples, this allows the AES to use more parsimoniously the genetic information for shaping cell clusters, because the intercellular communication allows the changing of, not cell after cell, but whole groups of cells to perform a given function. The concentration of a diffusing morphogen at a given position can alter cellular interactions leading to different cellular behaviours and, in the end, to different shapes. Another noteworthy property of the proposed approach is that it allows for continuously changing development. The artificial morphogenetic processes can be regulated in such a way that a cell performs a task quite similar, but slightly different than its neighbouring cells.

16.4.1 Linkage of genome and the physics between the cells

The cells are able to control their interactions with each other and their surroundings by either producing p.e. cell adhesion molecules or exerting forces by activating contractile fibres. These complex biological phenomena were approximated by simulating the cellular interactions by simple viscoelastic elements. If during cell differentiation a gene becomes active which produces a substance belonging to the set of 'cell adhesion molecules' (CAM), the properties of the interactions between its neighbouring cells is changed depending on the type and concentration of the 'CAM'. Depending on how fast these substances are produced and decay, the effects range from short contractions to long-lasting changes of adhesion between the cells. Although it is known that for morphogenetic processes the inertial forces can be neglected (Odell *et al.*, 1981), we did not change the equations as proposed there, but just increased the viscous coefficient, because the simulator is used not only for morphogenesis.

The differences with the work of Odell *et al.* (1981) is that the mechanics are linked with an artificial genome and that the model works in three dimensions. In Figure 16.7 the linkage of the

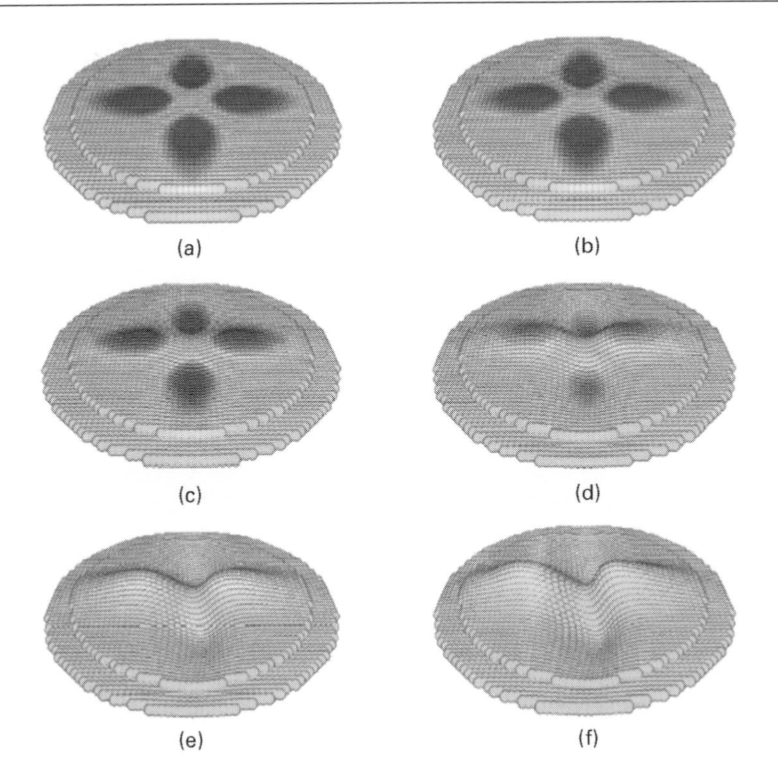

Figure 16.7 Linkage of the cells' genomes to the simulated physics. At the dark shaded places of the roundish cell sheets, an artificial morphogen is diffused. This factor will change the cells' adhesive properties and stresses and strains result due to the increased forces that the dark cells produce. The higher the tension is the darker the shade is. The deformation due to these stresses gets smaller from a to e, because the movement of the cells will in turn increase the tension to the pale grey cells and finally the cell sheet will reach an equilibrium. The shapes are initialized, but not determined by the morphogen. The resulting shape is the combined result of the physics (tensions and strains) and the properties of the cells (adhesion, which may vary in different directions).

cells' genomes to the physical simulator is illustrated. An artificial morphogen is diffused and builds up a gradient in a sheet of cells. Due to the activating effect of a gene in the cells a change of the spring properties proportional to the gene product is induced. These changes will lead to stresses and tensions in the cell sheet resulting in a physical deformation. More specifically the genome can change the links to all its six nearest neighbours individually by changing the spring constant k, the rest length L_0 and the viscosity factor μ. The location as well as the time of the changes depends on the intercellular communication between the cells, which is implemented as discussed above. The resulting shape is the combined result of physics and the properties of the cells such as adhesion properties and the cellular states determining them.

As the AES can only change these parameters indirectly, by the implemented developmental processes, the results of evolution can be analysed effectively compared to a direct encoding scheme, in which each parameter determines one single primitive. This is because in such a scheme the parameters are already the best description of the evolutionary result. In contrast, evolutionary processes combined with developmental processes allow us to analyse the reasons why something happened. To illustrate this claim I discuss a simple example of a shape change (Figure 16.8).

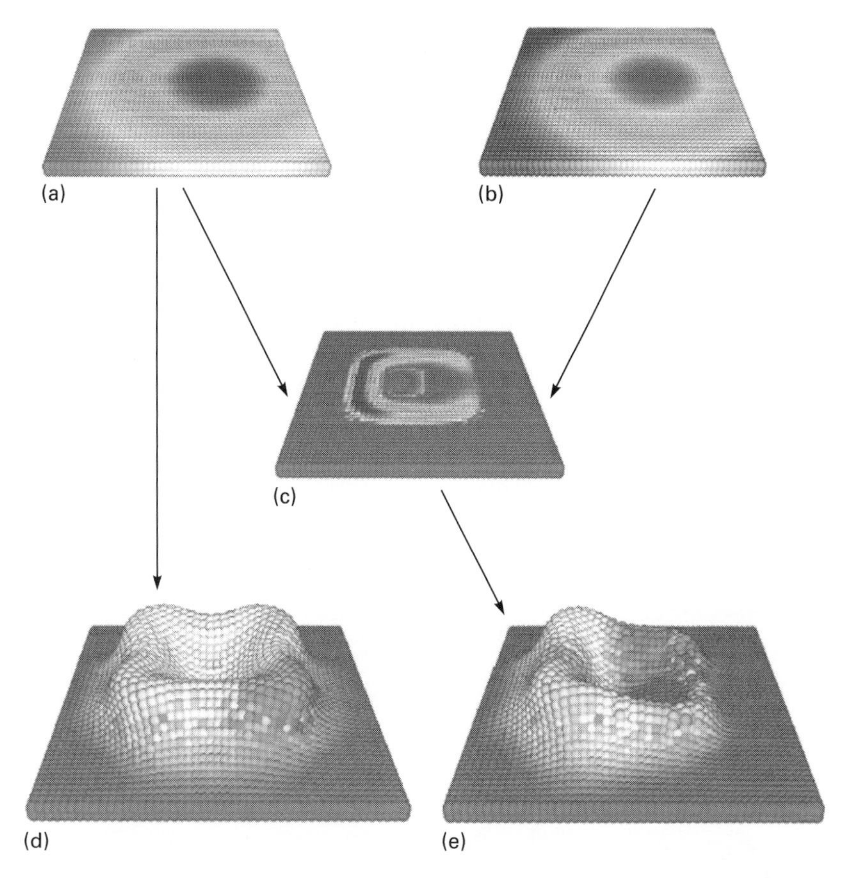

Figure 16.8 Interactions of genes, morphogens and the cellular physics. (a) Illustrations of the physical strains due to the default mechanism, which will produce a form illustrated in (d). (b) An additional morphogenetic gradient is diffused between the cells, which will downregulate the production of adhesion molecules. (c) In panel (c) the concentration of the adhesion molecules is illustrated. Due to the inhibiting effect of the second source the default shape (panel (d)) is changed to the shape in panel (e).

In this example a double layer of cells was prepared and a 'default' form (panel (d) of Figure 16.8) was selected by the designer during the evolutionary process. Initially two sources for positional information were given. In the first step the position of a second source was evolved by selecting for a given position. Remembering the illustration of the gene activity patterns in Figure 16.2, the regulatory mechanism easily found a regulatory gene that was only active in a band. (Another example of such a positioning can be found in Eggenberger and Dravid (1999), where we evolve butterfly wing patterns using a similar approach for gene regulation, but with a different set of structural genes.) One out of many different solutions to position a source is the following: the gene producing the second morphogen has four different regulatory units. Two are sensitive for one positional source and the other two are sensitive to the second source. A morphogen will inhibit the gene in high concentrations, activate the gene in medium concentrations and is unable to activate the gene in low concentrations. The same is true for the other couple of regulatory units for the second source of morphogen. By adding these two effects only the small region of intersection between the two bands of activities will be active and diffuse a morphogen. In other words the location of the source depends on the two sources and the genetic regulation of the gene producing the molecule of the second source. This molecule will diffuse and will have an impact on the other cells. The effect was that it regulated the gene activity for the unique cell adhesion molecule down and diminished the intercellular forces, which explains why the shape flattened out in regions where the molecule of the second source was present. By analysing the gene activities, the cell's products and interactions as well as the physics, one can finally understand what happens in the system. Often this can be done by visualizing the different gene activities and different concentrations of the products by different shades as is illustrated in Figure 16.8.

Another example was the simulation of invagination, which relied on the same principles explained above and which is illustrated in Figure 16.9.

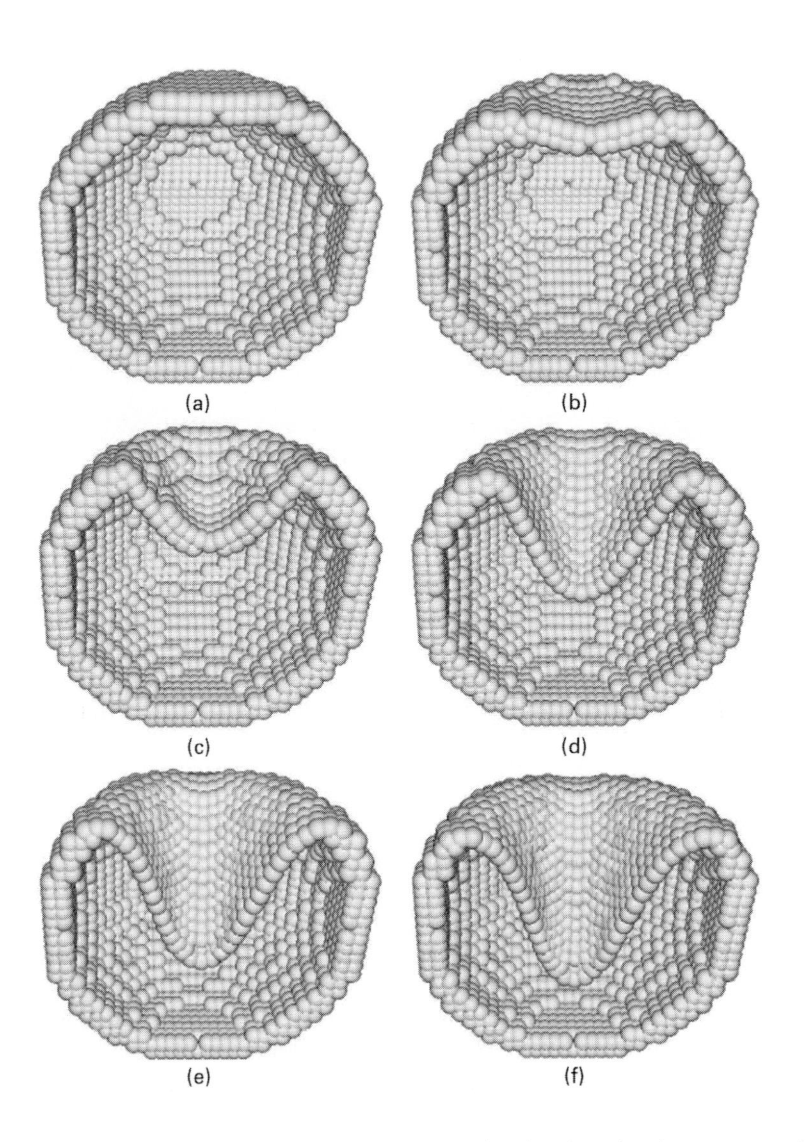

Figure 16.9 Six stages of simulated invagination. At a randomly selected point a source of morphogen was defined in the cellular ball and the AES had to find a parameter set which finally produced the following sequence of pictures. Note that the same mechanisms as above were used.

16.5 Discussion

Two main reasons led to the investigation of the developmental mechanisms producing different morphologies: to understand the possible reasons how the specific mechanisms generating different morphologies are used in nature and to investigate the possible advantage to combining artificial evolutionary techniques with principles of developmental biology. This chapter shows that more complex implementations of evolution than simple direct encoding schemes are feasible and, although the simulation of physics will increase the computational load, a reduction of the needed parameters to produce forms is achieved, because not every primitive (in this case the cells) has to be positioned in space using specific genetic information. The use of processes to construct a goal is more efficient genetically than a precise description.

Although the simulations are more complex than many of the older developmental approaches (Fleischer and Barr, 1992; Dellaert, 1995) or the direct encoding schemes, it is possible to evolve forms quite easily. As evolution is a process of ongoing adaptation with change of form and function, those biological mechanisms are selected that allow such changes without a decrease in fitness (Gerhart and Kirschner, 1997). By imitating such evolved mechanisms, it is reasonable to expect that the simulated investigation of these mechanisms will ameliorate the actual artificial evolutionary techniques. The following principles can be deduced. The developmental mechanisms are independent of the number of cells, in other words, the genetic information will not increase with the number of cells. The use of development allows us to extend easily these mechanisms by using new kinds of structural genes to control cell growth or cell migration. As shown in previous work not only forms, but also the patterning of butterfly wings (Eggenberger and Dravid, 1999) or the evolution of neural networks (Eggenberger, 1997, 2002) is possible. For large systems with more than a thousand elements, any direct encoding scheme will show an unnecessary increase in genome size and the more complex developmental methods will show their advantages. Another advantage is that the mechanisms do not depend critically on position addresses. The cells' positions depend on their behaviours and on their interactions (physics) and not every detail has to be explicitly specified. The combination of artificial evolutionary techniques and developmental processes allows for understanding why something happens. It is no longer just a random search in a property space, but a search for suitable developmental mechanisms which can be analysed and understood. Putting evolutionary techniques on firm ground, where the mechanisms can be understood is itself a major reason to investigate the potential of such systems.

References

Belew, R.K. (1993) Interposing an ontogenic model between genetic algorithms and neural networks. In *Advances in Neural Information Processing Systems (NIPS)*, Hanson, S.J., Cowan, J.D. and Giles, C.L. eds. Morgan Kauffman.

Bentley, P.J. and Kumar, S. (1999) Three ways to grow designs: a comparison of embryogenies for an evolutionary design problem. *Proceedings of the Genetic and Evolutionary Computation Conference (GECCO '99)*, Orlando, Florida, USA.

Bongard, J.C. (2002) Evolving modular genetic regulatory networks. In *Proceedings of IEEE 2002 Congress on Evolutionary Computation (CEC2002)*, pp. 1872–1877.

Bongard, J.C. and Pfeifer, R. (2001) Repeated structure and dissociation of genotypic and phenotypic complexity in artificial ontogeny. In Spector, L. and Goodman, E.D. *Proceedings of The Genetic and Evolutionary Computation Conference, GECCO-2001*, pp. 829–836.

Cangelosi, A., Parisi, D. and Nolfi, S. (1994) Cell division and migration in a 'genotype' for neural networks. *Network*, **5**, 497–515.

Cho, S.B. and Shimohara, K. (1998) Evolutionary learning of modular neural networks with genetic programming. *Applied Intelligence*, **9**, 191–200.

de Garis, H. (1991) *Genetic Programming:GenNets, Artificial Nervous Systems, Artificial Embryos* PhD thesis, Universitè Libre de Bruxelles, Belgium.

Dellaert, F. (1995) *Toward a Biologically Defensible Model of Development*, PhD thesis, Case Western Reserve University.

Edelman, G.M. (1998) *Topobiology: An Introduction to Molecular Embryology*. Basic Books.

Eggenberger, P. (1996) Cell interactions as a control tool of developmental processes for evolutionary robotics. In Maes, P., Mataric, M.J., Meyer, J.-A. *et al.* eds. *From Animals to Animats 4: Proceedings of the Fourth International Conference on Simulation of Adaptive Behavior*. pp. 446–448. MIT Press.

Eggenberger, P. (1997a) Creation of neural networks based on developmental and evolutionary principles. In Gerstner, W., Germond, A., Hasler, M. and Nicoud, J.D. eds. *Seventh International Conference of Artificial Neural Networks (ICANN'97)*, pp. 337–342. Springer.

Eggenberger, P. (1997b) Evolving morphologies of simulated 3d organisms based on differential gene expression. In Husbands, P. and Harvey, I. eds. *Fourth European Conference of Artificial Life*. MIT Press.

Eggenberger, P. (2000) Evolving neural network structures using axonal growth mechanisms. In *Proceedings of The Fifth Int. Symp. on Artificial Life and Robotics (AROB 5th'00)*, pp. 208–211.

Eggenberger, P. (2002) Evolving the morphology of a neural network for controlling a foveating retina – and its test on a real robot. In *Proceedings of The Eighth International Symposium on Artificial Life*, pp. 243–251. MIT Press.

Eggenberger, P. and Dravid, R. (1999) An evolutionary approach to pattern formation mechanisms on lepidopteran wings. In *Congress of Evolutionary Computation, Washington*, pp. 337–342.

Fleischer, K. and Barr, A.H. (1992) A simulation testbed for the study of multicellular development: The multiple mechanisms of morphogenesis. In Langton, C. ed. *Proceedings of the Workshop on Artificial Life (ALIFE '92)*, vol. 17 of Santa Fe Institute Studies in the Sciences of Complexity, pp. 389–416. Addison-Wesley.

Gerhart, J. and Kirschner, M. (1997) *Cells, Embryos and Evolution: Toward a Cellular and Developmental Understanding of Phenotypic Variation and Evolutionary Adaptability*. Blackwell.

Gruau, F. and Whitley, D. (1993) The cellular developmental of neural networks: the interaction of learning and evolution. *Technical Report 93-04*, Laboratoire de l'Informatique du paralléllisme, Ecole Normale Supérieure de Lyon, France.

Kaneko, K. (1992) Overview of coupled map lattices. *Chaos*, **2**(3), 279–282.

Kauffman, S.A. (1993) *The Origins of Order: Self-organization and Selection in Evolution*. Oxford University Press.

Kitano, H. (1994) Evolution of metabolism for morphogenesis. In Brooks, R. and Maes, P. eds. *Artificial Life IV: Proceedings of the Workshop on Artificial Life*, pp. 49–58, MIT Press.

Kodjabachian, J. and Meyer, J.-A. (1998) Evolution and development of neural controllers for locomotion, gradient-following, and obstacle-avoidance in artificial insects. *IEEE Transactions on Neural Networks*, **9**(5), 796–812.

McAdams, H.H. and Shapiro, L. (1995) Circuit simulation of genetic networks. *Nature*, **269**, 650–656.

Odell, G.M., Oster, G. Alberch, P. and Burnside, B. (1981) The mechanical basis of morphogenesis. *Developmental Biology*, **85**, 446–462.

Rechenberg, I. (1994) *Evolutionsstrategie '94*. Frommann-Holzboog.

Vaario, J. and Shimohara, K. (1997) Synthesis of developmental and evolutionary modeling of adaptive autonomous agents. In *Proc. of the International Conference of Artificial Neural Networks (ICANN '97)*, pp. 721–726. *LNCS*, Springer.

Wolpert, L. (1998) *Principles of Development*. Oxford University Press.

Evolution of differentiated multi-threaded digital organisms

17

TOM RAY and JOSEPH HART

17.1 Introduction

The work presented here consists of an exploration of the properties of evolution by natural selection in the digital medium. The evolving entities are self-replicating differentiated multi-threaded (emulated parallel) machine code programs. They live in a network of computers and are able to sense conditions on other machines and move between machines.

This work is explicitly not about the evolutionary origin of the differentiated condition, but rather about evolution that takes place just after that threshold has been crossed. This experiment begins with the most primitively differentiated condition: two cell types.

This is an extension of the work generally known as 'Tierra' (Ray, 1991, 1994a, b). The original Tierra was based on single-threaded (serial) machine code program living in a single computer. The original model was extended by Thearling and Ray (1994, 1997) to include multi-threaded programs, living on a sixty-four processor connection machine. However, these multi-threaded programs were of a single 'cell type' and never evolved into differentiated forms.

The seed program used by Thearling and Ray included a loop that was iterated many times. This loop was parallelized by using two threads, thus completing the work in half the time. Through evolution, the level of parallelism increased to as many as thirty-two threads. However, in the seed program and all programs that evolved from it in that experiment, all of the threads always executed the same code, thus there was no 'differentiation' between threads with respect to the code executed (genes expressed). This chapter extends the work of Thearling and Ray by starting with multi-threaded programs which are already differentiated (into sensory and reproductive threads).

17.1.1 Analogies

Here we are making analogies between some features of digital organisms and organic organisms. The objective of making these analogies is not to create a digital model of organic life, but rather to use organic life as a model on which to base our better design of digital evolution.

In organic organisms, the genome is the complete DNA sequence, of which a copy is found in each cell. Each cell is a membrane bound compartment and requires its own copy of the DNA, as the genetic information is not shared across the cell membranes. The entire genome includes many genes, which are segments of DNA that code for specific functions, mostly individual proteins. While each cell contains a complete copy of the genome, each individual cell expresses only a small subset of the genes in the entire genome. The specific subset of genes that are expressed in a cell determine the cell type. Groups of cells of the same type form a tissue. Different tissues are composed of cells that have differentiated in the sense that they express different subsets of the genes in the genome.

In our form of digital organisms, the genome consists of the complete sequence of executable machine code of the self-replicating computer program. Each thread of a multi-threaded process is associated with its own virtual CPU. These threads (CPUs) are considered analogous to the cells. However, the threads of a process all share a single copy of the genome, because they operate in a shared memory environment where the genetic information can easily be shared between CPUs. Duplication of the genome for each thread would be redundant, wasteful and unnecessary. In this detail, our digital system differs quite significantly from the organic system. Another difference is that here there is no spatial or geometric relationship between cells.

The genome of the digital organism includes several segments of machine code with identifiable functions, which are coherent algorithms or subroutines of the overall program represented by the entire genome. These individual algorithms can be considered analogous to the genes. Each thread (CPU) has access to the entire genome, yet each thread will execute only a subset of the complete set of genes in the genome. The specific subset of genes executed by a single thread determines its cell type. Groups of threads of the same cell type form a tissue. Different tissues are composed of threads that have differentiated in the sense that they execute different subsets of the algorithms (genes) in the genome.

17.1.2 Network Tierra

The work reported here is focused on the evolution of the differentiated multicellular condition. The multi-threaded digital organisms live in a networked environment where spatial and temporal heterogeneity of computational resources (most importantly CPU time) provides selective pressure to maintain a sensory system that can obtain data on conditions on various machines on the network, process the data, and make decisions about where to move within the network.

The experiment begins with a multi-threaded seed program that is already differentiated into two cell types: a sensory tissue and a reproductive tissue. The entire seed program includes about 320 bytes of executable machine code. However, no single thread executes all of this code, just as no cell in the human body expresses all of the genes in the human genome. The network ancestor genome has been somewhat arbitrarily labelled as composed of six genes, some of which have been further subdivided (Figure 17.1). Two of the genes are executed only during the

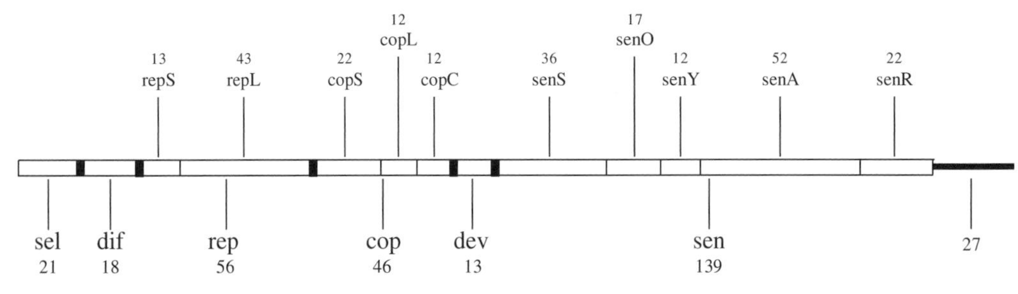

Figure 17.1 Ancestor genome. Lower labels indicate the six major genes and their sizes in bytes. Upper labels indicate subdivisions of the major genes and their sizes.

development from the single-celled to the mature ten-celled form (**sel**, **dif**). One gene is executed only by the reproductive tissue (**rep**), and one gene is executed only by the sensory tissue (**sen**). Two genes are executed by both tissues (**cop**, **dev**).

17.2 Methods

This section provides an outline of the experimental methods. The reader is encouraged to read Ray (1995) for details of the rationale and Charrel (1995) and Ray (1997, 1998) for technical details of the implementation.

17.2.1 The Tierra web

Tierra is another web on the internet. The Tierra web is created collectively as the result of running Tierra servers on many machines. The Tierra server is a piece of software written in the C language, which creates a virtual machine called Tierra. Tierra does not self-replicate, evolve or experience mutations. Tierra does not migrate on the net. In order to run a Tierra server, someone must download the software, install it and run it.

The collection of Tierra servers creates a subnet of the internet, within which digital organisms and Tierra browsers (Beagle) are able to move freely, accessing CPU cycles and the block of RAM memory that is made available by the server. Note that the digital organisms and Beagle cannot access other RAM on the machine, nor may they access the disk.

We can think of the web of Tierra servers as an archipelago of 'islands' (which we usually refer to as nodes or machines on the network) which can be inhabited by digital organisms. The digital organisms are mobile and feed on CPU cycles. Therefore, selection can potentially support the evolution of network foraging strategies.

In this experiment, we must create conditions under which selection will favour more complex migratory algorithms, over small highly optimized algorithms that only reproduce locally, such as evolved in non-network Tierra. Toward this goal we introduced the 'apocalypse' which at random intervals kills all organisms living on a single machine. This provides an absolute selection against non-migratory organisms, insuring that only migratory organisms can survive in the network environment.

Tierra runs as a low priority background process, like a screen saver, by using a 'Nice' value of 19. This causes the CPU cycles available to Tierra to mirror the load of non-Tierra processes on the machine (the speed of Tierra is high when the load from other processes is low). Thus the speed of Tierra will vary with the load on the machine. Also, when the user of a machine touches the keyboard or the mouse, Tierra immediately sleeps for ten minutes (from the last hit). We expect the heterogeneity in available CPU cycles to provide selective forces which contribute to maintaining cell differentiation.

The work reported here is based on a small-scale experiment conducted on a local-area network of about sixty Sparc stations running Unix.

17.2.2 Sensory system

Each Tierra server periodically sends a Tping data structure to all the other Tierra servers. In the current experiment, the structure contains the following entries (I32s is a 32 bit signed integer, I32u is a 32 bit unsigned integer):

```
struct TPingData /* data structure for Tping message */
{ I32s t; /* tag for message type */
  I32u address.node; /* IP address of node */
  I32u address.portnb; /* port number of socket */
  I32s cellID; /* unique identifier of organism in soup */
  I32s ranID; /* unique identifier, across network */
  I32s FecundityAvg; /* average fecundity at death */
  I32s Speed; /* average instructions/second */
  I32s NumCells; /* number of organisms on node */
  I32s AgeAvg; /* average inst age at death */
  I32s SoupSize; /* size of memory for Tierra soup */
  I32u TransitTime; /* in milliseconds */
  I32u Fresh; /* clock time at last refresh of this data */
  I32u Time; /* clock time at node */
  I32s InstExec; /* age of this Tierra process */
  I32s InstExecConnect; /* age while connected to net */
  I32s OS; /* operating system tag */
};
```

We will describe only those structure elements that are new in the current work, or which are mentioned elsewhere in this chapter.

address.node – is the 32 bit IP address of the machine from which these data came. These data are used by the organisms to specify the address of the machine that they will migrate to.
FecundityAvg – is the fecundity (number of offspring produced) at death or migration, averaged over all the organisms on the machine over the last million instructions executed.
Speed – is the speed of the virtual CPU in instructions per second executed, calculated over the last million instructions.
NumCells – is how many organisms are living on the machine at the time that the data structure is generated.
AgeAvg – is the age at death or migration, averaged over all the organisms on the machine over the last million instructions executed. The age is measured in virtual instructions executed by the individual organism.
InstExec, InstExecConnect – how many millions of instructions the Tierra process has been running, the age of the 'island'. InstExecConnect is how many millions of instructions the process has been running while connected to the network. The unix machines in our network are always connected to the network, so these two values are the same. They would differ on machines that are only intermittently connected to the network.

Each Tierra server maintains a 'map file' which is a list of Tping data structures from all the machines on the Tierra network. Digital organisms are born with a pointer into the list of Tping structures. The location of the pointer in the list is randomly initialized at birth. Each time the organism executes the **getipp** instruction, one Tping data structure is written into the soup at a location specified by a value in a CPU register and the pointer into the list is incremented, with wrap-around. Ray (1997) provides further details of the sensory mechanism.

17.2.3 Genetic operators

The central problem of the Tierra experiment is to find the conditions under which evolution can generate complexity. One primary consideration is to have a highly evolvable genetic language. The evolvability of a genetic language is not determined by its structure alone, but also by the nature of the genetic operators and the interaction between the two.

The genetic operators of the original Tierra do nothing more than flip bits in the linear genome. In order to enhance the power of genetic operators in Tierra, insertion, deletion and crossover have been added. In addition, the mutation operator has been enhanced to take two forms. One involves a bit flip, as in the original Tierra. The new form of mutation involves the replacement of a machine instruction with any other instruction chosen at random from the set of sixty-four instructions. The new genetic operations are performed on a daughter genome, just before it is born. In the runs described in this chapter, the rates of each of the different kinds of genetic operations were all set to the same values: each class of operation affects one in thirty-two individuals born.

17.3 Results

17.3.1 Genetic change

Table 17.1 illustrates the magnitude of genetic change in each of six major genes and ten subgenes, in each of seven genomes sampled from the end of seven runs ranging from 6 to 14 days. The changes are expressed as a percentage of the original gene. For example, if ten instructions are mutated (or inserted) in a 20 byte gene, the change will be 50%. If 30 bytes are inserted into a 20 byte genome, the change will be 150%.

Table 17.2 summarizes the source of the genetic changes, based on the same data as Table 17.1. Examination of the seven genomes of Table 17.1 revealed the following classes of genetic changes: **Mutation** – mutations are the result of flipping one bit in the six-bit machine instruction, or of replacing a machine instruction with one of the sixty-four instructions chosen at random. This analysis did not discriminate between the two types of mutation. **Single-byte-insertion** – the insertion of a single machine instruction into the genome. This kind of genetic change may be caused as a side effect of flaws in the increment and decrement instructions during the copying of the genome. **Single-byte-deletion** – the deletion of a single machine instruction from the genome. Like the single-byte-insertion, this may also be a side effect of flaws. **Multiple-byte-insertion** – the insertion of a sequence of more than one machine instruction into a genome. This could be caused by the insertion genetic operator. **Multiple-byte-deletion** – the deletion of a sequence of more than one machine instruction from the genome. This could be caused by the deletion genetic operator. **Rearrangement** – a change in the order of segments of the genome. This might be caused by some combination of insertor, deletion, or crossover genetic operators. **End-loss** – a couple of examples were seen in which a segment of code was lost from the end of the genome. This might be essentially the same process as the multiple-byte-deletion, or it might be a different process.

Mutation is by far the predominant source of genetic change (preserved by selection), in terms both of the number of genetic events and the amount of code affected. The next most common source of genetic change is multiple-byte-insertion, with an order of magnitude fewer events, but affecting more than half as much genetic code. The distribution of the various types of genetic change within the genome is very heterogeneous. For example, the 12 byte gene **copL**, makes up 4% of the genome, but contains 55% of the multi-byte-insertion events.

Table 17.1a and b Genetic change in each gene of seven genomes. Left columns are the run number and age of the genome in days. Top row is the name of each of the six genes and ten subgenes. Second row is the size of the gene in the ancestor. Remaining rows are the percentage change in the gene. * indicates that the gene is present in the genome, but is not expressed. – indicates that the gene has been lost from the genome

		sel	dif	rep	repS	repL	cop	copS	copL
Run	**Age**	**21**	**18**	**56**	**13**	**43**	**46**	**22**	**12**
1	11	0	0	27	0	35	11	5	25
2	6	0	0	25	0	33	28	0	100
3	8	10	6	23	0	30	20	14	17
4	9	14	0	61	8	77	13	0	50
5	6	29	6	27	15	30	50	9	158
6	6	10	0	13	0	16	4	0	8
7	14	19	0	29	0	37	35	14	50

		copC	dev	sen	senS	senO	senY	senA	senR
Run	**Age**	**12**	**14**	**144**	**41**	**17**	**12**	**52**	**22**
1	11	8	7	31	22	147	17	13	9
2	6	8	0	49	22	*59	*58	*63	*55
3	8	33	0	33	46	18	42	15	59
4	9	0	0	13	17	29	33	2	9
5	6	17	0	29	34	29	33	23	32
6	6	8	0	78	22	–	–	–	–
7	14	58	0	54	34	–	–	25	–

Table 17.2 Sources of genetic change

Genetic operation	Number of events	Bytes affected
Mutation	263	263
One-byte-insertion	9	9
One-byte-deletion	15	15
Multi-byte-Insertion	20	154
Multi-byte-deletion	11	64
Rearrangement	2	83
End-loss	2	125

17.3.2 Gene duplication

The insertion and crossover genetic operations cause segments of code to be moved about within or between genomes. In some instances, this results in a duplication of a segment of code within a genome. This duplicated code might or might not correspond to our arbitrary labelling of the code as genes or subgenes (see Figure 17.1).

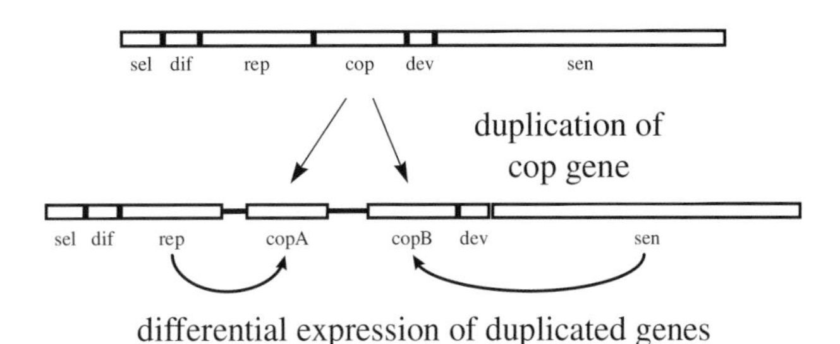

Figure 17.2 Gene duplication.

While we have observed many of these duplications, the most interesting examples have involved the complete duplication of functional algorithms which are called subroutines: either the **cop** gene, or the **dev** gene, or both together. We have observed instances of each of these duplications in which one copy of the duplicated gene is expressed in the reproductive tissue while the other copy is expressed in the sensory tissue (Figure 17.2).

At the time of duplication, both copies of the gene are generally identical. However, when the duplicated condition survives for prolonged periods of time, the two copies do diverge substantially in their structure and function.

17.3.3 Reproductive algorithm

The reproductive algorithm relies on a 12-byte copy loop (the **copL** gene) to perform a string-copy operation on the genome, resulting in the genetic code being copied from mother to daughter. The algorithm of the ancestor copies one byte for each iteration of the loop.

In the original Tierra experiments, it was observed that this algorithm sometimes evolved an optimization known as 'unrolling the loop', in which efficiency is increased by copying more than one byte in each iteration. In the original Tierra, the unrolled loops copied two or three bytes (Ray, 1994a). In the current experiment, we have observed loop unrollings of two, four and six bytes.

17.3.4 Developmental pattern

The development of the ancestor from the one-cell embryonic stage to the mature ten-cell stage is illustrated in Figure 17.3. The undifferentiated original cell splits into two cells. Soon after this first division, the differentiation event occurs (a conditional jump in the machine code), causing one cell to become a reproductive cell and the other to become a sensory cell. Subsequently, the reproductive cell divides once to form a two-celled reproductive tissue. The sensory cell goes through three division cycles to form an eight-celled sensory tissue.

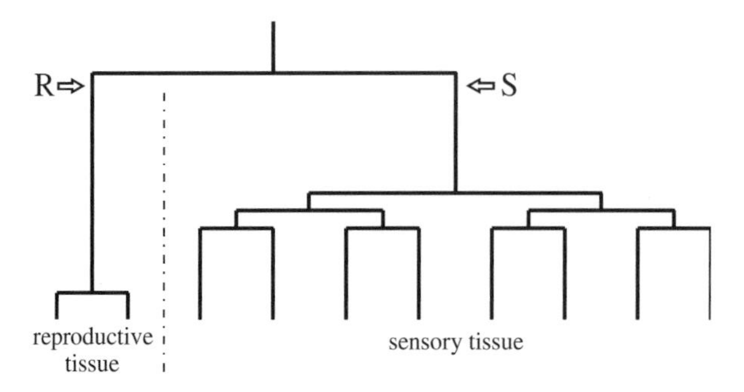

Figure 17.3 Developmental pattern.

Once the sensory tissue has reached the mature eight-cell form, it exhibits further developmental changes (Figure 17.4). Each of the eight sensory threads executes a **getipp** instruction to obtain a Tping data structure. These are then reduced to the single 'best' data through a series of three pair-wise comparisons (see the Sensory processing section below).

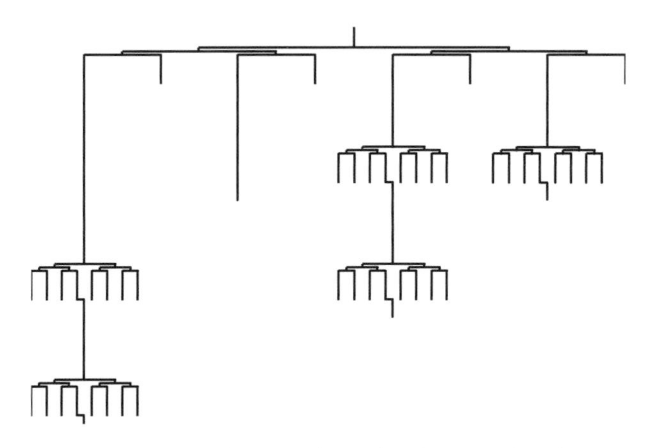

Figure 17.4 Sensory tissue developmental cycle.

Just before each pair-wise comparison, half of the threads halt (half of the cells die). The cells which remain alive compare two neighbouring data structures and if the one on the right is 'better' than the one on the left, the data are copied. The data are copied by calling the **cop** gene (which is also used by the reproductive tissue to copy the genome). The **cop** gene parallelizes its data copy function by splitting into multiple threads. When called from the sensory tissue, eight threads are used to copy the Tping data (if all four of the sensory threads doing the comparison should decide to copy the data, a total of thirty-two threads would be active simultaneously in the sensory tissue). After the data are copied, seven of the eight data copy threads halt.

At the end of the data reduction, only one of the eight sensory threads remains, but it splits into eight threads again to repeat the process, in an infinite loop. Similarly, after the genome has been copied by the two reproductive threads, one thread halts and the remaining thread executes the **divide** instruction, spawning the daughter as an independent process and potentially causing her migration. Then, the single reproductive thread splits into two threads again, and repeats the reproductive process in an infinite loop.

Table 17.3 presents a summary of the evolutionary changes in the configuration of the tissues. The first column lists the run number and the date of the sample in run–yymmdd format (for each of seven runs), or the name of the ancestral genome. The second column shows the configuration of the reproductive tissue, in the format: N×R, where N is the number of threads used to copy the genome, and R is the 'redundancy' of the reproductive tissue.

The reproductive tissue often manifested a redundancy of function. For example, a reproductive tissue might use eight cells to copy the genome, with each cell copying one-eighth of the genome. However, this entire configuration might be duplicated, so that there are actually sixteen reproductive cells, working as two groups, with each group of eight dividing the genome into eight parts in the same way. In this case, eight of the sixteen reproductive threads would be redundant. This case would appear in column two as: 8×2.

The third column shows the configuration of the sensory tissue, in the format: S×C, where S is the number of sensory threads which obtain Tping data (right part of Figure 17.3), and C is the number of threads used to copy the Tping data (middle of Figure 17.4) if the decision conditions (columns four and five) are met.

The first three rows of Table 17.3 show the structure of the three ancestral organisms used to seed the run: 960aad, 960aae and 960aaf. All three have the same configuration of tissues: 2×1 8×8. After listing the seed organisms, we show the result typical of all runs before November 1997, listed next to the date 971001.

Table 17.3 Thread and decision structure

Date	repro	sense	L dat	R dat	act
960aad	2×1	8×8	s/n	s/n	cd
960aae	2×1	8×8	f	f	cd
960aaf	2×1	8×8	s*f	s*f	cd
971001	256×1	0×0	–	–	–
1–971101	8×2	8×8	s/n	s/n	cd
1–971102	16×4	8×8	s/n	s/28	cd
1–971103	32×4	8×4	–	–	cd
1–971104	32×4	4×4	s	0	cd
1–971105	32×4	4×4	s	s*2	cd
1–971106	32×4	32×4	s	0	cd
1–971107	32×4	32×4	s	n	cd
1–971108	32×4	32×4	s	s*2	cd
1–971109	32×4	32×4	s	0	cd
1–971110	32×4	32×4	s	0	cd
1–971111	32×4	32×4	–	–	cd
2–971112	4×1	8×8	f	f	cd
2–971113	8×2	8×8	s	i/2	cd
2–971114	16×4	8×8	s	s/2	cd
2–971115	16×4	16×8	s	s/2	cd
2–971117	16×4	1×0	–	–	1 gt
3–971118	16×4	16×8	s*5	s*6	cd
3–971119	32×4	8×8	s*20	s*f	cd
3–971120	32×4	8×4	s	s	cd
3–971121	32×4	8×4	s	s	cd
3–971122	32×4	8×4	s	s	cd
3–971124	32×4	8×4	s	s	cd
3–971125	32×4	8×4	s*20	s	cd
4–971127	64×2	8×8	s*f	s*20	cd

4–971128	64×2	8×8	s*f	s*20	cd
4–971129	64×2	8×4	s*f	s*20	cd
4–971201	64×2	8×8	s*f	s*20	cd
4–971202	64×2	8×8	s*f	s*20	cd
4–971203	64×2	8×2	s*f	s*40	cd
4–971204	64×2	8×2	s*f	s*20	cd
4–971205	64×2	16×8	s*f	s*40	cd
5–971209	16×2	8×4	a	a-1	cd
5–971210	32×2	8×4	s	s-1	cd
5–971211	32×2	8×4	s	s-1	cd
5–971212	32×2	16×4	s	s-1	cd
5–971213	32×2	8×4	s	64	cd
5–971214	32×2	8×4	s	1	cd
6–971216	255×1	1×0	–	–	gt
6–971217	255×1	1×0	–	–	gt
6–971218	255×1	1×0	–	–	gt
6–971219	255×1	1×0	–	–	gt
6–971220	255×1	1×0	–	–	gt
6–971221	255×1	1×0	–	–	gt
7–971223	8×2	8×8	s	4096	cd
			n	28	
7–971224	8×2	1×0	s/n	28	gt
7–971227	64×2	1×0	s/n	65	gt
7–971230	64×2	1×0	s/n	28	gt
7–980102	64×2	1×0	s/n	128	gt
7–980105	64×2	1×0	s	784	gt
7–980110	64×2	1×0	s/n	64	gt
7–980114	64×2	1×0	s	7168	gt

In all runs before November 1997, the sensory tissue was completely lost and the reproductive tissue expanded to the limit of 256 cells. In order to migrate or send daughters to other machines on the network, the digital organism must suggest the IP address of the other machine. In early runs, we allowed any suggested IP address to be mapped to a valid address by finding the closest hamming-distance match in the map file. In these runs, loss of the sensory system and expansion of the reproductive tissue resulted in the 256×1 0×0 configuration shown for the 971001 date. In later runs, we required suggested IP addresses to be valid. In these runs, loss of the sensory system and expansion of the reproductive tissue resulted in the 255×1 1×0 configuration (which also occurred in run six).

At the end of October '97 some bugs were fixed which resulted in the survival of the sensory tissue through prolonged periods of evolution. An example of a bug that led to the selective elimination of the sensory system was the resetting of the pointer into the list of Tping data structures to zero, after its original random initialization. This had the consequence that all individuals in the population (of ancestral algorithms) could only sense the first fifteen machines on the net, regardless of the number of machines actually present in the network.

After fixing the bugs in the sensory system, the sensory tissue survived through prolonged periods of evolution in most runs. The structure of the developmental pattern and the resulting relative and absolute numbers of cells in the two tissues changed to the many forms listed in columns two and three of Table 17.3.

17.3.5 Sensory processing

The ancestral organism includes a 512 byte data area where it can hold sensory data. Each cell of the eight-cell sensory tissue reads a sixty-four byte Tping data structure into one of eight offsets into the data area. Each of the Tping structures contains data about the conditions on a different machine on the network. The sensory algorithm then undertakes a series of three pair-wise comparisons (Figure 17.5), to select the best machine to send the daughter to at the time of its birth. At the completion of the series of comparisons, the best looking data structure will be at the left-most position (zero offset) in the data area. The reproductive algorithm looks in this location for the IP address of the machine that it will send the daughter to.

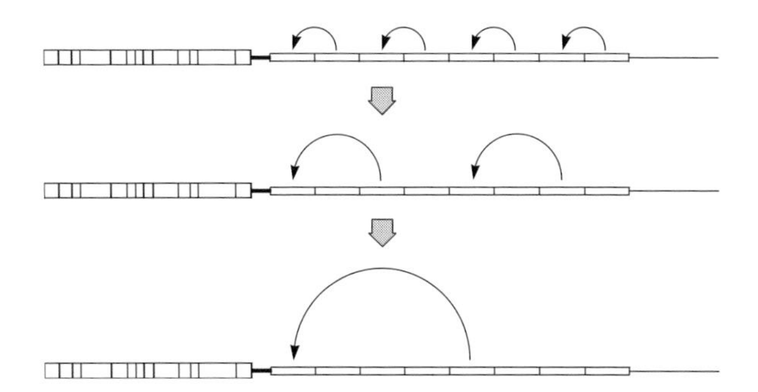

Figure 17.5 Sensory processing. The thick box on the left is the 320 byte genome marked with the divisions into genes and subgenes. The thin box to its right is the 512 byte data area marked with the eight 64 byte Tping data buffers.

The algorithm of the sensory tissue is an infinite loop, so that after the completion of the first cycle of three pair-wise comparisons, the entire sensory process repeats. After the first sensory cycle, only seven of the eight sensory threads write another Tping structure to the data area (the left-most data are preserved into successive cycles). During the time that it takes the reproductive tissue to copy the genome, the sensory system is able to complete two cycles, having collected and processed data from fifteen machines on the net.

The overall scheme of sensory data processing by multi-threaded sensory tissues tends to be preserved through evolution. However, there is a tendency for the reproductive algorithm to optimize, completing its function more quickly, with the result that the sensory system will be able to process fewer data before the results are needed by the reproductive tissue.

In some runs, the sensory processing algorithm evolved into a relatively simple form in which only a single buffer was used for storing the Tping data structure. In this case, the **getipp** instruction is used to read a structure into the buffer. Then a test such as 256 > Speed is performed and, if true, another **getipp** instruction is executed with the result that the previous data are replaced with new data.

17.3.6 Sensory data selection

The algorithm by which Tping data are selected is represented in columns four, five and six of Table 17.3. If the value in column four is less than or equal to the value in column five, then the action in column six is performed. Two different actions are represented in column six: cd – copy the Tping data on the right over the Tping data on the left; gt – get another Tping data structure from the map file list. The values listed in columns four and five include data from the Tping structures and constant values. The symbols used for the Tping entries are: s – Speed; n – NumCells; f – FecundityAvg; i – InstExecConnect; a – AgeAvg. In some cases two or more of these variables or constants are combined by the arithmetic operations of addition, subtraction, multiplication or division (+ – * / respectively).

In the studies reported here, all but the first of the seven runs were initiated with a mixture of three different ancestral genomes, using three different selection algorithms (top three rows of Table 17.3). 960aad copies the Tping data if Speed/NumCells <= Speed/NumCells; 960aae if FecundityAvg<=FecundityAvg; and 960aaf if FecundityAvg*Speed<= FecundityAvg*Speed.

It is likely that after a few generations of reproduction, an ecological process of competitive exclusion will result in a population that is entirely descended from only one of the three ancestors. Comparison of sequence similarity between the evolved organisms of Table 17.1 and the three ancestors reveals that in some runs, the population descended from the Speed/NumCells algorithm and in other runs from the FecundityAvg*Speed algorithm.

Evolution has also produced a diversity of sensory data selection algorithms. The element of the Tping data structure most commonly used by these algorithms is Speed. However, the algorithms also commonly integrate data other than elements of the Tping structure, such as some constant value. For example, copy data if Speed (on the left) <= 256.

The '-' symbol in columns four and five of the table indicates that the action in column six is performed unconditionally. The result in all of these cases is that the node to which the organism or its daughter migrate is chosen essentially at random. This is the situation found when the sensory system is completely lost through evolution (971001 and run 6). We call these organisms 'map-file-scanners', because they constantly get new IP addresses from the map file and then send the daughter (or migrate) to whatever address has been most recently accessed, by chance, when the reproductive process is completed. The notation '1 gt' in column six indicates that the program only gets one Tping data structure and uses its IP address as the migration destination. Since the pointer into the map file is initialized at random, this is another random method of node selection.

The selection mechanism for the 7–971223 organism is unique, in that it uses two conditionals, both from the data on the left. If s <= 4096 and n <= 28, the data on the right are copied over the data on the left.

17.3.7 Migration patterns

At the completion of the reproductive cycle, the network ancestor causes its daughter to be sent to another machine at birth. The IP address of the target machine is taken from the Tping data analysed by the sensory system. One change that commonly occurs through evolution is for the daughter to be born locally (on the machine where the mother lives) and for the mother to then immediately move to the machine whose IP address was recommended by the sensory system. Another common pattern is for the mother to send the daughter to another machine and then immediately to follow the daughter to that machine.

17.4 Discussion

17.4.1 Genetic change

The magnitude and source of genetic change preserved by selection varies greatly between genes. For example the developmental genes **dif** and **dev** are rarely altered, while **copL** which contains the critical code for copying data, often experiences large changes. In addition, the genetic operations predominantly responsible for the genetic changes vary widely between different parts of the genome.

The **copL** gene achieved higher levels of unrolling in this experiment than in the original Tierra. The large magnitude of change in this gene can be understood in terms of strong selection pressure for efficiency of copying data, combined with an accessible pathway for change with the genetic operators available (increasing levels of loop unrolling).

The sensory genes also show a high level of genetic change, but we suggest a different interpretation. It appears that the selective pressures on the sensory system are not as intense. The organisms can survive and reproduce without the sensory system, whereas the **copL** gene is essential for reproduction. Thus the high level of genetic change in the sensory genes may be due to lower selective pressures permitting higher levels of variation to survive.

These observations seem quite significant in the context of understanding the issue of evolvability (in fact like complexity, we do not have an adequate definition of evolvability). If we were to attempt to judge the evolvability of the system described here, we would reach very different conclusions from examining the changes exclusively in different parts of the genome. For example, evolvability seems to be high in the **copL** and **sen** genes, but low in the **dif**, **dev** and **repS** genes. At the same time, the causes of the high degree of genetic change in the **copL** and **sen** genes seem to be quite different.

17.4.2 Gene duplication

The phenomenon of gene duplication in which both copies of the gene are expressed, but by different tissues is surprising and quite interesting. Gene duplication and subsequent divergence of the sequence and function of the two copies of the gene is believed to be a primary mechanism for the increase in the complexity of genomes in organic evolution. It appears an analogous process has occurred, or at least begun, in this experiment.

17.4.3 Sensory data selection

The ancestor organisms were written with sensory data selection algorithms that seem 'smart' to their designers. Sometimes, apparently smart algorithms have been present in the later stages of evolution. However, it is often the case that the evolved algorithms appear to be less smart. However, they may be good adaptations to the environment in which the organisms live.

A predominant feature of the environment is the presence of other organisms and their behaviour. One of the most difficult problems in designing a sensory data selection algorithm is that if one specific algorithm comes to dominate the network-wide population, then most organisms will tend to make the same choices. This can result in a 'mob' behaviour (Figure 17.6).

Tests with the Speed/NumCells ancestor algorithm in a four node network (with genetic operations turned off to prevent evolution) revealed a severe problem. One of the four machines had a far faster processor, resulting in a consistently high value of Speed on that machine. The high Speed caused a high Speed/NumCells ratio, making it the machine of choice for the entire network-wide population. The result was that all daughters born on all machines were sent to this one machine. In effect, there was no birth or immigration on the other three machines. Thus the original organisms on those three machines lived indefinitely, accumulating very high fecundities (and associated Darwinian fitness).

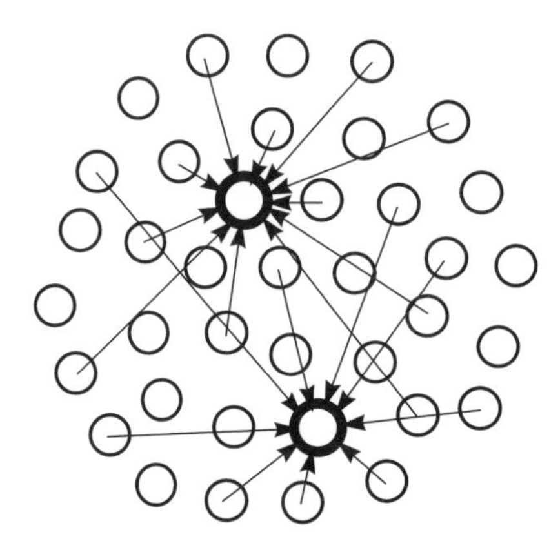

Figure 17.6 Mob behaviour.

Meanwhile, on the selected machine, there was a huge influx of immigration, in addition to the local reproduction. The result was a rapid flux of organisms (for each birth or immigration, the reaper must kill an existing organism to make space) such that few if any individuals survived long enough to reproduce. Thus the average fecundity (and Darwinian fitness) on the selected machine was near zero. Because the soup size on this machine was fixed, it was not possible for growth of the population to lower the Speed/NumCells ratio to a level comparable to the slower machines.

The consequence of the use of the Speed/NumCells algorithm throughout the small network was a mob behaviour that created a fitness landscape within which using the algorithm was the worst thing possible. Even random selection of machines would have been better. The severe mob behaviour seen in a small network is diffused somewhat in a larger network, because the individual organism does not have time to examine data from all machines. The ancestor is able to look at fifteen machines for each reproductive cycle. Because the pointers into the list of machines are initialized at random, each organism will look at a different list. However, there remains an underlying dynamic, in which some machines tend to be chosen by any organism that looks at them, generating some mob behaviour. In Figure 17.6 the favoured machines are represented by the heavy circles.

Selection algorithms such as 256 <= Speed appear to be relatively dumb, but they may have the selective advantage of reducing the mob effect by making the choice of machines more fuzzy.

17.4.4 Loss or degradation of sensory system

While this experiment demonstrated that the sensory system is able to survive long periods of evolution, some of the runs showed a complete loss or a serious degradation of the sensory system (see Tables 17.1 and 17.3). We believe that the primary selective factor for maintaining the sensory system is the temporal heterogeneity in the availability of CPU cycles to the Tierra process, due to the activity patterns of the human users of the machines in the Tierra network.

This experiment was conducted in a local-area network at ATR, where there are some fairly obvious patterns of human activity. ATR is in a somewhat remote location and most researchers commute by company bus. The bus service is available from 7:40 am to 10:00 pm on weekdays only. There is no bus service on weekends or holidays. Most researchers arrive between 8:00 am and 10:00 am and leave between 6:00 pm and 8:00 pm, on weekdays only. The data reported in this study cover the period of November 1 through January 14. In this period, weekends and holidays fell on: Nov 1–3, 8–9, 15–16, 22–24, 29–30; Dec 6–7, 13–14, 20–21, 23; Dec 27–Jan 4; Jan 10–11.

We can expect that an important component of the selective pressure for maintaining the sensory system will be relaxed on weekends, holidays, and weekdays from mid-evening to mid-morning. We suspect that this relaxation of selective pressures may partially explain the occasions of loss of the sensory system. It is worth noting that the sensory system was lost from the outset of run six, during the business week. However, this run was initiated in the mid-evening and probably lost its sensory system before ever experiencing the relevant selective pressures. We are preparing to test the pattern of loss of the sensory system against quantitative measurements of temporal heterogeneity in human activity in the network.

17.4.5 Migration patterns

There is an obvious benefit to the behaviour of the mother migrating after reproduction, rather than remaining on the local machine to attempt a second reproduction. When a creature moves to another machine, it enters the bottom of the reaper queue (Ray, 1991). By moving after reproduction, the mother effectively delays her death.

There are, however, some costs to the migration of a mature organism. If the Tierra process is sleeping on the target machine (due to user activity) the migrating genome can die as a result of having its packet(s) lost in the network. Furthermore, upon arrival, the formerly mature organism reverts to an essentially embryonic, one-celled state. It must then go through the developmental process leading to the mature ten-celled state, before it can begin the reproductive and sensory cycles. In addition, through migration, all sensory data are lost, whereas the mature organism which does not migrate would retain the selected sensory data in its left-most Tping data buffer.

17.5 Conclusions

The central objective of this project is to study the conditions under which evolution by natural selection leads to an increase in complexity of the replicators. For the purpose of this study, the primary quantitative measure of complexity is the level of differentiation of the multi-celled organism. The study begins with the most primitive level of differentiation: two cell types. There are two milestones in the study: (1) the differentiated state persists through prolonged periods of evolution; (2) the number of cell types increases through evolution.

In the work reported here, only the first of these two milestones has been achieved. There has been no sign of an increase in the number of cell types. However, the process of gene duplication with differential expression of the resulting genes is a kind of proto-differentiation event. This process offers some prospect of leading to new cell types.

Observations of a high degree of heterogeneity in the magnitude, source, and possible selective dynamics for genetic change in different parts of the genome provide raw data for our efforts to understand the nature of 'evolvability'. A practical understanding of evolvability, leading to an ability to design higher levels of evolvability into our synthetic evolving systems is crucial for progress in the area of evolutionary systems.

The ultimate imperatives in evolution are survival and reproduction. In the context of self-replicating computer programs, it is not obvious how selection can favour any behaviour beyond the efficient replication of the genome. However, in this experiment we demonstrate that selection can favour the ability to gather information about conditions in the environment, analyse that data and use the results of the analysis to control the direction of movements.

Digital organisms essentially identical to those of the original Tierra experiment, were provided with a sensory mechanism for obtaining data about conditions on other machines on the network; code for processing that data and making decisions based on the analysis, the digital equivalent of a nervous system; and effectors in the form of the ability to make directed movements between machines in the network. This sensory-nervous-effector system required 157 bytes of genetic code, compared to 136 bytes for the reproductive system alone. In addition, the sensory system required a data area almost twice the size of the entire genome. This sensory system is not 'hard-coded', in the sense that it is not essential for survival and reproduction in the network and it can be lost if selection does not maintain it in the face of degradation by genetic operations. Yet selection maintained this large burden of additional complexity due to the selective benefits of gathering, processing and acting upon information about the environment.

The migratory patterns of the digital organisms themselves become an important part of the fitness landscape in the network. The algorithms of the seed organisms generate an unfit (in the Darwinian sense) mob behaviour by causing all individuals in the network to migrate to the 'best' looking machines. Evolution resolves this problem by changing the algorithm to simply avoid poor quality machines.

Acknowledgement

This chapter is from Ray, T.S. and Hart, J. (1998) Evolution of differentiated multi-threaded digital organisms. In *Proceedings of Alife* 6, MIT Press.

References

Charrel, A. (1995) Tierra network version. *ATR Technical Report TR-H-145*. http://www.his.atr.co.jp/~ray/pubs/charrel/charrel.pdf

Koza, J.R. (1994) *Genetic programming II. Automatic Discovery of Reusable Programs*. pp. 746. MIT Press.

Ray, T.S. (1991) An approach to the synthesis of life. In Langton, C., Taylor, C., Farmer, J.D. and Rasmussen, S. eds. *Artificial Life II*, Addison-Wesley. http://www.his.atr.co.jp/~ray/pubs/tierra/

Ray, T.S. (1994a) Evolution, complexity, entropy, and artificial reality. *Physica D*, **75**, 239–263. http://www.his.atr.co.jp/~ray/pubs/oji/ojihtml.html

Ray, T.S. (1994b) An evolutionary approach to synthetic biology: Zen and the art of creating life. *Artificial Life*, **1**(1/2), 195–226. http://www.his.atr.co.jp/~ray/pubs/zen/

Ray, T.S. (1995) A proposal to create a network-wide biodiversity reserve for digital organisms. *ATR Technical Report TR-H-133*. http://www.his.atr.co.jp/~ray/pubs/reserves/

Ray, T.S. (1997) Selecting naturally for differentiation. In Koza, J.R., Deb, K., Dorigo, M. *et al.* eds. *Genetic Programming 1997: Proceedings of the Second Annual Conference*, July 13–16, 1997, Stanford University, pp. 414–419. Morgan Kaufmann. http://www.his.atr.co.jp/~ray/pubs/gp97/

Ray, T.S. (1998) *Continuing Progress Report on the Network Experiment*. Published only as a web page: http://www.his.atr.co.jp/~ray/tierra/netreport/

Thearling, K. and Ray, T.S. (1994) Evolving multi-cellular artificial life. In Brooks, R.A. and Maes, P. eds. *Artificial Life IV Conference Proceedings*, pp. 283–288. The MIT Press. http://www.his.atr.co.jp/~ray/pubs/alife4/alife4.pdf

Thearling, K. and Ray, T.S. (1997) Evolving parallel computation. *Complex Systems*, **10**(3), 229–237.

Section 5

Applications of Biologically Inspired Development

Artificial life models of neural development

18

ANGELO CANGELOSI, STEFANO NOLFI and **DOMENICO PARISI**

18.1 Introduction

Artificial neural networks are computational models of nervous systems. Natural organisms, however, do possess not only nervous systems but also genetic information stored in the nucleus of their cells (genotype). The nervous system is part of the phenotype which is derived from the genotype through a process called development. The information specified in the genotype determines aspects of the nervous system which are expressed as innate behavioural tendencies and predispositions to learn. When neural networks are viewed in the broader biological context of artificial life, they tend to be accompanied by genotypes and to become members of evolving populations of networks in which genotypes are inherited from parents to offspring (Parisi, 1997, Nolfi and Parisi, 2001).

Artificial neural networks can be evolved by using evolutionary algorithms (Holland, 1975; Koza, 1992; Schwefel, 1995). An initial population of different artificial genotypes, each encoding the free parameters of an individual neural network (e.g. the connection strengths and/or the architecture of the network and/or the learning rules), are created randomly. Each individual network is evaluated in order to determine its performance in some task (fitness). The fittest networks are allowed to reproduce (sexually or non-sexually) by generating copies of their genotypes with the addition of changes introduced by some genetic operator (e.g. mutations, crossover, duplication). This process is repeated for a number of generations until a network that satisfies the performance criterion (fitness function) set by the experimenter is obtained (for a review of methodological issue see Yao, 1993).

18.1.1 Evolution and development

The distinction between inherited genetic code (genotype) and the corresponding organism (phenotype) is a cornerstone of biology. What is inherited from the parents is the genotype. The phenotype is the complete individual that is formed according to the instructions specified in the genotype.

In simulations with evolving neural networks, the genotype might encode all the free parameters of the corresponding artificial neural network or only the initial value of the parameters and/or other parameters that affect learning. In the former case the network is entirely innate

 ISBN 0-12-428765-4

and there is no learning. In the latter networks change, both philogenetically across a succession of generations, and ontogenetically during the life of the individual, i.e. during the period of time in which they are evaluated.

Evolution is critically dependent on the distinction between genotype and phenotype and on their relation, i.e. the genotype-to-phenotype mapping. The fitness of an individual, which affects selective reproduction, is based on the phenotype. But what is inherited is the genotype, not the phenotype. Furthermore, while the genotype of an individual is one single entity, the organism can be considered as a succession of different phenotypes taking form during the genotype-to-phenotype mapping process, each derived from the previous one under genetic and environmental influences.

When the genotype-to-phenotype mapping process takes place gradually during an individual's lifetime we can talk of development. In this case, each successive phenotype, corresponding to a given stage of development, has a distinct fitness. The total fitness of a developing individual is a complex function of these developmental phases. Evolution must ensure that all these successive forms are viable and, at the same time, that they make a well-formed sequence where each form leads to the next one until a more or less stable (adult) form is reached. This puts various constraints on evolution but it also offers new means for exploring novelty. Small changes in the developmental rates of different components of the phenotype, for example, can have huge effects on the resulting phenotype. Indeed it has been hypothesized that, in natural evolution, changes affecting regulatory genes that control rate of development have played a more important role than other forms of change such as point mutations (Gould, 1977).

Although the role of the genotype-to-phenotype mapping and of development has been ignored in most of the experiments involving artificial evolution, there is now an increasing awareness of its importance. Wagner and Altenberg (1996) write:

> *In evolutionary computer science it was found that the Darwinian process of mutation, recombination and selection is not universally effective in improving complex systems like computer programs or chip designs. For adaptation to occur, these systems must possess* evolvability, *i.e. the ability of random variations to sometimes produce improvement. It was found that evolvability critically depends on the way genetic variation maps onto phenotypic variation, an issue known as the representation problem (p. 967).*

18.1.2 Artificial life approaches to modelling neural development

In the next sections, different approaches to modelling neural development in artificial life simulations will be presented. They range from simple direct genotype-phenotype encoding to more complex methods such as axonal growth, cellular encoding and regulatory models. Furthermore, we discuss some models of the interaction between evolution and learning. These models address a different type of plasticity in neural network development, i.e. the effects of ontogenetic learning in the overall evolutionary process. (For a review of neural network models of development, see Parisi, 1996 and Parisi and Nolfi, 2001.)

18.2 Genetic encoding: direct genotype-phenotype mapping

To evolve neural networks one decision that has to be taken is how to encode the network in the genotype in a manner suitable for the application of genetic operators. In most cases, all phenotypical characteristics are coded in a uniform manner so that the description of an individual at the level of the genotype assumes the form of a string of identical elements (such as binary or floating point numbers). The transformation of the genotype into the phenotypical network is called genotype-to-phenotype mapping.

In direct encoding schemes there is a one-to-one correspondence between genes and the phenotypical characters that are subjected to the evolutionary process (e.g. Miller *et al.*, 1989). Aside from being biologically implausible, simple one-to-one mappings have several drawbacks. One problem, for example, is scalability. Since the length of the genotype is proportional to the complexity of the corresponding phenotype, the space to be searched by the evolutionary process increases exponentially with the size of the network (Kitano, 1990).

Another problem of direct encoding schemes is the impossibility to encode repeated structures (such as network composed of several sub-networks with similar local connectivity) in a compact way. In one-to-one mappings, in fact, elements that are repeated at the level of the phenotype must be repeated at the level of the genotype as well. This affects not only the length of the genotype and the corresponding search space, but also the evolvability of individuals. A full genetic specification of a phenotype with repeated structures, in fact, implies that adaptive changes affecting repeated structures should be independently rediscovered through changes introduced by the genetic operators.

18.3 Growing methods

The genotype-to-phenotype process in nature is not only an abstract mapping of information from genotype to phenotype but it is also a process of physical growth (growth in size and in physical structure). By taking inspiration from biology, therefore, one can decide to encode growing instructions in the genotype. The phenotype is progressively built by executing the inherited growing instructions.

Nolfi *et al.* (1994b) used a growing encoding scheme (see also Nolfi and Parisi, 1995) to evolve the architecture and the connection strengths of neural networks that controlled a small mobile robot (for a similar method see Husbands *et al.*, 1994). These controllers are composed of a collection of artificial neurons distributed over a 2-dimensional space with growing and branching axons (Figure 18.1, left). Inherited genetic material specifies instructions that control the axonal growth and the branching process of neurons. During the growth process, when a growing axonal branch of a particular neuron reaches another neuron a connection between the two neurons is established. On the right of Figure 18.1 you can see the network resulting from the growth process displayed on the left of the figure after the elimination of non-connecting branches and isolated and non-functional neurons. However, axons grow and branch only if the activation variability of the corresponding neurons is larger than a genetically-specified threshold. This simple mechanism is based on the idea that sensory information coming from the environment has a critical role in the maturation of the connectivity of the biological nervous system and, more specifically, that the maturation process is sensitive to the activity of single neurons (see Purves, 1994). Since the actual sequence of sensory states experienced by the network influences the process of neural growth, in this model the developmental process is influenced not only by genetic factors but also by environmental factors.

This type of genotype-to-phenotype mapping allows the evolutionary process to select neural network topologies that are better suited to the task chosen. Moreover, by being sensitive to environmental conditions, the developmental process might display a form of plasticity. Indeed, as shown by the authors, if some aspects of the task are allowed to vary during the evolutionary

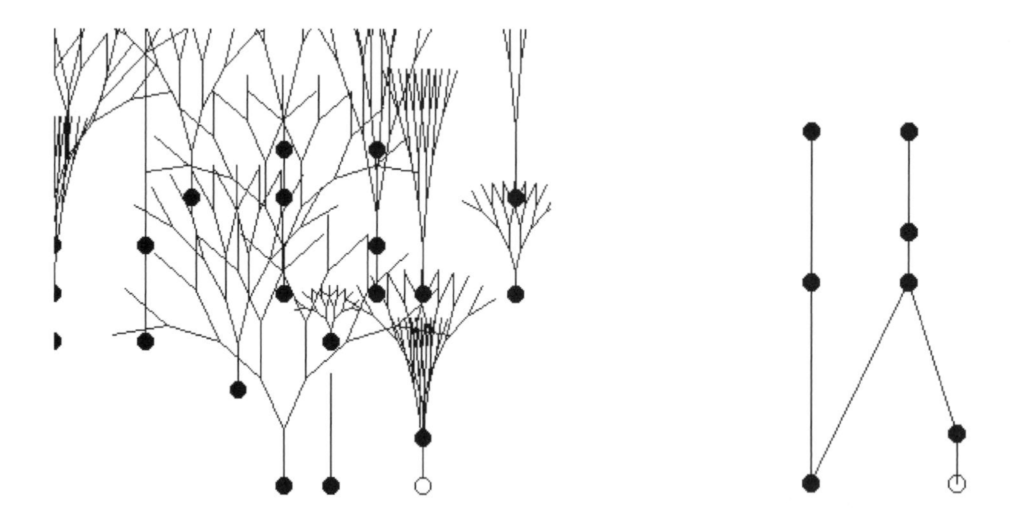

Figure 18.1 Development of an evolved neural network. Left: the growing and branching process of the axons. Right: the resulting neural network after removal of non-connecting branches and the elimination of isolated neurons and groups of interconnected neurons.

process, evolved genotypes display an ability to develop into different final phenotypical structures that are adapted to the current conditions.

18.4 Cellular encodings

In natural organisms the development of the nervous system begins with a folding in of the ectodermic tissue which forms the neural crest. This structure gives origin to the mature nervous system in a succession of three phases: the genesis and proliferation of different classes of neurons by cellular duplication and differentiation, the migration of neurons toward their final destination, and the growth of neurites (axons, dendrites). The growth process described in the previous section characterizes very roughly only the last of these three phases. A number of attempts have been made to include other aspects of neural development in artificial evolutionary experiments.

Cangelosi *et al.* (1994), for example, extended the model described in the previous section by adding a cell division and migration stage to the already existing stage of axonal growth. The genotype, in this case, is a collection of rules governing the process of cell division (a single cell is replaced by two 'daughter' cells) and migration (the new cells can move in 2D space). The genotype-to-phenotype process starts with a single cell which, by undergoing a number of duplication and migration processes, produces a collection of neurons arranged in a 2D space. At the end of this stage the neurons grow their axons and establish connections until a viable neural controller is formed (for a related approach, see Dellaert and Beer, 1994).

Gruau (1994) proposed a genetic encoding scheme for neural networks based on a cellular duplication and differentiation process. The genotype-to-phenotype mapping starts with a single cell that undergoes a number of duplication and transformation processes ending in a complete neural network. In this scheme the genotype is a collection of rules governing the process of cell divisions (a single cell is replaced by two 'daughter' cells) and transformations (new connections can be added and the strengths of the connections departing from a cell can be modified). In this model, therefore, connection links are established during the cellular duplication process.

In Gruau's model the instructions contained in the genotype are represented as a binary-tree structure as in genetic programming (Koza, 1992). During the genotype-to-phenotype mapping process, the genotype tree is scanned starting from the top node of the tree and then following each ramification. The top node represents the initial cell that, by undergoing a set of duplication processes, produces the final neural network. Each node of the genotype tree encodes the operations that should be applied to the corresponding cell and the two sub-trees of a node specify the operations that should be applied to the two daughter cells. The neural network is progressively built by following the tree and applying the corresponding duplication instructions. Terminal nodes of the tree (i.e. nodes that do not have sub-trees) represent terminal cells that will not undergo further duplications. Gruau also considered the case of genotypes formed by many trees where the terminal nodes of a tree may point to other trees. This mechanism allows the genotype-to-phenotype process to produce repeated phenotypical structures (e.g. repeated neural sub-networks) by re-using the same genetic information. Trees that are pointed to more than once, in fact, will be executed more times. This encoding method has two advantages: (a) compact genotypes can produce complex phenotypical networks, and (b) evolution may exploit phenotypes where repeated substructures are encoded in a single part of the genotype. Since the identification of substructures that are read more than once is an emergent result of the evolutionary process, Gruau defines this method automatic definition of neural subnetworks (ADNS) (Gruau, 1994).

18.5 Heterochrony in neural development

The existence of variable and plastic ontogenetic development is strictly related to the evolution of regulatory genotypes, i.e. genotypes whose main role is to control the functioning of simple ontogenetic events. Even though some genes directly encode structural molecules, most genetic products consist of regulatory elements such as enzymes. These regulatory genes act as ON-OFF switches on the complex chain of biochemical events that constitute the three main phenomena of cellular development: mitoses, cell differentiation and migration. A regulatory ontogenetic development consists of a variety of interactions between the growing organism and its environment.

In such a regulatory development, the timing of the events, i.e. their temporal activation/inhibition, and their rate, i.e. the frequency of occurrence of the phenomena, both have a strong impact. The temporal co-occurrence of two or more events can prove essential for allowing the activation of a biological phenomenon. Even the spatial relation between substructures of the developing organism is a key factor. The spatial interaction between cells can induce the phenomena of cell differentiation or cell migration. These classes of interactions, especially the temporal relations occurring during the organism's development, constitute the phenomenon known as heterochronic change. Heterochrony (McKinney and McNamara, 1991) is the study of the effect of changes in timing and rate of the ontogenetic development in an evolutionary context. In particular, heterochronic classifications are based on the comparison of ontogenies that differ in terms of (1) onset of growth, (2) offset of growth and (3) rate of growth of an organ or other biological traits. These three kinds of change correspond respectively to the following couples of heterochronic phenomena: predisplacement and postdisplacement for an anticipated and postponed growth onset, hypermorphosis and progenesis for a late and early offset, and acceleration and neoteny for a faster and slower rate of growth (see also Gould, 1977).

Cangelosi and Elman (1995; Cangelosi, 1999) have developed a model of development that simultaneously simulates many biologically-inspired phenomena for the development of neural networks in artificial organisms. They use a regulatory genotype in which most of the genes produce elements whose role is to control the activation, inhibition and delay of the developmental events. The phenomena occurring during neural network development (cell duplication, differentiation, migration, axonal growth and synaptogenesis) are directly inspired by their real biological functioning (Purves and Lichtheim, 1985). At the beginning of neural development the organism's neural system consists of a single egg cell with its own genome and a set of elements present in the intercellular environment. Some of these elements act as 'receptors' for extracellular signalling. Others are 'structural' elements for the activation and execution of developmental events. Others are pure 'regulatory' elements for the modulation of gene expression and do not play any direct role in development. The structural elements can regulate gene expression while the receptors cannot.

The physical environment in which the egg cell will grow consists of a 2D grid of 7*20 cells. The grid has a polarized orientation in the y dimension. The upper pole corresponds to the organism's muscle tissue side and the lower pole to the sensory tissue side. The initial intracellular elements are considered to be inherited from the parent organism. Their distribution, i.e. the initial amount of each element, will function as the zygote's pattern formation mechanism. During development, the amount of these elements, together with the other environmental conditions, will determine the activation, inhibition or delay of developmental events. Moreover, these elements act also as regulators of gene expression.

Five developmental events cyclically occur during the neural network's growth: cell duplication, cell differentiation, cell migration, axonal growth and synaptogenesis. For example, the cell duplication process consists of the replacement of the mother cell with two new daughter cells. The physical displacement of the new cells and their differentiation (i.e. the splitting of the mother cell's elements), is determined by the environment available around the mother cell and by the amount of the two elements responsible for mitosis. In this case, the choice of the two elements is inspired by the role played by cyclin and the kinase enzyme, two major regulatory proteins for mitosis (Marx, 1989).

The feasibility of regulatory development for adaptation to environmental changes is the hypothesis tested using this model. A two-stage, two-task simulation setting was used. In the first evolutionary stage, the organisms are selected according to their performance in a foraging task. In the second stage, dangerous elements are introduced in the environment, together with food. This task requires that organisms adapt their food approaching strategy to a new behavioural pattern for approaching only foods and avoiding dangers. To do this, organisms need to restructure their neural network, for example by adding or readapting some sensory and hidden neurons to the new processing needs. The way to re-adapt the neural network is by modifying its architecture.

Analysis of the distribution of neurogenetic changes that allow organisms to successfully re-adapt for the food and danger task shows that all five developmental events, except migration, are involved in re-adapting ontogenesis for coping with new behavioural requests. The events related to axon growth and synaptogenesis are the ones most frequently used as a re-adaptation strategy. Even small adjustments of the connectivity pattern can prove very functional for the evolution of good networks.

In the simulation different examples of heterochronic changes were observed. For example, a case of adaptive local progenesis and contemporaneous cell–cell induction effect due to spatial interaction was observed. Few mutations in an offspring were enough to cause significant changes in its neural development and to allow the organism to adapt to the new environment. Figure 18.2 shows the morphogenetic tree of the ancestor and descendant

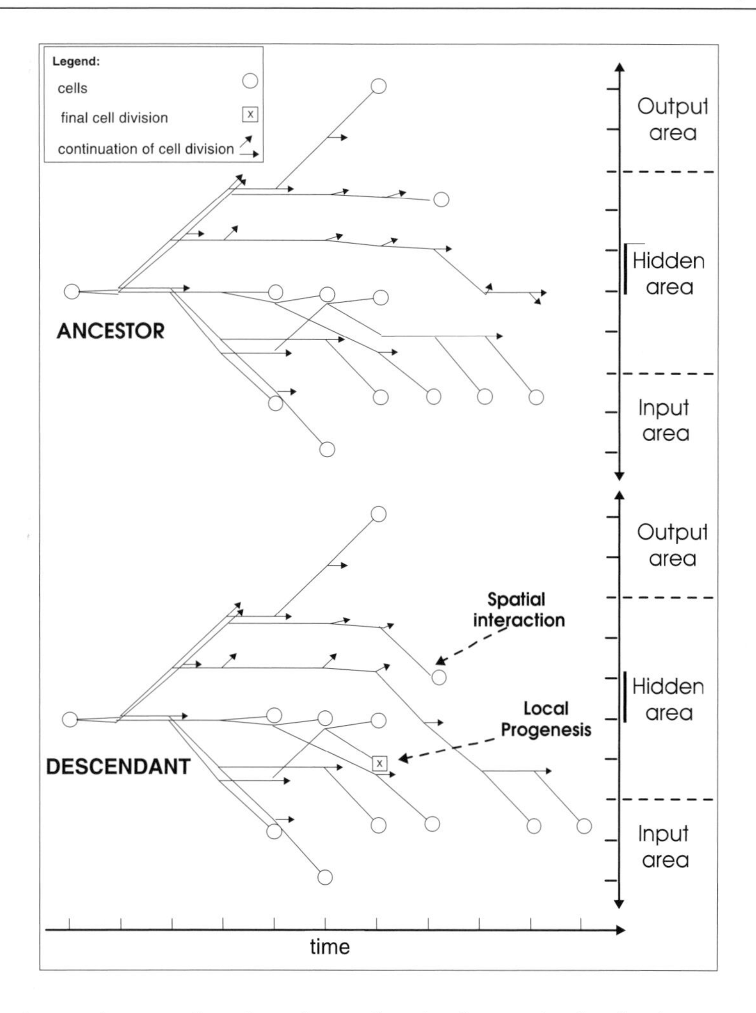

Figure 18.2 Morphogenetic trees that show heterochronic changes in the development of a neural network (see text for explanation) (from Cangelosi, 1999).

organisms where such changes happened. The morphogenetic tree is a graphic representation of a cell duplication tree using the two dimensions of time and space. It facilitates the understanding of the developmental events and their temporal and spatial interactions (Arthur, 1984). In the ancestor organism, two sensory cells for food location input originate from a common founder cell. In the descendant individual, this cell stops duplicating early, leaving two free spots in the sensory area of the developing grid. This is a case of local progenesis, because the offset of the mitotic sequence is anticipated. At the same time, there is a change in the cell displacement of other cell duplication branches. In the upper side of the developmental grid, two new cells, coming from a different mitosis branch, occupy the space left free in the sensory area. What happens in the descendant is that, in later stages of development, a newly formed cell changes position moving to the lower row. This new displacement will induce a dividing cell to place one of its daughters in the lower input area. Because of these spatial interactions, the progeny of this cell ends up in the two spots left free by the progenesis.

18.6 Evolution and learning

Evolution and learning are two forms of adaptation that operate on different time scales. Evolution is capable of capturing relatively slow environmental changes that might encompass several generations. Learning, instead, allows an individual to adapt to environmental changes that are unpredictable at the generational level. Moreover, while evolution operates on the genotype, learning affects the phenotype and phenotypic changes cannot directly modify the genotype. Recently, the study of artificial neural networks that are subjected to both an evolutionary and a lifetime learning process has received increasing attention. These studies (see also Nolfi and Floreano, 1999) have been conducted with two different purposes: (a) looking at the advantages, in terms of performance, of combining two different adaptation techniques; (b) understanding the role of the interaction between learning and evolution in natural organisms. The general picture that emerges from this body of research is that, within an evolutionary perspective, learning has several different adaptive functions:

- It can help and guide evolution by channelling evolutionary search toward promising directions. For example, learning might significantly speed up the evolutionary search.
- It can supplement evolution by allowing individuals to adapt to environmental changes that, by occurring during the lifetime of the individual or within few generations, cannot be tracked by evolution.
- It can allow evolution to find more effective solutions and increase the ability to scale up to problems that involve large search space.

However, learning also has costs and, in particular, it might increase the unreliability of evolved individuals (Mayley, 1997). Since an individual's abilities are determined by the individual's learning experiences, learning individuals might fail to acquire the required abilities in unfavourable conditions.

18.6.1 How learning might help and 'guide' evolution

A simple and clear demonstration of how learning might influence evolution, even if the characteristics that are learned are not communicated to the genotype, was provided by Hinton and Nowlan (1987). The authors considered a simple case in which (a) the genotype of the evolving individuals consists of 20 genes that encode the architecture of the corresponding neural networks, and (b) only one architecture, i.e. only a single combination of gene values, confers added reproductive fitness. Individuals have a genotype with 20 genes that can assume two alternative values (0 or 1). The only combination of genes that provide a fitness value above 0 consists of all ones. In this extreme case, the probability of finding the good combination of genes would be very small given that the fitness surface looks like a flat area with a single spike in correspondence of the good combination.

The fitness surface is a metaphor often used to visualize the search space on an evolutionary algorithm. Any point on the search space corresponds to one of the possible combinations of genetic traits and the height of each point on the fitness surface corresponds to the fitness of the individual with the corresponding genetic traits. In the fitness surface of Hinton and Nowlan's model, artificial evolution does not perform better than random search. Finding the right combination is like looking for a needle in a haystack.

The addition of learning simplifies evolutionary search significantly. One simple way to introduce learning is to assume that, in learning individuals, genes can have three alternative values [0, 1, and ?] where question marks indicate modifiable genes whose value is randomly selected within [0, 1] at each time step during an individual's lifetime. By comparing learning and non-learning individuals one can see that performance increases throughout generations much faster in the former than in the latter. The addition of learning, in fact, produces an enlargement and a smoothing of the fitness surface area around the good combination that, in this case, can be discovered much more easily by the genetic algorithm. This is due to the fact that not only the right combination of alleles but also combinations which in part have the right alleles and in part have unspecified (learnable) alleles might report an average fitness greater than 0 (fitness monotonically increases with the number of fixed right values because the time needed to find the right combination is inversely proportional, on the average, to the number of learnable alleles). As claimed by the authors, 'it is like searching for a needle in a haystack when someone tells you when you are getting close' (Hinton and Nowlan, 1987, p. 496). (For a variation of this model that has been used to study the interaction between evolution, learning, and culture, see Hutchins and Hazlehurst (1991).)

The Hinton-Nowlan model is an extremely simplified case that can be analysed easily but that makes several unrealistic assumptions: (1) there is no distinction between genotype and phenotype, (2) learning is modelled as a random process that does not have any directionality, and (3) there is no distinction between the learning task (i.e. the learning function that individuals try to maximize during their lifetime) and the evolutionary task (i.e. the selection criterion that identifies the individuals that are allowed to reproduce). Further research conducted by Nolfi *et al.* (1994a) showed how, when these limitations are released, learning and evolution display other forms of interactions that are also mutually beneficial.

Nolfi *et al.* (1994a) studied the case of artificial neural networks that 'live' in a grid world containing food elements. Networks evolve (to become fitter at one task) at the population level and learn (a different task) at the individual level. In particular, individuals are selected on the basis of the number of food elements that they are able to collect (evolutionary task) and try to predict the sensory consequences of their motor actions during their lifetime (learning task).

The genotypes of the evolving individuals encode the initial weights of a feedforward neural network that, each time step, receives sensory information from the environment (the angle and the distance of the nearest food element and the last planned motor action), determines a given motor action selected within four options (move forward, turn left, turn right or stay still) and predicts the next sensory state (i.e. the state of the sensors after the planned action will be executed). Sensory information is used both as input and as teaching input for the output units encoding the predicted state of the sensors – the new sensory state is compared with the predicted state and the difference (error) is used to modify the connection weights through back-propagation. As in the case of the Hinton-Nowlan model, modifications due to learning are not transferred back into the genotype.

The experimental results show two results. First, after a few generations, by learning to predict, individuals increase their performance during life not only with respect to their ability to predict but also with respect to their ability to find food. Secondly, the ability to find food increases evolutionarily faster and achieves better results at the end of evolution in the case of learning populations than in the case of control populations in which individuals are not allowed to learn during lifetime. Further analyses demonstrate that the first result can be explained by considering that evolution tends to select individuals that are located in regions of the search space where the learning and evolutionary tasks are dynamically correlated (i.e. where changes due to learning that produce an increase in performance with respect to the learning task also produce positive effects, on the average, with respect to the evolutionary task). The second result can be explained by considering that, since learning tends to channel evolution toward solutions in which the learning task and the evolutionary task are dynamically correlated, learning allows individuals to recover from deleterious mutations (Nolfi, 1999).

Consider for example two individuals, *a* and *b*, which are located in two distant locations in weight space but have the same fitness at birth, i.e. the two locations correspond to the same height on the fitness surface (Figure 18.3). However, individual *a* is located in a region in which the fitness surface and the learning surface are dynamically correlated, i.e. a region in which movements that result in an increase in height with respect to the learning surface also cause, on average, an increase with respect to the fitness surface. Individual *b*, on the other hand, is located in a region in which the two surfaces are not dynamically correlated. If individual *b* moves in weight space it will go up in the learning surface but not necessarily in the fitness surface. Because of learning, the two individuals will move during their lifetime in a direction that improves their learning performance, i.e. in a direction in which their height on the learning surface tends to increase. This implies that individual *a*, which is located in a dynamically correlated region, will end up with a higher fitness than individual *b* and, therefore, will have a better chance to be selected. The final result is that evolution will have a tendency to select progressively individuals which are located in dynamically correlated regions. In other words, learning forces evolution to select individuals which improve their performance with respect to both the learning and the evolutionary task.

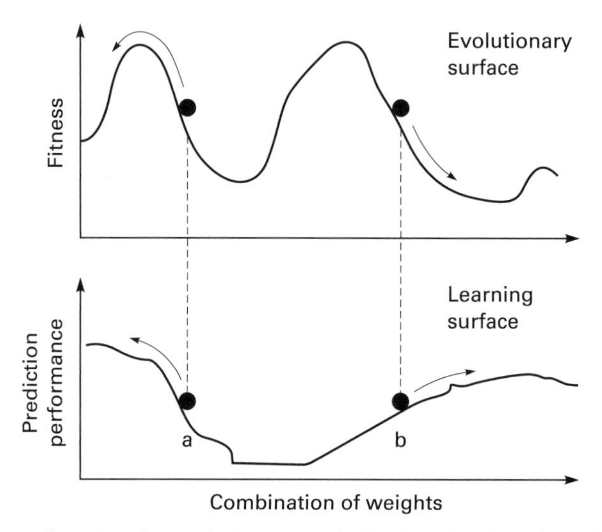

Figure 18.3 Fitness surface for the evolutionary task (finding food) and performance surface for the learning task (sensory prediction) for all possible weight matrices. Movement due to learning is represented as arrows.

18.6.2 Adapting to changing conditions on the fly

As we claimed above, learning might complement evolution by providing a means to master changes that occur too fast to be tracked by the evolutionary process. However, as we will see in this section, the combination of learning and evolution deeply alter both processes. In individuals that evolve and learn, adaptive characteristics emerge as the result of the interaction between evolutionary and lifetime adaptation and cannot be traced back to only one of the two processes.

Nolfi and Parisi (1997) evolved neural controllers for a small mobile robot that was asked to explore an arena of 60 × 20 cm surrounded by walls. The robot was provided with eight infrared sensors that could detect walls up to a distance of about 4 cm and two motors that controlled the two corresponding wheels. The colours of the walls switched from black to white and vice versa each generation. Given that the activity of the infrared sensors is highly affected by the colour of the reflecting surface (white walls reflect more than black walls), to maximize their exploration behaviour evolved robots should modify their behaviour on the fly. In the environment with dark walls, in fact, robots should move very carefully when sensors are activated given that walls are detected only when they are very close. In the environment with white walls, on the contrary, robots should begin to avoid walls only when the sensors are strongly activated in order to explore also the area close to the walls.

Individuals learn during their lifetime by means of self-generated teaching signals. The genotype of the evolving individuals encodes the connection strengths of two neural modules: a teaching module that each time step receives the state of the sensors as input and produce a teaching signal as output, and an action module that receives the state of the sensors as input and produce motor actions as output. The self-generated teaching signal is used to modify the connection strengths of the action module (for a similar architecture, see Ackley and Littman, 1991). This implies that not only the initial behaviour produced by the evolving individuals but also what individuals learn is the result of the evolutionary process and is not determined by the experimenter.

The results of the simulation show that evolved robots display an ability to discriminate the two types of environments and to modify their behaviour accordingly, thus maximizing their exploration capability. An analysis of the obtained results reveals that this ability results from a complex interaction between the evolutionary and the learning process. For example, evolved individuals display an inherited ability to behave so as to enhance the perceived differences between the two environments. This in turns allows the learning process progressively to modify the behaviour of the robots in such a way that they adapt to the different environmental conditions.

More generally, this and other research show that evolution, in the case of individuals that are able to change during life as a result of learning, does not tend to develop directly an ability to solve a problem. Rather, they tend to develop a predisposition to acquire such ability through learning.

Other experiments conducted by co-evolving two competing populations of predator and prey robots (Nolfi and Floreano, 1998) emphasized how lifetime learning might allow evolving individuals to achieve generality, i.e. the ability to produce effective behaviour in a variety of different circumstances. Predators consist of small mobile robots provided with infrared sensors and a linear camera with a view angle of 36° with which they could detect the prey. Prey consist of mobile robots of the same size provided only with infrared sensors but that have a maximum available speed set to twice that of the predators. Each individual is tested against different competitors for 10 trials. Predators are scored with 1 point for each trial in which they were able to catch the prey while prey are scored with 1 point for each trial they were able to escape predators.

What is interesting about this experimental situation is that, since both populations change across generations, predators and prey are facing everchanging and potentially progressively more complex challenges. The authors observe that, in this situation, evolution alone displays severe limitations and progressively more effective solutions can be developed only by allowing evolving individuals to adapt on the fly through a form of lifetime learning. Indeed, any possible fixed strategy is able to master only a limited number of different types of competitors and therefore only by combining evolution and learning was it possible to synthesize individuals able to deal with competitors adopting qualitatively different strategies. Indeed, only by evolving learning individuals was it possible to observe the emergence of predators able to detect the current strategy adopted by the prey and to modify their behaviour accordingly.

18.6.3 Evolving the learning rules

Floreano and Urzelai (2000) conducted a set of experiments in which the genotype of the evolving individuals encoded the learning properties of the neurons of the corresponding neural network (see also Belew *et al.*, 1991). These properties included one of four possible Hebbian learning rules, the learning rate and the sign of all the incoming synapses of the corresponding neuron. When the genotype is decoded into a neural controller, the connection strengths are set to small random values. As reported by the authors, after some generations, the genetically specified configuration of learning rules tends to produce changes in the synaptic strengths that allow individuals to acquire the required competencies through lifetime learning. By comparing the results obtained with this method with a control experiment in which the strength of the synapses were directly encoded into the genotype, the authors observed that evolved controllers able to adapt during lifetime can solve certain tasks faster and better than standard non-adaptive controllers. Moreover, they demonstrated that their method scales up well to large neural architectures.

The authors applied their method in order to evolve neural controllers for mobile robots. Interestingly, the analysis of the synaptic activity of the evolved controllers showed that several synapses did not reach a stable state but keep changing all the time. In particular, synapses continue to change even when the behaviour of the robot became rather stable.

Similar advantages have been reported by Husbands *et al.* (1999) who evolved a type of neural network in which neurons, that were distributed over a 2D surface, emitted 'gases' that diffused through the network and modulated the transfer function of the neurons in a concentration-dependent fashion, thus providing a form of plasticity. Finally, in the experiments performed by Di Paolo (2000) it has been shown how learning can play the role of a homeostatic process whereby evolved neural networks adapt in order to remain stable in the presence of external perturbations.

18.7 Discussion

All changes that occur during the life of an individual organism, especially those that concern the organism's nervous system and the resulting behaviour, are due both to the influence of the information contained in the organism's inherited genotype and to the influence of the individual's specific experience in the specific environment. When the first type of influences (genetic) are prevalent, one talks about maturation. Learning is the term used when the second type of influence (environmental) is predominant. When both are equally important the term most frequently used is development.

Simulations using neural networks of the 'classical' type (Rumelhart and McClelland, 1986) tend to concentrate on learning and to ignore the organisms' genotypes and the evolutionary process at the population level which results in those genotypes. This makes it difficult to study development using 'classical' neural networks. Neural networks viewed in an artificial life perspective (Parisi, in press), on the contrary, are used in simulations in which what is simulated is not only the nervous system of organisms but also their body, environment and genotype. Furthermore, the object of any particular simulation is not a single individual but a population of different individuals which reproduce selectively and evolve across of a succession of generations. In this framework it becomes possible to study neural development, i.e. changes that occur in an individual organism's nervous system (neural network) during the individual's lifetime and that are due to both the individual's inherited (evolved) genotype and the individual's experience in the specific environment.

In the past decade a number of simulations using neural networks have been conducted which have attempted to model how genetic information can be mapped in the organism's phenotype, how genetic information can determine changes in an individual's neural network and therefore in the individual's behaviour, and how genetic information and information from the environment can interact in determining these lifetime changes. In many circumstances this interaction results in better, more rapid, more flexible adaptation of the organisms to the environment in which they happen to live. Therefore, these simulations shed some light on the question why learning has emerged as an evolutionary adaptation.

Of course, like all simulations using neural networks these simulations greatly simplify everything, from the genetic encoding to the genotype-to-phenotype mapping, from the genetically-based changes that occur in an individual's neural network during the individual's life to the role of learning in modulating and directing these changes. These simplifications are all the more regrettable given the steadily increasing corpus of new and detailed knowledge that research in 'real' genetics and developmental biology is accumulating in these years. However, simulations using neural networks in an artificial life perspective can progressively incorporate new, although always highly selected, knowledge generated by the 'real' sciences. In any case they can play a useful role for testing in detailed and controllable ways ideas on general and specific mechanisms underlying neural and behavioural development in organisms.

References

Ackley, D.H. and Littman, M.L. (1991) Interaction between learning and evolution. In Langton, C.G. *et al.* eds. *Proceedings of the Second Conference on Artificial Life*. Addison-Wesley.

Arthur, W. (1984) *Mechanisms of Morphological Evolution*. Wiley.

Belew, R.K., McInerney, J. and Schraudolph, N.N. (1991) Evolving networks: using the genetic algorithm with connectionist learning. In Langton, C.G., Farmer, J.D., Rasmussen, S. and Taylor, C.E. eds. *Artificial Life II*. Addison-Wesley.

Cangelosi, A. (1999) Heterochrony and adaptation in developing neural networks. In Banzhaf, W. *et al.* eds. *Proceedings of GECCO99 Genetic and Evolutionary Computation Conference*. Morgan Kaufmann.

Cangelosi, A. and Elman, J.L. (1995) Gene regulation and biological development in neural networks: an Exploratory Model. *Technical Report*. CRL-UCSD, University in California at San Diego.

Cangelosi, A., Nolfi, S. and Parisi, D. (1994) Cell division and migration in a 'genotype' for neural networks. *Network – Computation in Neural Systems*, **5**, 497–515.

Dellaert, F. and Beer, R.D. (1994) Toward an evolvable model of development for autonomous agent synthesis. In Brooks, R. and Maes, P. eds. *Proceedings of the Fourth Conference on Artificial Life* MIT Press.

Di Paolo, E.A. (2000) Homeostatic adaptation to inversion in the visual field and other sensorimotor disruptions. In Meyer, J.-A., Berthoz, A., Floreano, D. *et al.* eds. *From Animals to Animats 6. Proceedings of the VI International Conference on Simulation of Adaptive Behavior*, MIT Press.

Floreano, D. and Urzelai, J. (2000) Evolutionary robots with on-line self-organization and behavioral fitness. *Neural Networks*, **13**, 431–443.

Gould, S.J. (1977) *Ontogeny and Phylogeny*. Harvard University Press.

Gruau, F. (1994) Automatic definition of modular neural networks. *Adaptive Behavior*, **3**, 151–183.

Hinton, G.E. and Nowlan, S.J. (1987) How learning guides evolution. *Complex Systems*, **1**, 495–502.

Holland, J.J. (1975) *Adaptation in Natural and Artificial Systems*. University of Michigan Press.

Husbands, P., Smith T., Jakobi, N. and O'Schea, M. (1999) Better living through chemistry: evolving gasnets for robot control. *Connection Science*, **3–4**, 185–210.

Husbands, P., Harvey I., Cliff, D. and Miller, G. (1994) The use of genetic algorithms for the development of sensorimotor control systems. In Gaussier, P. and Nicoud, J.-D. eds. *From Perception to Action*. IEEE Press.

Hutchins, E. and Hazelehurst, B. (1991) Learning in the cultural process. In Langton, C., Taylor, C., Farmer, J.D. and Rasmussen, S. eds. *Artificial Life II*. Addison-Wesley.

Kitano, H. (1990) Designing neural networks using genetic algorithms with graph generation system. *Complex Systems*, **4**, 461–476.

Koza, J.R. (1992) *Genetic Programming: On the Programming of Computers by Means of Natural Selection*. MIT Press.

Marx, J.L. (1989) The cell cycle coming under control. *Science*, **245**, 252–255.

Mayley, G. (1997) Landscapes, learning costs, and genetic assimilation. *Evolutionary Computation*, **4**, 213–234.

Miller, G.F., Todd, P.M. and Hedge, S.U. (1989) Designing neural networks using genetic algorithms. In Nadel, L. and Stein, D. eds. *Proceedings Third International Conference on Genetic Algorithms*. Morgan Kaufmann.

McKinney, M.L. and McNamara, K.J. (1991) *Heterochrony: the Evolution of Ontogeny*. Plenum Press.

Nolfi, S. (1999) How learning and evolution interact: the case of a learning task which differs from the evolutionary task. *Adaptive Behavior*, **2**, 231–236.

Nolfi, S. and Floreano, D. (1998) Co-evolving predator and prey robots: Do 'arm races' arise in artificial evolution? *Artificial Life*, **4**, 311–335.

Nolfi, S. and Floreano, D. (1999) Learning and evolution. *Autonomous Robots*, **1**, 89–113.

Nolfi, S. and Parisi, D. (1995) Genotypes for neural networks. In Arbib, M.A. ed. *Handbook of Brain Theory and Neural Networks*. MIT Press.

Nolfi, S. and Parisi, D. (1997) Learning to adapt to changing environments in evolving neural networks. *Adaptive Behavior*, **1**, 75–98.

Nolfi, S., Miglino, O. and Parisi, D. (1994b) Phenotypic plasticity in evolving neural networks. In Gaussier, D.P. and Nicoud, J.-D. eds. *Proceedings of the International Conference From Perception to Action*. IEEE Press.

Nolfi, S., Elman, J.L. and Parisi, D. (1994a) Learning and evolution in neural networks. *Adaptive Behavior*, **4**, 311–335.

Parisi, D. (1996) Computational models of developmental mechanisms. In Gelman, R. and Au, T.K. eds. *Perceptual and Cognitive Development*. Academic Press.

Parisi, D. (1997) Artificial life and higher level cognition. *Brain and Cognition*, **34**, 160–184.

Parisi, D. (in press) Neural networks and artificial life. In Amit, D. and Parisi, G. eds. *Frontiers of Life*. Academic Press.

Parisi, D. and Nolfi, S. (2001) Development in neural networks. In Patel, M., Honovar, V. and Balakrishnan, K. eds. *Advances in the Evolutionary Synthesis of Intelligent Agents*. MIT Press.

Purves, D. (1994) *Neural Activity and the Growth of the Brain*. Cambridge University Press.

Purves, D. and Lichtheim, J.W. (1985) *Principles of Neural Development*. Sinauer Ass.

Rumelhart, D.E. and McClelland, J.L. (1986) *Parallel Distributed Processing. Explorations in the Microstructure of Cognition. Volume I. Foundations*. MIT Press.

Schwefel, H.P. (1995) *Evolution and Optimum Seeking*. Wiley Press.

Wagner, G.P. and Altenberg, L. (1996) Complex adaptations and the evolution of evolvability. *Evolution*, **50**, 967–976.

Yao, X. (1993) A review of evolutionary artificial neural networks. *International Journal of Intelligent Systems*, **4**, 203–222.

Evolving computational neural systems using synthetic developmental mechanisms

19

ALISTAIR G RUST, ROD ADAMS, MARIA SCHILSTRA
and HAMID BOLOURI

19.1 Modelling biological neural development

19.1.1 Why should development be of interest to non-biologists?

Biological development is highly complex, beginning with an egg and resulting in a complete, living organism (Purves and Lichtman, 1985). Development is essentially sequential, establishing a gross structure which becomes progressively more complex over time (Goodwin, 1991). This refinement of structure and function/behaviour operates across many different levels of the biological scale, from molecules to cells to tissues and organs. On each level of scale there is interactive self-organization between the constituent elements (Goodwin, 1996). Neural development is an example of these processes which leads to the development of a nervous system and associated functions. For an abstract view of neural development see Figure 19.1.

Modelling the nervous system as computational artificial neural networks has long been the source of interest to engineers and computer scientists (McCulloch and Pitts, 1943). Mathematical models have been widely used to model the functions and behaviour of the nervous system and, from an engineering perspective, to make use of the powerful, parallel nature of the nervous system to solve complex problems. However, few models have explored the potential of using neural development to automate the process of designing artificial networks, bypassing the need for hand-coding the initial architectures. But why should biological development be of interest to computer scientists and engineers who design artificial neural networks/systems? From a systems perspective, biological neural development encompasses many powerful ideas including:

Variation. Development automatically produces a huge variety of neural systems, using a hierarchy of compactly encoded programs. Dynamic, self-organizing interactions between developmental mechanisms result in more complex systems than through single mechanisms alone (Goodwin, 1996). In other words, the sum is greater than the whole.

Adaptation. Different neural systems, tailored to specific tasks, can be crafted by adjusting the genes that modulate development.

Regulation. Complex developmental programs, involving multiple interactions and relations, may be controlled by relatively small numbers of genes (Raff, 1996; Davidson, 2001). System development can hence be effectively controlled by a minimal set of input parameters.

 ISBN 0-12-428765-4

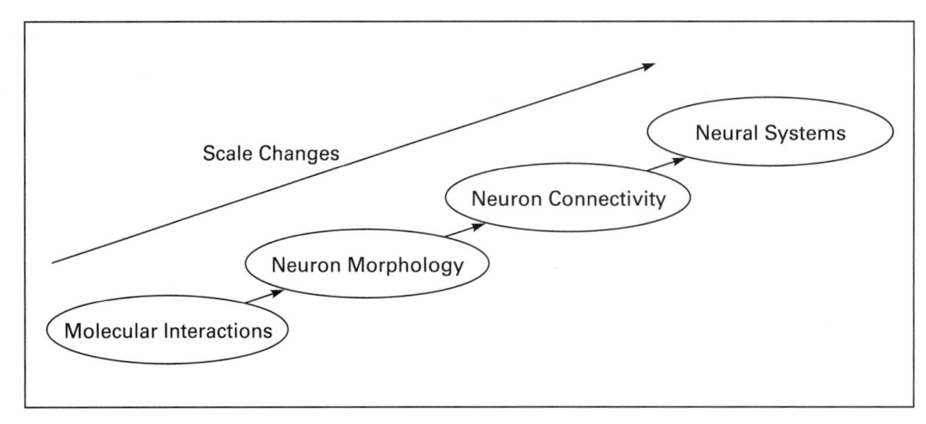

Figure 19.1 Developmental self-organization across scales of neural structure. Molecular interactions include the formation of chemical gradients within the embryo which guide neurons to their positions (Goodwin, 1991). Morphological characteristics of neurons are the result of interactions between cell lineage and the developmental environment, determining which growth programs a neuron follows (Purves and Lichtman, 1985). Neuron connectivity encompasses many complex mechanisms including cell migration, axon extension and connection formation (Purves and Lichtman, 1985; Hall, 1992). The complex interplay of the preceding mechanisms then give rise to the interconnected neural modules and their associated functions on a systems level (Hilgetag *et al.*, 1996).

Modularity is an intrinsic feature of development to create repeated structures and modules (Dawkins, 1989).

Robustness. Developmental programs are robust to variations in the developmental environment. This enables complex systems to be created without the need for an exact encoding or 'blueprinting' of the final system.

The assertion made in this chapter, is that to harness fully the potential of synthetic neural systems, more cues should be sought from the mechanisms of neural development. We are aiming to create a generic set of synthetic developmental mechanisms, capable of creating neural systems with a wide variety of different architectures and functionalities and not limited to stereotypical features. The long-term aim of the work is to develop computational neural systems that are more dynamic and adaptive.

19.2 Synthetic neuroscience

19.2.1 The appropriate level of modelling

Choosing the appropriate level at which to simulate neural systems is an open question and is principally determined by the problem under investigation. Broadly speaking, the computational simulation of neural systems is approached in two different ways: computational neuroscience and artificial life.

Computational neuroscience

The goal of computational neuroscience is to understand the fundamental processes of brain function through computer simulations. Biological neurosystems provide the data against which the modelled systems are measured for their accuracy and correctness. Modelling predominantly focuses on particular facets of biological systems, such as specific chemical and molecular interactions or the effects of neuron physiologies.

The field of computational neuroscience demonstrates that by reducing the levels of complexity found in biology to mathematical and computational models (Bush and Sejnowski, 1993), biologically useful insights can be achieved (Bower, 1998). For example, detailed compartmental membrane models of neurons can be accurately simulated by software packages such as GENESIS (Bower and Beeman, 1995) and NEURON (Hines and Carnevale, 1997).

Artificial life (ALife)

In ALife, biological neural systems are simulated as abstract models, to explore self-organization or to solve a specific application, such as an engineering problem. Unlike the computational neuroscience approach, biological plausibility is less of an issue. Abstract algorithms are used to capture a sufficient level of detail whereby the problem can be addressed. Included in this approach to simulation are parts of the disciplines of Artificial Intelligence (AI) and Artificial Life (ALife).

ALife models have shown much promise in the simulation of biological systems (Langton, 1989). Researchers in evolutionary robotics (Harvey *et al.*, 1993; Nolfi *et al.*, 1994) for example, have been using artificial developmental models to create artificial neural controllers for mobile robots (Dellaert and Beer, 1994; Kodjabachian and Meyer, 1995; Parisi and Nolfi, 1998). Development is simulated as abstract processes where the biological plausibility of models varies considerably.

19.2.2 Previous and complementary work

A number of different approaches have been reported by way of simulating neural development in order to create neuron morphologies and networks. Fleischer and Barr developed an extensive simulator which incorporated many self-organizing biophysical phenomena (Fleischer and Barr, 1994; Fleischer, 1995). The simulator represented a powerful tool with the potential to develop highly plausible biological structures, neuron morphology included. Attempts were made at tailoring the neural morphology but these proved to be non-trivial and inconclusive due to the large parameter search space involved (Fleischer, 1995).

Modelling of neural morphology using re-writing rules (in particular L-systems (Lindenmayer, 1968)) has been explored by a number of groups (McCormick and Mulchandani, 1994; Burton *et al.*, 1999; Ascoli, 1999). Ascoli (1999), in particular, has been using rules whose grammar is derived from observations of developing biological neural morphology (e.g. typical branching angles and rate of diameter reduction at branch points). The neuron models do not, however, allow interactions between a growing neuron and its environment. The development of a neuron and hence its visual form are dependent on the effectiveness and implementation of the hand-crafted, re-writing rules.

Other approaches have been reported by Nolfi and Parisi (1991), Dellaert and Beer (1994), Kodjabachian and Meyer (1994), Astor and Adami (2000). While these approaches have provided valuable insights they are characterized, among other things, by producing artificial neural networks which have stereotyped architectures.

In order to approach the computational complexity of biological neural systems, we argue some intermediate position between the computational neuroscience and artificial life approaches is necessary.

19.2.3 An overview of the chapter

Our work to date is briefly reviewed in this chapter. It begins by presenting the developmental mechanisms that have been modelled in software and illustrating the range of neural morphologies which may be generated. To examine the adaptability of the model we have combined it with an evolutionary algorithm to evolve neurons and networks with specific connectivity and functionality. Using this computational framework we have also investigated the feasibility of incorporating computational models of neural activity into the developmental model. A number of experiments are summarized before future directions are presented.

19.3 A modelling framework for neural development

19.3.1 Implemented developmental rules

We have been implementing a 3-dimensional (3D) simulation of biological development, in which neuron-to-neuron connectivity and neuron morphology is created through interactive self-organization (Rust, 1998). The goal of the modelling process has been to encapsulate key mechanisms of neural development, using a minimal, compact set of mathematical descriptions.

The key phase of neural development which is modelled is that of neuron outgrowth. Briefly to summarize the key facets of neural systems, neurons consist of three key structural elements: (a) a soma, or the cell body through which electrical pulse signals are collected and processed; (b) dendrites, which can form large tree-like structures and collect inputs from neighbouring neurons and (c) axons, a single growth from a neuron which acts effectively as the output from the soma, transmitting information onwards to other neurons. Axons from a single neuron can branch and may make widespread connections with the dendritic trees of its neighbours. Connections are made at specialized junctions called synapses, where the transmission of the electrical activity is mediated by a multitude of chemical feedback mechanisms.

The processes through which axons and dendrites branch to form connections, thereby creating the structure of individual neurons and networks, has been one focus of our modelling. Development is modelled in a number of overlapping stages, which govern how neurons extend axons and dendrites (collectively termed neurites). Neurites grow within an artificial, embryonic environment, into which they emit local chemical gradients of neurotrophins. A growth phase begins with the placement of neurons within the developmental environment. Neurons can be placed in arbitrary positions or within planes from which layered networks can develop. Growth occurs in discrete time steps, where each active neurite tip evaluates the gradients within its local environment before performing an action. An extensive set of synthetic neural developmental rules have been implemented. A detailed set of the implemented mathematical models are included in an appendix at the end of this chapter. These rules can be grouped into classes or phases of development.

Neurite outgrowth rules

These rules enable neurites to extend and explore their local environments within the simulation. Rules encode how the growing tips branch in response to sensed gradient conditions. Growth is based on a simple attraction/repulsion model. The gradients of chemicals produced by dendrites attract axons and vice versa. At the same time, dendrites repel other dendrites and axons repel other axons.

Two principal mechanisms of neurite branching have been implemented. Intrinsic branches are ones that occur at genetically determined times, but where the directions taken post-branching can be mediated by the local chemical gradients. Interactive splitting of a neurite is induced solely by the local environment, for example, a neurite may branch in response to chemical sources being emitted at right-angles to its current direction of growth.

Spontaneous neural activity rules

Once connections have formed between neurons, phases of spontaneous neural activity are used to regulate the growth rate in subsequent time steps. These particular rules are inspired by mechanisms observed during development of the mammalian retina and visual system (Shatz, 1994).

Pruning rules

Once growth has been completed, extraneous connections and neurons are removed based on the cumulative effects of the interspersed activity phases. Overgrowth and the creation of too many initial synpases is a common phenomenon in development, representing a built in mechanism for error correction through the removal of redundant neural circuitry (Hall, 1992; Davies, 1994). The model seeks to mimic such characteristics.

While the developmental rules have been subdivided into different phases, it should be stressed that the mapping from the specification of the rules (genotype) to the resulting neurons/networks (phenotype) is a complex, non-linear process due to self-organizing interactions between the rules.

19.3.2 Examples of diffusion gradients and simple growth

The effects of chemical diffusion within the synthetic developmental environment are illustrated in Figure 19.2 by the growth of a simple 3-neuron network. For ease of display, the network was grown using a 2D rather than a 3D model to obtain the plot in Figure 19.2(a). In 3D, diffusing chemicals form spherical gradients around growing neurites and neuron somas. Figure 19.2(a) illustrates both how the diffusion of emitted chemicals decays as a result of a diffusion law ($1/\text{distance}^n$) and that diffusion is restricted to a particular area. The diffusion of the sources of

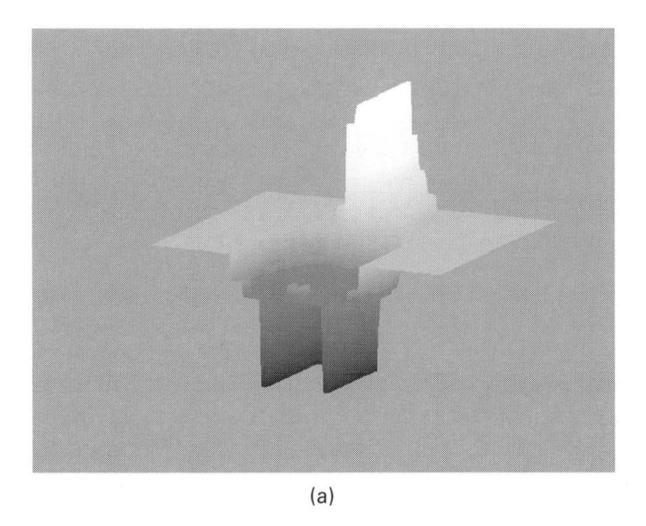

(a)

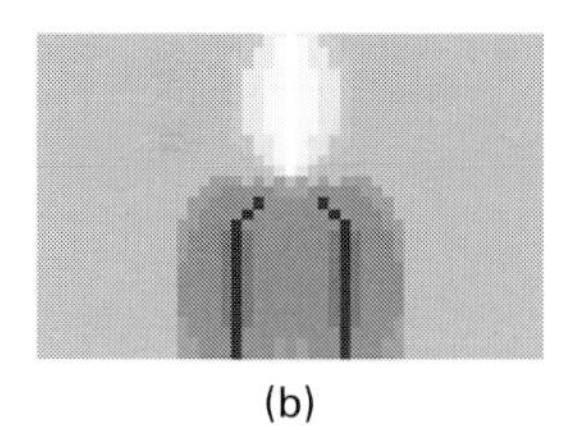

(b)

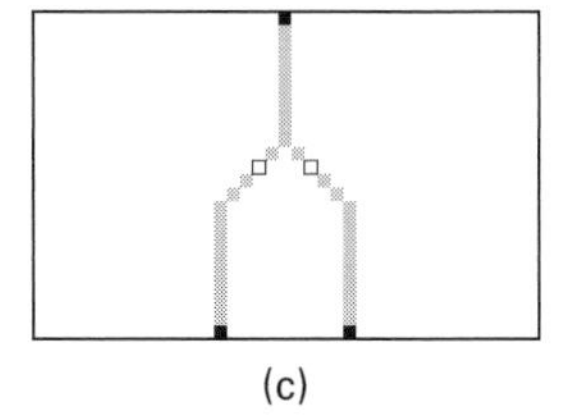

(c)

Figure 19.2 Chemical diffusion within the developmental environment resulting from the development of a 3-neuron network. (a) A 3D representation of the resultant chemical gradient produced by the growing neurites. (b) A plan view of the gradients. The highest, positive gradients are shown as white while the largest negative gradients are shown as black. (c) The resultant network structure with gradients removed. Neuron bodies are depicted as black squares, synapses as white squares, and axons and dendrites as grey lines.

chemicals is illustrated by the peaks and troughs of Figure 19.2(a). The diffusing sources interact destructively within the immediate locality of the synapses, resulting in gradient levels approaching the same level as the background chemical level.

The concept of neurites growing incrementally with time is illustrated with the sequence of snapshots in Figure 19.3. (The somas of neurons are represented by spheres while synapses are diamond shaped, unless otherwise stated.) Branching occurs as the result of a number of intrinsic splits and this sets the initial gross morphologies of the neurons. Up to Figure 19.3(d) the branches extend widely into the surrounding environment. After this point in developmental time, the tips of the respective neurites become attracted to each other and grow together, due to the effects of the invisible chemical gradients.

Synapses have formed in Figure 19.3(e) and more form before outgrowth is complete in (f). Figure 19.3(f) also illustrates how extraneous neurites develop, which will necessitate pruning once growth is completed.

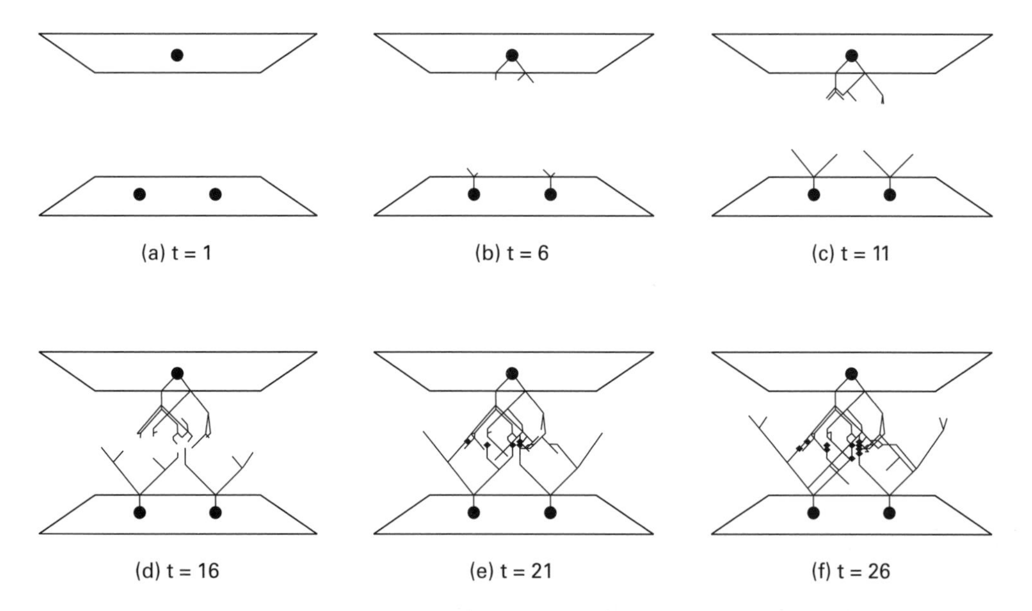

Figure 19.3 Snapshots of neural outgrowth of a two layer network at various stages of developmental time (t).

19.3.3 Neuron and network morphologies

The developmental rules are controlled by parameters, in the same way as gene expression levels can be thought of as parameters for biological development. A wide variety of neuron and network morphologies can be achieved by varying these parameters, some examples of which are illustrated in Figure 19.4. The creation of individual neuron morphologies or networks does not, however, have to be the result of a single phase of growth. Figure 19.4(f) was created by growing the initial parallel fibres in the lower plane of the figure. These parallel fibres then acted as static gradient sources which attracted the descending dendrites to create the final complex structure.

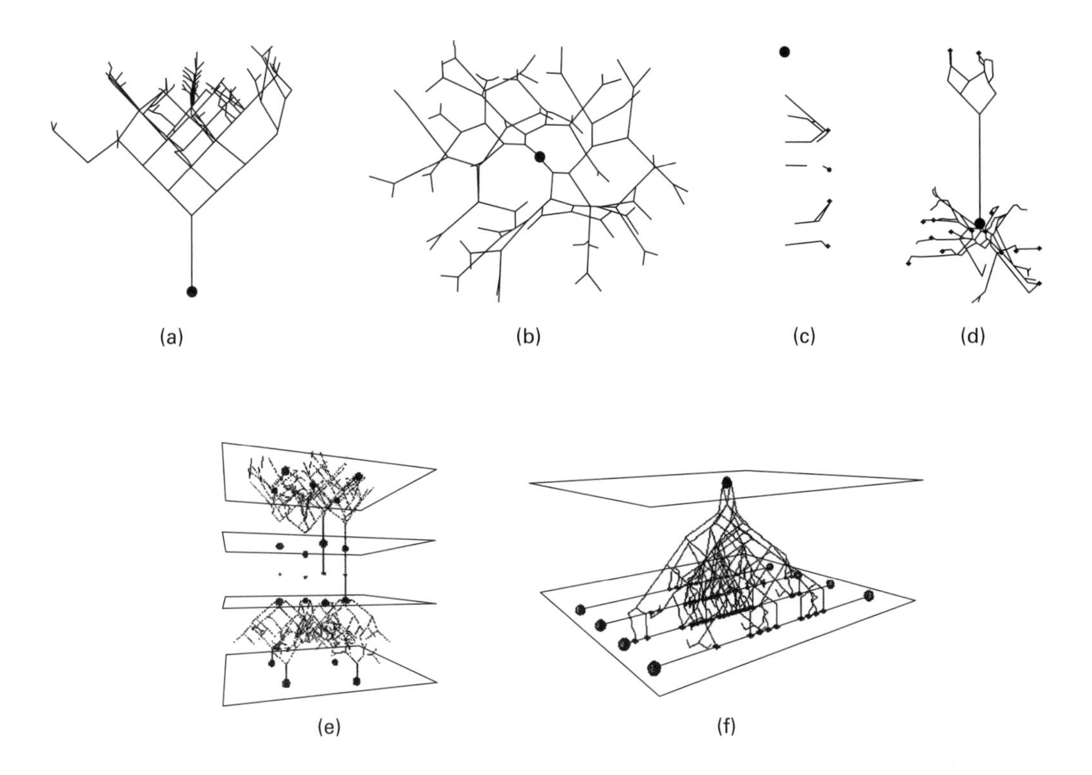

Figure 19.4 Examples of individual neuron and network morphologies. (a) Dendritic tree. (b) Widely-branching dendritic arbour. (c) Axon-like process. (d) Dendritic arbour of a pyramidal neuron. (e) A 4-layer network architecture. (f) A novel neuron architecture where the morphology of the top-most neuron has been shaped by the previously grown parallel fibres.

19.4 Evolutionary neural development

19.4.1 Optimizing the developmental parameters

Given the wide range of potential morphologies which the simulator can develop, the next step was to evaluate whether it was possible to direct the model to grow specific target morphologies with possible desired functionality. Essentially the capability of the model to create varieties of structure derives from the way in which the developmental rules are governed by changes in parameter values. Therefore, the task of using the developmental simulator to achieve particular architectures or functionality becomes equivalent to searching for optimal sets of developmental parameters.

The optimization process involves searching a multidimensional space of all developmental parameters. The size of the search space is determined by the number of parameters under investigation and the range of values that these parameters can take. Within this space there is a hoped for region, or possibly regions where, for a given task, a set of suitable parameter values may be found. For one particular application this will be one particular region, while for another application it may be another different region. If the number of parameters is large and their potential ranges great, then the potential search space is large. The larger and more complex the search space, the longer an optimization method will run for.

For the developmental model, although the set of parameters does not explicitly set the sequence of the developmental rules, a number of parameters can determine whether a rule is activated and the frequency with which it is used. For example, if the parameter controlling the strength of localized gradients is large, then this may encourage a greater frequency of neurite branching. Hence, although the primary focus is on searching for optimal rule parameters for a particular task, this also indirectly affects the expression of the sequence of rules. The genetic algorithm (GA) (Holland, 1975) was chosen over other optimization techniques (such as simulated annealing, hill climbing and conjugate gradient) as the tool with which to search the developmental parameter space. GAs are thought to offer the best results when search spaces are large and real-valued (Mitchell, 1996). The developmental search space can also be extremely rugged, containing many discontinuities. Other optimization techniques are less likely to be able to traverse such spaces. The GA is incorporated into the developmental process as described and illustrated in Figure 19.5.

An off-the-shelf GA implementation, GENESIS (Grefenstette, 1990), was used in the evolutionary experiments. By current standards GENESIS is a simple implementation of the GA but it does include the principal evolutionary mechanisms of crossover and mutation. We specifically chose a simplistic GA to verify that it was the in-built capabilities of the developmental model which would lead to desired architectures and that it was not due to a critical reliance on the optimization model. In essence, the GA was simply being used to fine-tune the self-organizing mechanisms.

Simulations to verify adaptability of the developmental model have been undertaken incrementally. Initial experiments verified that the developmental model could in fact be adapted to satisfy a particular application or function. Studies in these cases were to identify architectures and desired functionality in networks of neurons, where the functionality was similar to classical, computer-science based artificial neural networks. A second set of experiments was then undertaken which were inspired by the computational models of real-world neural activity.

19.4.2 Evolution of network architectures

As illustrated in Figure 19.5, the initial application to which the developmental model was tailored, was the evolution of synthetic artificial mammalian retinas. The chosen task was the evolution of neural architectures to mimic the function of edge-detection. These experiments verified the relationship between function and form, or connectivity in this case. Namely, that if the underlying connectivity of the evolved networks was correct then their function would be correct. Incorrect connectivity leads to incorrect functionality.

In keeping with the incremental strategy of experimentation, initial simulations were tailored to evolve specific connectivity in a 3-layer neural network system (Rust *et al.*, 1998; Rust, 1998). If the correct connectivity was achieved, then by mapping functionality onto the neurons in the network, the desired edge-detecting functionality was obtained. Similar experiments were also performed where the evolved networks were assessed only on a mathematical model of a desired edge-detection response, without any direct analysis of the connectivity of the network (Bolouri *et al.*, 1998). In both sets of experiments sets of optimal developmental parameters were identified which were able to create networks that performed edge-detection.

One facet of the developmental model which was explored during the network experiments was the robustness of the rules to noise, or to the perturbation of the initial positions of neurons within the network. If the rules were incapable of adapting to positional variation, then this would indicate that the model was brittle and susceptible to small changes in conditions. If connectivity and functionality could still be retained in the presence of noise, then the model

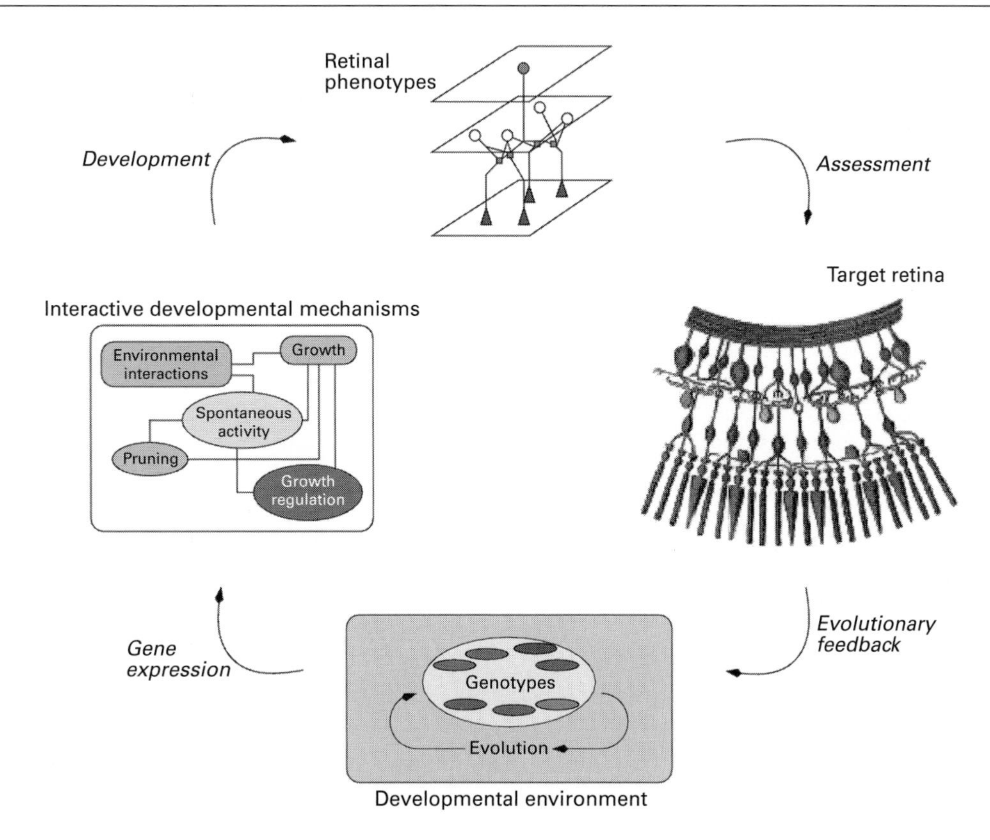

Figure 19.5 An overview of a typical evolutionary development scenario. A population of genotypes, or sets of different developmental parameters, specifying the creation of networks or neurons, exists within an artificial developmental environment. Under the control of the developmental rules, the genotypes are decoded to form 3D neuron architectures. Each developed network, or phenotype, is then measured against a target, possibly based on a specific architecture or a desired functionality. In this example, the target architecture is a model inspired by edge-detection circuitry within the mammalian retina. This assessment of the performance of the developmental program provides feedback to the evolutionary process which subsequently modifies the population of genotypes. Those genotypes, or sets of developmental parameters, that produce networks with the best performances are said to be the fittest and are retained. This subset of genotypes is then bred using algorithms akin to crossover and mutation, to form a new population of prospective genotypes. The evolutionary cycle is then repeated until the evolved phenotypes hopefully meet the desired target, in this case the target retina architecture.

would demonstrate inherent robustness, a desired feature. A summary of the investigation into the effects of environmental perturbations on the developmental model is presented in Figure 19.6 and explained below.

In the absence of any initial neuron placement noise, developmental rules which only encoded genetically-determined branching (referred to previously as intrinsic branching rules) can result in retina structures whose functional response matches the desired response (Figures 19.6(e) and (b)). However, if the positions of neurons are perturbed, the best parameter set for the branching only rules fails to produce adequate functionality, as illustrated in Figure 19.6(f).

The developmental program was made more robust to noise by incorporating rules that permit growing neurons to produce extra branches via interactions with the local developmental environment. The addition of interactive branching rules results in a significant improvement in

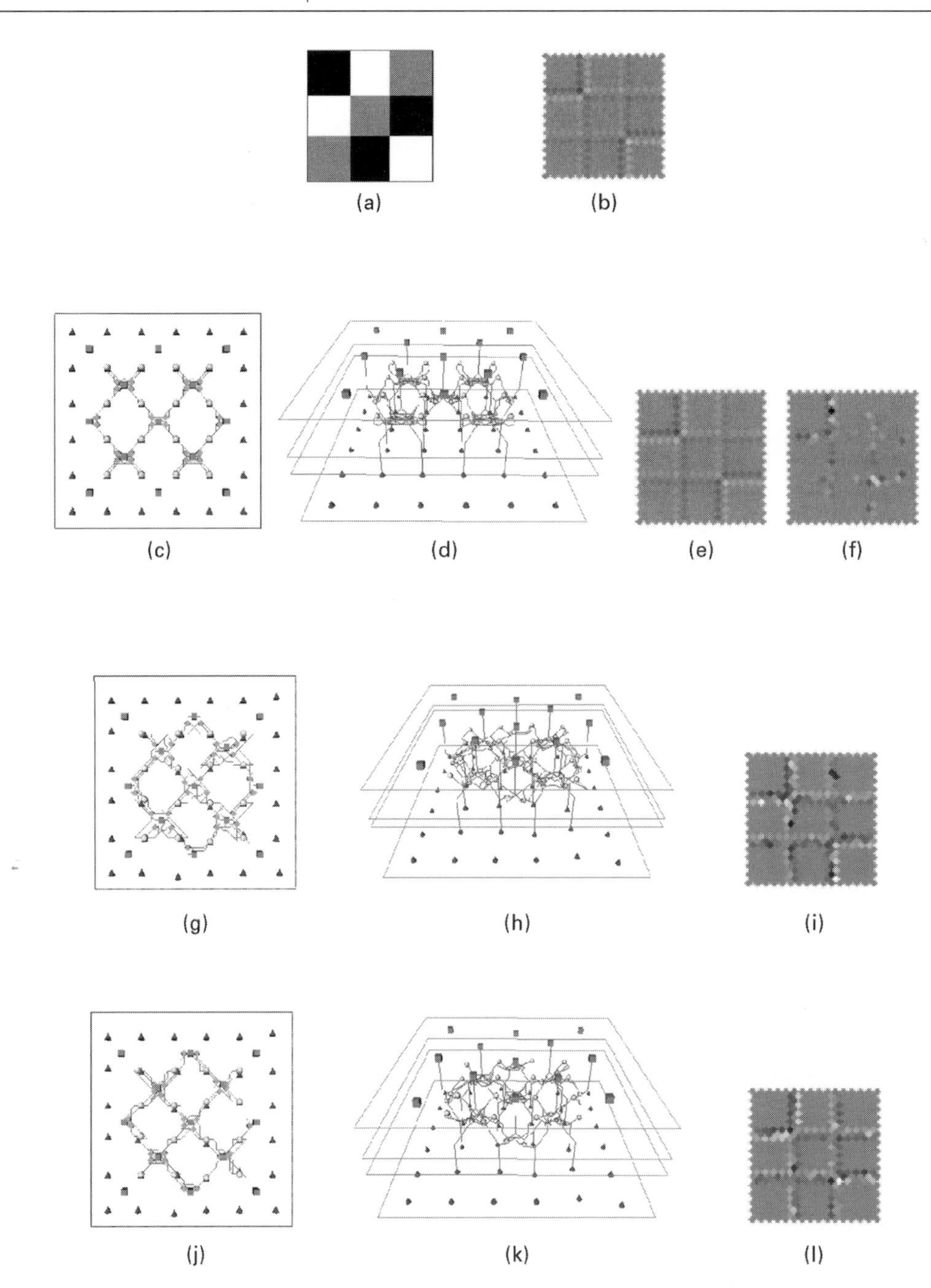

Figure 19.6 Architectures and functionality of evolved retinas. (a) Input image. (b) Target output image showing edge-detection. (c) and (d) are side and top views of a portion of the best evolved network upon unperturbed, symmetric neuron placement using intrinsic branching rules only. (e) is the output that is produced. However, (f) shows the output of a grown retina using the same developmental parameter values but where the initial positions of the neurons are perturbed. (g) and (h) are the side and top views of a portion of the best evolved network using perturbed neuron placement and interactive branching rules. (i) Output image from the best evolved network. (j) and (k) are the side and top views of a portion of the best evolved network using perturbed neuron placement, interactive branching and pruning rules. (l) Output image from the best evolved network.

structure and hence functionality over the intrinsic only branching rules (Figures 19.6(g) to (i)). The target response is far more distinguishable in Figure 19.6(i) compared to Figure 19.6(f). However, completely black and white pixels in the output image indicate that neurons are being saturated due to over-connectivity. Performance is further improved by incorporating pruning rules with the branching rules, to modify the effects of overgrowth. With the addition of the pruning parameters, the effects of noise on the functionality of the retina are reduced, as illustrated in Figure 19.6(l).

In these experiments evolution was carried out in stages, incorporating parameter values from previous stages and evolving them alongside new parameters. For example, the intrinsic branching rule parameters are evolved in the first stage and the values obtained are used to seed the next stage of development, including the interactive branching rules. In this second stage, both the seeded branching rules and the newly incorporated interactive branching rules are co-evolved. The evolved values of these developmental parameters are then used to seed the final stage which included the pruning rules.

The search space of developmental parameters is thus progressively enlarged in an orderly manner, permitting the evolutionary process to explore incrementally and sequentially. This contrasts to other models where all the developmental rules are co-evolved simultaneously. Presenting such large, global search spaces can cause evolution to stall in the early generations (Dellaert and Beer, 1994; Nolfi, 1997).

If the potential search space is large then all the networks in a population may have the same, low fitness value. Evolution is therefore given no clear trajectory along which to progress. Splitting the evolutionary development process into phases, we feel is more easily generalized and more biologically plausible.

19.4.3 Evolution of multi-compartmental neurons

Single biological neurons, let alone networks and systems, exhibit highly dynamic, adaptive computational capabilities (Koch and Segev, 1998). One of the key determinants of these capabilities is the relationship between a neuron's morphology and its function (Mainen and Sejnowski, 1996). The form or geometry of a neuron directly influences how action potentials are propagated along its axons and dendrites. Accurate and detailed simulations of neural signalling is undertaken within the domain of computational neuroscience using mathematical, compartmental models. Compartmental models of neurons mimic the transmission of neural activity pulses by modelling the ion channels within neurites, typically as sets of ordinary differential equations (ODEs). In electrical engineering terms, these models can be idealized as cable models of electrical transmission. Each compartment of the neuron has associated miniature circuits of resistance and capacitance, which are chained together to form cables. Parameters assigned to the components of each compartment determine how activity signals, or spike trains, are both generated and transmitted.

One computational neuroscience model which explicitly sought to examine the relationship between function and form was created by Mainen and Sejnowski (Mainen and Sejnowski, 1996). Using an ODE model and mapping it onto different morphologies of real neurons which had been extracted and measured, they showed that different neural activity spike train patterns would be obtained depending on the specific neuron morphology.

Inspired by Mainen and Sejnowski (1996) we investigated the relationship between synthetic neural morphology and functionality using multi-compartmental neuron models (Rust and Adams, 1999). The process involved growing individual neurons, mapping Mainen and Sejnowski's compartmental model ODE into the neurons and then evaluating the resulting response to an induced activity pulse. In this scenario the fitness of evolved neurons was determined by analysing the frequency and spacing of generated spikes against those pulse trains measured in biological neurons. Using this approach we have used the developmental simulator to evolve neurons with specific spike trains, as illustrated in Figure 19.7.

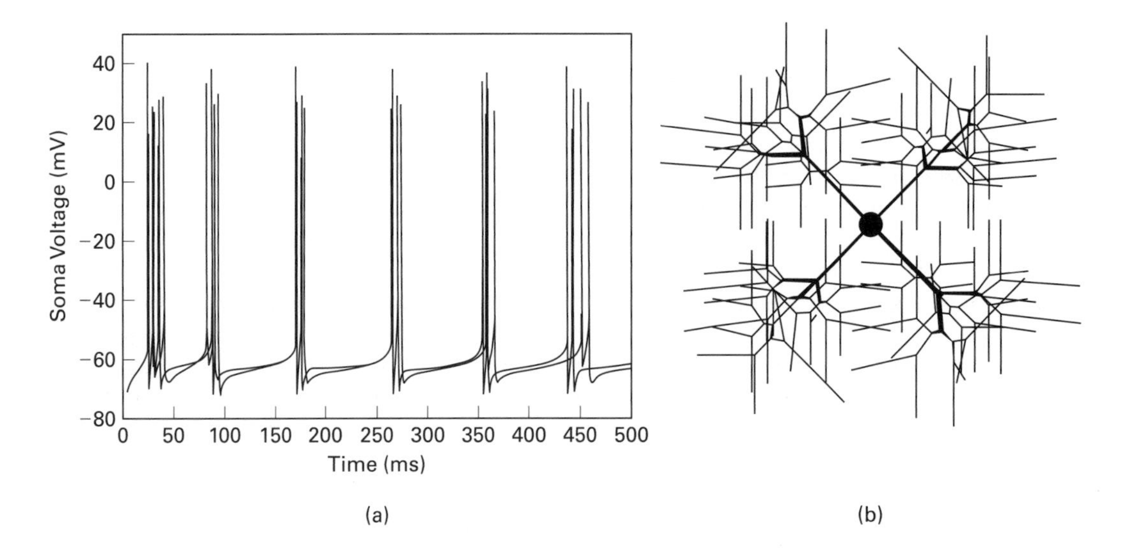

Figure 19.7 Evolution of activity spike train patterns in single compartmental neurons. (a) Output spike trains: target spike train and evolved spike train are virtually on top of each other. (b) Morphology of one of the best evolved neurons.

Computational neuroscience models, however, tend to be computationally intensive due to the need to solve large sets of ODEs. This computational bottleneck therefore permits only a handful of neurons to be simulated at any one time, reducing the potential to study networks of spatially extended neurons. Evolving networks of synthetic, compartmental neurons becomes unfeasible owing to the number of evaluations which must be performed with each evolutionary cycle. Therefore, we have been developing a simplified neural compartmental model which is based upon a relatively simple finite state automaton (FSA) model (Schilstra *et al.*, 2002). The FSA is behaviourally equivalent to current ODE models, where the validity of such models to retain the characteristics of more complete, compartmental models has been demonstrated by Gerhardt *et al.* (1990a, b). Effectively the FSA is able to reduce a multiequation ODE to a more simplified and computationally faster algorithm. Figure 19.8 illustrates the FSA mapped onto different neuronal morphologies. With such a model it is hoped eventually to permit faster evaluation of 3D neuronal networks.

19.5 Discussion

As stated previously there is no single appropriate way to model neural development. Ultimately it is the application for which the modelling is required that will dictate the exact implementation details, e.g. an engineering application versus the understanding of a biological phenomenon. In this chapter we have presented an engineering approach to the modelling of neural

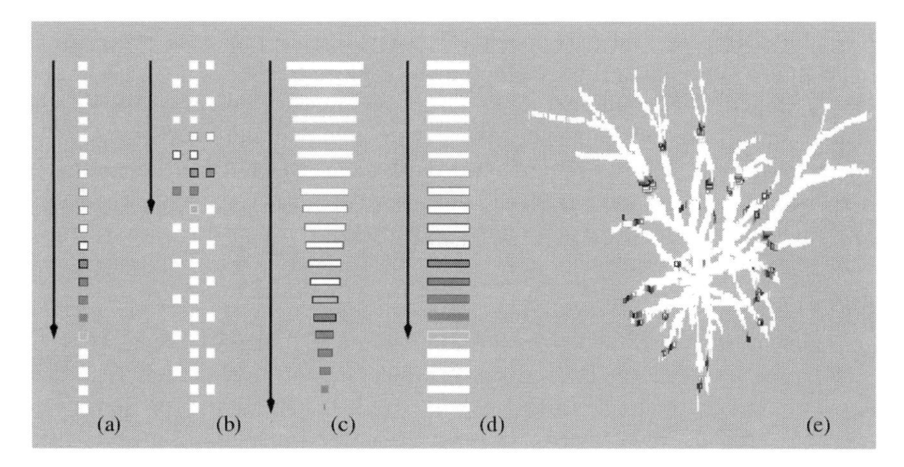

Figure 19.8 An FSA model of neural activity mapped onto differing neural morphologies. The relative velocities of the neural activity waves are determined by the branching pattern and the degree of tapering of the branches. Structures a, b, c, and d are 'neurites' that have received a short (1 time step) excitatory pulse into their top compartments. All structures were updated 50 times and the resulting excitation wave moved from the top of the figure downwards. The length of all compartments is 1.0, their diameters 1.0 (a, b), 5.0 (d) or vary from 10.0 at the top to 0.5 at the bottom compartment (c). The darker the fill of a compartment, the higher the level of excitation. The darker its outline, the higher the recovery level. In the uniform structures a and d, the front of the excitation wave is at compartment 16, showing that the width of the neurite has no effect on wave velocity. However, in the highly branched structure b, wave velocity is significantly lower (the wave front is at compartment 9), whereas the wave has moved faster through the tapering structure (wave front at compartment 20). (e) Shows an excitation wave, back-propagating from the soma, along the dendrites of a layer 2/3 neocortical pyramidal neuron modelled with our FSA (compartmentalization as in Mainen and Sejnowski (1996)). The snapshot was taken 300 updates after a brief excitatory pulse was applied to the dendrite compartments directly adjacent to the soma (at the centre of the image).

development. It aimed at capturing the key mechanisms of neural outgrowth and was shown to be capable of creating a wide range of neuron and network morphologies. Reviewing the model in these terms highlights the framework within which the modelling approach sits and points towards further areas of research, across different levels of system modelling.

Gene interaction networks

From a simplistic viewpoint, the developmental rules may be considered as genes whose effects are modulated by the developmental parameters. Increasing a modelling parameter can have the knock-on effect of activating a growth rule, which in turn may cause a cascade of other interactions. This is hence a very simplified genetic regulatory network. It is thought that certain neural characteristics, such as branching patterns, may be controlled by single genes. Therefore, with more detailed analysis of potential candidate scenarios, the implemented developmental environment could be adapted into a testbed for simplistic gene interaction networks, where neuron morphology is the observed behaviour.

Neuron functionality

The morphology-evolving experiments of the compartmental neurons only investigated the adaption of morphology with a static ODE model. Adapting both the neuron and ODE simultaneously would enable the interrelationship between the intraneurite processes of neural signalling and that of gross neural physiology to be explored.

The optimization of ODE parameters has been previously attempted but only for single neurons (Vanier and Bower, 1999; Eichler-West *et al.*, 1998). Vanier and Bower (1999) argue that parameter optimization need not be limited to single neurons but acknowledge that to optimize networks of neurons, the complexity of the model must be reduced. The FSA model of neural activity hence provides a more tractable entry point.

Network functionality

Combining 3-dimensional morphology from the simulator and the FSA model will enable more neural models to be investigated, beyond simple 'integrate-and-five' models. The reduction in computational complexity of the FSA model will permit the simulation of sparsely connected networks of neurons. This leads to areas of potential interest including active dendrite models and the effects of the spatial arrangement of synapses on the firing patterns of neurons (Mel, 1994).

Neuron libraries

Evolving classes of neurons with specific geometries and functionality will allow such classes to be simply called into the implementation process of neural systems. The classes can provide basic neuron primitives with which to instigate synthetic system designs. Developmental self-organization within the implemented rules will enable the basic classes of neurons to adapt and fine tune themselves to their environment. This provides a stepping-off point to investigate the evolution of modular neural systems where the morphology and connectivity of neurons within separate modules can be different.

References

Ascoli, G.A. (1999) Progress and perspectives in computational anatomy. *Anatomical Record,* **257**(6), 195–207.

Astor, J.C. and Adami. (2000) A developmental model for the evolution of artificial neural networks. *Artificial Life*, **6**, 189–218.

Bolouri, H., Adams, R., George, S. and Rust, A.G. (1998) Molecular self-organisation in a developmental model for the evolution of large-scale neural networks. In Usui, S. and Omori, T. eds. *Proceedings of the International Conference on Neural Information Processing and Intelligent Systems (ICONIP'98)*, **II**, 797–800.

Bower, J. and Beeman, D. (1995) *The Book of GENESIS: Exploring Realistic Neural Models with the GEnesis NEural SImulation System*. Springer-Verlag.

Bower, J. ed. (1998) *Computational Neuroscience: Trends in Research, 1998*. Kluwer Academic/Plenum Publishers.

Burton, B.P., Chow, T.S., Duchowski, A.T. *et al.* (1999) Exploring the brain forest. *Neurocomputing*, **26–27**, 971–980.

Bush, P.C. and Sejnowski, T.J. (1993) Reduced compartmental models of neocortical pyramidal cells. *Journal of Neuroscience Methods*, **46**, 159–166.

Davidson, E.H. (2001) *Genomic Regulatory Systems: Development and Evolution*. Academic Press Ltd.

Davies. A.M. (1994) Intrinsic programs of growth and survival in developing verebrate neurons. *Trends in Neurosciences*, **17**(5), 195–198.

Dawkins, R. (1989) The evolution of evolvability. In Langton, C.G. ed. *Artificial Life: The Proceedings of an Interdisciplinary Workshop on the Synthesis and Simulation of Living Systems*. Addison-Wesley.

Dellaert, F. and Beer, R. D. (1994) Toward an evolvable model of development for autonomous agent synthesis. In Brooks, R. A. and Maes, P. eds. *Artificial Life IV: Proceedings of the Fourth International Workshop on the Synthesis and Simulation of Living Systems,* pp. 246–258. MIT Press.

Eichler-West, R.M., De Schutter, E. and Wilcox, G.L. (1998) Using evolutionary algorithms to search for control parameters in a nonlinear partial differential equation. In Davis, L., De Jong, K., Vose M. and Whitley, L.D. eds. *Institute for Mathematics and its Applications Volume on Evolutionary Algorithms and High Performance Computing*. Springer-Verlag.

Fleischer, K. (1995) *A Multiple-Mechanism Developmental Model Defining Self-Organizing Geometric Structures*. PhD dissertation, California Institute of Technology.

Fleischer, K. and Barr, A. H. (1994) A simulation testbed for the study of multicellular development: the multiple mechanisms of morphogenesis. In Langton, C.G. ed. *Proceedings of Artificial Life III*, pp. 389–416. Addison-Wesley.

Gerhardt, M., Schuster, H. and Tyson, J.J. (1990a) A cellular automaton model of excitable media. II, curvature, dispersion, rotating and meandering waves. *Physica D*, **46**, 392–415.

Gerhardt, M., Schuster, H. and Tyson, J.J. (1990b) A cellular automaton model of excitable media including curvature and dispersion. *Science*, **247**, 1563–1566.

Goodwin, B. (1996) *How the Leopard Changed Its Spots: The Evolution of Complexity*. Simon and Schuster.

Goodwin, B.C. (1991) *Development*. Hodder and Stoughton and the Open University.

Grefenstette, J.J. (1990) GENESIS Version 5: ftp://www.aic.nrl.navy.mil/pub/galist/source-code/ga-source.

Hall, Z.W. (1992) *An Introduction to Molecular Neurobiology*. Sinauer Associates.

Harvey, I., Husbands, P. and Cliff, D. (1993) Issues in evolutionary robotics. In Meyer, J.A., Roitblat, H. and Wilson, S. eds. *From Animals to Animats 2: Proceedings of the 2nd International Conference on Simulations of Adaptive Behaviour SAB '92*, MIT Press/Bradford Books.

Hilgetag, C.-C., O'Neil, M.A. and Young, M.P. (1996) Indeterminate organization of the visual systems. *Science*, **271**, 776–777.

Hines, M.L. and Carnevale, N.T. (1997) The NEURON simulation environment. *Neural Computation*, **9**(6), 1179–1209.

Holland, J.H. (1975) *Adaptation in Natural and Artificial Systems: an Introductory Analysis with Applications to Biology, Control and Artificial Intelligence*. University of Michigan Press.

Koch, C. and Segev, I. (1998) *Methods in Neuronal Modeling: From Ions to Networks*. MIT Press.

Kodjabachian, J. and Meyer, J.-A. (1994) Development, learning and evolution in animats. In Gaussier, P. and Nicoud, J. D. eds, *Proceedings of PerAc'94: From Perception to Action,* pp. 96–109. IEEE Computer Society Press.

Kodjabachian, J. and Meyer, J.-A. (1995) Evolution and development of control architectures in animats. *Robotics and Autonomous Systems*, **16**, 161–182.

Langton, C.G. (ed.) (1989) *Proceedings of an Interdisciplinary Workshop on the Synthesis and Simulation of Living Systems*. Addison-Wesley.

Lindenmayer, A. (1968) Mathematical models for cellular interactions in development. *Journal of Theoretical Biology,* **18**, 280–299.

Mainen, Z.F. and Sejnowski, T.J. (1996) Influence of dendritic structure on firing pattern in model neurocortical neurons. *Science*, **382**, 363–366.

McCormick, B.H. and Mulchandani, K. (1994) L-System modeling of neurons. In *Visualisation in Biomedical Computing 1992*, pp. 693–701.

McCulloch, W. and Pitts, W. (1943) A logical calculus of the ideas immanent in nervous activity. *Bulletin of Mathematical Biophysics*, **5**, 115–133.

Mel, B.W. (1994) Information processing in dendritic trees. *Neural Computation*, **6**(6), 1031–1085.

Mitchell, M. (1996) *An Introduction to Genetic Algorithms*. MIT Press.

Nolfi, S. (1997) Evolving non-trivial behaviours on real robots: a garbage collecting robot. *Robotics and Autonomous Systems*, **22**, 187–198.

Nolfi, S., Floreano, D., Miglino, O. and Mondada, F. (1994) How to evolve autonomous robots: different approaches in evolutionary robotics. In Brooks, R. and Maes, P. eds. *Proceedings of the Fourth International Workshop on the Synthesis and Simulation of Living Systems*. MIT Press.

Nolfi, S. and Parisi, D. (1991). Growing neural networks. *Technical Report PCIA-91-15*, Institute of Pyschology, Rome.
Parisi, D. and Nolfi, S. (1998) Development in neural networks. In Patel, M.J. and Honavar, V. eds. *Advances in Evolutionary Synthesis of Neural Systems*. MIT Press.
Purves, D. and Lichtman, J. W. (1995) *Principles of Neural Development*. Sinauer Associates.
Raff, R.A. (1996) *The Shape of Life: Genes, Development, and the Evolution of Animal Form*. University of Chicago Press.
Rust, A.G. (1998) *Developmental Self-Organisation in Artificial Neural Networks*. PhD thesis, Department of Computer Science, University of Hertfordshire.
Rust, A.G. and Adams, R. (1999) Developmental evolution of dendritic morphology in a multi-compartmental neuron model. In *Proceedings of the 9th International Conference on Artificial Neural Networks (ICANN'99)*, **1**, 383–388. IEE.
Rust, A.G., Adams, R., George, S. and Bolouri, H. (1998) Developmental evolution of an edge detecting retina. In Niklasson, L., Boden, M. and Ziemke, T. eds. *Proceedings of the 8th International Conference on Artificial Neural Networks (ICANN'98)*, pp. 561–566, Springer-Verlag.
Rust, A.G., Adams, R., George, S. and Bolouri, H. (2001) Towards computational neural systems through developmental evolution. In Wermter, S., Austin, J. and Willshaw, W. eds. *Emergent Neural Computational Architectures based on Neuroscience*, pp. 188–202. Springer-Verlag.
Schilstra, M., Rust, A.G., Adams, R. and Bolouri, H. (2002) A finite state automaton model for multi-neuron simulations. *Neurocomputing*, **44–46**, 1141–1148.
Shatz, C.J. (1994) Role for spontaneous neural activity in the patterning of connections between retina and LGN during visual system development. *International Journal of Developmental Neuroscience*, **12**(6), 531–546.
Vanier, M.C. and Bower, J.M. (1999) A comparative survey of automated parameter-search methods for compartmental neural models. *Journal of Computational Neuroscience*, **7**(2), 149–171.

Appendix: modelled developmental rules

The following tables contain the set of mathematical rules used to encapsulate the developmental mechanisms which have been modelled. A more detailed set of rules 1 to 14 were previously published in Rust *et al.* (2001). They are reproduced here, in a more concise form, to improve the readability of the subsequent rules.

The rules are presented in three columns arranged by content as: (1) a mathematical description of the rule; (2) illustrative diagrams and accompanying explanations, where applicable; and (3) any developmental parameter(s) used in the rule, with the range of values used in evolutionary experiments and an example of a typical parameter value.

Table A1 Initial developmental configuration rules

<table>
<tr><td colspan="3">1. Developmental environment</td></tr>
<tr><td>$(x, y, z) \in Z^3$</td><td>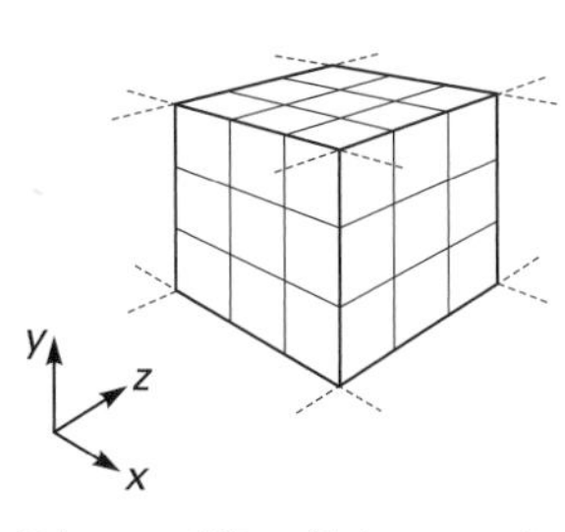
Discrete 3D cellular matrix.</td><td></td></tr>
<tr><td colspan="3">2. Discrete directional vectors from one grid point to 26 neighbouring grid points ($3 \times 3 \times 3$ cube)</td></tr>
<tr><td>$\vec{v}_1 \in$ 26 unit vectors $=$
$V = \{(a, b, c):$
$a, b, c, \in \{-1, 0, +1\},$
$(a, b, c) \neq (0, 0, 0)\}$</td><td>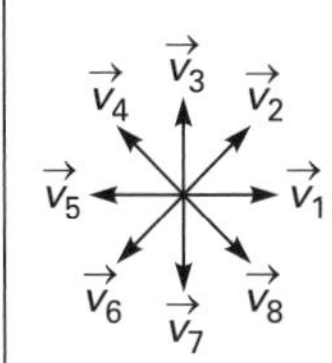
e.g. the 8 unit vectors in the xy-plane
<table><tr><td>$\vec{v}_1$</td><td>(1, 0, 0)</td></tr><tr><td>$\vec{v}_2$</td><td>(1, 1, 0)</td></tr><tr><td>$\vec{v}_3$</td><td>(0, 1, 0)</td></tr><tr><td>$\vec{v}_4$</td><td>(−1, 1, 0)</td></tr><tr><td>$\vec{v}_5$</td><td>(−1, 0, 0)</td></tr><tr><td>$\vec{v}_6$</td><td>(−1, −1, 0)</td></tr><tr><td>$\vec{v}_7$</td><td>(0, −1, 0)</td></tr><tr><td>$\vec{v}_8$</td><td>(1, −1, 0)</td></tr></table></td><td></td></tr>
<tr><td colspan="3">3. Chemical concentration at a source</td></tr>
<tr><td>$\mathcal{C}(\vec{S})$ is the concentration at $\vec{S} \in \mathbb{Z}3$</td><td></td><td>$10 \leq \mathcal{C} \leq 1000$
Typical:100</td></tr>
<tr><td colspan="3">4. Chemical concentration at a point away from source $\vec{S}$</td></tr>
<tr><td>$\mathcal{C} = \frac{\mathcal{C}(s)}{d^{\omega}}$ where
$d = \sqrt{(x - x_s)^2 + (y - y_s)^2 + (z - z_s)^2}$
and $\omega =$ decay factor</td><td>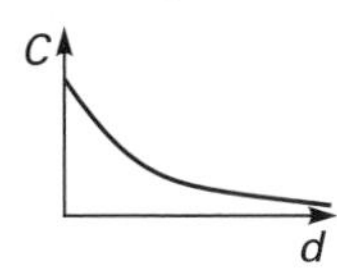
Concentration slope $\propto \omega$</td><td>$\omega \in \{1, 2, 3, 4\}$
Typical : 1</td></tr>
<tr><td colspan="3">5. Grid resolved directional concentration gradients</td></tr>
<tr><td>Gradients in directions $\vec{v}_i = \nabla\mathcal{C} \cdot \vec{v}_i$</td><td></td><td></td></tr>
<tr><td colspan="3">6. Background concentration gradient</td></tr>
<tr><td>$\vec{g}_b = (0, g_b, 0)$</td><td>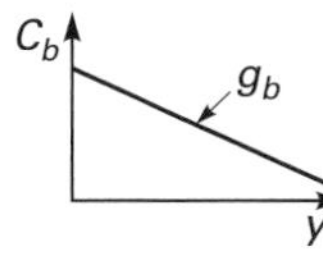
$\vec{g}_b = 0$ along x, and z, and constant along y.</td><td>$0.001 \leq \vec{g}_b \leq 1.00$
Typical: 0.01</td></tr>
</table>

Table A2 Developmental rules that determine neurite navigation

7. Growth cone location		
$\vec{\Lambda} = (x, y, z)$	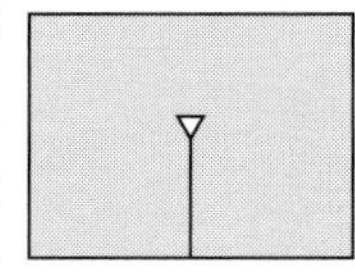 A single growth cone at the tip of a growing neurite.	
8. Chemical gradient sensed by a growth cone at $\vec{\Lambda}$		
$\vec{G}_r = \left(\sum_{n=1, n \neq r}^{att} \nabla C_n - \sum_{m=1}^{rep} \nabla C_m \right) \Big\|_{A}$ where *att* is the number of attractant sources and *rep* is the number of repulsive sources.	Each $C_n(s), C_m(s)$ can be different. Hence each attractor does not necessarily have equal status. Further, growth cones are not necessarily sensitive to all chemical concentration sources.	
9. Growth cone gradient resolution		
$g_m = \max_i \{\vec{G}_r \cdot \vec{v}_i\}$ for $i = 1$ to 26		
10. Growth cone movement		
$\vec{\Lambda}(t+1) = \vec{\Lambda}(t) + \vec{v}_j \Delta l$ where $\vec{v}_j = \begin{cases} \overrightarrow{v_m} : (g_m - g_c) > \Gamma \\ \overrightarrow{v_c} : (g_m - g_c) \leq \Gamma \end{cases}$ $\overrightarrow{v_m} = \overrightarrow{v_i} \in V \; s.t. \overrightarrow{G_r} \cdot \overrightarrow{v_i} = g_m$. g_c is the concentration gradient in the current direction, $\overrightarrow{v_c}$ is the unit vector in the current direction, Γ is a turning threshold and $\Delta l \in \{1, \sqrt{2}, \sqrt{3}\}$	Growth cones hill-climb within the gradient landscape. 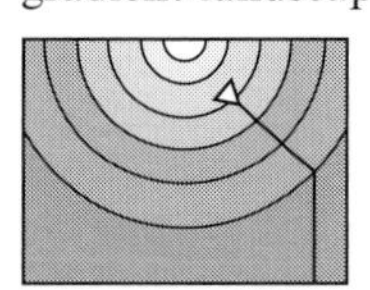 Γ low: growth cone instantly responds to gradient changes. 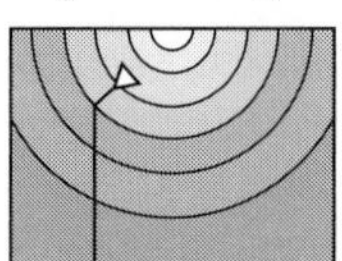 Γ high: growth cone responds only when close to the concentration source. 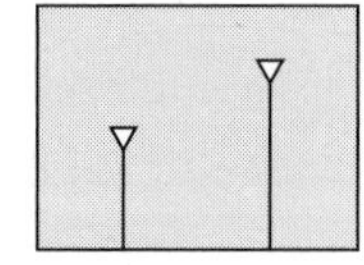 Neurites can extend at different rates, implemented using time delays.	$0.005 \leq \Gamma \leq 10.0$ Typical: 0.05

Table A3 Developmental rules that control neurite branching

11. Interactive branching		
Prob branch $= P(\vec{\Lambda}(t)) = \dfrac{g_m \times g_{m'}}{\tau + g_m^2}$ where $g_{m'} = \max\{\overrightarrow{G_r} \cdot \overrightarrow{v_1}, \overrightarrow{G_r} \cdot \overrightarrow{v_i} \neq g_m\}$ where g_m and $g_{m'}$ are the largest and second largest gradients respectively, and τ is a concentration gradient threshold.	t $t+1$	$0.05 \leq \tau \leq 10.0$ Typical: 1.0
12. Intrinsic branching		
Branch times $\subset \mathbb{Z}^+$	Growth cones branch at specific times.	

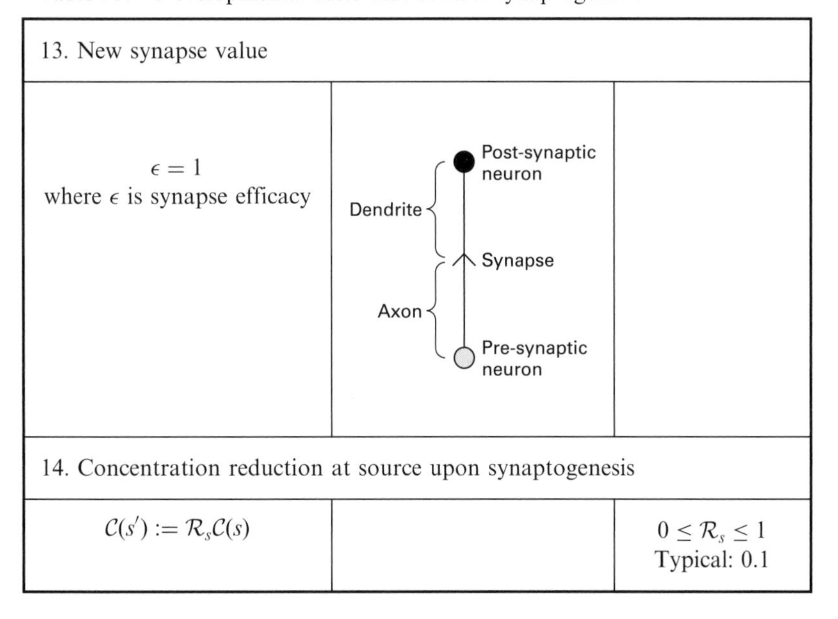

Table A4 Developmental rules that control synaptogenesis

13. New synapse value		
$\epsilon = 1$ where ϵ is synapse efficacy	Post-synaptic neuron; Dendrite; Synapse; Axon; Pre-synaptic neuron	
14. Concentration reduction at source upon synaptogenesis		
$\mathcal{C}(s') := \mathcal{R}_s\mathcal{C}(s)$		$0 \leq \mathcal{R}_s \leq 1$ Typical: 0.1

Table A5 Spontaneous neural activity rules

15. Spontaneous activity initiation		
$Initiate\_Activity ==$ $(Rand < Act_{prob})$ $AND\ (Number\_synapses > Syn_{th})$		$0 \leq Act_{prob} \leq 1$ Typical : 0.5 $0 \leq Syn_{th} \leq 100$ Typical : 10
16. Origin of spontaneous activity		
$\overrightarrow{P_a} = (x_a, y_a, z_a)$ s.t. $(x_a - x_l)(x_a - x_u)(z_a - z_l)(z_a - z_u) = 0$ for $x_l \leq x_a \leq x_u$ and $z_l \leq z_a \leq z_u$ and y_a is the position of the layer	Plan view of pre-synaptic layer 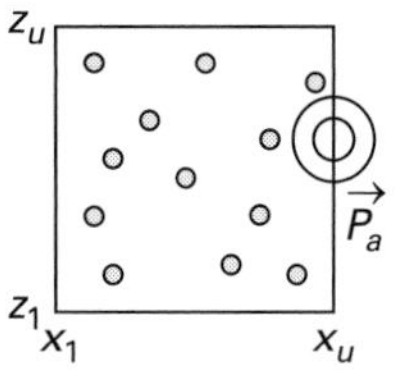 Select a point of activity from the boundary of the pre-synaptic layer.	
17. Time of induced activity pulse at a pre-synaptic neuron		
$t_a = \frac{w}{c}$ where w is the distance of the cell soma from $\overrightarrow{P_a}$ and c is the propagation velocity of activity	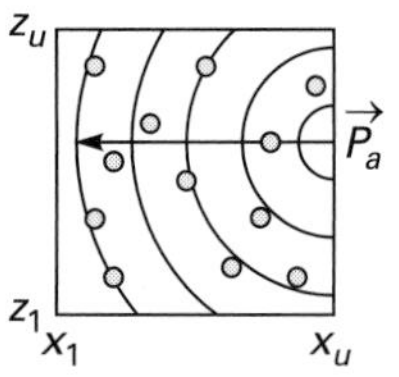	$1 \leq c \leq 5$ Typical : 1
18. Time of first induced activity pulse at any pre-synaptic neuron		
t_{first}	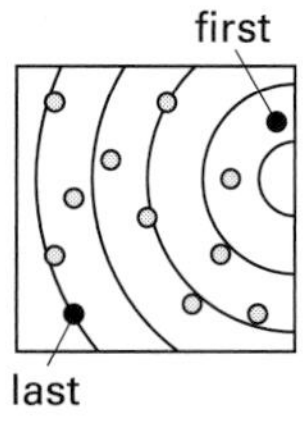	

Table A6 Spontaneous neural activity rules (continued)

19. Activity pulse		
$height = \mathcal{A}_{init}$ $width = 1$	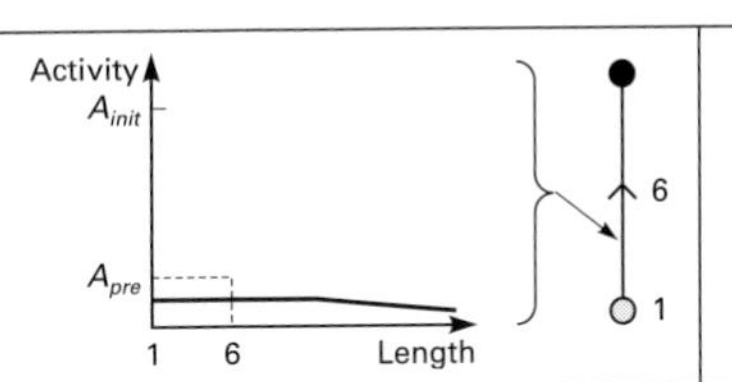	$0.5 \leq \mathcal{A}_{init} \leq 10$ Typical : 1.0
20. Activity pulse propagation to synapse		
$\mathcal{A}_{prc} = \frac{\mathcal{A}_{init}}{l_a}$ where l_a is axon length	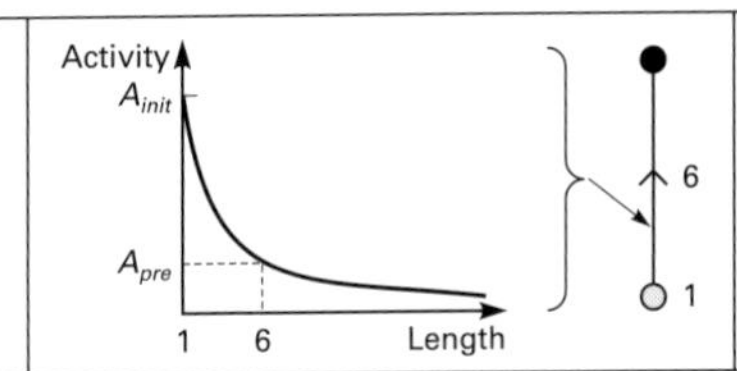	
21. Activation level at synapse		
$\alpha_{syn}(t_0) = \mathcal{A}_{pre}$		
22. Activity pulse transfer at synapse		
$\mathcal{A}_{post} = \epsilon\, \mathcal{A}_{pre} = \frac{\epsilon \mathcal{A}_{init}}{l_a}$	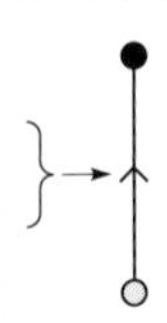	
23. Activation decay at synapse		
$\frac{d\alpha_{syn}}{dt} = d_{syn} < 0$	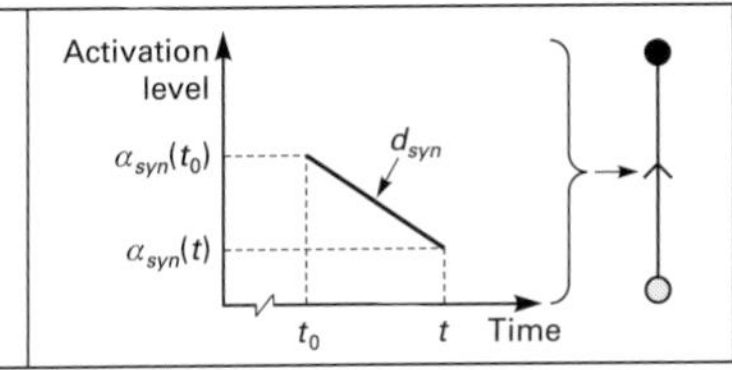	$-0.1 \leq d_{syn} \leq$ -0.0001 Typical : -0.0005
24. Activity arriving at post-synaptic soma		
$\mathcal{A}_{som} = \frac{\mathcal{A}_{post}}{l_d} = \frac{\epsilon \mathcal{A}_{init}}{l_a l_d}$ where l_d is dendrite length	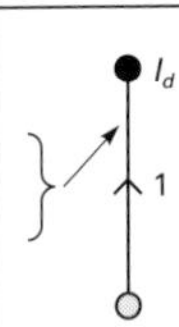	
25. Initial activation level at post-synaptic soma		
$\alpha_{som}(t_0) = \sum_{i=1}^{p} \mathcal{A}_{som,i}$ where p is the number of pulses arriving during time step t_0		

Table A7 Spontaneous neural activity rules (continued)

26. Decay of activation at post-synaptic soma		
$\frac{d\alpha_{som}}{dt} = d_{som} < 0$	c.f. Figure in 27.	$-0.1 \leq d_{som} \leq -10^{-3}$ Typical: -0.00025
27. Activation level at post-synaptic soma		
$\alpha_{som}(t+1) = (\alpha_{som}(t) + d_{som}) + \sum_{i=1}^{p} \mathcal{A}_{som,i}$ where p is the number of pulses arriving during one time step	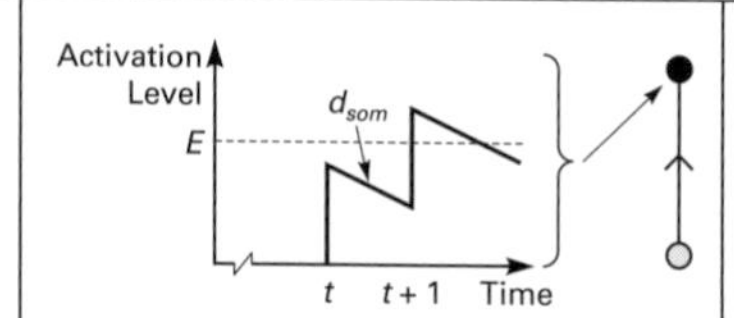	
28. Generation of reinforcement signal at post-synaptic soma at time t		
$\mathcal{N} = \begin{cases} \mathcal{N}_{init} & : \alpha_{som}(t) \geq E \\ 0 & : \alpha_{som}(t) < E \end{cases}$ where E is an activation threshold	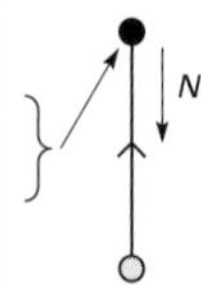 also see Figure in 27.	$0 \leq \mathcal{N}_{init} \leq 2$ Typical: 1.0
29. Reinforcement signal reduction		
$\mathcal{N}_{branch} = \frac{\mathcal{N}_{trunk}}{2}$	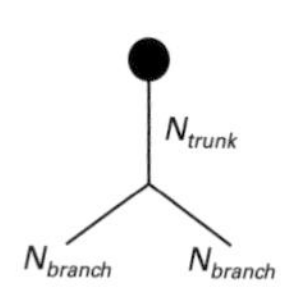	
30. Synapse reinforcement upon a reinforcement signal arriving at a synapse		
$\Delta\epsilon = \begin{cases} \mathcal{N}_d \alpha_{syn}(t) & : \alpha_{syn}(t) > 0 \\ 0 & : \alpha_{syn}(t) \leq 0 \end{cases}$ where $\mathcal{N}_d$ is the level of the reinforcement signal arriving at a synapse	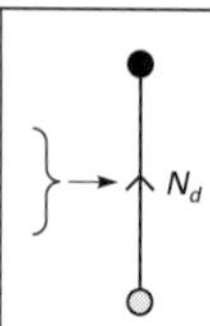 c.f. Figure in 23.	
31. Initial accumulated reinforcement level at pre-synaptic soma		
$\eta(t_0) = \sum_{i=1}^{p} (\mathcal{N}_d)_i = \mathcal{N} t_0$ where p is the number of pulses arriving during t_0		
32. Accumulated reinforcement signal at pre-synaptic soma		
$\eta(t+1) = \eta(t) + \mathcal{N}_{t+1}$	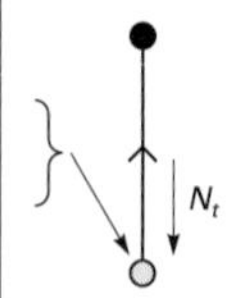	

Table A8 Spontaneous neural activity rules (continued)

33. Time of the last reinforcement signal to reach a pre-synaptic soma		
t_{last}		
34. Duration of an activity phase		
t s.t. $t_{first} \leq t \leq t_{last}$		
35. End of growth and activity phases		
t_{end}		

Table A9 Growth rate regulation rules

36. Dendritic growth regulation due to activity level		
Dendritic growth rate reduction $= \psi \mathcal{D}_d$ where $\mathcal{D}_d$ is the maximum growth rate reduction and $$\psi = \begin{cases} \frac{1-e^{-\mu(\Pi-\kappa)}}{1+e^{-\mu(\Pi-\kappa)}} & : \Pi-\kappa > 0 \\ 0 & : \Pi-\kappa \leq 0 \end{cases}$$ where $\mu > 0$ is a slope constant, Π is the actual frequency of neurotrophin firing events during an activity phase and κ is a threshold frequency value	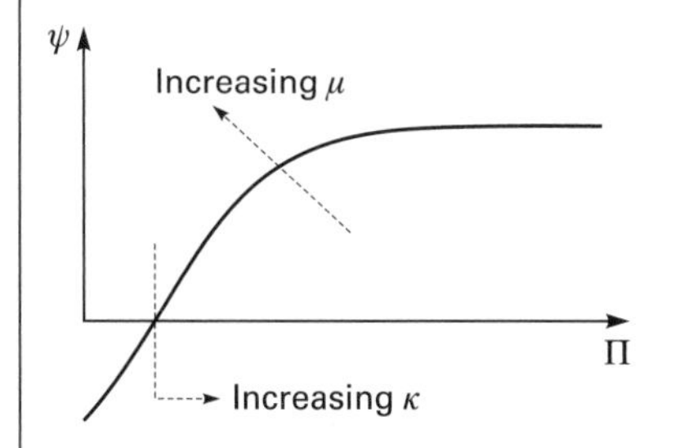 Growth regulation applies to the whole neuronal tree.	$0 \leq \mathcal{D}_d \leq 10$ Typical: 3 $0.5 \leq \mu \leq 5.0$ Typical: 2.0 $1 \leq \kappa \leq 50$ Typical: 20.0
37. Axonal growth regulation due to activity dependent reinforcement signals		
Axonal growth rate reduction $-u\mathcal{D}_a$ where $\mathcal{D}_a$ is the maximum growth rate reduction and $$u = \begin{cases} \frac{1-e^{-\xi(\mathcal{N}_t-\sigma)}}{1+e^{-\xi(\mathcal{N}_t-\sigma)}} & : \mathcal{N}_t-\sigma > 0 \\ 0 & : \mathcal{N}_t-\sigma \leq 0 \end{cases}$$ where $\xi > 0$ is a slope constant, $\mathcal{N}_t$ is the actual value of reinforcement signal received during an activity phase and σ is a threshold value of reinforcement signal	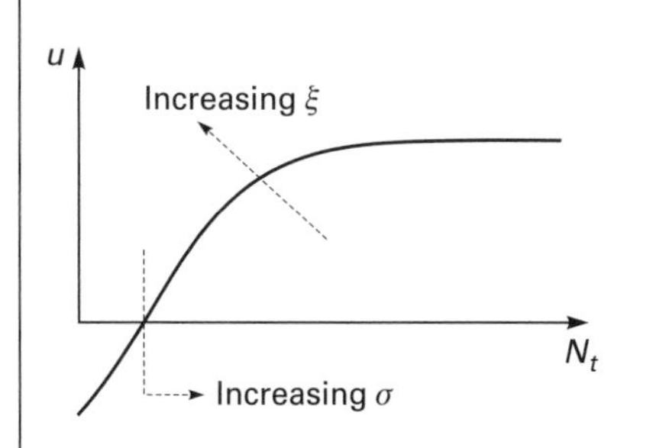 Growth regulation applies to the whole neuronal tree.	$0 \leq \mathcal{D}_a \leq 10$ Typical: 3 $0.5 \leq \xi \leq 5.0$ Typical: 2.0 $0.5 \leq \sigma \leq 10.0$ Typical: 5.0

Table A10 Pruning rules

38. Pruning of unterminated branches		
if $t = t_{end}$ *AND* $\vec{\Lambda}(t)$ *AND* $synapse = false$ while ($synapse = false$ *AND* $branch = false$) $\vec{\Lambda}(t) = \vec{\Lambda}(t-1)$	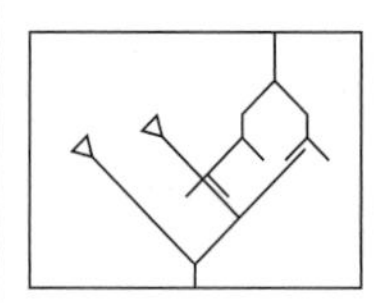 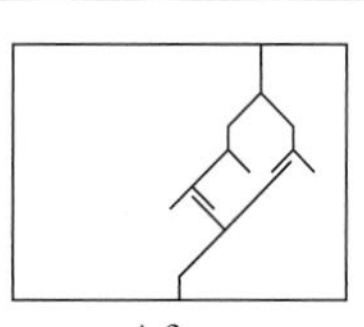 Before After	
39. Pre-synaptic neuron death:		
39a. Pre-specified if $\eta(t_{end}) < \eta_{th}$ $neuron = remove$		$0 \leq \eta_{th} \leq 100$ Typical: 30.0
39b. Global average $$if\ \eta(t_{end}) < \frac{\sum_{i=1}^{n} \eta_i(t_{end})}{n}$$ $remove(neuron)$ where n the number of pre-synaptic neurons		
40. Synapse pruning:		
40a. Specific for each post-synaptic neuron connected to a pre-synaptic neuron via synapses with efficacies $\epsilon_1 \ldots \epsilon_h$ $$\epsilon_{avg} = \frac{\sum_{i=1}^{h} \epsilon_1}{h}$$ for each synapse, if $\epsilon_i < \epsilon_{avg} remove(\epsilon_i)$	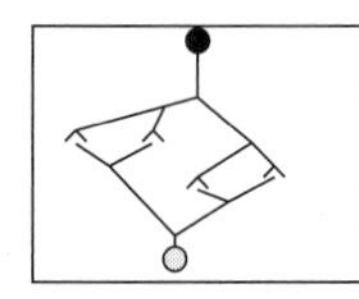 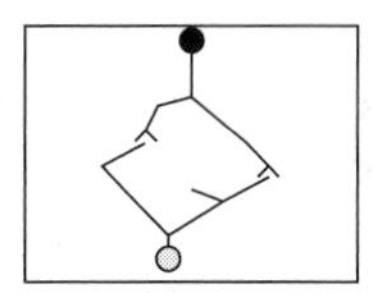 Before After Incorporating 38.	
40b. Global average for each post-synaptic neuron connected to pre-synaptic neurons $a_1 \ldots a_p$ via synapses with efficacies $\epsilon_{11} \ldots \epsilon_{1m_1} \ldots \epsilon_{pm_p}$ $$\epsilon_{avg} = \frac{\sum_{i=1}^{p} \sum_{j=1}^{m_i} \epsilon_{ij}}{\sum_{k=1}^{p} m_k}$$ for each synapse, if $\epsilon_{ij} < \epsilon_{avg}$ $remove(\epsilon_{ij})$	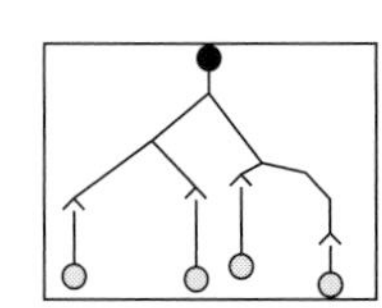 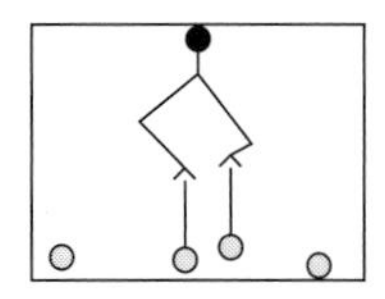 Before After Incorporating 38.	
41. Neuron death		
for each pre- and post-synaptic neuron if $number\_synapses = 0$ $remove(neuron)$		

A developmental model for the evolution of complete autonomous agents 20

FRANK DELLAERT and **RANDALL D BEER**

20.1 Introduction

Development is an important and integral part of biological evolution. Genetic changes are not directly manifested in phenotypic changes, as is often assumed in population genetics, as well as in most autonomous agent work involving artificial evolution. Rather, complex developmental machinery mediates between genetic information and phenotype and this has many consequences. It provides robustness by filtering out genetic changes (i.e. some genetic changes make little or no difference to the final phenotype; there is an equifinality to development). It also provides a natural way to try out a spectrum of mutations, i.e. the same type of genetic mutation can produce anything from no effect to a very large effect in the phenotype, depending on when the affected gene acts during development. Furthermore, development provides a compact genetic encoding of complex phenotypes, allows incremental building of complex organisms and supports symmetry and modular designs.

For these reasons, there is a growing interest in modelling development (Lyndenmayer and Prusinkiewicz, 1989; Wilson, 1989; Mjolsness *et al.*, 1991; deBoer *et al.*, 1992; Fleischer and Barr, 1994; Kitano, 1994). Many ongoing efforts aimed at including simple developmental models in evolutionary simulations can be found in the literature (Belew, 1993; Cangelosi *et al.*, 1993; Gruau and Whitley, 1993; DeGaris, 1994; Kodjabachian and Meyer, 1994; Nolfi *et al.*, 1994; Sims, 1994; Jakobi, 1995).

Much of the latter work has focused on modelling neural development, but in biology, bodies and nervous systems co-evolve. Body morphology and nervous systems can constrain and shape one another. Somatic and genetic factors can interact and this occurs not only during development, but on an evolutionary scale as well. Problems posed by evolution can be solved by a combination of body and neural changes. Allowing both body and nervous system to co-evolve can provide a smoother and more incremental path for substantial changes.

Also, much of the existing work is highly abstracted from biological development (e.g. using grammars). While there are good reasons for this (familiarity, simplicity, computational speed, emphasis on performance not biology, etc.), too little is currently understood about development to know what are the right abstractions to make. Development completely transforms the structure of the space that is being searched. If we are lucky, this transformation will allow us to evolve interesting agents more easily. But if we are unlucky, we could actually make the search problem harder. Because so little is currently understood about the overall 'logic' of development, it is important that we explore many different levels of abstraction to get a sense of the tradeoffs involved. An important aspect of this exploration should be to explore developmental models that are more biologically realistic in their basic structure than the highly abstract models that have currently been explored.

Thus, in this chapter we present a model of development that has been used to evolve functional autonomous agents, complete with a morphological structure and a neural control system. Earlier work that involved a more biologically defensible but more complex model is contrasted with a new and simplified approach that performs surprisingly better. In the next section we give a brief overview of the general approach that we have adapted to model a developmental process for autonomous agents. It is common to both models we discuss in the chapter. Section 20.3 presents a biologically defensible model of development, complete with a mechanism for the emergence of a nervous system inspired by axonal growth cones. The expressiveness of the model is demonstrated by means of a hand-designed genome, able to direct the development of a functional and complete agent that can execute a simple task in a simulated world. In Section 20.4, we discuss a simplified model that addresses some of the problems of the earlier model and show that it can be used to evolve functional agents from scratch. We show the example of an agent evolved to execute a line following task. Section 20.5 discusses some of the lessons we learned from this work and suggests some avenues for further research.

20.2 Overview of the developmental model

In nature it is the performance of an adult organism in its environment that will determine whether its genetic material is propagated. However, in sharp contrast with the model usually assumed in the genetic algorithm literature, genes do not directly specify the traits of an animal. Rather, they specify the developmental sequence by which an animal grows out of a single egg cell to a fully developed phenotype. It is only after this process is complete that specific traits or behaviour can be selected for or against.

For reasons specified in the introduction and elsewhere (Dellaert and Beer, 1994a; Dellaert, 1995), we believe that using a developmental model in conjunction with the genetic algorithm (GA) can help us better to evolve autonomous agents. Thus, we have attempted to implement this in simulation, where a genetic algorithm will supply us with some genetic material, the *genotype*, which is then transformed by a developmental model into a fully-grown organism, the *phenotype*. It is the performance of the phenotype that will determine whether its genotype is selected inside the GA.

In particular, we have implemented a model of development for simple simulated organisms that start out as a single cell but 'grow' into multicellular organisms. An example of one such developmental sequence is shown in Figure 20.1. The fully developed agent will then be eval-

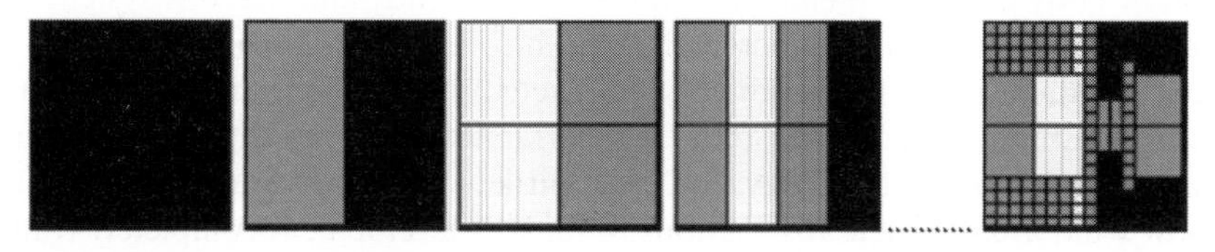

Figure 20.1 Starting with an 'egg cell' the developmental simulation yields a multicellular square as the adult organism.

uated on how well it performs a simple task in a simulated world. In contrast to the complex nature of real biological cells that are able to move and change their shape, our simulated cells are modelled as simple, rigid two-dimensional squares. This choice allows us to implement cell division in an efficient way, keeping the computational cost of the simulation acceptable. Indeed, after two rounds of divisions, the resulting cells are again square cells. Our implementation does allow for more complex cell models to be substituted in place of the square cell model should that need arise in the future.

We explicitly model the existence of different cell types. In biology, each cell possesses the same identical copy of the genome, a sequence of DNA unique to that particular animal. Yet, not all the cells have the same characteristics or behaviour: a range of different cell types coexists in any fully-grown biological organism. Although the cells contain identical genes, different subsets of genes are expressed in different cells, giving them different properties. In our model we have implemented a similar arrangement: each cell has an identical copy of a simulated genome inside it, but the subset of the 'genes' active at any given moment determines what type of cell it is and how the cell will behave during development. Exactly how this is implemented is at the core of both models we will present and will be discussed in detail below.

During biological development, the subset of genes expressed in each cell is not static but rather in constant flux. The cell responds to its environment and to the instructions coded in its genome by changing its composition continuously, until it has differentiated fully into one of the 'adult' cell types. In turn, the instructions given by the genome are a function of the state of the cell. Thus, cell state and genome comprise an interwoven dynamical system, a genetic regulatory network, and it is the collective unfolding of the dynamics of many such systems – one for each cell – that constitutes development. On a higher level of abstraction, development can be seen as the sequence of events by which the cells in the body differentiate to perform the various functions inside the animal.

The heart of our developmental model is formed by exactly one such genome state dynamical system that lives inside each of our model cells. The active subset of model genes inside a cell will be regarded as the cell state, and the genome, the current state and the environment in which the cell finds itself govern the possible state transitions from that state. Qualitatively, this picture is similar to what we see in biological cells, albeit quite simplified in the details. Our developmental simulation can now similarly be viewed as the sequence of events by which the state of the cells differentiate from the initial 'egg cell' state – and from one another – to form a particular cell type arrangement that will suit a particular task.

As explained in the introduction, we are interested in having both a morphological component, i.e. the development of the physical extent of a simulated organism, and a neural component, i.e. how the nervous system of an organism develops. Together, these components should form a complete autonomous agent. To this end, we have associated certain simulated cell types with particular functions, e.g. sensors, interneurons or actuators. We then also provide a model of how these control components get wired up in a working neural network, i.e. the neural developmental component of the model.

This high level specification of the model needs to be complemented with quite a few implementation issues and choices to make it work. We need to specify which states will lead to cells dividing. Since we have chosen to update all the cell states synchronously, we need to break the symmetry after the first division to get interesting dynamical behaviour. Some mechanism of intercellular communication must be implemented to make it possible for cells to influence each other's state. Meaning must be assigned to the different possible genes, so that we can interpret state as cell type. Finally, the detailed mechanism for neural development needs to be fully specified. In the following we will present two different implementations of the developmental model: a complex one and a simple one. The complex model is more biologically defensible, but suffers from its complexity. The simple model is further removed from biological reality in its implementation details, but has the advantage that it is computationally more tractable and easier to analyse. We will discuss both models in the next sections.

20.3 A complex model

The first implementation models simplified genome and cytoplasm entities inside each cell, and has model 'proteins' that are produced by the genome and are collected inside the 'cytoplasm'. The set of proteins in each cell determines what kind of events the cell reacts to and which signals it emits while the developmental sequence unfolds. In addition, there is an elaborate model of neural development based on a growth-cone model that detects the presence of proteins in the cells and grows accordingly.

20.3.1 Genome–cytoplasm model

In the complex model, each cell contains a 'cytoplasm' and a 'genome'. The cytoplasm contains 'proteins', and the proteins present in each cell determine its capabilities, i.e. cells with a different set of proteins can be thought of as having a different cell type. In this regard our model proteins can be thought of as having a similar role as biological proteins. We represent each model protein by a unique integer and implemented the cytoplasm as a set of integers.

The genome consists of a set of 'operons', an example of which is shown in Figure 20.2. Each operon is made up of a set of input tags, a Boolean function and a set of output tags. In each time step of the simulation, the input tags determine the input to the Boolean function and, if the output of the Boolean function is evaluated to TRUE, the protein corresponding to the output tag is injected into the cytoplasm. An input tag will give a 1 to the Boolean function if its corresponding protein is present, and 0 otherwise.

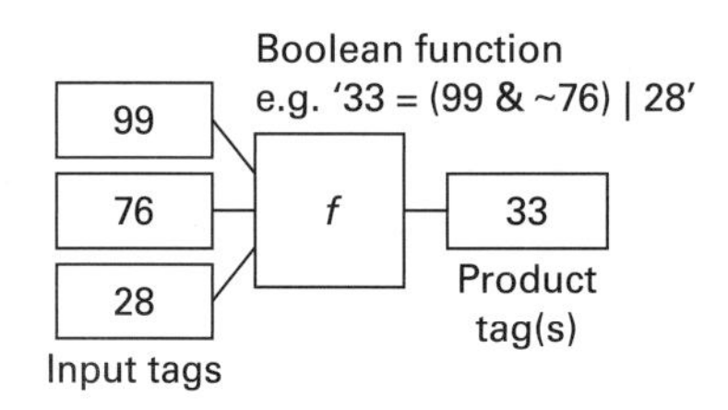

Figure 20.2 An artificial genome consists of artificial 'operons', one of which is shown here.

Thus, the content of the cytoplasm can be seen as the state of the cell or its cell type and the genome will determine how this state will change over time. It is the interplay between genome and cytoplasm that determines how the cell's type changes over time. This interplay is summarized graphically in Figure 20.3.

20.3.2 Development

With the description of the cell's internals in the previous section, we can explain how the developmental simulation will proceed. The organism starts out as a single cell, with its genome

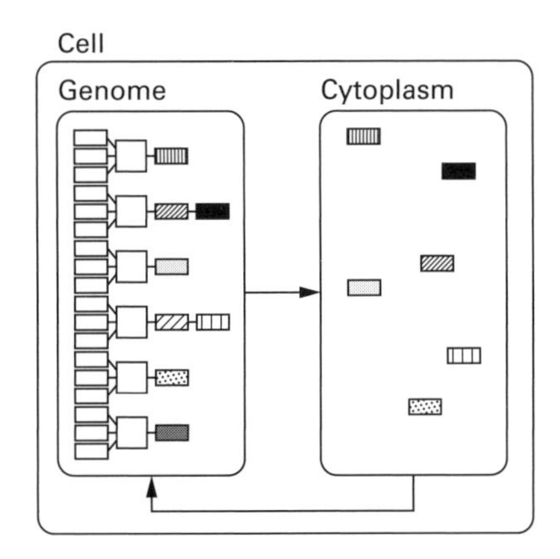

Figure 20.3 Schematic summary of the cytoplasm–genome model.

given by the GA. The cytoplasm is initially the empty set. Then, for a constant number of iterations, all cells in the organism go through a simulated cell cycle in parallel. This 'cell cycle' consists of two phases. (1) During 'interphase' the cytoplasm is updated by evaluating all the operons in the genome, as explained above. (2) During mitosis, each cell checks for the presence of a special protein (which we will denote tDividing, where the leading t stands for 'protein tag') and divides if found. The simulation ensures that the first cell always goes through division by injecting the tDividing protein into the cell's cytoplasm prior to starting the cell cycle. Thus, after the first division we end up with two cells, each with identical cytoplasm and genome that it inherited from their parent cell. This presents a problem: since all the cells obey the same deterministic rules set out above, the remainder of the simulation will yield identical cell types at each time step and no interesting organisms will emerge.

Therefore, we add two additional mechanisms, symmetry breaking and intercellular communication, to ensure that the developmental process exhibits interesting dynamics. Symmetry breaking happens just after the first division event and is implemented by injecting a special protein in only one of the first two daughter cells. This needs to happen only once: from now on these cells' descendants will also differ, as their dynamical trajectories start from different initial conditions. Cell communication, on the other hand, can happen at every cell cycle and consists of a mechanism that allows one cell to cause the introduction of a protein into another cell. This is modelled after the biological mechanism of induction and involves proteins that represent morphogens, receptors and intracellular messenger proteins. The implementation details can be found in Dellaert and Beer (1994a, b) and Dellaert (1995).

The model for intercellular communication is also used to introduce two supplementary features that establish a perimeter and midline on the organism that can be detected by the cells, which can lead to local differentiation of cells as seen in Figure 20.4.

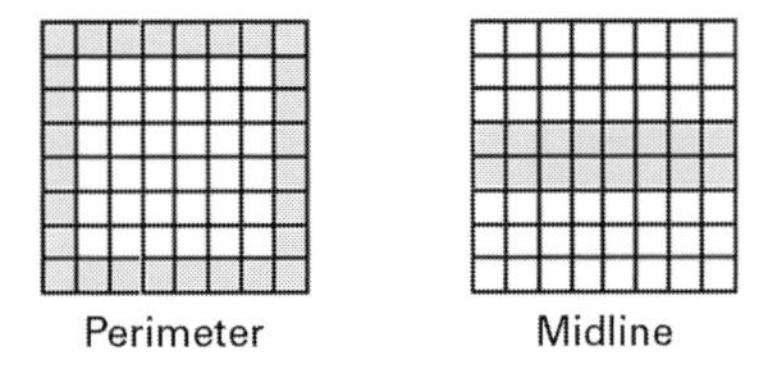

Figure 20.4 Two special receptors can lead to the differentiation of cells at the perimeter and adjacent to the midline of the organism, respectively.

20.3.3 A detailed neural developmental model

When the development of the agent morphology has settled into a configuration in which divisions no longer occur (but the cell state can still continue to evolve), a control architecture or 'nervous system' develops on top of the arrangement of cells. This happens because specialized cells, i.e. those that express a specific protein (tAxon), will send out axons that will innervate other cells and, as such, establish a neural network architecture. Only cells expressing the tTarget protein will be innervated and axons will only grow on top of cells that express tCAM proteins. This is modelled directly after one postulated mechanism in biological development, in which neurons emit growth cones that can detect the presence of certain molecules (cellular adhesion molecules or CAMs) on the surface of cells and adjust their direction of growth accordingly.

The central feature of this model is the 'growth cone' model, illustrated in Figure 20.5. The black rectangle represents a growth cone. It is linked back to the cell from which the axon originated by 'link' elements, and it sends out 'flanks' that sample the neighbourhood ahead by means of 'spikes'. The flank whose spikes detect more tCAM proteins will be promoted to a growth cone in the next time step, with the axon splitting in different directions in case of a tie. This fairly intricate finite state model is modelled on the workings of a real biological growth cone, albeit quite simplified. After the process of axon growth is complete, a dynamical neural network (Beer and Gallagher, 1992) is instantiated that connects sensor, interneuron and actuator cells according to the connections made during the neural developmental phase. Time constants and biases of the organism are global and are specified separately in the genome. Thus, emergence and placement of sensor and actuator cells needs to be coordinated with appropriate neural developmental events to lead to interesting behaviour. Sensor cells that did not send out an axon or actuators that were not innervated will have no effect on the agents' behaviour.

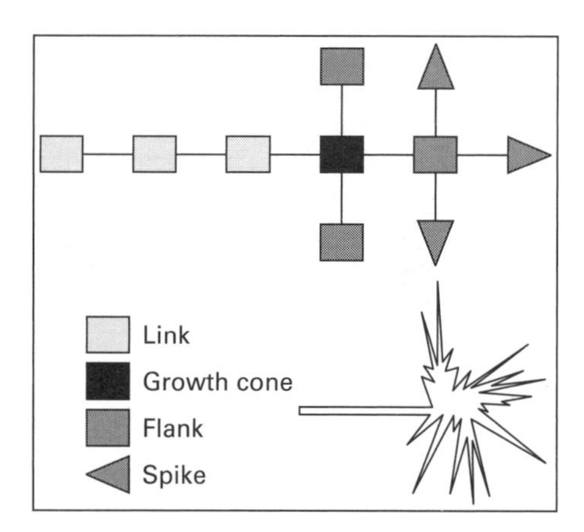

Figure 20.5 The different axon-element states.

20.3.4 A Braitenberg 'hate vehicle'

In order to explore the expressiveness of our developmental model, we have hand-designed a genome capable of directing the development of both the body and nervous system of a simple Braitenberg-style 'hate vehicle'. This agent executes a simple avoidance task in a simulated world. Figure 20.6 shows the adult form of the hand-designed organism and its behaviour on the task; sensors are at the frontal side of the organism (on the right in the figure), and a simple network relays their activation to patches of actuator cells (on the left). Figure 20.7 shows the expression domains of the different proteins in the final developmental stage of the organism. A complete analysis of the developmental sequence is beyond the scope of this chapter, but can be found in Dellaert (1995).

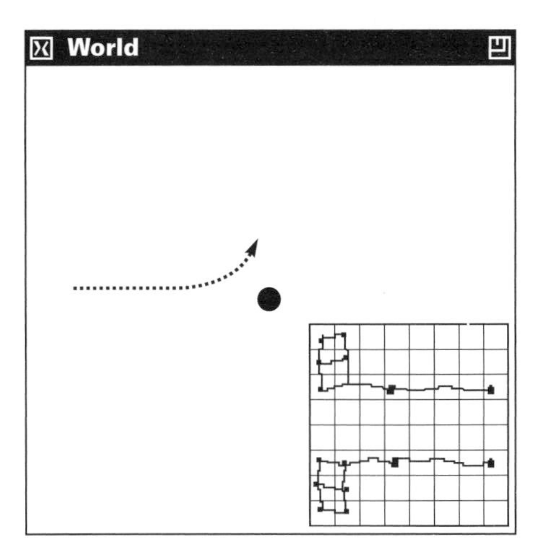

Figure 20.6 The behaviour of the hand-designed Braitenberg hate vehicle in a simulated world.

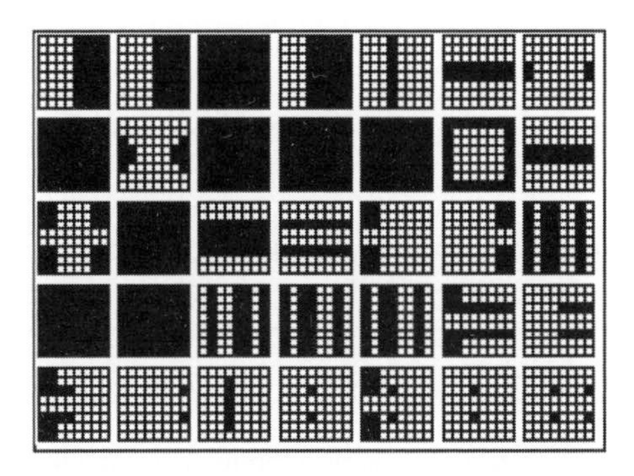

Figure 20.7 The different 'protein expression domains' in the adult organism. Each square in the matrix shows in which cells each of the 35 proteins is present. For example, the bottom right inset where the tAxon protein is expressed, and corresponds to the cells in Figure 20.6 that send out an axon.

It is a demonstration of the power of the model that the initial genetic specification can direct the simultaneous development of both morphology and nervous system, leading to a complete autonomous agent. We want to stress that only the genome has been manually specified (i.e. a set of fully specified operons) and that all subsequent development follows from the model without intervention. Note also that no learning takes place in the agent. All the behaviour it exhibits is solely a function of its evolved architecture.

Although we were unable to evolve such an agent from scratch, we were able to improve its performance significantly through incremental evolution. We started out with the hand-designed genome as a primer for the genetic algorithm and used a performance function to evaluate each agent's aptness for the avoidance task. The result was an agent that was markedly better at executing the task, as shown in Figure 20.8. The consecutive steps in the development of this incrementally evolved agent are shown in Figure 20.9. In this figure, the big black dots represent axon-emitting cells, whereas the smaller black squares represent innervation of target cells. Time goes from top left to bottom right, the last square representing the adult organism.

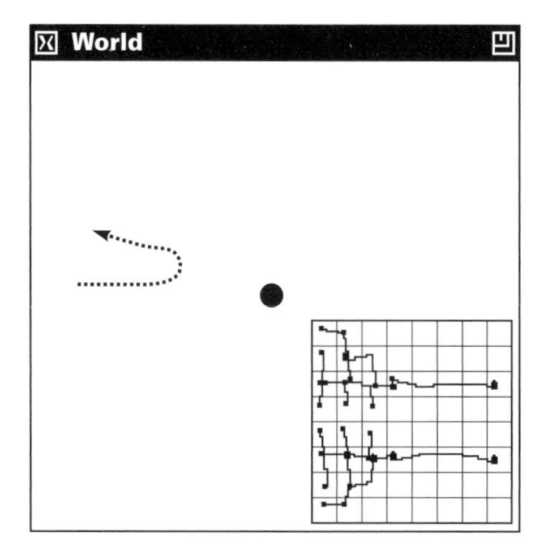

Figure 20.8 The behaviour of the incrementally evolved agent in a simulated world.

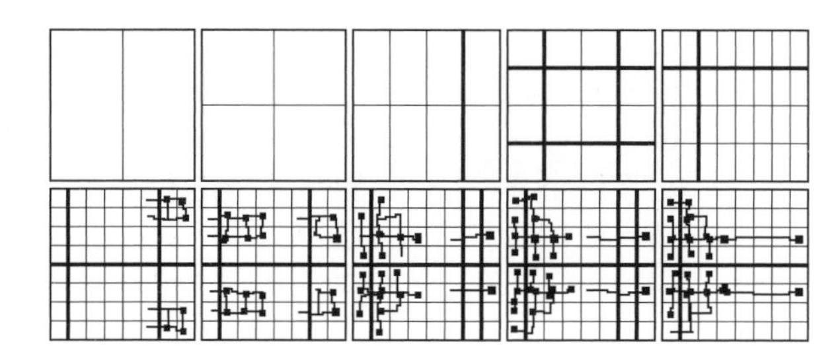

Figure 20.9 The consecutive steps in the development of the incrementally evolved agent.

20.3.5 Lessons learned

The expressiveness of the developmental model as illustrated in the previous paragraphs is also its greatest disadvantage. There is a vast space of possible genomes with their associated phenotypes. Many of the protein expression domains need to be tightly coordinated with one another to have a working nervous system emerge. Thus, both the size and structure of the search space make the problem hard. Although the Braitenberg example convinces us that viable organisms (with respect to the task) exist, it does not give us a path to them starting from random genotypes. Although we were able to evolve incrementally better performing agents starting with the hand-designed genome (see also Dellaert and Beer, 1994b), we have not been able to obtain convincing results when starting evolution from scratch. In addition, the complexity of the model makes it quite hard and error-prone to implement. To cope with both these problems, we have experimented with a simplified model.

20.4 A simplified model

The simplified model uses a *random Boolean network* (RBN) as an abstraction for the genome, where the state of each cell is equal to the state of its RBN. The topology and rules of the RBN are the same for each cell and can be regarded as the genetic specification of the organism. The model for neural development has been simplified considerably and is now based on the range and position of the interacting cells. In this section we will discuss each of these aspects in more detail.

20.4.1 The random Boolean network model

RBNs were first thought of as an abstraction for genetic regulatory networks by Kauffman (1969, 1993) and extended by Jackson *et al.* (1986) to systems of multiple, communicating networks as needed in the context of development. A random Boolean network can be represented by a graph, for example the simple RBN of Figure 20.10. In an N node network, each node is defined by K incoming edges, defining a pseudo-neighbourhood and a particular Boolean function 2. The edges can be recurrent, i.e. nodes can connect to themselves. In addition, each node has an associated state variable assuming a value of 0 or 1. Each node synchronously updates its state in discrete time steps, and the state of a node at time t+1 is the value of the Boolean function using the state of the input nodes at time t.

An RBN can then be used as an abstraction of the genetic regulatory network inside a cell. The state of the cell can be equated to the state of the RBN and each node can be seen as equivalent to a particular 'protein', although we will not use that terminology here. The 'random' refers to the fact that each node can have a different Boolean function and pseudo-neighbourhood, in contrast to cellular automata which represent a special case of RBNs. The

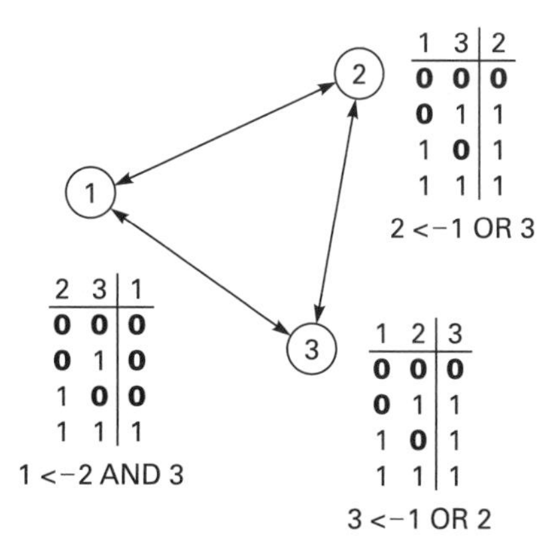

Figure 20.10 A simple RBN. Boolean functions are specified as truth tables. Example from Kauffman (1993).

topology and the particular Boolean functions that specify the RBN will then determine how the state evolves over time and thus correspond to the 'genome' of the complex model. The incoming edges for each node indicate which other nodes influence its state and can be seen to correspond to the input tags of an operon in the earlier model. If we complement this with a mechanism for intercellular communication, allowing for edges to occur between cells, we have a very similar picture to the more complex model sketched above.

An appealing property of RBNs is that their dynamics can be made explicit using phase portraits (Wuensche, 1993, 1994). This provides us with a powerful tool to analyse the dynamics of development (Dellaert, 1995). An example is shown in Figure 20.11. In the figure, each small circle represents one of the 128 states of the 7-node RBN and the edges between them represent state transitions. The transitions always occur in the direction of the state attractors, in this case state cycles respectively with period 5 and 2.

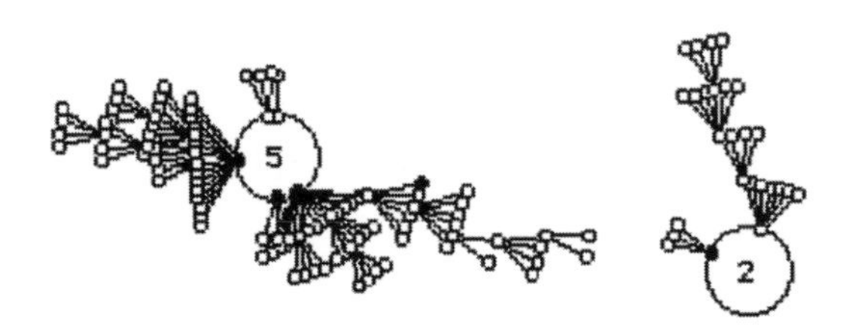

Figure 20.11 The phase portrait of a random Boolean network with N = 7 and K = 3. See text for an explanation.

20.4.2 Development

The developmental simulation unfolds similarly as with the more complex model. Each cell contains an identical copy of the RBN, but the state of the RBN can vary between cells. Instead of a given protein signalling division, a cell will now divide if an *a priori* specified bit in the state vector is set. This bit is set in the 'egg cell' to ensure at least one division event. Symmetry breaking occurs by deterministically perturbing the state (by flipping one bit) of one of the daughter cells resulting from this first division. Intercellular communication is accomplished by calculating neighbourhood state vectors that serve as external inputs to the intracell RBNs, perturbing their phase portrait. Details can be found in Dellaert and Beer (1994a); the final result is a developmental sequence that is qualitatively similar to the ones obtained by the more complex model, but at a considerably reduced computational cost.

20.4.3 Simplified neural development

To cope with the problems that arose when using the complex model described above, we have implemented a much-simplified neural developmental model. As in the earlier model, the final differentiation of a cell determines whether it will send out an axon or not, and/or whether it is a target for innervation. This is done simply by associating a prespecified node of the RBN with these respective properties. In addition, three bits in the RBN state vector determine whether the cell will be a sensor, an actuator or an interneuron and one bit specifies whether any innervation will be inhibitory or excitatory.

The development of the neural network is simple and straightforward: each cell that has the 'axon bit' set will innervate all target cells (with the 'target bit' set) within its range. The constants specifying the connection strength and the range at which cells innervate each other are evolved together with the genome and are identical for all cells. Thus, unlike before, the developmental process does not need to specify elaborate pathways of CAM molecules on which the axons can grow, but merely needs to make sure that it places cells that need to be connected within each other's range. In addition, it can adjust the architecture by specifying for each cell individually the sign of the connection and the cell type. The range and weight factors can be co-evolved with the placement of the cells to obtain the desired behaviour.

20.4.4 A line follower

Using this simpler model, we have not only evolved agents that execute the earlier avoidance task quite well, but have also succeeded in evolving agents that perform more difficult tasks, in this case following a line made up of circular segments. In this experiment, each agent in a genetic algorithm population (steady state GA with tournament selection, population size 100, mutation rate 2.5%, crossover rate 100%, tournament size 7, run for 10 000 evaluations) was put into a simulated world and evaluated on how well it could follow the curve sketched in Figure 20.12, made up of two semicircles. In the neural developmental phase, standard static neurons were used instead of the dynamical neurons of before.

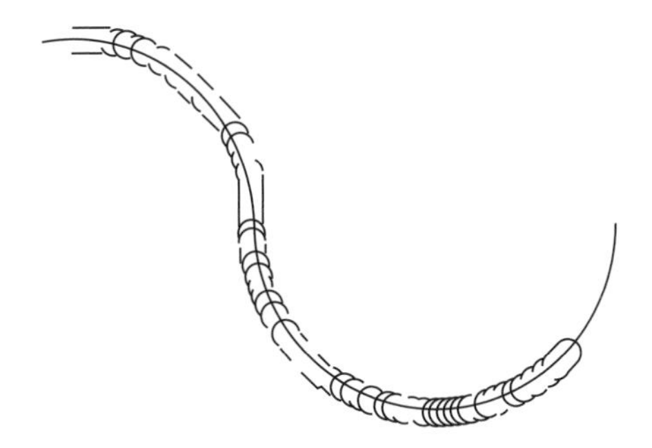

Figure 20.12 The outline of the line following agent as it executes the task. The agent moves from left to right.

The evaluation function used was the integrated squared error to an ideal position along the curve. This can be easily calculated since the velocity of each agent is kept constant: only the steering angle varies. The response of each sensor cell in the agent is a Gaussian function of its closest distance to the line. The output of the actuators is averaged for each side, added together and multiplied by a constant factor to become a steering output. In the following paragraph we describe the structure of one such agent, the best of one particular run of the GA, whose behaviour in the simulated world is shown in Figure 20.12.

The developmental sequence of the evolved agent is shown in Figure 20.13. In this figure, the dark cells represent sensors while the lighter ones are the actuators. As you can see, development unfolds asymmetrically as not all cells divide an equal number of times. Also, you can see that the sensor cell type is being induced by the perimeter (as shown before in Figure 20.4). No interneuron type cells were present.

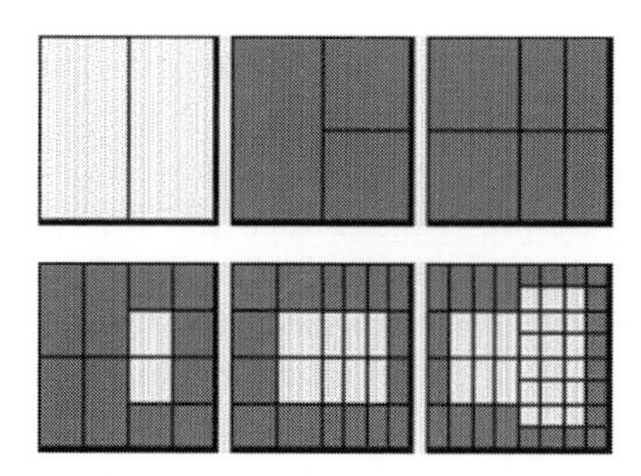

Figure 20.13 The developmental sequence, from the two-cell stage onwards, of the line-following agent.

If we look at the detailed innervation of the cells, we get the picture sketched in Figure 20.14, where all the connections turned out to be excitatory. This makes sense when you consider how movement is implemented. By this innervation, the agent steers in the direction where it can sense the line. Note that this particular innervation depends on the range factor: with a bigger innervation range more cells would be innervated. With this picture in mind, it is easy to understand Figure 20.15, which shows the activation of each neuron/cell during the execution of the task. You can see that the actuators not innervated by any sensors remain inactive throughout the whole task. When evaluating the evolved line-followers on novel lines of different curvatures, their control strategy turned out to be unstable. They were able to follow the line for a while, but they eventually overshot and lost track of the line. Given that they have not been presented with different types of lines during the time that they were evolved it is reasonable that they experience these problems. It is an open question whether one can evolve dynamically stable controllers by exposing them to many different line-following tasks during evolution.

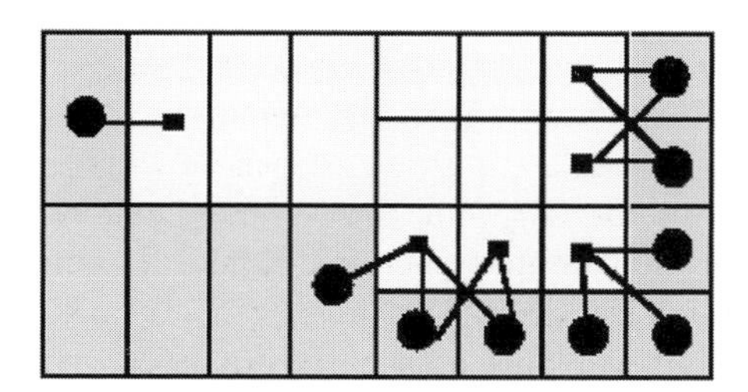

Figure 20.14 The innervation of actuators by sensors in detail. Only the bottom half of the agent is shown.

20.5 Discussion

In this chapter we made an attempt to incorporate a developmental model in the artificial evolution of complete autonomous agents, i.e. with a 'body' and a 'nervous system'. We presented two models that operated at different levels of abstraction from the biological phenomenon of development, although we feel that both possess many of the properties that are at the core of development, most notably the dynamic interplay between genome and cell state. In addition, they are both able, albeit in quite different ways, to account for the emergence of a nervous system on top of a developed multicellular body arrangement.

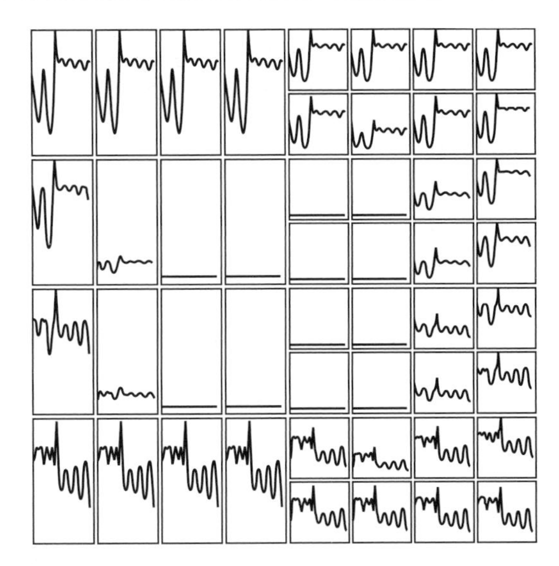

Figure 20.15 The activation of each neuron/cell during the execution of the task. Time is on the x-axis of each cell, the bottom and top of the cell represent activation of 0 and 1.

The first and most complex model is more biologically defensible in its details. It represents an initial attempt at extending our earlier work (Dellaert and Beer, 1994a) to include neural development. We were able to demonstrate its expressiveness by showing that it can account for the development of agents complete with a neural control architecture, capable of executing a simple avoidance task in simulation. We were also able to evolve incrementally better agents starting from that hand-designed agent.

The second and simplified model was adopted in the hope that we could use it to evolve agents from scratch. It used a model for the genome and cell state that was somewhat more removed from the biological example, random Boolean networks. These have the advantage of being computationally cheaper and they lend themselves to analysis using tools from dynamical systems theory. In addition, the second model used a considerably simplified model for neural development. We have used it successfully to evolve agents from scratch that can execute simple tasks. The line following agent was presented as an example.

In the domain of autonomous agents, one could extend this work to investigate whether more complex tasks are within reach of the model. A simple extension would be to evolve line followers with a stable controller for any type of line, given some smoothness constraint. We have also experimented with introducing learning into the process. If an agent is capable of learning during its lifetime, we might benefit from the Baldwin effect to speed up evolution (Whitley *et al.*, 1994). Finally, it would be of interest to investigate the behaviour of the model under selection of an implicit fitness function, i.e. where agents are placed in a world in which their only task is to outsmart their peers. Here, we expect the complexity of the simulated world to be reflected in the complexity of the agents and perhaps this brings out the potential of a developmental model better than using explicitly designed fitness functions.

However, perhaps the most intriguing outcome of this work is that such models can conceivably be used in the quest to understand more about the logic of actual, *biological* development. In the case of the RBN model (and possibly also for the genome-cytoplasm model), it is possible to use analysis tools like a phase portrait to visualize what processes unfold during the simulated developmental process. We might one day be able to use the abstractions that are explored here to visualize the dynamical trajectories being traversed by cells in actual organisms. In addition, we can try to understand more by trying to synthesize observable aspects of development using these models. Some of this has already been explored in Dellaert (1995). We have, among other things, examined the use of phase portraits to visualize the dynamics of interconnected RBNs (as used in our model for intercellular communication) and we have succeeded at synthesizing pattern formation mechanisms like those found underlying the development of compound insect eyes.

In summary, development is an important, powerful and integral element of biological evolution. It is our hope that explorations such as those we have presented here, will contribute to its understanding, both in its own right and as an element of autonomous agent research.

Acknowledgements

We would like to extend special thanks to James Thomas, Katrien Hemelsoet and Shumeet Baluja for their helpful comments. This work was supported in part by grant N00014-90-J-1545 from the Office of Naval Research.

This chapter is from a paper Dellaert, F. and Beer, R.D. (1996) A developmental model for the evolution of complete autonomous agents. In Simulation of Adaptive Behaviour, SAB 1996 Conference Proceedings: Maes, P., Mataric, M., Meyer, J.-A. *et al.* eds. From Animals to Animats 4. MIT Press/Bradford Books.

References

Beer, R.D. and Gallagher, J.C. (1992) Evolving dynamical neural networks for adaptive behaviour. *Adaptive Behaviour*, **1**, 91–122.

Belew, R.K. (1993) Interposing an ontogenic model between genetic algorithms and neural networks. In *Advances in Neural Information Processing Systems (NIPS) 5*, Hanson, S.J., Cowan, J.D. and Giles, C.L. eds. Morgan Kauffman.

Cangelosi, A., Parisi, D. and Nolfi, S. (1993) Cell division and migration in a 'Genotype' for neural networks, *Technical Report PCIA-93*, Institute of Psychology, C.N.R., Rome.

de Boer, M.J.M., Fracchia, F.D. and Prusinkiewicz, P. (1992) Analysis and simulation of the development of cellular layers. In *Artificial Life II*. Langton, C.G. *et al.* eds. Addison-Wesley.

De Garis, H. (1994) CAM-Brain: the genetic programming of an artificial brain which grows/evolves at electronic speeds in a cellular automata machine. ICNN'94, Int'l Conf. on Neural Networks, June 1994, Orlando, Florida.

Dellaert, F. (1995) *Toward a Biologically Defensible Model of Development*, Masters Thesis, Case Western Reserve University.

Dellaert, F. and Beer, R.D. (1994a) Toward an evolvable model of development for autonomous agent synthesis. In *Artificial Life IV, Proceedings of the Fourth International Workshop on the Synthesis and Simulation of Living Systems*. Brooks, R. and Maes, P. eds. MIT Press.

Dellaert, F. and Beer, R.D. (1994b) Co-evolving body and brain in autonomous agents using a developmental model, *Technical Report CES-94-16*, Dept. of Computer Engineering and Science, Case Western Reserve University.

Fleischer, K. and Barr, A.H. (1994) A simulation testbed for the study of multicellular development: the multiple mechanisms of morphogenesis. In *Artificial Life III*, Langton, C.G. ed. Addison-Wesley.

Gruau, F. and Whitley, D. (1993) The cellular development of neural networks: the interaction of learning and evolution, *Research 93-04*, Laboratoire de l'Informatique du Parallélisme, Ecole Normale Supérieure de Lyon.

Jackson, E.R., Johnson, D. and Nash, W.G. (1986) Gene networks in development. *J. Theor. Biol.*, **119**(4) 379–396.

Jakobi, N. (1995) *Harnessing Morphogenesis*, Technical Report School of Cognitive and Computing Sciences, University of Sussex.

Kauffman, S. (1969) Metabolic stability and epigenesis in randomly constructed genetic nets. *J. Theor. Biol.*, **22** 437–467.

Kauffman, S.A. (1993) *The Origins of Order*. Oxford University Press.

Kitano, H. (1994) Evolution of metabolism for morphogenesis. In *Artificial Life IV, Proceedings of the Fourth International Workshop on the Synthesis and Simulation of Living Systems*. Brooks, R. and Maes, P. eds. MIT Press.

Kodjabachian, J. and Meyer, J.-A. (1994) Development, learning and evolution. In *Proceedings of PerAc '94: From Perception to Action, Lausanne*. Gaussier, P.N., Nicoud, J.-D. eds. IEEE Comput. Soc. Press. pp. 96–109.

Lyndenmayer, A. and Prusinkiewicz, P. (1989) Developmental models of multicellular organisms: a computer graphics perspective. In *Artificial Life*. Langton, C.G. ed. Addison-Wesley.

Mjolsness, E., Sharp, D.H. and Reinitz, J. (1991) A connectionist model of development. *J. Theor. Biol.*, **152**(4) 429–453.

Nolfi, S., Miglino, O. and Parisi, D. (1994) Phenotypic plasticity in evolving neural networks. In *Proc of the First Conference from Perception to Action*. Lausanne. Gaussier, P.N. and Nicoud, J.-D. eds. IEEE Computer Soc. Press.

Sims, K. (1994) Evolving 3D morphology and behaviour by competition. In *Artificial Life IV, Proceedings of the Fourth International Workshop on the Synthesis and Simulation of Living Systems*. Brooks, R. and Maes, P. eds. MIT Press.

Whitley, D., Gordon, V.S. and Mathias, K. (1994) Lamarckian evolution, the Baldwin effect and function optimization. In *Parallel Problem Solving from Nature – PPSN III*. Davidsor, Y., Schwefel, H.-P. and Manner, R. eds. Springer-Verlag.

Wilson, S.W. (1989) The genetic algorithm and simulated evolution. In *Artificial Life*. Langton, C.G. ed. Addison-Wesley.

Wuensche, A. (1993) Memory, far from equilibrium. In *Proceedings, Self-Organization and Life: From Simple Rules to Global Complexity, European Conference on Artificial Life (ECAL-93)*, pp. 1150–1159. MIT Press.

Wuensche, A. (1994) The ghost in the machine: basins of attraction of random Boolean networks. In *Artificial Life III*, Langton, C.G. ed. Addison-Wesley.

Harnessing morphogenesis 21

NICK JAKOBI

21.1 Introduction

Many people have noted the difficulties involved in designing control architectures for robots by hand (Brooks, 1991; Husbands and Harvey 1992). As robots and the behaviours we demand of them become more complicated, these difficulties can only increase. Evolutionary robotics is an attempt to overcome some of these difficulties by automating the design process (see Harvey *et al.*, 1994 for an example). Artificial evolution is employed to evolve control architectures using a genetic algorithm (Goldberg, 1989). In the experiments reported in this chapter, a population of initially random genotypes (character strings) are developed into phenotypes (artificial neural networks) and assigned a fitness value based on their ability to control a small mobile robot at a simple task. A new generation of genotypes is then 'bred' from the old population by probabilistically selecting fit parent genotypes and applying the genetic operators of crossover and mutation to create offspring genotypes. Over many generations, average fitness gradually improves until individuals evolve that can perform the task satisfactorily.

Both developmental biology and evolutionary robotics, then, deal with processes by which genotypes become phenotypes. This is where the similarity between the two disciplines ends at the moment. With a few notable exceptions (among them Dellaert and Beer, 1994; Nolfi *et al.*, 1994; Vaario, 1994), the crude bit-string to neural network transformations normally used in evolutionary robotics are clumsy parodies of the subtle mechanisms of morphogenesis that developmental biologists are only beginning to uncover. In itself this is no bad thing, but since no entirely satisfactory method of encoding really complicated neural networks in a compact way has been discovered, this chapter represents a step towards exploiting the encoding scheme used in nature.

The inspiration behind the encoding scheme described in this chapter comes from biology, but it was designed with evolutionary robotics firmly in mind and not as a defensible model of biological development. The overall picture to be gleaned from the model, described in the next section, will at least be reminiscent to the developmental biologist of what actually takes place during morphogenesis, but where precise functional details of a particular developmental mechanism were unavailable, or where the biological details appeared to conflict with the goals of evolvability and flexibility of design necessary to any encoding scheme, alternative

 ISBN 0-12-428765-4

mechanisms were implemented. Biological terms are used extensively throughout the chapter. Unless stated otherwise they should be seen as explanatory tools rather than strong references to their biological counterparts.

21.2 An overview of the developmental model

This section gives a general overview of the encoding scheme in three stages. First it describes what takes place at the level of the genome, then how this controls the behaviour of individual cells and finally how a multicellular 'organism' develops from a single cell. In the current scheme, a genome consists of a single string of what might be thought of as base pairs of nucleic acids (i.e. one of four characters). The start of each gene on the genome is identified by a certain pattern of preceding characters, similar to the TATA box on a real genome. Genes are of fixed length. Each gene is responsible for the production of a particular protein and in turn is regulated by certain combinations of proteins within the cell. What is important to realize at this stage is that genes create proteins that regulate genes (see Ptashne, 1986 for a beautifully clear example of this), forming genomic regulatory networks (GRNs) that control the entire behaviour of each cell during development. These are best thought of as independent dynamical systems within each cell, capable of being knocked from one basin of attraction into another by internal and external stimuli, following independent trajectories through state space[1].

The proteins within each cell are divided into different classes. Each has a unique effect on the gross behaviour of the cell. The ability of a given protein to perform its role within its class depends on its 'shape'. Signal proteins diffuse out of one cell and into another, allowing cells to influence each other at a distance. The direct consequences of this influence may take one of two forms, either perturbing the cell's internal dynamics (turning certain genes 'on' and others 'off') or applying a force on the cell body towards or away from the protein source. In practice, there are many different signal proteins, from many different sources, entering each cell at any given time and each cell will only respond to a subset (depending on the state of its internal dynamics). This means that different cells may behave differently in response to identical external stimuli and also that different cells may behave identically in response to different external stimuli. Initially a single cell is placed in an environment containing a number of strategically placed extracellular signal protein sources. These fixed sources provide a reference for the developmental process. After the initial cell begins to divide, individual cells divide and move, as part of one big dynamical system coupled together by signal protein interactions, until, eventually, they differentiate.

When this occurs, a number of dendrites grow out from each cell guided by 'growth cones' sensitive to unique combinations of signal proteins. On contact with another cell, a synaptic connection is formed. After every cell has differentiated and every dendrite has either connected or died, thresholds and weights are assigned to each cell and dendrite respectively and inputs and outputs are assigned to cells which lie in specific regions of the developmental environment, to form a neural network ready for testing.

[1] For a fuller account of this interpretation, see Kauffman (1993).

21.3 Preliminary experiments

Before going into the details of the encoding scheme, I will first outline the results of preliminary experiments. This is done so that the reader may understand the general ideas behind the work reported here before becoming too bogged down with the finer points. Two initial experiments were carried out to test the encoding scheme. Both were carried out using a simulation (Jakobi *et al.*, 1995) of the Khepera miniature mobile robot (K-Team, 1993) of the correct level of detail to ensure transfer of evolved behaviours from simulation to robot. Both experiments involved a population of one hundred character strings, each of which was 5000 (initially random) characters long. The genetic algorithm in both cases was based on a simple generational model with rank-based selection. Crossover and mutation were the only genetic operators used. Crossover happened at every 'breeding' and the mutation rate was set at 0.002 mutations per character on the genome (about 12 mutations per genome on average).

In the first experiment, the robot was placed at one end of a long corridor with many bends and neural networks were evolved that could guide it down the corridor without crashing into the walls. The fitness function (similar to that used in Floreano and Mondada, 1994) returned the normalized product of three terms: one for going as fast as possible, one for going as straight as possible and one for staying away from walls. By generation 10, networks had evolved that could guide the robot down the corridor. The network displayed in Figure 21.1, which was the fittest individual of generation 50, could successfully guide the robot down the corridor without it touching the walls. The experiment was run for a total of 500 generations but no real improvement was made after the fiftieth generation. In the second experiment, the robot was placed in a rectangular environment containing many small cylinders. The fitness function was the same as that used above. Again, by generation 10, neural networks had evolved which evoked obstacle avoiding behaviour in the robot. The network displayed on the right hand side of Figure 21.1 is again the fittest member of the 50th generation. This network proved to be very good at obstacle avoidance. However, since it only makes use of two of the available eight sensors on the robot, the robot has several 'blind spots', including straight ahead for small objects.

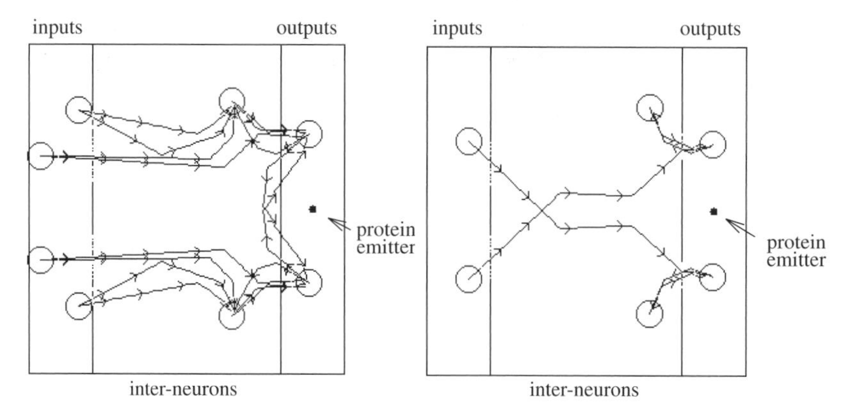

Figure 21.1 Two evolved neural networks. The network on the left exhibits corridor following behaviour, while that on the right exhibits obstacle avoiding behaviour. The input region of the developmental environment was divided into eight subregions, from top to bottom, corresponding to the eight infrared light sensors on the Khepera robot. The output region was divided into two subregions, corresponding to the left and right motors of the Khepera.

21.4 The encoding scheme in detail

The reader who is not interested in the details of the algorithm that converts character strings to artificial neural networks may skip this section. They should, however, be warned that a full appreciation of the issues raised in Section 21.5 is not possible without a more involved understanding of the encoding scheme.

21.4.1 The genomic regulatory network (GRN)

Proteins regulate genes which produce proteins. In other words, proteins regulate other proteins. Each unit in the GRN corresponds to exactly one protein in the cell and the pattern and nature of links between units is defined by which protein(s) regulate which other protein(s). The activity of each unit equals the intracellular concentration of its particular protein (which is equal in turn to the activity of the gene that produces it: maximum 1.0, minimum 0.0). In reality, it may (and usually does) take the presence of several proteins and the absence of several others to activate a particular gene (thus increasing the cellular concentration of the protein it encodes for). There are usually several links fanning into each unit in the GRN, each with a weight between 2.0 and −2.0 and each unit has a unique threshold that sets the lower bound of the linear threshold activation function.

21.4.2 Preprocessing the genome

Various forms of template matching routines are used to work out the nature of the GRN and indeed all protein interactions, including those that have no direct influence on genomic activity. These routines are computationally expensive but, since the GRN and all protein interactions remain fixed for any given genome, the genome may be exhaustively preprocessed before development begins and then discarded. This leaves the GRN (which may now be treated exclusively as a recurrent neural network with linear threshold activation functions[2]) and various relations between the units of the GRN, extracellular stimuli and internal variables that have a direct bearing on the gross behaviour of the cell.

The makeup of a gene is shown in Figure 21.2. As with a real gene there is a short regulatory region that contains several fields and a longer coding region. The 'TATA' region marks the start of the gene, the threshold region defines the amount of stimulation needed to switch the gene on, the link template controls which proteins affect the gene's activity and the type region defines what class of protein the gene codes for. A protein is translated from the coding region by taking triplets of characters to produce a string whose characters derive from a 64 letter alphabet. The ends of each protein string are then joined together to create a circle. There are two types of template matching, protein-to-genome and protein-to-protein. These routines are used in three different ways: to find relationships between proteins and the genome, to find relationships between proteins and other proteins, and to find relationships between proteins and fixed class-specific templates (designed to elicit specific information about the functionality of that protein within its class). Each of the 64 characters that may occur in a circular protein string can be regarded as a chemical with a unique affinity for each of the four chemicals (characters) that

[2] Care should be taken to distinguish the GRN from the phenotype as both are, in fact, recurrent neural networks. It is important to keep in mind that one (the GRN) controls the development of the other (the phenotype).

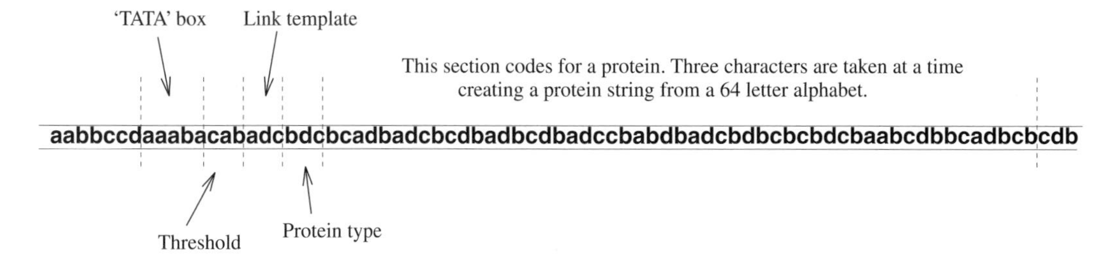

Figure 21.2 The layout of a gene.

make up the genome and with a unique affinity for each of the other 63 chemicals that may constitute a protein. All affinities are values between 0 and 1. Any single character match (protein-to-protein or protein-to-genome) will, therefore, result in one of 64 values between zero and one. These routines, in the current implementation, usually operate on templates of three contiguous characters. These matches return one of 192 values between 0 and 2.

The lower bound of the linear threshold activation function of each unit in the GRN is read directly from the threshold region of the corresponding protein's gene. The triplet of characters is converted from base 4 to a decimal number between 0 and 64 and then to a decimal between plus and minus one. In the nucleus of a real cell, which and how many genes any given protein regulates depends on that protein's shape. This shape is arrived at by a complicated folding process, the biological intricacies of which are poorly understood. In the model described in this chapter, each protein is a circular string. To work out whether any given protein plays a part in the regulation of any given gene (and hence its product), the protein is rotated over the link template of the gene until, if at all, the value of the match between the template and the protein string segment it is opposite goes over a threshold (Figure 21.3). If the threshold is surpassed

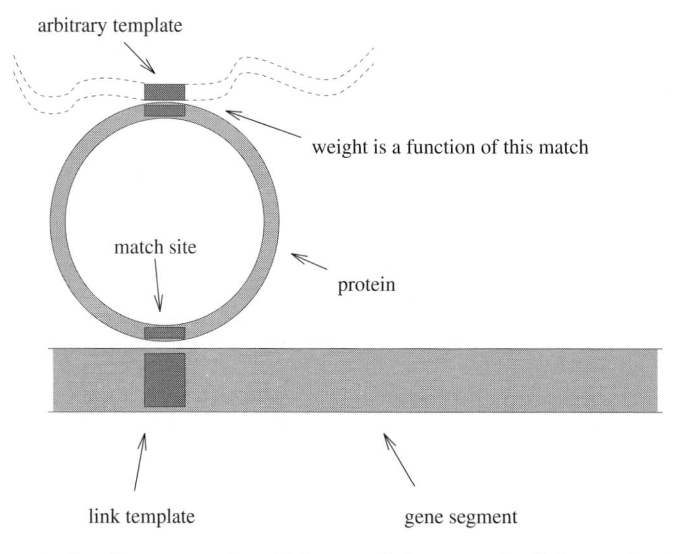

Figure 21.3 A protein has been rotated until the match between the link template on the gene and the corresponding protein segment goes over a certain threshold. The contribution this protein makes to the regulation of the gene is then calculated from matching the diametrically opposite side of the protein with an arbitrary fixed template.

then the protein plays a role in the regulation of the gene and a link is formed in the GRN between the unit corresponding to the protein and the unit corresponding to the gene's product. The weight on this link is found by matching three characters of the circular protein string, diametrically opposite to where the link template matched, against an arbitrarily defined but fixed template that might conceivably be thought of as part of an RNA polymerase molecule.

Also at the preprocessing stage, all proteins are assigned to a class depending on the base 4 number expressed by the type regions on their genes. The functionality of a protein within its class is found in similar ways to its regulatory role, as described above. There is one special class of protein that is worth mentioning here as it has a direct influence on the nature of the GRN. These are the signal proteins that may be emitted from a cell or extracellular source and may enter through the wall of a cell to affect the dynamics of its GRN. A signal protein plays no regulatory role within the cell it is produced in, but only within other cells it has entered through the cell wall. The amount of signal protein diffusing out of a particular cell is calculated directly from the state of that cell's GRN. However, the activations of those genes (and hence the corresponding units in the GRN) within that cell that are regulated by signal proteins are a function of the amount of signal protein diffusing into that cell through the cell wall.

21.4.3 Protein classes

There are six classes to which a protein may belong: signals, movers, dendritics, splitters, differentiators and threshold proteins. Each plays a crucial role in the determination of individual cell behaviour and hence to the overall shape and functionality of the final organism. The following is a description of the role of each class within the cell.

Signals

These proteins diffuse uniformly out from the cell within which they are produced, creating a chemical field (there is no time factor involved) in the space around them. They infuse into cells lying on the resultant chemical gradient wherein they may influence the internal dynamics of the GRN, turning some genes on and others off, or interact with mover proteins.

Movers

Mover proteins control the way in which individual cells move. Each mover protein responds to a subset, particular to that mover protein, of the available signal proteins. This subset is found (during preprocessing) by finding all signal proteins that have a contiguous 3-character string segment that matches a similar 3-character string segment somewhere on the mover protein, with a value that exceeds a certain threshold. This is akin to finding an orientation of two proteins whereby they 'fit together'. If a signal protein matches a mover protein in this way then the affinity of the mover protein (treated as a weight) for the signal protein is defined as the value returned by matching 3 characters on the signal protein, diametrically opposite to the match site with the mover protein, to a fixed class-specific template.

Dendritics

Dendritic proteins control the way in which dendrites grow. Each dendritic protein also responds to a subset, particular to that dendritic protein, of the available signal proteins. This subset, and the affinity of a dendritic protein for each of the signal proteins in its subset, is found in exactly the same way as for mover proteins, described above. In addition, dendritic proteins also have a threshold associated with them, found by taking the lesser of the maximum match values of two class-specific templates rotated around the protein. This threshold represents the concentration of dendritic protein that must be present in a cell (at differentiation) for the particular dendrite to grow. The dendrite will be excitatory or inhibitory depending on which of the two templates has the lesser maximum match.

Splitters

A cell may split in one of two ways, either parallel to the direction of the total force acting on the cell or at right angles to it. Which, if either, of these two options occurs depends on which of two internal variables goes over a certain threshold first. Splitter proteins have two weights between 0 and 1 attributed to them, one for each of these variables, found by taking the maximum match values of two class-specific templates rotated around the protein. At each time step the weighted sum of splitter protein concentrations in the cell is added to each variable; if it splits, both variables are reset.

Differentiators

Differentiator proteins work in much the same way as splitter proteins. Each cell contains an internal variable which is not reset if the cell splits. At each time step the weighted sum (each differentiator protein also has a weight between 0 and 1 attributed to it, found in the same way using a class-specific template) of differentiator protein concentrations is added to this internal variable until it, also, goes over a certain threshold and the cell starts to differentiate.

Threshold proteins

After development has ceased, the finished structure is translated into a neural network, with cells becoming units in the network and dendrites becoming the links between these units. Threshold proteins are responsible for setting the thresholds on the units in the final neural network. At differentiation, these thresholds are calculated from the normalized sum of the total concentration of threshold protein within each cell.

21.4.4 Initial development

After the genome has been preprocessed, development may begin. A single cell is placed in a two-dimensional environment with a number of strategically placed extracellular signal protein sources. These protein sources emit chemicals in exactly the same way as a cell does and should not be regarded as somehow having more control over the developmental process than individual cells that emit signal proteins. The concentration of signal protein in the environment drops off with distance from the source. This means a concentration of 1 at the signal source drops to 0 at a distance of 100 (the developmental environment used in Section 21.3 was a 100 by 100 square) with a non-linearity caused by the $\sqrt{d}$ term which ensures that all signal protein maxima will be located at signal protein emitters (important for dendrite growth).

At each time step the concentrations of every signal protein in the local environment of each cell is calculated. This is done by iterating through all signal protein sources and adding the effects of those sources that emit the same chemical. The GRNs of each cell may then be updated synchronously. First, new inputs for all nodes in each GRN are calculated from the weighted sum of recurrent connections and any external signal protein inputs and then new outputs are calculated using a linear threshold function with lower bounds set by the threshold on each node. All internal variables are then updated.

Forces on each cell are calculated by the application of mover proteins to the signal proteins in each cell's locality. Each mover protein causes a force proportional to, and in the direction of the maximum gradient of, the weighted sum of the local chemical concentrations of the members of its particular subset of signal proteins. The weights in this sum correspond to the mover protein's affinities for each of the members of its subset. The cell moves in the direction of the resultant of these forces.

If any internal variable (for splitting or for differentiation) has gone over its threshold, then the appropriate action takes place. Note that when a cell splits, each daughter cell is identical to the mother cell in all but a slight positional displacement at right angles to the axis of cleavage. When the internal differentiation variable within a cell goes over a threshold, the cell differentiates. At this point the cellular concentrations within the cell become fixed, including the levels of signal proteins output from that cell.

21.4.5 Differentiation and dendrite growth

At differentiation the number of dendrites that will grow from a cell is calculated. This is the number of dendritic proteins whose intracellular concentration lies above the threshold associated with that protein (see Section 21.4.3). As with mover proteins, each dendritic responds to its own particular subset of signal proteins. Growth starts from the side of the cell where the weighted sum of the concentrations of each signal protein in this subset is greatest.

Dendrites are guided by a trident-shaped 'growth cone' with three sensors, one at the tip of each fork, each responsive to the dendritic protein's subset of signal proteins. At each time step the concentrations at each sensor are calculated and the position of the base of the trident is updated to that of the trident tip with the highest concentration. In this way the dendrite is steered, at a fixed speed, towards the local maximum of the weighted sum. This will usually be another cell, at which point the dendrite forms a connection with that cell and stops. However, since the sum is weighted, the dendrite may actively steer away from signal protein sources as well as towards them. This creates two sorts of problems. One, dendrites may grow off to infinity, and two, they may go into orbit around local chemical minima. For this reason there is a maximum length to which a dendrite may grow before it is said to be dead.

21.4.6 Interpreting a neural network

Once all cells have differentiated and all dendrites have either connected or died, the finished structure may be interpreted as an artificial neural network (ANN) which can then be used to control a robot. The architecture and connection matrix of the ANN is taken directly from that of the developed organism. Activation is interpreted to flow in the direction of dendrite growth. Each link is either inhibitory or excitatory (depending on the results of a template match). The weight on each link is calculated from the concentrations of the signal protein subset attributed to the dendritic protein responsible. The weighted sum of these concentrations, at the point where the growth cone connects, is divided by the maximum possible value of this weighted sum (i.e. as if all the concentrations were 1.0) to give a number between 0 and 1. This is then scaled appropriately.

Thresholds on the units in the ANN are calculated directly from the concentration levels of threshold proteins in the corresponding cells. Input and output units to the ANN correspond to those cells that end up in certain regions of the developmental environment.

21.5 Issues of evolvability

There are two properties essential to an encoding scheme capable of evolving complex control architectures for robots. First, it must be robust with respect to the genetic operators used by the genetic algorithm (crossover, mutation, translocation, genome growth etc.), and secondly, it must be capable of 'subroutining', in the sense of encoding for repeated structure in a compact way. The encoding scheme described in this chapter displays both of these properties, which are described below and it is this that makes it potentially very powerful indeed.

21.5.1 Robustness to the genetic operators

If an encoding scheme is to work at all (i.e. to provide the basis for something more than random search) then it must be robust with respect to the genetic operators it is used with. It is crucial, if evolution is to progress, that fit phenotypic traits are not destroyed by the breeding process. Ideally, with the exception of mutation, all genetic operators should cause as little phenotypic disruption as possible. The crossover operator is a special case, since it acts on two genotypes as opposed to one, but I would argue, along with Harvey (1992), that it is only meaningful, in an evolutionary robotics context, to use crossover in conjunction with converged populations.

The encoding scheme, reported here, is robust with respect to operators involving translocation and genome growth. This allows two things: first, genes that work in tandem (e.g. to create a fit subnetwork in the phenotype) may be relocated next to each other on the genotype, thus minimizing the chance of their separation by crossover. Secondly, open-ended evolution becomes possible, where phenotypic complexity is not restricted by the original size of the genome. This robustness follows from the simple reason that, using template matching, parts of the genotype address other parts of the genotype (to form the links of the GRN) by the character sequences that occur there and not, as in many encoding schemes, by a function of the region's location on the genome. If these sequences of characters are relocated on the genotype, or if the genotype is allowed to grow or shrink, it will have minimal effects on the shape and nature of the GRN.

Ideally, mutating any genotype should, in the main, provide only minor changes to the phenotype (of a degree proportional, in some sense, to the changes made to the genotype). However, especially in the case of converged population genetic algorithms, small mutations to the genotype should occasionally result in large changes to the phenotype[3]. Because of the way a developmental scheme works, small scale changes to the GRN that controls development may, nevertheless, produce large scale changes in the phenotype (whole new portions of network may grow, or substructures may get repeated, see below). Also a small change to the GRN may do nothing but slightly alter one parameter of the phenotype. Slight mutations to the GRN, then, provide the right range of changes in the phenotype.

What is less obvious is why slight changes to the genotype, in the encoding scheme reported here, should result in slight changes to the GRN. Section 21.4.2 explains how the GRN is decoded from the genome using a variety of template matching operations. The problem is

[3] Since we are, in effect, in charge of the physics under which development occurs, it is also possible to restrict the form that large scale changes may take. This may well be desirable since some forms of large scale change, such as the insertion of a totally random subnetwork in an ANN, are less likely to be useful than, say, the repetition of an already existing subnetwork in the same ANN. In Goodwin (1990), it is argued that something like this sort of restriction plays a major role in evolution.

that if two or more templates overlap then a single mutation at this point may result in a disproportionately large change to the GRN. In order to minimize this possibility, the template matching routines were designed to extract the maximum amount of information from templates that were as short as possible. Also, the ratio of protein length to template length (e.g. the link templates on the gene and the class specific templates used in preprocessing) was made as large as possible. Inevitably, though, a certain amount of overlap will occur.

However, in practice, evolution very quickly favours those genotypes that display low amounts of overlap, since mutation of these genomes is far less disruptive in general and fit individuals are more likely to have fit offspring. The epistasis caused by this overlap is then almost entirely 'bred out' of the population.

21.5.2 Encoding repeated structure

The positions of the extracellular signal protein sources within the developmental environment restricts the final neural network architectures that are possible. For example, in the experiment detailed in Section 21.3, development unfolds in conjunction with only a single extracellular signal protein source, ensuring bisymmetry in the resultant neural networks. It must be stressed that, with the addition of more extracellular protein sources (in asymmetric positions), fully developed networks would not necessarily be asymmetric. This is because the signal protein produced by any given source may well not template match anywhere on the genome and thus would not affect the dynamics of any cell within its chemical field.

During the developmental process, each cell will only respond to a subset of the signal proteins within which it is immersed. Different cells at different times will respond to different subsets, depending on the particular positions of their GRNs in dynamical state space. This allows the developmental process to buy in or out of symmetry and the same applies to buying in or out of repeated structure. For example, imagine two cells on opposite sides of a developmental environment which is asymmetric with respect to signal protein concentrations. Now, even though each cell is immersed in a different set of signal proteins, it so happens that they both respond to the same subset and that the environment is symmetrical with respect to this subset. If both of the cells' GRNs are at the same position in dynamical state space at this time, then they will display the same behaviour, since, in effect, their internal and external influences are identical, and they may split, move and differentiate to produce identical subnetworks on either side of the developmental environment. However, cells elsewhere in the developmental environment may respond to different signal proteins subsets and may, therefore, develop asymmetrically. The final developed network will be asymmetric as a whole but with two identical subnetworks.

21.6 Conclusions

This chapter represents an attempt to harness the developmental power of morphogenesis to the still young discipline of evolutionary robotics. A biologically inspired encoding scheme has been outlined, capable of encoding repeated structure and symmetry in a compact way, which is robust to the wiliest of genetic operators. It has been successfully used to evolve two simple behaviours for a small mobile robot.

It is the author's hope that with time, a descendant of the current encoding scheme will make possible the evolution of far more complicated behaviours than those outlined in Section 21.3, and maybe even robots themselves (see Dellaert and Beer, 1994). The scheme's ability to cope with open ended, incremental evolution (Harvey, 1992), while building up a library of useful subnetworks on the genome, may finally provide the evolutionary robotocist with the means to compete (in terms of time and resources) with other more traditional methods of control architecture design. This is a necessary step if evolutionary robotics is to fulfil its early promise.

21.6.1 Postscript November 2002

Much of the artificial life mobile robotics work of the 1990s centred around new and more complex ways of evolving simple obstacle avoidance. My work (performed in 1994) has to represent one of the most complex methods of achieving this simple behaviour to date. Its main motivation was an engineering one, to find a more powerful way of designing robot control architectures using artificial evolution. There was a feeling in the community at the time that some of the difficulties in evolving really complex behaviours stemmed from the simplicity of the developmental schemes being used to map genotype to phenotype. I therefore set out to make a more complicated developmental scheme, stealing as many ideas as I could from biology. The results were ultimately unsuccessful in terms of my engineering aims, but in hindsight they had quite a lot to say about the nature of evolutionary search.

As pointed out in the chapter, a good encoding scheme for evolving structures such as neural networks for robots must be able to: (a) respond to mutation sometimes in a small fashion and sometimes in a big fashion, (b) be capable of encoding for repeated structure at all levels of complexity. The encoding scheme described in the chapter is capable of doing this and yet, in practice, it was actually very hard to get it to evolve anything but the simplest behaviours. The reason for this is that there is another property of successful encoding schemes, including those present in the natural world, that the chapter does not deal with: an offspring must be able to differ in *interesting* ways from its parents. Ideally, we would like mutation always to make a small or large change in the robot's behaviour, but hardly ever a lethal one. In the encoding scheme described in the chapter, however, mutation either produced no change at all or such a large and totally unpredictable change that the resultant network did nothing at all when downloaded onto the robot. Thus the entire methodology described in the chapter actually represented a fairly poor way of exploring behaviour-space which was ultimately what it was created to do.

There are several ways of understanding the search mechanism described here. One is as a genetic algorithm which is searching genotype space for genotypes of high fitness, where fitness is measured according to a function that incorporates a complicated developmental process, and the genetic operators are moving the focus of search from one genotype to another. However, this is not a particularly helpful way of looking at things. Much better is to think of the process as an evolutionary algorithm in a broader sense acting directly on phenotype (i.e. behaviour) space. In this space the move operators that transform one phenotype into another are the genetic operators transformed through the developmental process and it is the properties of these operators that we are concerned with. For engineering purposes, we would like to be able to design and tune these phenotypic move operators so that they search the space as efficiently as possible. With a developmental process as complicated as the one in the chapter, tuning these operators in practice is very hard if not impossible. In the natural world, of course, the developmental process itself has evolved so that the phenotypic search that results from genotypic changes is performed efficiently.

In my current work I have changed my emphasis. I now do commercial combinatorial optimization, still using evolutionary methods, in domains such as vehicle routing, timetabling and scheduling. Instead of encoding potential solutions as bit strings, I operate directly on the solution itself. There is no developmental process. Instead the move operators within the evolutionary process are carefully designed to provide the type of variation *directly* that is likely to search solution space efficiently.

Some may feel that this takes all the fun out of it in that it is possible beforehand to define the class of structures that the system is able to create and that evolution is therefore unable to 'discover' new and surprising structures. However, this sense of discovery only follows from the computational intractability of working out what the phenotypic move operators, which result from transforming genetic mutation operators through a complicated developmental process, actually look like. It does not necessarily result in more efficient evolutionary search.

The direct approach works well for the type of commercial combinatorial optimization problems I deal with where efficient domain-specific move operators are often easy to design. However for certain types of move operator, like for instance an operator that is able to repeat structure at any level of complexity, it is hard to see how such an operator could be made to work *without* a developmental process.

Acknowledgements

During this research Nick Jakobi was supported by a postgraduate bursary from the School of Cognitive and Computer Science at the University of Sussex.

Many thanks to Inman Harvey for allowing me, and encouraging me, to build on the proposed encoding scheme outlined in Harvey (1993).

References

Brooks, R. (1991) Intelligence without representation. *Artificial Intelligence*, **47**, 139–159.

Dellaert, F. and Beer, R.D. (1994) Co-evolving body and brain in autonomous agents using a developmental model. *Technical Report*, (CES-94-16).

Floreano, D. and Mondada, F. (1994) Automatic creation of an autonomous agent: genetic evolution of a neural network driven robot. In Cliff, D., Husbands, P., Meyer, J.A. and Wilson, S. eds. *From Animals to Animats, volume 3*

Goldberg, D.E. (1989) *Genetic Algorithms in Search, Optimization and Machine Learning*. Addison-Wesley.

Goodwin, B.C. (1990) The evolution of generic form. In Maynard Smith, J. and Vida, G. eds. *Organizational Constraints on the Dynamics of Evolution, Proceedings in Non-linear Science*, pp. 107–117. Manchester University Press.

Harvey, I. (1992) Species adaptation genetic algorithms: the basis for a continuing saga. In Varela, F.J. and Bourgine, P. eds. *Toward a Practice of Autonomous Systems: Proceedings of the First European Conference on Artificial Life*, pp. 346–354, MIT Press/Bradford Books.

Harvey, I. (1993) *The Artificial Evolution of Adaptive Behaviour*. PhD thesis, University of Sussex.

Harvey, I., Husbands, P. and Cliff, D. (1994) Seeing the light: artificial evolution, real vision. In *Proceedings of the Third International Conference on Simulation of Adaptive Behavior*, pp. 392–401. MIT Press/Bradford Books.

Husbands, P. and Harvey, I. (1992) Evolution versus design: controlling autonomous robots. In *Integrating Perception, Planning and Action: Proceedings of the Third Annual Conference on Artificial Intelligence, Simulation and Planning*, pp. 139–146. IEEE Press.

Jakobi, N., Husbands, P. and Harvey, I. (1995) Noise and the reality gap: The use of simulation in evolutionary robotics. In Moran, F., Moreno, A., Merelo, J.J. and Chacon, P. eds. *Advances in Artificial Life: Proc. 3rd European Conference on Artificial Life*, pp. 704–720. Springer-Verlag.

K-Team (1993) *Khepera Users Manual*. EPFL, Lausanne, June 1993.

Kauffman, S. A. (1993) *The Origins of Order*. Oxford University Press.
Nolfi, S., Miglino, O. and Parisi, D. (1994) Phenotypic plasticity in evolving neural networks. In *Proceedings of the PerAc'94 Conference*. IEEE Computer Society Press.
Ptashne, M. (1986) *A Genetic Switch*. Cell Press.
Vaario, J. (1994) *From Evolutionary Computation to Computational Evolution*. Informatica.

Evolvable hardware: pumping life into dead silicon

22

PAULINE HADDOW, GUNNAR TUFTE
and PIET VAN REMORTEL

22.1 Introduction

The design productivity gap in the electronic industry is a well-known fact. How can the design community utilize the design capacity that technology is offering and at the same time ensure its correctness? To find solutions to the problem of developing large and complex designs new design paradigms are required (Semiconductor Industry Association, 1997).

One possible solution is to turn away from traditional design techniques following the various design and testing phases and instead allow hardware to *evolve* until a correct solution is found. This technique is termed hardware evolution or equally, evolvable hardware. Evolvable hardware may be considered to be a subset of evolutionary computation, where the evolved solution is represented in hardware instead of software.

In this chapter the field of evolvable hardware is introduced along with some of the limitations which prevent evolution of complex circuits. The process of natural development is analysed with respect to the features which may be represented in a developmental process for circuit development. The technology itself is also studied to identify features which development may exploit. We present some initial experiments for circuit development based on L-systems, a mathematical formalism for development (Lindenmayer, 1968). The results of these experiments along with our analysis of development and our technology lead us to describe some of the challenges we face in our bid to develop electronic circuits.

22.2 Evolving circuits

22.2.1 Evolvable hardware

In evolvable hardware we wish to use the power of evolution to search for solutions both within and outside the traditional design space. Within the traditional design space we can look for more optimal solutions, whereas outside the traditional design space we can explore for solutions to problems unsolvable using traditional design techniques. To achieve evolvable hardware we need an evolutionary algorithm, a circuit problem to be solved and a technology. In general, the technology is usually a reconfigurable analogue or digital array, i.e. an FPAA (a

 ISBN 0-12-428765-4

reconfigurable chip known as a field programmable analogue array) or an FPGA (field programmable gate array).

There are three main methods for achieving evolvable hardware: extrinsic, intrinsic and complete hardware (on-chip) evolution, as shown in Figure 22.1. In extrinsic evolution, the entire evolution process, including fitness evaluation of the individuals, is implemented in software. Extrinsic evolution uses a software simulation of the underlying hardware to evaluate the fitness value of each individual. This may be an advantage if technology independence is required or if the technology is not available for on-chip/on-board testing.

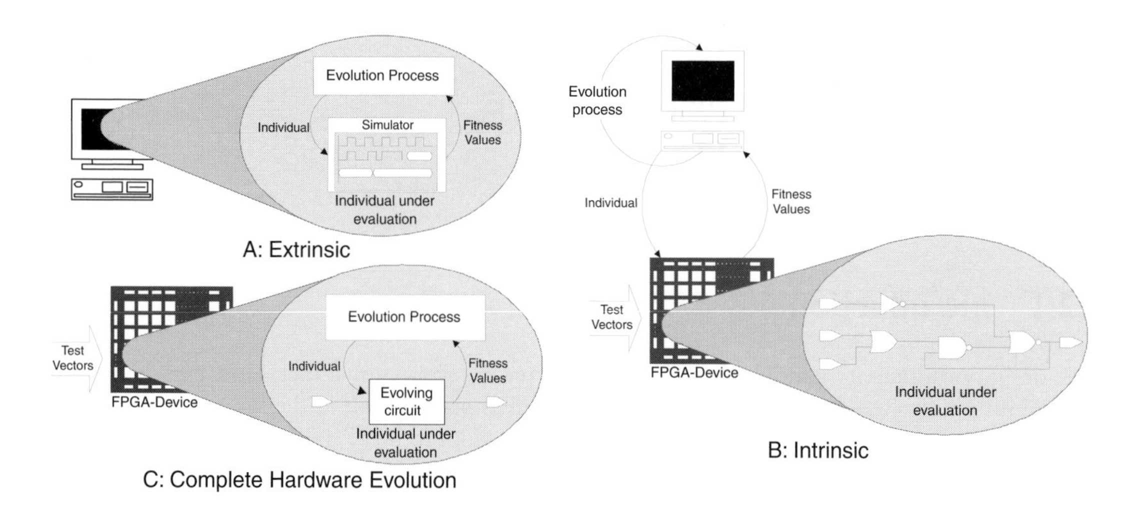

Figure 22.1 Evolving circuits.

On the other hand, if technology is the goal then more accurate fitness values may be obtained from a physical implementation. Testing of each individual on the desired technology itself is termed intrinsic evolution. As shown, the evolution process is still implemented in software but fitness evaluation requires that each individual be downloaded (programmed) onto the FPGA for testing. The resulting fitness values are then fed back to the evolution process to complete fitness evaluation of the given individual.

The final and less used form of evolution is an on-chip strategy, also termed complete hardware evolution (CHE) (Tufte and Haddow, 1999; Macias, 1999), where the evolution process is located on the same chip as the evolving circuit. This, as in intrinsic evolution, enables technology specific evaluation of the evolving circuit, but the on-chip GA speeds up the evaluation phase since only on-chip communication is needed. Another similar setup is that of an on-board processor holding the evolutionary algorithm (Murakawa *et al.*, 1999).

22.2.2 Limitations in today's evolvable hardware

The field of evolvable hardware (EHW) promises many possibilities within optimization and exploration of new circuit designs apart from one missing factor: scalability. Scalability is the property of a method or solution to keep on performing acceptably when the problem size increases. In our case, acceptable performance may be said to be a non-exponential resource increase and performance decrease. Our problem domain is the evolution of electronic circuits.

In recent years, small electronic circuits have been shown to be evolved successfully (Thompson, 1995; Higuchi *et al.*, 1996; Koza, 1997) and, although complex circuits have been evolved (Thompson, 1996), these may be said to be *structurally* complex only. Evolution of complex functionality is still beyond our reach.

The introduction of evolutionary design methods may be said to be a move from designing the actual hardware to finding a good representation for the circuit: the genotype, and designing a specification for evaluating its functionality: fitness. Let us consider some of the inherent problems.

Direct mapping

Generally in EHW, a one-to-one mapping, i.e. a direct mapping, has been chosen for the genotype–phenotype transition. This means that the genotype needs to include all the information required for the phenotype, i.e. configuration data to program an FPGA. The larger the circuit to be evolved, the more logic and routing information (configuration data) required. As the complexity of the genotype increases, so does the computational and storage requirements of the evolutionary process.

Indirect mapping

Moving away from a direct mapping to an indirect mapping we have the possibility of introducing a smaller genotype for the large phenotype. However, by such a move, we move complexity from the evolution process over to the mapping process, i.e. the genotype–phenotype mapping.

Fitness values are used by the selection mechanism to find good and bad solutions in the population, consisting of individuals in their genotype form. Due to the indirect mapping, the genotype is now structurally different and will include less information than that of the phenotype. However, fitness is an evaluation of the phenotype. As such, we are evaluating one way to represent a circuit, i.e. the phenotype, to provide feedback as to the quality of a different representation of the same circuit. As such we have an inconsistency. How do we then provide useful information on good and bad genes in the genotype by evaluating a phenotype?

Representation of genes in the genotype

It is hard to make a good representation for circuits with wiring and gates and avoid the problem that one small change in a genotype may change an almost working circuit (phenotype) to a circuit with no measurable functionality (Miller and Thomson, 1998). This is often the case since fitness is based on measurements or simulations of one single point: the output signal of the circuit. If the change affects the connection to the output we have no way of measuring how good the rest of the circuit is since we have no way of testing the circuit. Although test vectors are fed into the inputs of a circuit, we need to read their effect on the circuit at the output(s).

A similar problem may be seen when we make a change to a gene in the genotype. We would wish that the degree to which the functionality of the circuit is affected is similar no matter which gene is changed. Unfortunately this is hard to achieve. Destructive mutation describes the case where a simple change in a gene causes a major difference to the functionality of the circuit. A similar effect may be seen by destructive crossover where one or two offspring are made with lower fitness than those of the parents. That is we move away from our goal.

Underlying technology

If evolution aims to reduce the constraints of traditional design, is it possible to reduce the constraints of the technology itself by introducing technologies better suited to evolution (Haddow and Tufte, 2000)? In digital EHW today, FPGAs may be seen to be the main target technology. They not only provide a reconfigurable platform, essential for realistic evaluation of individuals, but also chips are commercially available. However, are today's FPGAs 'evolution friendly'? Do they exhibit features which enable us to utilize fully the power of evolution?

To answer this question we need to look at the characteristics of evolvable hardware. Four important characteristics are the size of the phenotype required, flexibility, speed of configuration and robustness of the evolved design.

Today's FPGAs offer more and more resources, requiring more and more configuration data, thus requiring a large phenotype. In addition, there is a relatively slow interface to today's FPGAs when regarded with respect to the fact that evolution of large circuits may require a larger population of individuals and many evolutionary runs, i.e. a large number of reconfigurations. As such, evaluation is very slow. Partial reconfiguration in the newer FPGAs simplifies, but does not remove this problem.

Fitness specification

In the search for solutions for a given task or problem, the problem has to be specified. The solution to the problem is often dependent on the specification of the problem as this steers the implementation of the solution. Moving from traditional design techniques to evolution, specification relates to specifying the genotype and the fitness function. As complexity of the functionality of circuits that we want to evolve increases, so the complexity of the fitness function increases.

22.3 Improving evolvable hardware: shrinking the genotype

In the past two to three years, researchers in the field of both software and hardware evolution have looked to biology to gain better insights into improving existing techniques. Inspired by biological development where, from a single cell a complex organism can develop, we are interested in finding ways in which development may be introduced to genetic algorithms so as to solve the genotype challenge. This challenge may be expressed in terms of shrinking the genotype. We need to move away from a direct genotype–phenotype mapping so as to enable evolution to evolve large complex electronic circuits.

As stated, one of the reasons for this interest in biology is to improve the resource effectiveness of evolutionary techniques. Another reason is a move from optimization to exploration (Bentley, 2000), i.e. from improving existing solutions to finding novel solutions. In the context of circuit design these novel solutions might be novel due to unique solutions to unsolved problems of a reasonable complexity, i.e. of the complexity level solvable using today's techniques. On the other hand, they may be novel due to solution of more complex problems than those solvable today.

Shrinking the genotype effectively moves the complexity problem over to the genotype–phenotype mapping thus increasing the complexity of the mapping. Artificial development is this mapping and, as such, we need a development process that can handle such a complex mapping. To simplify the problem, we have chosen to split the mapping into two simpler mappings by introducing a virtual technology between the genotype and phenotype.

The mapping from the virtual technology – a virtual evolvable hardware (EHW) FPGA – to a physical FPGA is a relatively simple mapping as both architectures are based on FPGA principles. The development process itself is the first stage of the mapping. Instead of developing to our complex organism (phenotype) we are developing to a simpler organism, the intertype, i.e. the configuration data for our virtual EHW FPGA.

22.3.1 Decreasing development complexity – introducing an intertype

In digital EHW, FPGAs may be seen to be the target technology as they provide a reconfigurable platform and chips are commercially available. The main elements of FPGA chips are configurable logic blocks (CLBs) connected together in a grid format and configurable routing resources. In addition, configurable input/output blocks (IOBs) are connected to the grid at the perimeter of the chip.

Configuration can be categorized into non-partial or partial configuration. The former may be applied to any FPGA and enables the complete chip to be configured (programmed). The latter is offered on some devices and allows a portion of the chip to be reconfigured while the remaining parts of the chip retain their current configurations.

To offer the possibility for complex designs, FPGAs are both expanding in size, i.e. increasing the number of combinational logic blocks per chip, increasing the complexity of these blocks and introducing vast and varied routing resources. The effect of these features on the configuration data, our phenotype, is a large increase in data, i.e. a more complex phenotype.

As shown by Thompson (1996), evolution may be used to optimize designs in ways unachievable by means of traditional techniques and, in fact, can often result in designs that are not understandable to the human designer. As such, we can exploit fewer resources for achieving complex designs by using evolution. Therefore, these vast resources are not only unnecessary for evolution but, in fact, reduce its effectiveness since there are so many unnecessary resources that must be processed by the evolution process.

The virtual EHW FPGA

The virtual EHW FPGA contains blocks (Sblocks) laid out as a symmetric grid where neighbouring blocks touch one another (Haddow and Tufte, 2001). There is no external routing between these blocks except for the global clock lines. Input and output ports enable communication between touching neighbours.

Each Sblock neighbours onto Sblocks on its four sides. Each Sblock consists of both a simple logic/memory component and routing resources. The Sblock may either be configured as a logic or memory element with direct connections to its four neighbours or it may be configured as a routing element to connect one or more neighbours to non-local nodes. By connecting together several nodes as routing elements, longer connections may be realized.

Figure 22.2 illustrates the internal routing of the Sblock. Each Sblock is responsible for processing input signals and routing signals to its neighbouring Sblocks. There are four pairs of unidirectional wires at each interface. The input wires are attached through routing logic to the input routing channel which provides a dedicated wire for each input. This channel is then fed into the logic/memory block. Output from the logic/memory block is a single wire which feeds the output routing ring. All the four outputs of the Sblock are attached to this output routing ring through the output routing logic.

A more detailed view of the logic/memory block is shown in Figure 22.3. Inputs from neighbouring Sblocks and a feedback from the output are connected to a 5 input look up table (LUT). The LUT may be configured to hold a function. When 'Don't Care' (DC) bits are placed at a given input, then that neighbouring Sblock is not connected to it. In this way the LUT is programmed not only for functionality but external connectivity of the Sblocks. Therefore, to alter the functionality and connectivity of an Sblock only the LUT content need be reprogrammed. When the Sblock is used for routing, all inputs apart from the incoming input to be routed are 'don't care' inputs.

Figure 22.4 illustrates a routing Sblock (R) which takes a west input and routes it to the North. The logic block (OR) then reads its south input as well as its north input and forwards its

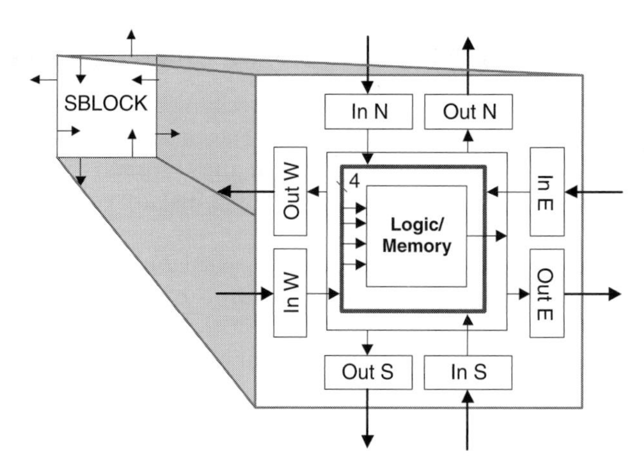

Figure 22.2 Sblock – routing and logic/memory block.

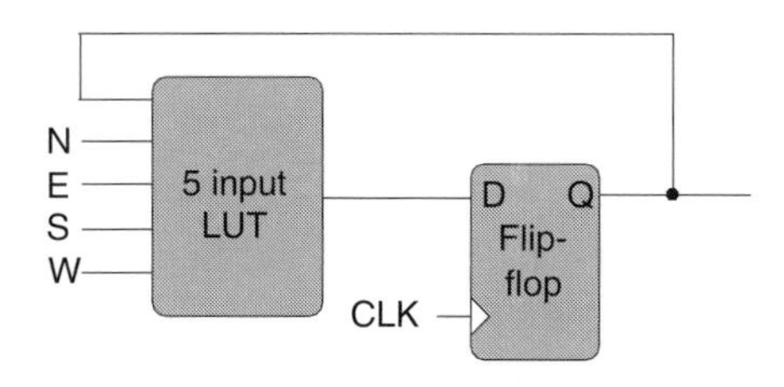

Figure 22.3 Sblock logic/memory.

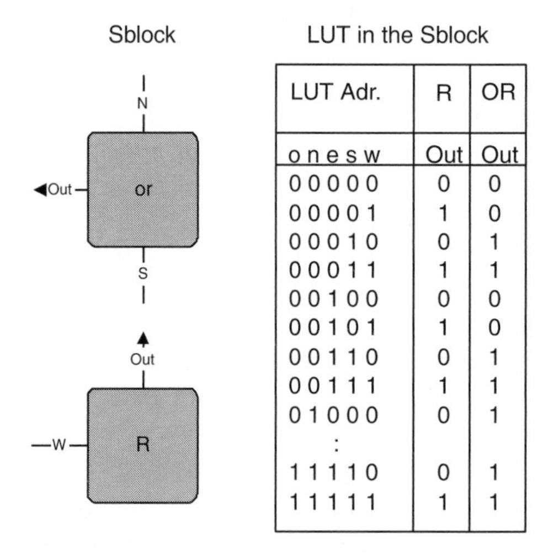

LUT Adr.	R	OR
o n e s w	Out	Out
0 0 0 0 0	0	0
0 0 0 0 1	1	0
0 0 0 1 0	0	1
0 0 0 1 1	1	1
0 0 1 0 0	0	0
0 0 1 0 1	1	0
0 0 1 1 0	0	1
0 0 1 1 1	1	1
0 1 0 0 0	0	1
:		
1 1 1 1 0	0	1
1 1 1 1 1	1	1

Figure 22.4 Sblock as a routing module.

result to its west output. If we look at the LUT output for the routing module (R) we see that the output is identical to the west input (w). As such, the other inputs are treated as 'don't cares'. The LUT for the logic module (OR) will most likely have different input values but, for simplicity, we consider the same LUT inputs and consider the output values for an OR module. The OR column shows that in this case the north and the south inputs have been Or-ed and the other inputs are treated as 'don't cares'. It is worth noting, however, that for clarity we have chosen to highlight only the relevant outputs in this figure but, of course, outputs are available from a given Sblock in all directions.

Development and evolution friendly features

One of the key features of the virtual EHW FPGA that makes it more suitable to evolution is the vastly reduced configuration data needed to program the device, i.e. the intertype. This is mainly due to the lack of external routing which accounts for a significant part of today's FPGAs. In addition, the logic blocks themselves – Sblocks, are relatively simple.

When we consider development we may say that this virtual EHW FPGA is just a small step in the right direction. Long term we can expect that evolvable hardware and developmental techniques will help drive the need for better technologies where these techniques can be further exploited. However, to begin with let us look at what the virtual EHW FPGA offers (Tufte and Haddow, 2003).

Artificial development could, as in nature, use: pattern formation, morphogenesis, cell differentiation and growth as interrelated processes. Biologically we are looking at a cell-based organism where the maximum size of our organism is the maximum grid size of the technology we are using. Each Sblock is a cell in the organism and can communicate with other cells through local communication – although limited to four neighbours due to the grid structure. An organism (circuit) should grow from a single cell (an Sblock) by programming a neighbour cell as part of the organism. Growth should only affect neighbours, i.e. growth causes our growing organism to expand cell-wise within the grid of our technology. In addition, our development algorithm will need to allow for cell death, which would mean deactivating an Sblock by reprogramming it to the default value of non-organism cells.

Cell differentiation may be achieved by reprogramming a given cell. When we consider the case of a single Sblock as a cell, we have a structure limited to communication with the nearest four neighbours and a fixed internal structure. Varying the structure through differentiation would mean reprogramming the Sblock to connect to or disconnect to individual neighbours. Varying the functionality through differentiation involves programming the cell's internal functionality, i.e. its look-up table. It should be noted that programming the cell's connectivity and functionality are intertwined through the look up table, as described earlier.

The development algorithm should control pattern formation within the cells based on the cell types. However, what is a cell's type with respect to Sblocks? If as in biology, the type is the same functionality, e.g. a muscle cell, then perhaps we refer to the functionality of the given block, e.g. a 3-input AND gate. The type will then describe how the cell functions, i.e. how it reacts to various input signals.

If we only consider cells as simple Sblocks then morphogenesis will not be achieved as each cell has a fixed size in the technology. However, it is possible to abstract slightly from the technology and not only view the grid as an organism but perhaps view the cell itself as an expanding body, consisting of one or more cells.

As such, we see features in the virtual EHW FPGA which may be termed 'development friendly'. However, the challenge is to find an artificial developmental algorithm which will take some form of electronic circuit representation and develop it to a functionally correct electronic circuit.

22.4 Combining development with a genetic algorithm

To enable development, one solution is to follow the working of DNA and combine rules into our representation. That is, the genotype includes rules telling how, where and which gene, i.e. component in a component representation, should grow to develop a solution (Bentley and Kumar, 2000). However, no matter how the development algorithm is expressed, it may be useful to use evolution to fine tune the process. For our purposes we wish to integrate development with a genetic algorithm.

The genetic algorithm itself is a standard GA (Holland, 1975). The GA works on a population of individuals where each individual is represented in its genotype form. For a rule based development system, the genotype form may include a start string and a set of rules. Genetic operations are applied to a selected individual and the individual is placed in the fitness module in its genotype form. Development appears in the evaluation of the individual, i.e. during fitness evaluation. Development is a mapping of the genotype, represented in the GA, to a phenotype.

Figure 22.5 illustrates the process of fitness evaluation, including development. The individual is given to the development process in its genotype form. In the case of a rule based genotype, the main operation of the development process is to start up the ring of rules on the starting string and allow the ring to continue until there are no more rules to fire. When the ring is complete the genotype will have developed into a phenotype, in our case a circuit design solution. Figure 22.5 illustrates intrinsic evolution where the phenotype is then down-loaded into the FPGA and fitness values arising from testing of the developed circuit are fed back to the GA.

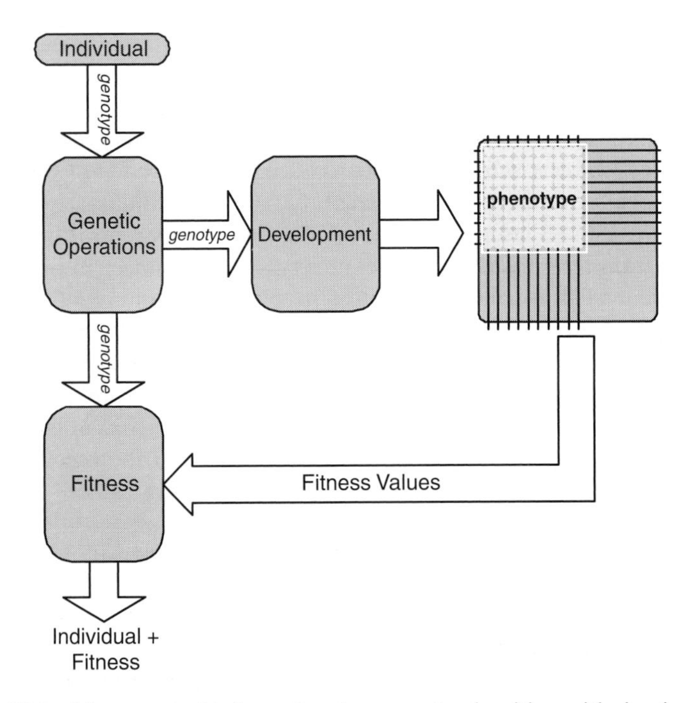

Figure 22.5 Movement of information in a genetic algorithm with development.

22.5 Development in practice: L-systems

As a first case study, we consider L-systems which may be said to be based on differentiation through cell lineage, i.e. line of descent. That is, development originates from a single cell through differentiation expressed as rules for growth. In addition, further specialization may be caused by a cell's internal state, expressed as rules for change.

In this work, L-systems have been adapted to developing digital circuits. A genetic algorithm is used in combination with L-systems to retune the rules. The work presented is a first attempt at using L-systems in combination with a genetic algorithm to achieve development on a digital platform. A more detailed description of this work may be found in Haddow *et al.* (2001).

22.5.1 L-systems introduction

An L-system is a mathematical formalism used in the study of biological development. One of the main application areas is the study of plant morphology. Phenotypes are branching structures attained through the derivation and graphical interpretation of the development process, described by a set of rules. The rules describe how the plant should grow and specialize and interpretation of the development itself enables morphogenesis to be studied.

An L-system is made up of an alphabet, a number of ranked rules and a start string or axiom. Applying a rule means finding targets within the search string which match the rule condition. This condition is a pattern on the left-hand side (LHS) of the rule. This condition is also a string and the string is made up of elements from the alphabet. Firing the rule means replacing the targets, where possible, with the result of the rule, i.e. the right-hand side (RHS) of the rule. Firing of the rules continues until there are no targets found for any rule or until the process is interrupted.

22.5.2 L-systems for circuit development

The work described applies L-system principles to the development of electronic circuits. As such, although the principles are the same the rules and their ring sequence have been adapted to the application in hand, i.e. circuit design, rather than the more common plant development. This work may be said to be of a very experimental nature where the main goal is to find positive signs that a rule based system such as L-systems might have qualities that would make development of circuits possible.

As stated earlier, development is an indirect mapping from a smaller genotype (axiom plus rules) to a larger phenotype (circuit) with the inherent problems of obtaining reliable feedback on good and bad solutions from the developed phenotype. To at least simplify this problem slightly, it was decided that both the form of the rules themselves and the application of these rules would be designed with determinism in mind. As such, a given genotype will always develop to a given phenotype. In addition, to find possible rules for development it was important to involve evolution to help us to refine an initial set of rules.

We have added technology constraints where our intertype technology is the virtual EHW FPGA. Following the principles of biological development and treating an Sblock as a cell, then growth steps are limited to 32 bit Sblocks, i.e. a complete cell. As in any other developing organism we wish to introduce shape and our goal is the shape of an Sblock architecture – a grid. Therefore, growth is limited to the grid size chosen for our virtual EHW FPGA. Change rules both affect connectivity of the architecture – in biological terms communication between cells – as well as functionality – specialization.

Change and growth rules

In the described system there are two types of rules: change and growth rules. Figure 22.6 illustrates part of the rule list used in the experiments herein. Change rules have an RHS string of equivalent length to their LHS string. This is to avoid any growth due to the application of a change rule. As such, it represents specialization due to the cell's internal state. Growth rules introduce not only new cells but cells with different properties to that of their ancestors.

```
1  - 21200110101212 -> 10110100001110
2  - 000012110111   -> 100001110011
3  - 010121121222   -> 100100010100
4  - 2220021022     -> 0111001110
5  - 01001 ->  001011100111011001100000000100111
6  - 10111 ->  100101110001110101010001100101110
7  - 01011 ->  101011001111110111011001000111100
8  - 10101 ->  010111101110011001000010111101010
9  - 00011 ->  111110110010000110010110011010110
10 - 01011 ->  101111010010100100110111100001 01
11 - 01110 ->  110100010010001101010110010000 01
12 - 01100 ->  111111001011001110001010101110 01
```

Figure 22.6 Change and growth rules.

The rules are ranked from the most to the least specific, as shown. The growth rules are given a random priority and ranked accordingly. In this example change rules have been ranked before growth rules. It may be noted that since these rules are randomly generated and then evolved, more than one rule may have the same LHS and therefore the rule ring sequence will control which of these rules will be read first. A 'don't care' feature, typical of digital design, has been introduced to the change rule concept. 'Don't cares' are represented by a 2, as shown. To retain determinism 'don't cares' are only allowed on the LHS of a change rule.

Through change rules, the contents of one or more Sblocks are changed. A change rule targets the locations in the intertype where the string on the LHS matches. Firing the rule means replacing this string at the different targets with the RHS of the rule. The LHS string may be found within a single Sblock or overlapping two or more Sblocks as illustrated in Figure 22.7.

As illustrated in Figure 22.7, each Sblock has an associated seed. If a seed in the array matches the LHS of a growth rule then this means that the associated Sblock is a target for the growth rule. If there is no available space to grow into, i.e. the Sblock is surrounded by configured Sblocks, then there will be no growth at this point. Firing the growth rule means placing an Sblock in the first free location according to the priority rule: north, south, west and then east. In introducing a new seed it is important that determinism is maintained. Therefore, the new seed is given the first four bits of the Sblock which was targeted and its bits 5 to 8 are used to replace its own seed. As shown, the new Sblock is configured by the RHS of the growth rule.

Rule firing sequence

To ensure fairness, we propose firing the rules in batches. The rules are ordered from the most to the least specific but all rules will have a chance of firing before a given rule can fire again. Figure 22.8 illustrates this firing sequence. Once any rule in the batch has fired on a target then no other rule can fire on the same target area. As such, these target areas are reserved for the remainder

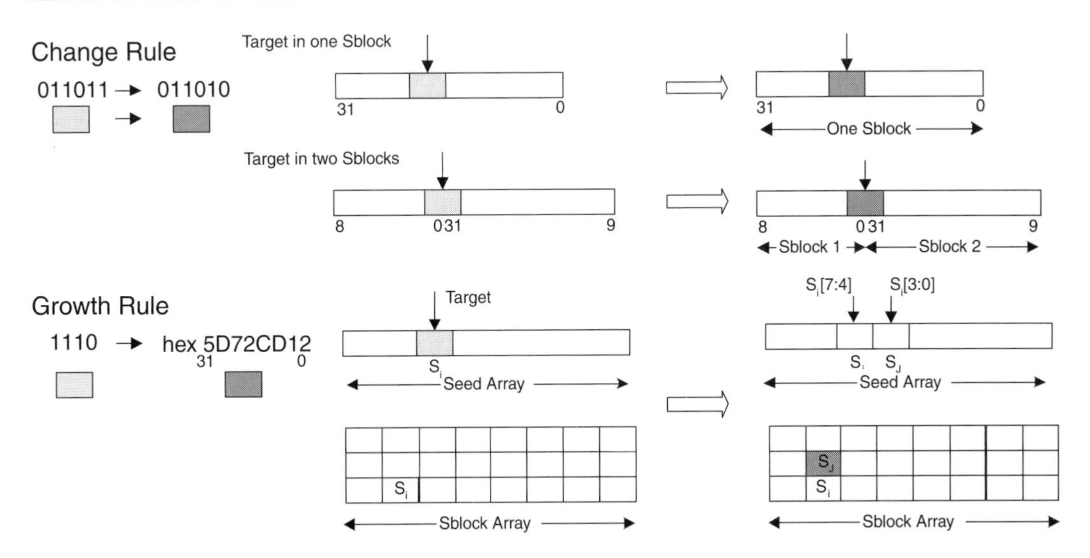

Figure 22.7 Applying change and growth rules.

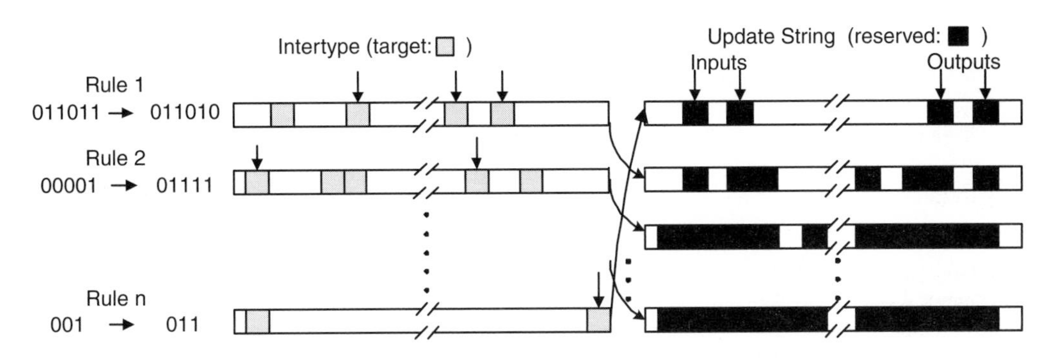

Figure 22.8 Rule firing sequence.

of the current ring batch, as illustrated in what is termed the update string. The default value of the update string is the configuration of the input and output Sblocks.

We assume, as shown, that rule 1 (the top ranked rule) has found targets in the intertype, i.e. matching strings to its LHS. However, checking the update string, one of its targets cannot fire as the target overlaps an input Sblock. The rule fires on the other targets in the normal way by writing its RHS string over the target bits at each target point. In addition, it sets the equivalent bits in the update string. Rule 2 finds five targets but only two are free. Again the rule fires and sets the respective bits in the update string. The last rule of the ranked list finds one target and therefore can fire on that target. The update string can now be reset to its default value and the firing process continues, beginning again with the most specific rule. In the example illustrated in Figure 22.6, growth rules are ranked after change rules and therefore do not affect the firing of the change rules within a given batch. However, their affect will be seen at the next batch where the update string will reflect the size of the grown intertype. The firing process may be stopped either when there are no more rules to fire or when the developed string meets some other chosen condition.

22.5.3 Experimental platform

Figure 22.9 illustrates the workings of our experimental platform. As shown, our platform has been designed with intrinsic evolution in mind but also with the possibility to run extrinsic experiments. The genotype may consist of one or more 32 bit axioms and a ranked list of rules. The process of fitness evaluation is similar to that described earlier apart from the introduction of the virtual EHW FPGA and the possibility to simulate this technology to provide extrinsic fitness feedback.

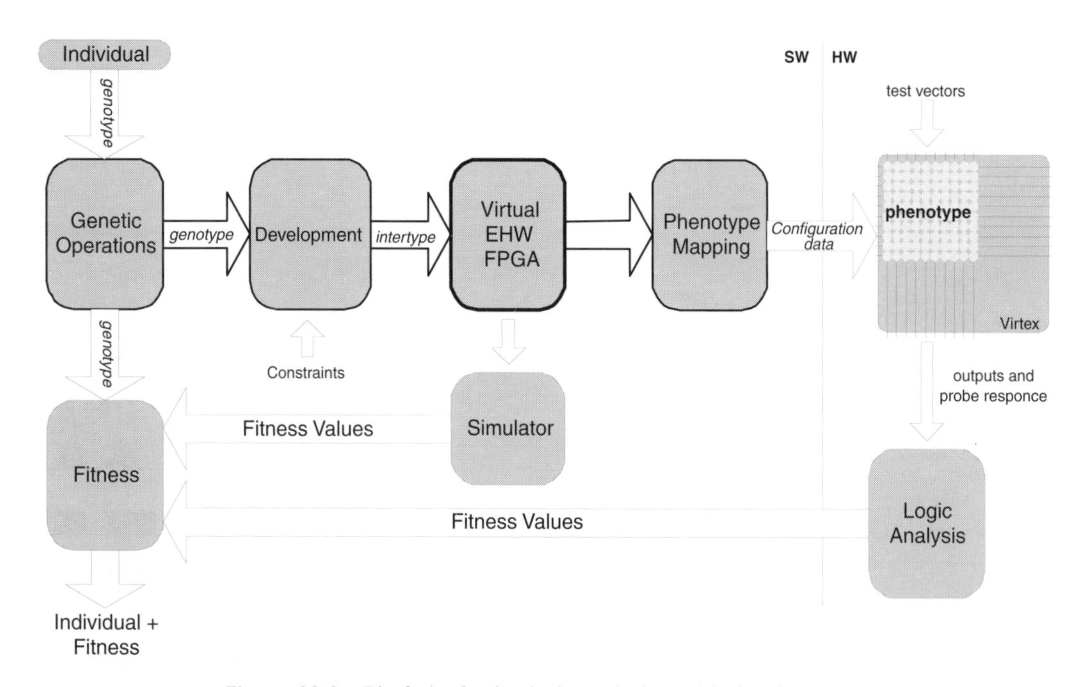

Figure 22.9 Platform for intrinsic evolution with development.

The development process is given the constraints of the required design. These include the technology constraint – Sblock grid size and the design features – the number and position of inputs and outputs. The Sblocks specified in the design features are protected from further changes during the development process. The axiom(s) is allocated to an Sblock location(s). The development process may then begin firing rules, according to the method described above and, as such, develop the genotype to the intertype string. Note that the introduction of the virtual EHW FPGA means that we no longer develop to the phenotype representation but to the intertype representation. A simulator is available to generate fitness values based on the intertype representation. This simulator enables initial studies of the interplay between the L-system rules and the GA to be conducted at the intertype level. Fitness values generated are fed back to the fitness module.

The next stage involves mapping the virtual EHW FPGA design (intertype) to a circuit description for FPGA (phenotype). The phenotype mapping is a translation process from the configuration data expressed in the intertype, to Xilinx Virtex (the FPGA device used) configuration data format. More information about this mapping may be found in Haddow and Tufte (2001). The circuit downloaded onto the FPGA is tested using a set of test vectors and the results of these tests are fed through what is termed a logic analyser which is a tool that helps analysis of data from such a source. The analyser contains a program that generates fitness values from these data and these are forwarded to the fitness module of the GA.

22.5.4 L-systems: experimentation

The experimental goal may be expressed as follows. The 'circuit' has no specified function. From a single axiom in the centre of the Sblock grid, an Sblock solution consisting of north, south, west or east routing modules is grown. The maximum grid size is 16 by 16.

The experiments were conducted using the development platform in extrinsic evolution mode. That is, fitness values were generated from the simulator and not from the FPGA chip itself. The GA parameters were as follows: population 300; roulette wheel selection with elitism; crossover 0.8 or 0 (2 cases); mutation 0.1 and maximum number of generations 150.

Although our goal is to evolve routing Sblocks, fitness evaluation also provided information about how close non-routing blocks are to routing blocks. That is low-input Sblocks are credited more than high input Sblocks except for the case of zero-input Sblocks. These are given a low weighting as they do not offer any routing possibilities. Fitness is expressed:

$$F = 15^{*}R + [6^{*}(C - R) - (6^{*}6in + 5^{*}5in + 4^{*}4in + 3^{*}3in + 2^{*}2in + 1^{*}N1in)]$$

where R is the number of routing Sblocks; C the number of configured Sblocks; Xin the number of blocks with X inputs and $N1in$ the number of inverter Sblocks. A pure router Sblock has only one input and no inverter function.

Little growth was achieved in the initial experiments before the rules stopped firing. To aid growth and ensure that firing continued, we both increased the ratio of growth to change rules and made the change rules less specific, i.e. shorter LHS. A significant improvement in growth was seen but the results presented no clear trend and seemed almost random.

Randomness observed in the results seemed to be caused by the crossover operator. To test our assumptions further, we removed the crossover operator which we believed was forcing the results to jump around, rather than progress steadily towards higher fitness. As a result, the population gradually found better and better solutions and the average fitness improved gradually. Figure 22.10 illustrates one of our runs where a typical pattern of fitness improvement is seen. Decreasing number of inputs is illustrated from dark grey to white. White boxes with a cross are router modules. Black boxes indicate either zero-input Sblocks or non-configured Sblocks, i.e. not grown. In general our solutions achieved 30 to 50% routers in less than 100 generations. Limitations in our current simulator have not allowed us to run enough generations to study further improvements.

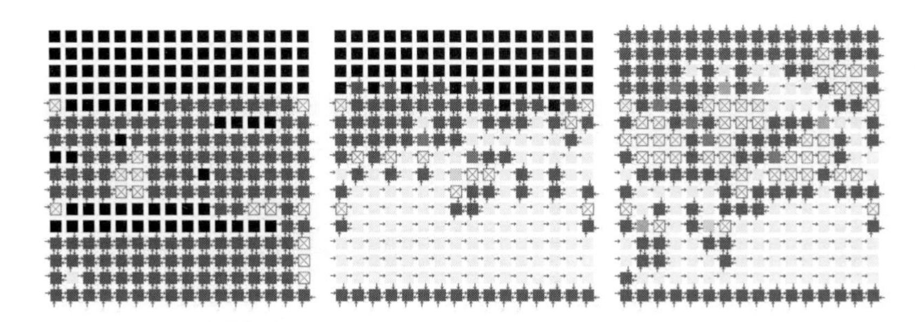

Figure 22.10 Configured Sblocks after 3, 23 and 57 generations.

The achievement of a trend towards increasing routing modules indicates that the combination of the GA and L-systems can at least approach a solution. This solution was achieved through what was intended as a knowledge-poor representation. That is no knowledge was introduced to the rules as to logic or structures expected in the phenotype. However, the experimental results indicate that the current representation might not be as knowledge-poor as intended. Figure 22.10 shows that most router blocks achieved are west router blocks, indicating that west router blocks were favoured over other router blocks. In an earlier set of experiments where zero-inputs were omitted from the fitness calculation, zero-inputs were favoured over the poorly weighted high-input blocks resulting in grids with a large number of zero-inputs. In both cases, the fitness function did not give special credit to the favoured configurations. As such, this tendency towards a specific type of Sblock must lie in the representation, i.e. in the rules.

22.6 Challenges for artificial development

In the search for suitable artificial developmental processes for developing electronic design, the L-system work presented herein was used as a first case study to apply the principles of an existing developmental model to a new application area, i.e. that of electronic design. The goal was to learn how to apply the principles of this model towards the application and to find weaknesses in the use of such a model. This work and continuing work has highlighted a number of challenges that face us within the field of artificial development, both in general and towards the development of electronic circuits.

Since development is an indirect mapping between a genotype and a phenotype, there is less information in the genotype than in the phenotype. Figure 22.11 illustrates how a 128 bit genotype is developed to a 3200 bit phenotype as well as the relationship between the genotype search space and the phenotype solution space. A 128 bit genotype can search a solution space of 2^{128} different phenotypes of length 128 bit, if we assume a direct mapping. If we assume a deterministic development mapping, as shown, then we still search 2^{128} different phenotypes but this time of length 3200 bits as development has extended the phenotype to 3200 bits. As such, the search space for the 128 bit genotype is only a very small part of the phenotype search space.

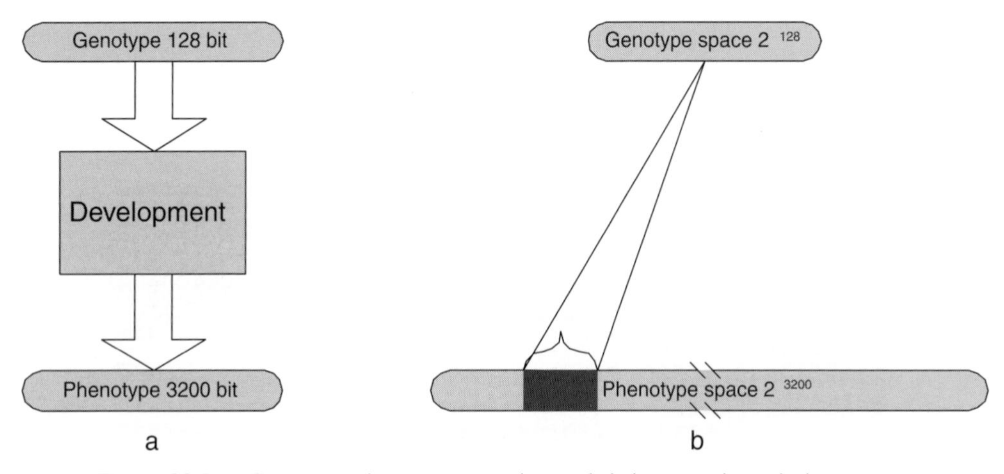

Figure 22.11 Genotype–phenotype mapping and their respective solution space.

How do we represent our 128 bit genotype such that it may be developed to a 3200 bit phenotype? Using development, we may assume that the genotype does not include information on the actual phenotype to be developed but more a building plan of how to build the functionality that we wish to achieve in the phenotype. However, how do we express this?

Using our L-systems work, as an example, our building plan was made up of an axiom and a set of rules. Although to a limited extent we managed to fine tune our rules to achieve a trend towards increasing west routing modules we did not reach the desired phenotype. The main problem may be said to lie in the representation used and in the fitness evaluation (to be discussed later). As stated, our representation included a set of rules which are the description of our building plan on how to develop the axiom. However, designing rules that are suitable for such a purpose and that will, in fact, give rise to the desired phenotype is quite a challenge, even with the help of evolution.

For each new individual with its given axiom and set of rules, the rules were allowed to fire until no more rules could fire. Then a fitness value was assigned to the resulting phenotype. As such, fitness evaluation of each individual included the complete development of an individual from genotype, through the rule firing process, to phenotype and a fitness assessment of the phenotype developed. Thus each individual's fitness evaluation required significant resources. This implies high resource usage for the evolution process since fitness is a part of the process. In fact we have exactly the problem that we are trying to avoid through the use of development.

A further problem is the design of the rules themselves. To provide evolution with greater freedom to explore for solutions, a knowledge-poor representation was chosen (Bentley, 2000). In an L-system context this may be interpreted as the rules having no chosen meaning with respect to improving connectivity or logic. Also the rules are not context-sensitive, relying on neighbouring Sblock information. To begin with the rules in the L-system experiments were randomly generated, evolved, fine tuned to a certain extent through experimental results and then re-evolved. However, with little knowledge in the rules other than random strings it was hard to interpret individual rule effects on the final phenotype.

Fitness evaluation

As we move towards more complex functionality so the challenge to express fitness increases. Unfortunately, the introduction of development to the evolutionary process cannot be said to reduce this challenge but, in fact, perhaps increase it.

When development is added to the evolution process, each individual must be developed before fitness evaluation. As such, fitness is still a measure of the quality of the resulting phenotype. However, to be able to use this quality as a useful feedback to the evolution process, we need to know what parts of the building plan are good or bad with respect to the resulting phenotype. In other words, if we use a genotype consisting of rules, then the fitness function must be able to detect good rules or rule combinations to be able to exploit the genetic material towards a good result. However, how do we detect such features?

Considering the L-system work described we see that feedback from fitness evaluation was limited to feedback over how good the set of rules were, rather than feedback on individual or combination of rules. As such, it was not possible to identify good building blocks. In addition, certain events hid what might be good qualities within given rules. For instance, growth rules caused the system to grow thus increasing the random probability of routing blocks in the developed system. This thus hid any possibility of identifying the effect of change rules themselves. Also, a change rule may have made a routing block but been overwritten by another change rule and as such, the effect of the good change rule is hidden.

Developmental stages

In a biologically inspired developmental process, one might include stages like pattern formation, morphogenesis, cell differentiation and growth. During a certain stage of the development process, the evolution process might give a better fitness score to a given rule or set of rules which are very suitable to that given stage. This rule/rules would then have a higher chance of being selected. This also means that those with lower scores would have less chance of being selected thus losing some of this genetic material. Once this genetic material is lost, i.e. no longer in the population, future stages are unable to use this genetic material. Therefore fitness evaluation in stages, a type of incremental evolution, might be a dangerous way to go.

On the other hand, is it necessary to incorporate all developmental stages in our developmental process? Our goal is not to model and understand the principles of biological development. It is our intention to perform problem solving, in which the observation of biological phenomena in the model is of little real importance and instead the 'raw' performance of the model in fulfilling our goal is of essence, i.e. the development of electronic circuits.

In the case of L-systems, biological development stages implemented are limited to differentiation and growth and cell identity is decided by lineage. On the other hand, cellular automata (Von Neumann, 1966) may also be said to be based on differentiation and growth but cell identity is decided by induction, i.e. a cell's identity depends on its own identity (state) and the states of its neighbours. Cell systems (Lantin and Fracchia, 1995) may be said to combine lineage and induction in its development interpretation. As such, these models although not describing the full stages of development, extract relevant properties which are appropriate for the computational models in mind. However, which properties of development are relevant for developing electronic circuits?

Functionality and structure

If we want to evolve hardware, both functionality and structure are important. Functionality is the task we want our circuit to execute and structure is how the circuit is put together. In EHW, the main goal has been to achieve a functionally correct circuit. There may be a number of structures that can give rise to a given functionality. If we want to find the most effective structure then fitness can include some measure of effectiveness, such as number of gates, as well as a measure of functionality. That is fitness measures both functionality and structure. However, if we are just looking to solve a given function then fitness may measure functionality alone.

However, when we move to some form of development we are looking at a building plan, i.e. a description of activities that will lead to the final organism's structure and functionality. Again we might have the freedom of various building plans to achieve a given functionality, i.e. differing activities leading to different structures with the same functionality. However, our genotype is based on the building plan and, as such, fitness will need to give both structural and functional feedback. It may be said, therefore, that development increases the need to give fitness feedback on structure as well as functionality.

If we consider the development process to be a deterministic mapping then the translation of a certain genotype must always result in exactly one phenotype, i.e. identical structure and functionality. When using evolution to fine tune the development process this has the advantage that fitness assigned to the genotype will always have the same value after development to the phenotype.

However, in nature development may be said to be non-deterministic. That is a dandelion is still a dandelion, i.e. it has the same DNA, if it grows on a hillside or in a valley. However, those that grow on the hillside tend to be smaller, i.e. a different structure, than those that grow in the valley. As such, the DNA reacts in some way to the environment to produce the same functionality with some diversity in structure.

Considering this in the context of electronic circuits could non-determinism be an advantage? If we consider traditional design then many different solutions may be found for a given problem. That is many different designs, i.e. structures, may be found to solve a given problem – a given function. This means that if non-determinism can be used to develop functionality and this functionality may be achieved through different structures, i.e. different phenotypes, then maybe this would improve the chance of development finding a solution to a given problem. When we consider evolution's role in evolving the genotype for development, this structural freedom enhances the possibility for neutral mutations, i.e. mutations that have no measurable effect on the phenotype. Neutral mutations are often believed to be an important mechanism for evolutionary optimizations (Kimura, 1983; Vesselin and Miller, 2000) and therefore may help evolution to evolve appropriate rules for development. However, fitness will need to be more general than that required for deterministic development as fitness will deem a range of phenotypes fit for a given genotype.

22.7 Approaching these challenges

If we wish to reduce resource usage we need to simplify our genotype representation in a way so as to reduce the resources used in the development process and thus in evolution as well. We need to simplify the development process such that it consumes fewer resources. Two possible improvements to our L-system concept are the introduction of limited types and the introduction of rules related to this typing, rather than random bit strings as used in the L-system experiments described herein.

With respect to type reduction, we can use the Sblock as an example. In the experiments described, an Sblock may have up to 2^{32} different formats, i.e. 2^{32} types. The number of types may be limited by, for example, only allowing specified strings, such as those describing 3-input NAND gates and 2-input NOR gates. To limit development to these types, no cell may hold a bit string which does not describe one of these types. This means that both the axiom and the rules in the genotype are limited to strings of these types. However, can we find a development algorithm that can develop a correct phenotype using such a limited typing?

In our former L-system implementation, we used what may be termed as a knowledge-poor representation. However, following the guidelines above and restricting gene types in the genotype by limiting the axiom and rule elements to the allowed for SBlock types, we in fact build more knowledge into our genotype representation. The Sblock types in the developing intertype will be limited to those that one would expect to find in the desired phenotype.

When more knowledge is built into the genotype, it may be expected that this would allow more knowledge to be gained from a study of the development process that has led to a given phenotype. The challenge of being able to provide useful feedback will be simplified as knowledge is now available to be explored. We may study the development process in detail by, for example, studying the activity pattern of genes leading to the phenotype. That is, providing fitness evaluation based on not only the phenotype itself but also the development process. Another approach might be to consider only significant events in the development of a given phenotype.

A further improvement on the development process itself may be achieved by building local knowledge into the development process as well as into the developed phenotype. If we include this feature of natural development, what we are looking for is that a cell in the developing organism is not just developing through a set of rules applying to its own state but the state of its neighbours as well, i.e. induction. In the case of L-system development towards the virtual EHW FPGA, there are four neighbours and therefore induction will be similar to that seen in cellular automata. Another possibility is to free the neighbourhood from the restriction of four neighbours but update a given cell's state based on a dynamic neighbourhood where the number of neighbours varies over time. This is of course not attuned to today's technology but may be an interesting approach to the development of circuits on a more futuristic technology. The work into evolution and development might drive the creation of future technologies to exploit these methodologies to achieve complexity not achievable with today's design techniques and technology.

The above discussion has focused on development itself and, in particular, the resource problems encountered through experimentation. However, before we enter the development stages we might first consider whether a pre-stage to development might be useful. One thing we do know from biology, is that in *Drosophila* the mother sets up four different systems (or regions) in the egg. The fertilized egg then develops from these four pre-setup systems. This pre-stage could be compared to a well-known computing technique: 'divide and conquer'. That is divide the problem into different parts and then solve, or in this case develop, the individual parts. This may be a way to reduce resource usage in the development process as the total resource usage of developing each of the separate parts may be less than that of developing the complete system.

Research into development leaves engineers of developmental models for specific tasks, such as evolvable hardware, with the choice of a range of mechanisms to choose from and a number of ways in which these mechanisms may be implemented. The choice may be a balance between expressiveness and computational efficiency and simplicity and this choice may well vary considerably from application to application.

References

Bentley, P.J. (2000) Exploring component-based representations – the secret of creativity by evolution. In *Proc. of the Fourth International Conference on Adaptive Computing in Design and Manufacture*, pp. 161–174.

Bentley, P.J. and Kumar, S. (2000) Three ways to grow designs: a comparison of embryogenies for an evolutionary design problem. In *Proc. of the Genetic and Evolutionary Computation Conference (GECCO 2000)*, pp. 35–43.

Haddow, P.C. and Tufte, G. (2000) An evolvable hardware FPGA for adaptive hardware. In *Congress on Evolutionary Computation* (CEC00), pp. 553–560.

Haddow, P.C. and Tufte, G. (2001) Bridging the genotype-phenotype mapping for digital FPGAs. In *3rd NASA/DoD Workshop on Evolvable Hardware*, IEEE Computer Society, pp. 109–115.

Haddow, P.C., Tufte, G. and van Remortel, P. (2001) Shrinking the genotype: L-systems for EHW? In *4th International Conference on Evolvable Systems (ICES01)*, Lecture Notes in Computer Science, pp. 128–139. Springer.

Higuchi, T. *et al.* (1996) Evolvable hardware and its application to pattern recognition and fault-tolerant systems. In Sanchez, E. and Tomassini, M. eds. *Towards Evolvable Hardware: the Evolutionary Engineering Approach*, volume 1062 of Lecture Notes in Computer Science, pp. 118–135. Springer.

Holland, J.H. (1975) *Adaptation in Natural and Artificial Systems*. The University of Michigan Press.

Kimura, M. (1983) *The Neutral Theory of Molecular Evolution*. Cambridge University Press.

Koza, J.R., Bennett, F.H. III, Andre, D., Keane, M.A. and Dunlop, F. (1997) Automated synthesis of analog electrical circuits by means of genetic programming. *IEEE Transactions on Evolutionary Computation*, **1**(2), 109–128.

Lantin, M. and Fracchia, F.D. (1995) Generalized context-sensitive cell systems. In *International Workshop on Information Processing in Cells and Tissues (IPCAT'95)*, pp. 42–54.

Lindenmayer, A. (1968) Mathematical models for cellular interactions in development. *Journal of Theoretical Biology*, pp. 280–299.

Macias, N. (1999) Ring around the PIG: a parallel GA with only local interactions coupled with a self-reconfigurable hardware platform to implement an o(1) evolutionary cycle for evolvable hardware. In *Proceedings of the 1999 Congress on Evolutionary Computation (CEC99)*, pp. 1067–1078. IEEE.

Miller, J.F. and Thomson, P. (1998) Aspects of digital evolution: evolvability and architecture. In *Fifth International Conference on Parallel Problem Solving from Nature (PPSN'98)*, pp. 927–936. Springer Verlag LNCS.

Murakawa, M. Yoshizawa, S., Kajitani, I. *et al.* (1999) The GRD chip: genetic reconfiguration of DSPs for neural network processing. *IEEE Transactions on Computers*, **48**(6), 628–639.

Semiconductor Industry Association (1997) *The National Technology Roadmap for Semiconductors.*

Thompson, A. (1995) Evolving electronic robot controllers that exploit hardware resources. In Advances in Artificial Life, *Proc. of the 3rd European Conference on Artificial life (ECAL95)*, pp. 640–656. Springer Verlag LNAI 929.

Thompson, A. (1996) An evolved circuit, intrinsic in silicon, entwined with physics. In *1st International Conference on Evolvable Systems (ICES96)*, Lecture Notes in Computer Science, pp. 390–405, Springer.

Tufte, G. and Haddow, P.C. (1999) Prototyping a GA pipeline for complete hardware evolution. In *The 1st NASA/DoD Workshop on Evolvable Hardware*, pp. 18–25. IEEE Computer Society.

Tufte, G. and Haddow, P. (2003) Building Knowledge into Development Rules for Circuit Design. In *Proc. of the Fifth International Conference on Evolvable Systems (ICES 005)*, pp. 69–80. Springer Verlag LNCS 2606.

Vesselin, K.V. and Miller, J.F. (2000) The advantages of landscape neutrality in digital circuit evolution. In *3rd International Conference on Evolvable Systems (ICES03)*, Lecture Notes in Computer Science, pp. 252–263. Springer.

Von Neumann, J. (1966) *Theory of Self-Reproducing Automata.* University of Illinois Press.

Glossary

Computer science

Algorithm A generic description of a computer program or computational procedure, written to be human-readable.

Artificial life A field of computer science related to evolutionary computation, focusing on the modelling and simulation of adaptive behaviour observed in biology.

Binary coding/binary encoding In contrast to *real coding*, binary coded genes are represented using strings of binary 1s and 0s.

Cellular automata A computer algorithm that updates the status of cell-states in a grid according to the state of each cell's neighbours.

Code A set of computer program instructions. Another word for computer program.

Compiler A computer program used to convert the code written in a programming language into native machine code.

Computational development The field of research devoted to modelling development processes by computer.

Encoding The way in which data are represented for a particular algorithm in a computer, e.g. *genetic encoding* or *binary encoding*.

Evolutionary algorithm A stochastic population based algorithm, such as evolutionary programming, evolutionary strategies, genetic programming or genetic algorithm, that evolves solutions to problems.

Evolutionary computation The field of research devoted to study and application of evolutionary algorithms.

Evolvable hardware A sub-field of evolutionary computation that focuses on the evolution of electronic circuit designs, often using FPGAs.

Execute Running a computer program.

 ISBN 0-12-428765-4

Fitness function Typically used in evolutionary algorithms to provide a measure of the fitness of a phenotype, corresponding to how well the phenotype satisfies the problem objective.

FPGA (field programmable gate array) A silicon chip designed to be reconfigurable into alternative electronic circuit designs.

Genetic encoding/genetic coding The way in which 'genes' are represented in an implementation of an evolutionary algorithm, e.g. real coding or binary coding.

Global maximum/global optimum The global maximum is the highest peak of a function (best overall value if maximizing the function).

IF statement A conditional instruction in most computer programming languages that takes the form: 'IF condition THEN action' or 'IF condition THEN action1 ELSE action2'.

Local maximum A peak in a multimodal fitness function below the global maximum/global optimum.

Loop A programming structure where the code contained within the loop may be continuously re-executed until the loop ends. Examples include 'FOR' loops, 'WHILE' loops and 'REPEAT..UNTIL' loops.

Machine code The low-level binary instructions executed by a computer processor, typically generated by a compiler.

Multimodal A function with multiple modes or peaks.

Neural network A computer algorithm based on the operation of interconnected neurons.

Programming language A language such as 'C', 'Prolog', or 'Java', in which computer programs are written, before being compiled into native machine code and executed by the computer.

Real coding/real encoding In contrast to *binary coding*, real-coded genes are represented using parameters that store values in decimal.

Recursion A subroutine that calls itself is recursive.

Subroutine (or function) A self-contained 'module' of code that can be executed multiple times by calling the function instead of repeating the code in the main program.

Biology

Adipogenic Able to form fat or adipose tissue.

Allele One of several alternative forms of a gene occupying a given locus on a chromosome.

Annelid A segmented worm.

Anterior-posterior axis Axis which defines the head (anterior) and tail (posterior) end of the animal.

Apoptosis Often referred to as programmed cell death, occurs widely during development. Unlike necrosis apoptosis does not cause damage to surrounding cells.

Asymmetric division Cell divisions in which daughter cells are different from each other due to the unequal distribution of some cytoplasmic determinants.

Base pair A partnership of A with T or of C with G in a double helix.

Blastula A hollow ball of cells, composed of an epithelial layer of small cells enclosing a fluid-filled cavity, the blastocoele.

5-Bromodeoxyuridine (BrdU). A base analogue of thymidine, which is often used experimentally to label dividing cells.

Cardiomyocyte A muscle cell of the heart.

Cell differentiation A process of development that causes cells to become functionally and structurally different from each other and become distinct cell types.

Cell cycle The sequence of events by which a cell duplicates itself and divides in two.

Chromosome A discrete unit of the genome carrying many genes.

Chondrogenic Able to form cartilage.

Cleavage A rapid series of cell divisions, without growth, dividing the embryo into smaller cells, occurring after fertilization.

Codon A triplet of nucleotides that represents an amino acid or a termination signal.

Coding region Is the part of the gene whose DNA encodes a polypeptide or functional RNA.

Control region The region of a gene to which regulatory proteins bind determining whether or not the gene is transcribed.

Cryogenic infarction Obstruction of the blood supply as a result of extremely low temperatures.

Cytokinesis The process of cytoplasmic division.

Cytoplasm Material between the plasma membrane and the nucleus.

Cytoplasmic factors Proteins, RNA, etc.

Degeneracy Refers to the genetic code and the lack of an effect of many changes in the third base of the codon on the amino acid that is represented.

DNA microarrays Devices that are used to analyse complex nucleic acid samples by hybridization. They make it possible to quantitate the amount of different nucleic acid molecules that are present in a sample of interest.

Dedifferentiation Loss of the structural characteristics of a differentiated cell, which may result in the cell then differentiating into a new cell type.

Dendrites Extensions from the body of a nerve cell that receive stimuli from other nerve cells.

Determinants Cytoplasmic factors in the egg and in embryonic cells that can be asymmetrically distributed at cell division and so influence how the daughter cells develop.

Determination Implies a stable change in the internal state of a cell such that its fate is now fixed, or determined.

Diploid Cells contain two sets of homologous chromosomes, one from each parent and thus two copies of each gene.

Dorsoventral axis Axis which defines the relation of the upper surface or back (dorsal) to the under surface (ventral) of an organism or structure.

Ectoderm The germ layer that gives rise to the epidermis and the nervous system.

Endoderm The germ layer that gives rise to the gut and associated organs, for example, the lungs.

Enhancers DNA sequences to which regulatory proteins bind to control the time and place of transcription of a gene. Enhancers can be many thousands of base pairs away from the gene's coding region.

Epidermis The outer layer of cells, in vertebrates and insects, that forms the interface between the organism and its environment.

Epigenetic Changes of this type influence the phenotype without altering the genotype. They consist of changes in the properties of a cell that are inherited but that do not represent a change in genetic information.

Eukaryote A single- or multicellular organism whose cells contain a distinct membrane-bound nucleus.

Exocrine cell A cell that makes up part of an exocrine gland, which discharges its secretion through a duct.

Exons Any segment of an interrupted gene that is represented in the mature RNA product.

Fate Describes what a cell will normally develop into.

G2 The phase of the cell cycle through which cells progress after S phase but before M phase.

Gastrula The stage in animal development when the endoderm and mesoderm of the blastula move inside the embryo.

Gastrulation The process in animal embryos in which the endoderm and mesoderm move from the outer surface of the embryo to the inside, where they give rise to internal organs.

Genotype A description of the exact genetic constitution of a cell or organism in terms of the alleles it possesses for any given gene.

Gene The segment of DNA involved in producing a polypeptide chain. It includes regions preceding and following the coding region as well as intervening sequences (introns) between individual coding segments (exons).

Genetic code The correspondence between triplets in DNA or RNA and amino acids in protein.

Germ layers Those regions of the early animal embryo that will give rise to distinct types of tissue. Most animals have three germ layers: ectoderm, mesoderm and endoderm.

Giant cells Large multinucleated cells that are thought to result from the fusion of macrophages.

Green fluorescent protein An autofluorescent protein that was originally isolated from the jellyfish *Aequorea victoria*. It can be genetically conjugated with proteins to make them fluorescent. The most widely used mutant, EGFP, has an emission maximum at 510 nm.

Haploid Cells contain only one set of chromosomes and thus only one copy of each gene.

Heterochrony An evolutionary change in the timing of developmental events.

Heterokaryon A cell that contains two nuclei in a common cytoplasm.

Hepatocytes The parenchymal cells of the liver that are responsible for the synthesis, degradation and storage of a wide range of substances.

Induction The process in which one group of cells signals to another group of cells in the embryo and so affects how they will develop.

Interphase The phase of the cell cycle where DNA replicates and proteins are synthesized before and after mitosis.

Intron A segment of DNA that is transcribed, but removed from within the transcript by splicing together sequences (exons) on either side of it.

Invagination The local inward movement of a sheet of embryonic epithelial cells to form a bulge-like structure.

Kinase An enzyme that phosphorylates (adds a phosphate group) to a substrate. The substrates for protein kinases are amino acids in other proteins.

Lateral inhibition The mechanism by which cells inhibit neighbouring cells from developing in a similar way to themselves.

Locus Position on the chromosome at which the gene resides.

Maternal factors Proteins and RNAs that are deposited in the egg by the mother during oogenesis. The production of these maternal proteins and RNAs is under the control of so-called maternal genes.

Meiosis Special type of cell division that occurs during formation of sperm and eggs and in which the number of chromosomes is halved from diploid to haploid.

Mesenchyme Immature connective tissue that consists of cells that are embedded in extracellular matrix.

Messenger RNA (mRNA) The RNA molecule that specifies the sequence of amino acids in a protein; produced by transcription of the DNA.

Mesoderm The germ layer that gives rise to the skeletomuscular system, connective tissue, the blood and internal organs.

Metazoan Refers to the kingdom Animalia (animals), which comprises ~ 35 phyla of multicellular organisms.

Microtubule A hollow tube, 25 nm in diameter, that is formed by the lateral association of 13 protofilaments, which are themselves polymers of α- and β-tubulin subunits.

Mitosis Division of the nucleus during cell division resulting in both daughter cells having the same diploid complement of chromosomes as the parent cell. Mitosis itself consists of a number of stages.

Morphogen Any substance active in pattern formation whose spatial concentration varies and to which cells respond differently at different threshold concentrations.

Morphogenesis Collective term for the processes involved in bringing about changes in form in the developing embryo.

Myelin Proteins that are produced by Schwann cells or oligodendrocytes that cause adjacent plasma membranes to stack tightly together.

Myofibre A skeletal muscle fibre that consists of one long multinucleate cell.

Myofibril The structural unit of striated muscle fibres. Several myofibrils make up each fibre.

Myotube The structure comprising multiple nuclei that is formed by the fusion of proliferating myoblasts (undifferentiated precursors to muscle cells) and is characterized by the presence of certain muscle-specific marker proteins.

Nomarski optics Also known as differential interference contrast microscopy, this technique forms images of high contrast and resolution in unstained cells using birefringent prisms and polarized light.

Notochord Refers to the rod-like structure, in vertebrate embryos, that runs from head to tail and lies centrally beneath the future central nervous system. It is derived from the mesoderm.

Ontogeny Refers to the development of an individual organism.

Oogenesis Process of egg formation in the female.

Operon A unit of bacterial gene expression and regulation, including structural genes and control elements in DNA.

Osteoclast A mesenchymal cell that can differentiate into a bone degrading cell.

Pattern formation The process by which cells in a developing embryo acquire identities that lead to well-ordered spatial pattern of cell activities.

Phenotype The appearance or other characteristics of an organism, resulting from the interaction between its genetic constitution and the environment.

Phylogeny Evolutionary history that is sometimes represented by the hypothesized ancestor-descendant relationship of a group of organisms.

Pleiotropic gene Gene affects more than characteristic of the phenotype.

Pluripotent cell A stem cell that can give rise to more than one differentiated cell type.

Planarian Describes free-living members of the invertebrate phylum platyhelminths.

Prokaryotic Organisms (bacteria) lacking a membrane-bound nucleus or membrane-bound orgonelles, and having DNA not organized into chromosomes.

Promoter A region of DNA close to the coding sequence to which RNA polymerase binds to begin transcription of a gene.

Positional information In the form of a gradient of an extracellular signalling molecule can provide the basis for pattern formation. Cells acquire a positional value that is related to their position with respect to the boundaries of the given field of positional information. The cells then interpret this positional value according to their genetic constitution and developmental history and develop accordingly.

Receptor A trans-membrane protein located in the plasma membrane that binds a ligand in a domain on the extracellular side and as a result has a change in activity of the cytoplasmic domain.

Reaction-diffusion mechanisms Produce self-organizing patterns of chemical concentrations which could underlie periodic patterns.

Regeneration The ability of a fully developed organism to replace lost parts.

Regulation The ability of the embryo to develop normally even when parts are removed or rearranged. Embryos capable of regulation are termed regulative.

Regulatory gene A gene that codes for an RNA or a protein whose function is to control the expression of other genes.

Rotifera A small phylum of microscopic multicellular organisms. They have a wheel-like ciliated organ (from which they derive their name) that they use for swimming and feeding.

Schwann cell A cell that produces myelin and ensheathes axons in the peripheral nervous system.

Segmentation The division of the body of an organism into a series of morphologically similar units or segments.

Signal transduction The process by which a receptor interacts with a ligand at the surface of the cell and then transmits a signal to trigger a pathway within the cell.

Somites In vertebrate embryos are segmented blocks of mesoderm lying on either side of the notochord. They give rise to body and limb muscles, the vertebral column and the dermis.

Structural gene A gene that codes for any RNA or protein product other than a regulator.

Synapse The specialized point of contact where a neuron communicates with another neuron or a muscle cell.

Symmetric division Cell division where daughter cells are the same as each other. Contrast with asymmetric division.

S (synthesis) phase The phase of the eukaryotic cell cycle in which DNA is synthesized.

Transcription Refers to the synthesis of RNA on a DNA template.

Transcription factor (TF) A regulatory protein required to initiate or regulate transcription of a gene into RNA. TFs act within the nucleus of a cell by binding to specific regulatory regions in the DNA.

Transdifferentiation The process by which a differentiated cell can differentiate into a different cell type.

Translation The process by which messenger RNA directs the order of amino acids in a protein during protein synthesis.

Turbellarians A class of platyhelminths that comprises mostly aquatic and free-living organisms. They have a ciliated epidermis for locomotion and a simple gut.

Urodele An order of the class Amphibia, which comprises newts and salamanders, which have elongated bodies, short limbs and a tail.

Zygote The fertilized egg.

Index

Acetabularia acetabulum development, 189–90
Activator-inhibitor models, 18–19, 163
  pattern formation, *see* Pattern formation
Activin, 50
  Xenopus gene activation, 150
Adenylyl cyclase, 68, 76
Adrenaline, 76, 77
Amino acid sequence, 7, 182, 183
Amphioxus, 57
Anabaena, 141
Angelfish markings, 193–4
Animal movement, 195–9
Annelid segmentation, 247, 248
Anterior–posterior axis, 3
  Drosophila development, 52–4, 126
  Hox gene expression, 51, 52
Apical dominance, 144
Apoptosis, 79, 168
  evolutionary developmental system (EDS), 35
  see also Cell death
Arabidopsis thaliana:
  flower morphogenesis model, 262–3
  shoot apical meristem (SAM), 156
Arthropod segmentation, 247, 248
Artificial genomes, 256, 265–76
  artificial evolutionary system for shapes, 313–14
  biological genome similarities, 271, 274
  dynamics, 268–9
    chaotic, 269
    complex, 269
    controlled differentiation, 271
    cyclic gene expression, 269, 271
    homeostasis, 270–1
    multiple expression patterns, 269–70, 271
    ordered, 268
  evolutionary experiments, 272–3, 275
    fitness function, 273
  expression graphs, 268
  gene regulation modelling, 271–6
  model, 265–7
    new implementations, 275–6
  template matching principle, 265, 266
Artificial neural networks, *see* Neural development modelling
Assembly construction, 203
Asymmetric cell division, 4, 6, 62
  artificial evolutionary system for shapes, 310
  morphogenesis modelling, 170–2
    biomorphs (*Blind Watchmaker* software), 248–9
Attractor, 189
Autocatalysis, 136, 137
Autocrine signalling, 5
Automatic definition of neural subnetworks (ADNS), 343
Autonomous agents development, 182, 377–90
  complex model implementation, 380–5, 389
    Braitenberg 'hate vehicle', 382–4
    developmental simulation, 380–1
    genome–cytoplasm model, 380
    incremental evolution, 384, 385
    neural network development, 382
    problems, 384–5
  developmental model, 378–80
  fitness functions, 389
  learning, 389
  simplified model implementation, 385–8, 389
    developmental simulation, 386
    line follower, 387–8

Autonomous agents development (*cont'd*)
neural development, 386–7
random Boolean network model, 385–6
Axonal outgrowth, 356
autonomous agents developmental model, 382
simplified model implementation, 386
neural development simulations, 341–2, 344
see also Dendrite growth
Axons, 356

Bacillus subtilis sporulation:
forespore, 122
genetic regulatory network, 122
modelling, 110, 123–5, 129, 130
phosphorelay, 123
mother cell, 122
Bacteriophage lambda, 115
Baldwin effect, 62
Base pairing, 7
'Behavioural' genes, 27
Belousov-Zhabotinskii reaction, 194–5
Beta-catenin, 141
hydra homologue, 145, 146, 147
Bicoid morphogen gradient, 52–3, 54
Bifurcation, 188
Biochemical switching behaviour, 148
Bipedal gait, 196
Bird flight, 199
Blastema, 95
myotube de-differentiation assay, 96–7
Blastula, 3
Blind Watchmaker, 239
artificial embryology, 241–4
lines, 242
mirror algorithms, 242
pixel peppering, 241–2
recursive trees, 242–4
emergent biomorph phenotypes, 244–52
scaling gradients, 247–8
segmentation, 247, 248
symmetry/asymmetry, 246–7, 248–9
procedures, 241
Body plan, 3
Bongard gene regulation model, 261, 304
Brachyury, hydra homologue, 146
Braitenberg 'hate vehicle', autonomous agents developmental model, 382–4
Branchial arches, 55
Brusselator reaction, 143
Budding, 170

Caenorhabditis elegans, 79
Calcium signalling, 68, 69, 70–1, 77
Acetabularia acetabulum development, 190
cAMP (cyclic adenosine monophosphate), 59, 68, 69, 71, 76
Dictyostelium discoideum signalling, 166, 167
cAMP-dependent protein kinase (PKA), 76, 77
Canalization, 87, 88, 182
Cardiomyocyte regeneration, 93, 95
Cartesian genetic programming, 279–81
developmental programming (cell evolution), 281–5
modulo operation, 283
node programmes, 282–3, 285
node transformation, 284–5
seed node, 282
two-dimensional maps evolution, 285–9
fitness function, 281, 300
mutation, 281
two-dimensional cellular maps evolution:
cell representations, 286–7
chemicals representation, 286–7
experimental parameters, 288
experiments/results, 289–97
French Flag model task, 288–9
genotype analysis, 297–8
genotypes, 288
Catastrophe theory, 182
Cell adhesion, 214
cell sorting simulations, 222, 227–31
Dictyostelium discoideum morphogenesis, 166, 167
morphogenesis modelling, 170, 303
self-sorting functionality, 214
Cell adhesion molecules:
artificial evolutionary system for shapes, 306, 310–11, 313
autonomous agents neural developmental model, 382
morphogenesis simulations, 302
Cell cycle:
autonomous agents developmental model, 381
mouse myonuclei re-entry induction, 99
urodele regeneration studies:
cardiomyocytes, 93
iris pigmented epithelial cells, 93
limb myotubes, 95, 96, 98–9
pRb function, 98
thrombin function, 98–9
Cell death, 60, 79
artificial evolutionary system for shapes, 306
morphogenesis modelling, 168, 169, 303
virtual evolvable hardware, 410
see also Apoptosis
Cell differentiation, 4, 49, 59, 79, 320

autonomous agents developmental model, 379, 381
morphogenesis modelling, 168, 169, 303
neural development simulations, 342–3, 344
plasticity:
mammalian tissues, 95–6
urodele regeneration, 92, 93–5
see also Transdifferentiation
regulatory processes, 343
robot control architecture morphogenesis, 393, 398, 399
virtual evolvable hardware, 410
Cell division, 6, 214
artificial evolutionary system for shapes, 306, 308–10
asymmetric, *see* Asymmetric cell division
autonomous agents developmental model, 378, 380
evolutionary developmental system (EDS), 35
morphogenesis modelling, 168, 170–2, 303
neural development simulations, 342, 344
orientation, 6
regulatory processes, 343
robot control architecture morphogenesis, 393, 398, 399
symmetric, 6
see also Cleavage division
Cell membrane, 5
Cell migration:
morphogenesis simulation, 303
neural development simulations, 342, 344
regulatory processes, 343
robot control architecture morphogenesis, 397, 399
Cell movement (protrusion formation), 152
Cell program evolution, *see* Cartesian genetic programming
'Cell receptor' genes, 26–7
Cell signalling, 4, 5–6, 59, 64–80, 208
artificial evolutionary system for shapes, 306–7
autonomous agents developmental model, 379, 381
simplified model implementation, 386
Cartesian genetic developmental programming, 294–7
defects, 79–80
developmental role, 78–80
electrical signals, 71
mechanisms:
amplification of signal, 77
phosphorylation, 74–5
structural change, 73
morphogenesis simulation, 303
neural networks, 78
pathways (signal cascades), 76–7
complex interactions, 77
primary/secondary messengers, 66
receptors, 64, 71–3, 76
robot control architecture morphogenesis, 398
shoot apical meristem (SAM), 156–61
signal transduction, 65
see also Signalling molecules/signalling proteins
Cell sorting, 165, 214–15
Dictyostelium discoideum morphogenesis model, 165–8
differential cell adhesion simulations, 222, 227–31
Cells, 4–6, 204–5, 278, 279
autonomous agents developmental model, 380
evolutionary developmental system (EDS), 35
evolutionary programme, *see* Cartesian genetic programming
scale, 208
surface tension effects, 213
cell sorting, 214–15, 216
Cellular encoding, 19–20
Cellular networks, 209–13
Central pattern generator (CPG), 197
cGMP (cyclic guanosine monophosphate), 71, 72
Chaos, 189
Chemotaxis, 59, 166, 167, 300
Cis-regulatory regions, 8, 48, 49
evolutionary developmental system (EDS), 32, 34
CLAVATA1 (CLV1), 156, 157, 159–60
CLAVATA3 (CLV3), 156, 157, 159–60
Cleavage division, 3
artificial evolutionary system for shapes, 309–10
division plane regulation, 171–2
Coding regions, 257
artificial genomes, 265
evolutionary experiments, 273
evolutionary developmental system (EDS), 32, 34
Codon, 8
Complementary base pairing, 7, 8
Complex adaptive systems, 12, 182
Computational development, 2, 9–22
advantages/disadvantages, 10–11
definition, 13–15
embodiment within environment, 15–16
evolvability, 16–18
modules, 14–15
review of background, 18–21
Computational neuroscience, 354–5

Computational neuroscience (*cont'd*)
  see also Neural development modelling
Computer models, 1, 2
Connectionist modelling approach, 21
Construction, 9–10, 203–5
Contact inhibition, 20
  lateral inhibition simulations, 222–7
    cell state equations, 223–5
    importance of temporal sequence, 225–6
Conus marmoreus shell patterns, 152
Convergent extension, 170, 216
Couette-Taylor flow, 190
Creep mutation:
  evolutionary developmental system (EDS), 36
  fractal protein modelling approach, 28
Creep in solids, 211
Crypsis, 87
Crystal networks, 210–12, 213
Cubitus interuptus (*ci*), 144, 263
Cumulative selection, 254
Cyclin, 344
Cytokines, 70, 73
Cytokinesis, 6
Cytoskeleton, 59

Dappled patterns, 191
Decapentaplegic (Dpp), 50, 263
Delta-Notch, 21, 51, 222, 227
Dendrite growth, 356
  robot control architecture morphogenesis, 393, 397–8, 399
Dendrites, 356
Dendritic trees, 356
Development, 2–9, 10
  autonomous agents, *see* Autonomous agents development
  cell signalling pathways, 78–80
  definition, 2
  environmentally-induced variation, 86–8
  evolution, 54–62, 88–9
  evolutionary algorithms, 12–13
  fractal protein modelling approach, 27
  genotype–phenotype relationship, 82–3
  modelling 12, 377–8
    see also Cartesian genetic programming; Computational development
  modularity, 83–6
    definition, 83
  processes, 2–4
    morphogenesis simulation, 168
  relation to genetics, 239–41
  symmetry-breaking cascades, 189
  virtual evolvable hardware, 411
    challenges, 418–21
    developmental stages, 420
    functionality, 420–1
Developmental reaction norms (DRN), 87
Diabetes, 79
Diacylglycerol (DAG), 71, 77
Dictyostelium discoideum, 165
  life cycle, 165
  modelling behaviour, 166–8
    differential cell adhesion, 166, 167
    long-range cell signalling, 166, 167
Differentiation:
  computational models, 20, 21
  evolutionary developmental system (EDS), 35
  see also Cell differentiation
Diffusion gradients, 3
  neural development modelling, 357–8
  positional information, 50
  reaction-diffusion systems, 18
  see also Morphogen gradients
Digital organism evolution, 319–34
  developmental pattern, 325–8
  experimental methods, 321
    genetic operators, 323
    selection against non-migratory organisms, 321
    sensory system, 322
  gene duplication, 325, 331
  genetic change 323–4, 330–1
  genome/genes, 320
  migration patterns, 330, 333
  organic organism analogies, 319–20
  reproductive algorithm, 325
  results, 323–30
  sensory data selection, 329–30, 331–2
    mob behaviour, 331, 332
  sensory processing, 329
    loss/degradation of system, 332–3
  see also Tierra
Digits evolution, 56, 57
Discrete Dynamics Lab software, 259–60
Dissociation of modules, 84
Divergence:
  meristic changes in morphology, 85
  modules, 84–5
DNA (deoxyribonucleic acid), 7–8, 181, 182, 183, 191, 193, 199–200, 319, 320
Dorsal–ventral axis, 3
  Drosophila development, 54
Drosophila:
  leg disc formation modelling, 263
  neurogenesis modelling, 264–5
  pair rule stripes, 48

pattern formation during development, 52–4, 141, 163, 191
positional information, 50, 51
segmentation, 143–4
gap gene cross-regulatory interactions qualitative modelling, 126–9, 130–1
genetic regulatory networks modelling, 110, 115, 125–9
loss-of-function mutations simulation, 129, 130
Turing-like patterns in hair development, 193
Duplication:
fractal protein modelling approach, 28
meristic changes in morphology, 85
modules, 84–5
see also Gene duplication

E. coli cell division, 152
Ectoderm, 3
EGFs, 51
Egg:
evolution, 60
internal asymmetry, 140
relevance to evolvability, 60–1
Eggenberger gene regulation model, 260–1, 304
Elastic forces, 208
see also Viscoelastic forces
Electrical signalling, 71
Electrostatic forces, 208
Elongation, 170
Endocrine signalling, 5, 70
signalling pathway, 76–7
Endoderm, 3
Engrailed (*en*), 143, 144
Enhancers of transcription, 257
artificial genomes, 266, 268
Environmental adaptation, 346
'Environmental' genes, 27
Environmental variation, 86–8
Enzymes, 7
cellular receptor activity/receptor recruitment, 72
intracellular signalling, 68, 69, 73, 76, 77
Ephrins, 51
Epistasis, 17
Equilibrium states, 189
Ethylene, 70
Eukaryotic cells, 4, 5, 8, 58–9
Evolution:
co-option of modular structures/processes, 85–6
development, 54–62, 88–9
developmental biology, 47–54
duplication/divergence in regulatory genes, 85
egg, 60
evolutionary developmental system (EDS), 36
evolvability, 239–54
fractal protein modelling approach, 28
genetic system changes, 252–3
learning impact, 346–8
multicellularity, 58–60
neural development simulation, 359–64
learning process interactions, 348–50
learning rules evolution, 350
network architecture, 360–3
optimization of developmental parameters, 359–60
new structures, 54–8
processes:
modelling, 89
morphogenesis modelling, 168, 170
robot control architecture automated design (evolutionary robotics), 392
selection for developmental/mutational robustness, 172–3
selection pressure on embryo, 62
watershed events, 253
Evolutionary algorithms, 10, 11–12, 13, 303–4, 339
Cartesian genetic programming, 281
developmental modelling, 12–13
evolvability, 17
Evolutionary developmental system (EDS), 29, 31–9
cells, 35
components, 31–5
evolution, 36
genes, 32, 34–5
genetic regulatory networks, 37
isospatial coordinate system, 36
key features, 31
proteins:
creation/initialization, 31–2
destruction, 32
diffusion, 32
spherical embryo morphogenesis, 37
visualization, 36
Evolutionary programming (EP), 11
Evolutionary strategies (ES), 11
Evolvability, 16–18, 22, 60–2
evolution, 239–54
ontogenetic features, 88–9
robot control architecture automated design, 400–1
Evolvable hardware, 16, 21, 405–22
artificial development, 411

Evolvable hardware (*cont'd*)
challenges, 418–21
developmental stages, 420
functionality, 420–1
genetic algorithm, 412
experimental platform, 416–17
extrinsic evolution, 406
fitness specification, 407, 408, 412, 416, 419
genotype complexity level, 408, 421
genotype-phenotype mapping, 407, 418–19
intertype representation, 409–11, 416
intrinsic evolution, 406
L-systems, *see* L-systems
limitations, 406–8
direct/indirect mapping, 407
genotype representation, 407
technological constraints, 407–8
methods, 405–6
on-chip (complete hardware) evolution, 406

FGFs, 51
Field programmable analogue array (FPAA), 405–6
Field programmable gate array (FPGA), 406, 407–8
configurable input/output blocks (IOBs), 409
configurable logic blocks (CLBs), 409
virtual evolvable hardware, 409–11, 416
development, 411
L-systems, 413, 417–18, 422
Finite state automaton (FSA):
neuron morphology–function relationship, 364
neuron network function investigation, 366
Fitness function, 17
artificial genome experiments, 273
autonomous agents development, 388
Cartesian genetic programming, 281, 300
evolvable hardware, 407, 408, 412, 416, 419
robot control architecture automated design, 392, 394, 402
Flower morphogenesis, 262
Foams, 209–13
Fractal protein modelling approach, 22–8
development, 27
evolution, 28
fractal proteins:
definition, 23–4
interactions, 24
fractal sampling, 27
genes, 26
'behavioural', 27
'cell receptor', 26–7
'environmental', 27
regulatory, 26
genetic operators, 28
mutation, 28
protein concentration levels:
calculation, 24
updating, 24–5
representation, 23
results, 28, 29, 30
Fucus eggs, 140

G protein signalling, 73, 76, 77
G-protein linked receptors, 59, 72, 76
Gaits:
bipedal, 196
central pattern generator (CPG) model, 197
networks of coupled oscillators, 197–9
phase relationships, 196, 197
quadrupedal, 196
as symmetry-breaking patterns, 195–9
Gap genes, 54
cross-regulatory interactions modelling, 126–9
differential equation models, 115
Gap junctions, 6, 66
Gastrula, 4
Gastrulation, 4, 61, 141
Gene activation, 148–50
biochemical switching behaviour, 148
pattern formation, 149–50
Gene duplication, 48, 49, 57, 85
digital organisms evolution, 325
Gene expression, 8–9, 65, 79
genetic regulatory networks, *see* Genetic regulatory networks
Gene interaction matrices:
Arabidopsis flower morphogenesis, 262, 263
Drosophila leg disc development, 263
Drosophila neurogenesis, 264, 265
Generalized logical models:
genetic regulatory networks qualitative analysis, 119–22
Drosophila gap gene cross-regulatory interactions, 127–9
GIN-sim, 121–2
Generative encoding, 21
Genes, 7–9, 181, 240
evolvable hardware representations, 407
model genomes, *see* Artificial genomes
mutation, *see* Mutation
parameter selection theory, 189
regulation, 8–9, 65, 256–7
models, 258–65
regulatory, 340
divergence in evolution, 85

fractal protein modelling approach, 26
heterochronic processes, 343
robot control architecture morphogenesis, 395
GENESIS, 355, 360
Genetic algorithm (GA), 11, 12, 304
artificial genome experiments, 272–3, 275
autonomous agents developmental simulation, 378, 381
evolutionary developmental system (EDS), 36
evolutionary robotics, 392, 402
evolvable hardware, 405, 408, 412
L-systems, 413
neural development modelling, 360
Genetic code, 8
Genetic drift, 17, 281
Genetic Network Analyser (GNA), 118, 123–4
Genetic programming (GP), 11
Genetic regulatory networks, 19, 21, 257
artificial evolutionary system for shapes, 305–8
autonomous agents developmental model, 385–6
dynamical modelling, 110–15
applications, 115
directed graphs, 111
ordinary differential equations (ODEs), 111–13
partial differential equations (PDEs), 113
stochastic models, 113–15
evolutionary developmental system (EDS), 37
fractal protein modelling approach, 22–8
results, 28, 29, 30
key features, 110–11
models, 258–65, 304
biological network simulations, 262–5
dynamics of regulatory systems, 258–62
morphogenesis modelling, 168, 302, 303
neural development modelling, 365
qualitative modelling/simulation, 109–31
asynchronous multivalued logic, 110
Bacillus subtilis sporulation initiation, 110, 122–5, 129, 130
generalized logical models, 119–22
piecewise-linear (PL) differential equations, 110, 116–19
segmentation during *Drosophila* embryogenesis, 110, 115, 125–30
robot control architecture design model, 393, 395, 398, 400–1
shoot apical meristem (SAM), 157
modelling, 157–60
see also Artificial genomes
Genome, 5, 7, 319–20
see also Artificial genomes
Genotype:
autonomous agents developmental model, 380
simplified model implementation (random Boolean networks), 385–6
evolvable hardware representations, 407, 416, 421
natural selection, 240
neural network simulations, 339–40
phenotype generation, 82–9
robot control architecture design model, 393
Genotype-phenotype mapping, 17, 340, 377
autonomous agents developmental simulation, 378
Blind Watchmaker, 241–4
cellular encodings, 342–3
direct encoding, 340–1
evolvable hardware, 407, 418–19
growing encoding scheme, 341–2
mechanical approach, 203–18
morphogenesis modelling, 170–1
neural development simulations, 340
robot control architecture design, 341, 392
Germ layers, 3
Glide reflections, 186
Glycogen breakdown, 77
Gradients:
biomorphs (*Blind Watchmaker* software), 247–8
see also Diffusion gradients; Morphogen gradients
Graftal patterns, 163–4
Gravity, 208, 209
Group theory, 186, 188
Growth, 4, 49, 50, 57
ancestry-based systems, 163
cell sorting behaviour, 215
genotype-to-phenotype mapping, 341
neural networks, *see* Neural development modelling
pattern formation:
graded concentration profile maintenance, 144
periodic, 142
shoot apical meristem (SAM) modelling, 158
Growth cone models, 382, 393, 399
Growth hormone, 73, 80
gt, 126

Haemoglobin, 57
Hardware evolution, *see* Evolvable hardware
Hedgehog family, 54
hedgehog (*hh*), 144
Heterochrony, 2
developmental modules, 84, 85

Heterochrony (*cont'd*)
  neural development simulations, 343, 344
    morphogenetic tree, 345
Heterotopy, 84
Hexapod movement, 199
Homeobox, 51
Homeobox gene family, *see* Hox genes
Homologous elements, 88–9
Hormones, 66, 70
Hox genes, 17, 48, 51–2, 57, 83, 86, 191, 261
  evolution by gene duplication, 57
Hoxc6, 52
hunchback (*hb*), 54, 126, 127, 129
Hydra, 141
  regeneration, 144, 145–7

Induction, 5
Inhibitors of transcription, 257
  artificial genomes, 266, 268
Inositol trisphosphate (InsPU3u), 71, 77
Insects:
  appendages, 51, 52, 86
  movement, 199
  pigmentation patterns, 86
Insulin, 68, 70, 73, 79
Interphase, 6
Invagination, 316
Ion-channels, 5
  cellular receptors, 72
  neuron activity modelling, 363
Iris pigmented epithelial cells:
  transdifferentiation, 93
  urodele regeneration studies, 93, 95

JAK/STAT transcriptional control, 59
Jaws, 47, 55

Khepera miniature mobile robot, 394
Kinases, 75, 77, 123, 125, 344
Knirps, 54, 126, 127
Kruppel, 54, 127, 129

L-systems, 19, 163, 164, 355
  evolvable hardware, 405, 413–18, 419, 421, 422
    change and growth rules, 414
    circuit development, 413–15
    developmental stages, 420
    experimentation, 417–18
    fitness evaluation, 419
    genetic algorithm, 413
    rule firing sequence, 414–15
Large Q Potts model, 165
Lateral inhibition, 4, 21
  simulations, 222–7
    cell state equations, 223–5
    importance of temporal sequence, 225–6
Leaf initiation, 152
Learning:
  autonomous agents development, 389
  neural development simulation, 346–50
    evolutionary processes interaction, 348–50
    impact on evolutionary search, 346–8
    learning rules evolution, 350
Lefty-2, 141
Lens:
  evolution, 85
  urodele iris pigmented epithelial cell
      transdifferentiation, 93, 95
Lindenmayer systems, 19
Line following tasks, 387–8, 389
Liver regeneration, 95–6

Mandelbrot set, 22–3
Mathematical modelling, 183–4
Meristematic growth, 169
Meristic changes, 85
Mesoderm, 3
Metal microstructure, 210–12
Microbial interactions, 87
Middle ear, 47, 55–6
Millipedes, 199, 247
Mitosis, 6
  see also Cell division
Modules, 354
  Bongard gene regulation model, 261
  definition, 83
  developmental, 83, 84
    Bacillus subtilis, 131
    co-opted functional change, 85
    Drosophila segmentation regulation, 131
    evolutionary consequences, 85–6, 88
    evolutionary models, 89
    heterochrony, 85
  Dictyostelium discoideum morphogenesis
      modelling, 168
  dissociation, 84
  divergence, 84–5
  duplication, 84–5
  evolutionary, 84
  external connectivity, 84
  formation/interaction, 14–15
  genetic regulatory networks modelling, 131
  homologous, 88–9
  patterns of change, 84
  process, 84
  structural, 83–4

types, 83–4
Mollusc shell pattern formation, 150–2, 195
Morphogen gradients, 3, 136
activator–inhibitor interaction, 137
artificial evolutionary system for shapes, 306–7, 314, 315–16
biochemical switching behaviour, 148
Drosophila development, 52–3
gene activation, 149–50
morphogen receptors, 50
Morphogenesis, 4, 16, 49, 163–5, 188–90, 191
artificial evolutionary system for shapes, 312–13
convergent extension, 216
evolutionary computation, 20–1
evolutionary developmental system (EDS), 37
modelling, 168–73, 302–17
adhesion molecules, 302
cell sorting model, 165
cellular approach, 303
Dictyostelium discoideum, 166–8
genetic regulatory networks, 302, 303
spherical embryo, 37
robot control architecture design, 391–402
visual models, 21
Morphogens, 182, 192, 195
autonomous agents developmental model, 380
see also Morphogen gradients
Msx-1, 101
Multicellularity, 208
evolution, 58–60
Mutation, 376
Cartesian genetic programming, 281
digital organisms evolution, 323–4
evolvable hardware, 407
fractal protein modelling approach, 28
robot control architecture morphogenesis, 400, 402
Myocyte enhancer factor 2 (MEF2), 98
Myoseverin, 101, 102
Myotubes:
differentiation:
assay of reversal in blastema, 96–7
phenotypic features, 96
mouse:
cellularization induction, 101–2
myonuclei cell-cycle re-entry induction, 99
urodele limb regeneration:
cell-cycle re-entry, 95, 96, 98–9
cellularization, 99–101
thrombin regulatory role, 98–9
transdifferentiation, 93–4, 95
Myriapod movement, 199

Natural forms, 205–7
Natural selection, 240, 253–4
Neural crest development, 342
Neural development modelling, 20, 21, 339–51, 353–66
artificial life approach, 355
autonomous agents developmental model, 379, 382
simplified model implementation, 386
cellular encodings, 342–3
automatic definition of neural subnetworks (ADNS), 343
chemical diffusion gradients, 357–8
complementary approaches, 355
computational neuroscience approach, 354–5
neuron function–morphology relationship, 363, 364
direct genotype-to-phenotype mapping, 340–1
evolutionary development, 341–2, 344, 359–64
multi-compartment neurons, 363–4
network architecture, 360–3
optimization of developmental parameters, 359–60
simplified compartment model, 364
features of biological systems, 353–4
finite state automaton (FSA) model, 364, 366
gene interaction networks, 365
growing encoding scheme, 341–2
heterochronic processes, 343, 344
morphogenetic tree, 345
implemented developmental rules, 356
neurite outgrowth, 356
pruning, 357
spontaneous neural activity, 357
interactive self-organization, 356
learning, 346–50
evolutionary processes interaction, 348–50
impact on evolutionary search, 346–8
learning rules evolution, 350
mathematical rules, 368–75
growth rate regulation, 375
initial developmental configuration, 369
neurite branching, 371
neurite navigation, 370
spontaneous neural activity, 372–5
synaptogenesis, 371
modelling approaches, 340
modelling framework, 356–8
network functionality, 366
network growth, 357–8
branching, 358

Neural development modelling (*cont'd*)
morphological parameters, 358, 359
synapse formation, 358
network morphology, 358, 359
neuron functional activity, 365–6
neuron libraries, 366
neuron outgrowth, 356
robot control architecture morphogenesis, 393, 394, 398, 399
encoding scheme, 395–9
Neurogenesis, 21
Neurogenetic learning (NGL), 20
NEURON, 355
Neuron outgrowth, 356
modelling, *see* Neural development modelling
see also Axonal outgrowth; Dendrite growth
Neuron signalling, 78
simulation, 363
Neuron structural elements, 356
Neurulation, 47
Nitric oxide, 70
Nodal, 141
Notochord, 54
Nuclear envelope, 5, 8
Nucleic acids, 7
Nucleotides, 7, 8

Obstacle avoidance tasks, 394, 402
Oliva porphyria shell patterns, 152
Oncogenes, 79, 80
Ontogeny, 47
selectability 88–9
'Ontogeny recapitulates phylogeny', 2, 61–2, 181–2
Ordinary differential equations (ODEs):
genetic regulatory networks, 111–13
neuron function–morphology relationship modelling, 363
Organizing regions, 5, 136, 141
hydra experimental studies, 145, 146

Pair-rule genes, 115, 126
Pair-rule stripes, 48
Paracrine signalling, 5, 70
Partial differential equations (PDEs):
activator–inhibitor interaction, 137
genetic regulatory networks, 113
Pattern formation, 135–53, 163–5, 181–2, 183
activator–inhibitor models, 18–19
hydra regeneration, 145–7
pattern regeneration following perturbation, 141
periodic patterns during growth, 142
source density, 144, 145
stripes/patches formation, 141–2
substrate depletion effects, 142–3
time-dependent patterns in tropical mollusc shells, 150–2
two-antagonist mechanisms, 151, 152
conserved developmental mechanisms, 48
contact-mediated lateral inhibition simulations, 222, 226, 227
developmental processes, 3–4, 49–54
Drosophila development, 52–4
French Flag model task, *see* Cartesian genetic programming
gene activation, 149–50
graded concentration profiles, 140–1
maintenance during growth, 144, 145
graftal patterns, 163–4
initiating stimulus, 140
local self-enhancement/long-range antagonistic effects, 136, 139–40
activator–inhibitor interaction, 136–7
organizing regions, 136
physical mechanisms, 206, 207
positional information, 49–51, 54
reaction-diffusion systems, 18, 148, 163
symmetry-breaking cascades, 189, 191–5
Turing patterns, 18, 139–40, 163
Periodic cycles, 189
Peripheral nerve regeneration, 96
Perturbatory bandwidth, 16
Phenotype:
environmentally-induced variation, 86–8
fitness, 340
generation from genotype, 82–9
natural selection, 240
plasticity, 87
genetic mechanisms, 87
selection, 87–8, 89
traits, 7
see also Genotype-phenotype mapping
Pheromones, 66
Phosphatases, 75, 123, 125
Phosphatidylinositol bisphosphate (PIPU2u), 71
Phospholipase C, 77
Phosphorelay, 123
Phosphorylase, 74, 77
Phosphorylase kinase, 77
Phosphorylation:
Bacillus subtilis sporulation initiation, 123
signal transduction, 74–6
Phylogeny, 47
Piecewise-linear (PL) differential equations:

genetic regulatory networks qualitative simulation, 110, 116–19
Bacillus subtilis sporulation initiation, 123–5
equilibrium inequalities, 118
Genetic Network Analyser (GNA), 118–19, 123–4
qualitative behaviour of system, 118
regulatory domains, 116
switching domains, 116
target equilibrium, 117
threshold inequalities, 118
transition graph, 118, 124, 125
Plant development, 21
cell signalling, 156–61
L-systems, 19
Plasmodesmata, 66
Pleiotropy, 17
Polygeny, 17
Polypeptides, 7
Pomacanthus imperator, 193
Position-dependent gene activation, 135–53
Positional information, 49–51, 136, 193
activator–inhibitor interactions, 137
developmental pattern formation, 3–4
Hox genes, 51–2
gene activation, 148
morphogenesis simulation, 303
regeneration studies, 92
Predation risk, 87, 88
Prokaryotic cells, 4, 5, 8
Promoter, 8, 257
artificial genomes, 265
Protein kinases, 59
Protein synthesis, 257
regulation, 8, 65
negative (repression), 9
positive (amplification), 8–9
Proteins, 5, 6–7, 181, 182–3
autonomous agents developmental model, 379
cell differentiation, 4, 79
conformational change during signal transduction, 73, 74, 75, 76
evolutionary developmental system (EDS), 31–2
functional aspects, 6
gene regulation, 7, 8–9, 65
see also Genetic regulatory networks
intracellular signalling, 73
robot control architecture design model, 393, 395, 396–7
Proto-oncogenes, 79
Purines, 7
Pyrimidines, 7
Quadrupedal gait, 196
Quasiperiodic dynamics, 189

Rana pipiens development, 190, 191
Random Boolean networks (RBNs), 19, 258–60, 304
artificial genomes comparison, 271
autonomous agents developmental model, 385–6
dynamic analysis, 259–60
phase portraits, 386, 390
state cycles (limit cycle attractors), 259–60
synchronous versus asynchronous updating, 260
Ras, 80
Rb, 98
Reaction-diffusion equations, 192
Reaction-diffusion systems, 18, 57, 148, 163
Receptors, 5, 6, 50, 64, 163
artificial evolutionary system for shapes, 306
autonomous agents developmental model, 381
cell signalling, 64, 71–3
classification, 72, 73
internalization, 69
ligand binding, 71
structural aspects, 71, 72
Reflections, 186
Regeneration, 92–3
cell-cycle re-entry, 93
induction in mouse myonuclei, 99
thrombin regulatory role, 98–9
urodele myotubes, 95, 96, 98–9
hydra, 144
mammalian liver, 95–6
mammalian peripheral nerve, 96
plasticity of differentiated cells:
mammalian cell studies, 95–6
skeletal muscle cells, *see* Myotubes
urodele amphibians, 93–5
see also Urodele regeneration
Regulatory genes, 257, 340
evolutionary experiments in artificial genomes, 273
fractal protein modelling approach, 26
heterochronic processes, 343
Retinal development modelling, 360–3
branching rules, 361, 363
edge-detecting function, 360
pruning rules, 363
robustness, 360–1
Rhodnius, 142
Ribosomes, 8
RNA (ribonucleic acid), 7, 8

RNA polymerase, 257, 397
Robot control architecture morphogenesis, 341, 392–403
  dendritic growth, 397–8, 399
  developmental model, 393
  encoding scheme, 395–9
    dendrite growth, 399
    dendritic proteins, 397
    differentiator proteins, 398
    genome preprocessing, 395–7
    genomic regulatory network, 395, 400
    initial development, 398–9
    mover proteins, 397
    neural network interpretation, 399
    regulatory proteins, 396–7
    signalling proteins, 397
    splitter proteins, 398
    template matching routines, 395, 396, 400–1
    threshold proteins, 398
  evolvability, 400–1
    encoding repeated structures, 401, 402
    robustness of genetic operators, 400–1, 402
  fitness function, 394, 402
  genetic regulatory network, 395, 398, 400–1
  neural network evolution, 394, 398
  preliminary experiments, 394
Robustness, 377
  evolutionary processes selection, 172–3
  retinal development modelling, 360–1
  robot control architecture evolvability, 400–1
Rotations, 186

Sand dunes, 187–8
Scaling, 279
  biomorphs (*Blind Watchmaker* software), 247–8
Screw (helical motion), 186
Sea urchin cis-regulatory regions, 48
Seashell markings, 150–2, 195
Seasonal variation, 87, 88
Segmentation, 17, 143
  biomorphs (*Blind Watchmaker* software), 247, 248
  Drosophila, see Drosophila
  as evolutionary watershed, 253
Selection:
  natural, 240, 253–4
  ontogenetic features, 88–9
  pressure on embryo, 62
Self-assembled structures, 13, 206, 207, 208
  crystal networks, 212, 213
  surface tension effects in organisms, 216–18
Self-repairing circuits, 300
Sensors, 5
Shapes evolution simulation, 302–17
  model, 304–12
    affinity between molecules, 308
    algorithm, 304–5
    cell adhesion, 306, 310–11, 313
    cell division, 306, 308–10
    evolutionary strategy, 312
    gene regulation, 305–8
      mechanism, 308
    signalling molecule gradient, 306–7
  simulation results, 312–16
    invagination, 316
    morphogen gradient, 314, 315–16
  viscoelastic cellular interactions, 310–11, 313
    linkage with artificial genome, 313–14
    physical simulator, 311–12
Shoot apical meristem (SAM), 156, 157
  central zone, 156
  genetic regulatory network, 157
  modelling, 157–60
    generic model, 157–8
    results, 159–60
    simulation, 158–9
  peripheral zone, 156
  rib meristem, 156
Sign flip mutation, 28
Signal transduction, 59, 65
  co-opted functional change, 85–6
  pathways, 6
    modular features, 84, 85
  phosphorylation, 74–6
  protein conformational change, 73, 74, 75
  trimeric G proteins, 73
Signalling molecules/signalling proteins, 51, 54
  composition, 70
  cross-talk, 77
  diffusion, 208
  enzymatic synthesis, 68
  intracellular, 70–1
  removal/reversal, 69–70
  robot control architecture morphogenesis, 393, 397, 398
  sequestered, 67–8, 69
  threshold levels, 70
  transfer mechanisms, 66–7
Single gene activity, 21
Sintering process, 213
Site-specific biological addressing, 304
Slime moulds, 21, 59, 60, 194–5
  aggregation patterns, 194–5
  life-cycle, 194
Soap froths, 209–10
Somites, 47

Sonic hedgehog, 50, 51
Species selection, 254
Spiral forms, 207
Spotted patterns, 191, 192, 193
Stem cell transdifferentiation, 94–5
Steroid signalling molecules, 70
  receptors, 73
Stochastic models:
  genetic regulatory networks, 113–15
    master equation, 113
    stochastic simulation methods, 114
Striped patterns, 141–2, 143, 163, 191, 192
  angelfish markings, 193–4
Structural coupling, 15
  ontology independence, 15, 16
Surface energy, 209
  cells, 214
Surface tension, 208–18
  crystal boundaries, 210–11
  microstructure formation, 210–12
  organism construction, 213–16
    self-assembly mechanisms, 216–18
  soap froths, 209–10
Symmetry, 185–6
  bilateral, 185
  biomorphs (*Blind Watchmaker* software), 246–7, 248–9
  breaking, 4, 18, 187–8
    animal gaits, 195–9
    autonomous agents developmental model, 381, 386
    bifurcations, 188
    morphogenesis, 189–90
    pattern formation, 191–5
    spontaneous, 187
  departures, 185
  fivefold rotational, 185
  mathematical modelling, 184
  transformations, 185, 186
    groups, 186
Synapses, 356
  neural development modelling, 358
  signalling, 5
Synaptogenesis:
  neural development simulations, 344
  robot control architecture design, 393
Synthetic biology approach, 220–34
  applications in archival model representation, 232–3
  cell sorting by differential cell adhesion, 227–31
  challenges, 232
  contact-mediated lateral inhibition simulations, 222–7
    cell state equations, 223–5
    importance of temporal sequence, 225–6
  evaluation criteria, 220–1, 231–2
  multiple-mechanism simulator, 221–2
    range of created models, 233–4

TATA box, 8, 257, 393, 395
  artificial genomes, 265
TATA-binding protein (TBP), 8
Tcf, 145, 147
Testosterone, 70
TFIID, 8
TGF beta family, 51, 54
Thrombin, 98–9
Tierra, 319, 320–1
  experimental methods, 321–3
  genome, 320–1
  multi-threaded seed programme, 320
  web, 321–3
  see also Digital organism evolution
Time-dependent pattern formation, 150–2
Timing of developmental processes, 58
Transcription, 8, 257
  gene expression regulation, 257
  see also Genetic regulatory networks
Transcription factors, 257
  artificial genomes, 265
  Drosophila development, 52, 53, 54
  evolutionary developmental system (EDS), 34, 35
  target sites, 8
    evolutionary change, 48
    evolutionary developmental system (EDS), 34
Transdifferentiation:
  definition, 93
  mammalian tissues, 94
  stem cells, 94–5
  urodele regeneration studies, 93–4
Translation, 7, 8
Translations, 186
Trichoplax, 61
Tubularia regeneration, 144
Tumour formation, 80, 86
Turing patterns, 18, 139–40, 163, 192, 193
  angelfish markings, 193–4
  seashell markings, 195
  slime mould aggregation patterns, 195

Urodele regeneration, 92–103
  heart, 93
  iris, 93
  limb, 93, 95

Urodele regeneration (*cont'd*)
  plasticity of differentiated cells, 93–5
  transdifferentiation, 93–4

Vertebral number/anatomy, 52, 86
Vertebrate development, 54–5
  pattern formation, 49–50
  phylotypic stage, 47, 54
  segmentation, 247, 248
Vertebrate jaws, 47, 55, 85
Vertebrate limb, 49, 50, 51, 52, 54, 56–7, 58, 83, 193
  pattern formation in development, 54
  urodele regeneration studies, 93, 95
Vesicles, 213–14
Viscoelastic forces, 208
  artificial evolutionary system for shapes, 310–11
    physical simulator, 311–12

Wingless (*wg*), 50, 54, 144
  hydra homologue, 145
Wnt, 51, 141
  hydra homologue, 146, 147
WUSCHEL (WUS), 156, 157, 159–60

Xbra, 150
Xenopus:
  activin-dependent gene activation, 150
  cell cycle regulation modelling, 115
  pattern formation, 54
  positional information, 50
Xgsc, 150

Yeast, 59
  cell cycle regulation, 115

Zebra fish fin development, 56–7
Zootype, 51
Zygote cleavage division, 3